EXAMKRACKERS MCAT®

BIOLOGY 2 SYSTEMS

10TH EDITION

JONATHAN ORSAY

Major Contributors:
Joshua Albrecht, M.D., Ph.D.
Kaitlyn Barkley
Jennifer Birk-Goldschmidt, M.S.
Stephanie Blatch, M.A.
Austin Mattox
Lauren Nadler
Laura Neubauer
Mark Pedersen, M.D.
Sara Thorp, D.O

Contributors:
Mark Alshak
Leena Asfour
Max Blodgett
David Collins
Erik Davies, M.Ed.
Claudia Goodsett
Marouf Hossain
Jay Li
Xintong Li, M.D.
Mohan Natrajan
Colleen Moran Shannon
Christopher Stewart
Darren Sultan
Steven Tersigni, M.D., M.P.H
Richmond Woodward

Art Director:
Erin Daniel

Designers:
Dana Kelley
Charles Yuen

Illustrators:
Stephen Halker
Kellie Holoski

Published in the United States of America by Osote Publishing, New Jersey.

ISBN 10: 1-893858-85-5 (Volume 2)
ISBN 13: 978-1-893858-85-5 (6 Volume Set)
10th Edition

To purchase additional copies of this book or the rest of the 6 volume set, call 1-888-572-2536 or fax orders to 1-859-305-6464.

Examkrackers.com
Osote.com

Printed and bound in the United States of America.

PHOTOCOPYING & DISTRIBUTION POLICY

Acknowledgements

Although I am the author, the hard work and expertise of many individuals contributed to this book. The idea of writing in two voices, a science voice and an MCAT® voice, was the creative brainchild of my imaginative friend Jordan Zaretsky. I would like to thank Jerry Johnson for lending his science talent and exceptional science skills to this project. I would like to thank nineteen years worth of Examkrackers students for doggedly questioning every explanation, every sentence, every diagram, and every punctuation mark in the book, and for providing the creative inspiration that helped me find new ways to approach and teach biology. Finally, I wish to thank my wife, Silvia, for her support during the difficult times in the past and those that lie ahead.

Introduction to the Examkrackers Manuals

The Examkrackers manuals are designed to give you exactly the information you need to do well on the MCAT® while limiting extraneous information that will not be tested. This manual organizes all of the information on the organ systems biology tested on the MCAT® conceptually. Concepts make the content both simple and portable for optimal application to MCAT® questions. Mastery of the organ systems biology covered in this manual will increase your confidence and allow you to succeed with seemingly difficult passages that are designed to intimidate. The MCAT® rewards your ability to read complex passages and questions through the lens of basic science concepts.

An in-depth introduction to the MCAT® is located in the Reasoning Skills manual. Read this introduction first to start thinking like the MCAT® and to learn critical mathematical skills. The second lecture of the Reasoning Skills manual addresses the research methods needed for success on 20% of questions on the science and psychology sections of the MCAT®. Research questions are the most difficult on the MCAT® and are the questions that help students beat the mean and get a high score. Once you have read those lectures, return to this manual to begin your review of the biology you will need to excel on the MCAT®.

How to Use This Manual

Examkrackers MCAT® manuals can be used with the Examkrackers Comprehensive MCAT® Course or as a tool for independent MCAT® study. Examkrackers MCAT® preparation experience has shown that you will get the most out of these manuals when you structure your studying as follows. If you are taking the Examkrackers Comprehensive MCAT® Course, read each lecture twice before the class lecture. During the first reading, you should not write in the book. Instead, read purely for enjoyment. During the second reading, highlight and take notes in the margins. Complete the twenty-four questions in each lecture during the second reading before coming to class. Do not look at the In-Class Exams before class. Immediately after class, read the lecture again, slowly and thoroughly.

If you are studying independently, read each lecture twice (as described above), completing the in-lecture questions during the second reading. Take the In-Class Exam, then read the lecture once more, slowly and thoroughly.

The In-Class Exams are designed to educate. They are similar to an MCAT® section, but are shortened and have most of the easy questions removed. We believe that you can answer most of the easy questions without too much help from us, so the best way to raise your score is to focus on the more difficult questions. The In-Class Exams are designed to help you prepare for test day when you will be faced with an intensity that is difficult to simulate. Technically, the Examkrackers In-Class Exams are timed to take as long as 35 minutes to complete. By practicing with a few minutes less, you will be ready for the pressure and pace of the real exam, and MCAT® day will feel more manageable. These methods are some of the reasons for the rapid and celebrated success of the Examkrackers prep course and products.

With each In-Class Exam you should see the number of questions you answer correctly increase. Do not be discouraged by poor performance on these exams; they are not meant to predict your performance on the real MCAT®. **The questions that you get wrong (or even guess correctly) are the most important ones. They represent your potential score increase.** When you get a question wrong or have to guess, determine why. If content was the problem, target those areas for review. If the issue was your approach to the question, make a commitment to a different approach for the next practice test. As you learn to think like the MCAT®, you will see your score increase.

In order to study most efficiently, it is essential to know what topics are and are not tested directly in MCAT® questions. This manual uses the following conventions to make the distinction. Any topic listed in the AAMC's guide to the MCAT® is printed in **red, bold type**. You must thoroughly understand all topics printed in **red, bold type**. Any formula that must be memorized is also printed in **red, bold type**.

If a topic is not printed in **bold and red**, it may still be important. Understanding these topics may be helpful for putting other terms in context. Topics and equations that are not explicitly tested but are still useful to know are printed in *italics*. Knowledge of content printed in *italics* will enhance your ability to answer passage-based MCAT® questions, as MCAT® passages may cover topics beyond the AAMC's list of tested topics on the MCAT®.

Features of the Examkrackers Manuals

The Examkrackers manuals include several features to help you retain and integrate information for the MCAT®. Take advantage of these features to get the most out of your study time.

- **The 3 Keys** – The Three Keys unlock the material and the MCAT® by highlighting the most important things to remember from each chapter. The Three Keys are listed at the beginning and end of each lecture with reminders from Salty throughout the text. Examine the Three Keys before and after reading a lecture to make sure you have absorbed the most important messages. As you read, continue to develop your own key concepts that will guide your studying and performance.

- **Signposts** – The MCAT® is fully integrated, asking you to apply the biological, physical, and social sciences simultaneously. The signposts alongside the text in this manual will help you build mental connections between topics and disciplines. This mental map will lead you to a high score on the MCAT®. The post of each sign "brackets" the paragraph to which it refers. When you see a signpost next to a topic, stop and consider how the topics are related. Soon you will begin making your own connections between concepts and topics within and between disciplines. This is an MCAT® skill that will improve your score. When answering questions, these connections give you multiple routes to find your way to the answer.
- **MCAT® Think** – These sidebars invite deeper consideration of certain topics. They provide helpful context for topics that are tested and will challenge you just like tough MCAT® passages. While MCAT® Think topics and their level of detail may not be explicitly tested on the MCAT®, read and consider each MCAT® Think to sharpen your MCAT® skills. The MCAT® Think sidebars provide essential practice in managing seemingly complex and unfamiliar content, as you will need to do for passages on MCAT® day.

I'm Salty the Kracker. Where you see purple text, that's me. I will show you why the content makes sense and help you develop your MCAT® intuition. My job is to make sure you 1. stay awake and 2. really understand and remember what you're reading. If you think I am funny, tell the boss. I could use a raise. If you get the munchies, reconsider... you'll want me around.

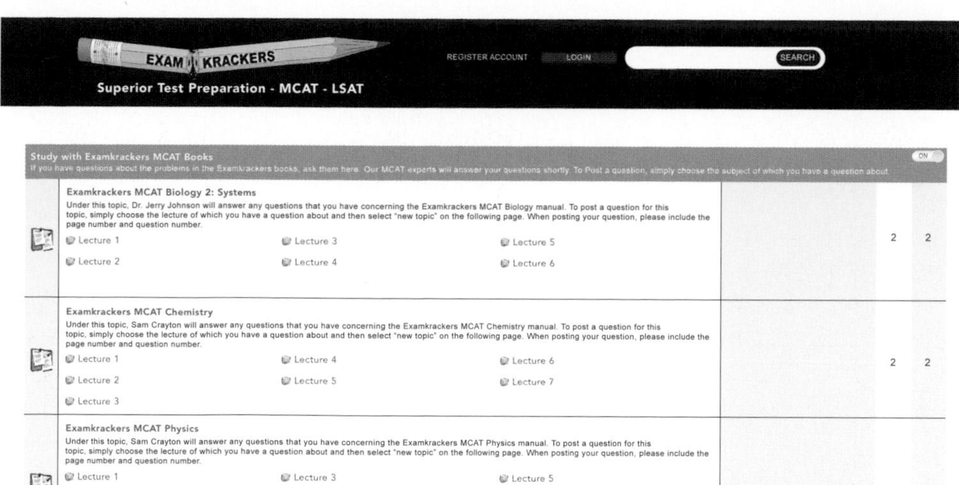

Additional Resources

If you find yourself struggling with the science or you want more practice, take advantage of the additional Examkrackers resources that are available.

Examkrackers offers a 9-week Comprehensive MCAT® Course to help you achieve a high score on the MCAT®. Whether in person or online, the course includes up to 120 hours with expert instructors, a unique course format, and regular full-length MCAT® exams. Each class includes lecture, a practice exam, and review, designed to help you develop essential MCAT® skills. For locations and registration please visit Examkrackers.com or call 1.888.KRACKEM.

EK-Tests® are Examkrackers full-length, online simulated MCAT® exams that match the AAMC MCAT® in style, content, and skills tested. Written by educators, high-scoring medical students, and physicians, EK-Tests® provide the best MCAT® practice available. To purchase EK-Tests®, please visit Examkrackers.com.

Your purchase of this book new will also give you access to the **Examkrackers Forums** at Examkrackers.com/mcat/forum. These bulletin boards allow you to discuss any MCAT® question with an MCAT® expert at Examkrackers. All discussions are kept on file so you can refer back to previous discussions on any question in this book. Once you have purchased the books you can take advantage of this resource by calling 1.888.KRACKEM to register for the forums.

Although we make every effort to ensure the accuracy of our books, the occasional error does occur. Corrections are posted on the Examkrackers Books Errata Forum, also at Examkrackers.com/mcat/forum. If you believe that you have found a mistake, please post an inquiry on the Study with Examkrackers MCAT® Books Forum, which is likewise found at Examkrackers.com/mcat/forum. As the leaders in MCAT® preparation, we are committed to providing you with the most up-to-date, accurate information possible.

Study diligently, trust this book to guide you, and you will reach your MCAT® goals.

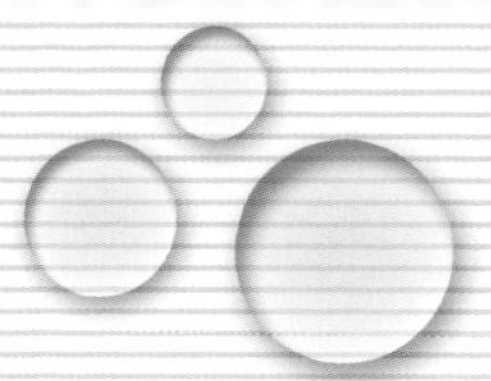

Table of Contents

LECTURE 1 The Cell 1

LECTURE 2 The Nervous System 37

LECTURE 3 The Endocrine System 77

LECTURE 4 The Circulatory, Respiratory, and Immune Systems 107

LECTURE 5 The Digestive and Excretory Systems 143

LECTURE 6 Muscle, Bone and Skin 175

30-Minute In-Class Exams 199

Answers & Explanations to In-Class Exams 247

Answers & Explanations to Questions in the Lectures 273

Photo Credits 293

Index 295

BIOLOGICAL SCIENCES

DIRECTIONS. Most questions in the Biological Sciences test are organized into groups, each preceded by a descriptive passage. After studying the passage, select the one best answer to each question in the group. Some questions are not based on a descriptive passage and are also independent of each other. You must also select the one best answer to these questions. If you are not certain of an answer, eliminate the alternatives that you know to be incorrect and then select an answer from the remaining alternatives. A periodic table is provided for your use. You may consult it whenever you wish.

PERIODIC TABLE OF THE ELEMENTS

1 **H** 1.0																	2 **He** 4.0
3 **Li** 6.9	4 **Be** 9.0											5 **B** 10.8	6 **C** 12.0	7 **N** 14.0	8 **O** 16.0	9 **F** 19.0	10 **Ne** 20.2
11 **Na** 23.0	12 **Mg** 24.3											13 **Al** 27.0	14 **Si** 28.1	15 **P** 31.0	16 **S** 32.1	17 **Cl** 35.5	18 **Ar** 39.9
19 **K** 39.1	20 **Ca** 40.1	21 **Sc** 45.0	22 **Ti** 47.9	23 **V** 50.9	24 **Cr** 52.0	25 **Mn** 54.9	26 **Fe** 55.8	27 **Co** 58.9	28 **Ni** 58.7	29 **Cu** 63.5	30 **Zn** 65.4	31 **Ga** 69.7	32 **Ge** 72.6	33 **As** 74.9	34 **Se** 79.0	35 **Br** 79.9	36 **Kr** 83.8
37 **Rb** 85.5	38 **Sr** 87.6	39 **Y** 88.9	40 **Zr** 91.2	41 **Nb** 92.9	42 **Mo** 95.9	43 **Tc** (98)	44 **Ru** 101.1	45 **Rh** 102.9	46 **Pd** 106.4	47 **Ag** 107.9	48 **Cd** 112.4	49 **In** 114.8	50 **Sn** 118.7	51 **Sb** 121.8	52 **Te** 127.6	53 **I** 126.9	54 **Xe** 131.3
55 **Cs** 132.9	56 **Ba** 137.3	57 **La*** 138.9	72 **Hf** 178.5	73 **Ta** 180.9	74 **W** 183.9	75 **Re** 186.2	76 **Os** 190.2	77 **Ir** 192.2	78 **Pt** 195.1	79 **Au** 197.0	80 **Hg** 200.6	81 **Tl** 204.4	82 **Pb** 207.2	83 **Bi** 209.0	84 **Po** (209)	85 **At** (210)	86 **Rn** (222)
87 **Fr** (223)	88 **Ra** 226.0	89 **Ac**= 227.0	104 **Unq** (261)	105 **Unp** (262)	106 **Unh** (263)	107 **Uns** (262)	108 **Uno** (265)	109 **Une** (267)									

*	58 **Ce** 140.1	59 **Pr** 140.9	60 **Nd** 144.2	61 **Pm** (145)	62 **Sm** 150.4	63 **Eu** 152.0	64 **Gd** 157.3	65 **Tb** 158.9	66 **Dy** 162.5	67 **Ho** 164.9	68 **Er** 167.3	69 **Tm** 168.9	70 **Yb** 173.0	71 **Lu** 175.0
=	90 **Th** 232.0	91 **Pa** (231)	92 **U** 238.0	93 **Np** (237)	94 **Pu** (244)	95 **Am** (243)	96 **Cm** (247)	97 **Bk** (247)	98 **Cf** (251)	99 **Es** (252)	100 **Fm** (257)	101 **Md** (258)	102 **No** (259)	103 **Lr** (260)

For your convenience, a periodic table is also inserted in the back of the book.

LECTURE 1

The Cell

THE 3 KEYS

1. The cell has compartments to separate chemical properties and processes.

2. A cell is like an organism – understand the function of organelles by treating them as organs.

3. Transport in and out of a cell occurs through diffusion down the concentration gradient, by pumps against the diffusion gradient, or via endocytosis/exocytosis.

1.1 Introduction

This lecture discusses the cell, the basic unit of life. Evolution is used as a framework to discuss key principles of viruses, prokaryotic cells, and eukaryotic cells.

The lecture will first cover the organizational and naming scheme of living organisms. Then, it will address evolution and how evolution explains changes in a genome over time, as well as how a species can become a new species altogether. Viruses, which have played a role in evolution and are dependent on living cells, will be described.

Next, the lecture will discuss the structure and reproduction of bacteria, medically relevant prokaryotes. Bacteria reproduce asexually but are able to exchange genetic material among themselves. The cell walls of bacteria protect them from the outside environment, so they are the target of many antibiotics.

Eukaryotic cells, including human cells, have two major types of organization. First, cells are organized by pathways. This lecture follows the flow of proteins after their initial synthesis; the Genetics Lecture in *Biology 1: Molecules* examines the flow of DNA and RNA; and the Metabolism Lecture in this same manual describes the flow of energy and biological molecules. Eukaryotic cells are also organized by chemical properties. Some compartments of the cell contain proteins, others lipids, and others acidic or basic environments. These compartments are surrounded by membranes that regulate the transport of substances.

The lecture concludes by considering how cells come together to form tissues and organs. Cells are held together by an extracellular matrix and communicate with each other so that the processes of these cells become synchronized. While an organ is more than the sum of its cells, the function of an organ depends on the functions of its cells. For example, organs that secrete proteins must have the organelles that make, synthesize, and secrete proteins. The remaining lectures in this manual will discuss systems of organs and how these systems work together to maintain the entire body.

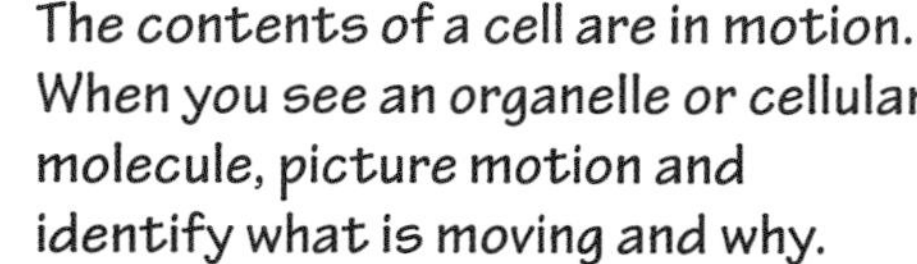
The contents of a cell are in motion. When you see an organelle or cellular molecule, picture motion and identify what is moving and why.

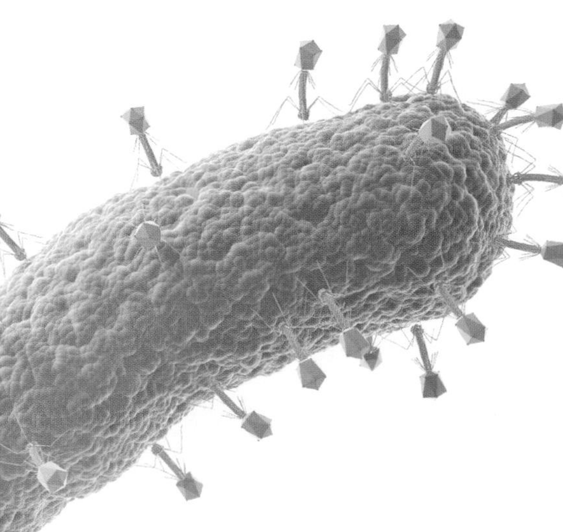

Everything in the cell is in flux. Proteins are constantly being synthesized, modified, transported, and degraded. When the cell can no longer maintain these processes, the cell dies.

A horse and a zebra would be of the same species only if they could reproduce to create fertile offspring. Horses and zebras are not of the same species, so this horse/zebra hybrid cross is probably infertile.

Remember the order of the classification system with the mnemonic: "Dear King Philip Cried Out For Good Soup."

1.2 Classification of Organisms

In biology, *taxonomy* is the organization of the naming system for living organisms. Most taxonomical classification systems are based upon genetic similarity resulting from shared phylogeny, or evolutionary history. Genetic similarity often allows scientists to predict the function of uncharacterized genes in humans based on their functions in closely related species. The most commonly used classification system contains increasingly specific groupings in the following order: **Domain**, *Kingdom, Phylum, Class, Order, Family, Genus, Species*. There are three domains: Bacteria, Archaea, and Eukarya. The kingdoms of Protista, Fungi, Plantae, and Animalia are in the domain Eukarya. All mammals belong to the class *Mammalia* and the phylum *Chordata*; all mammals likely share a common genetic ancestor that they do not share with birds, which are also in the phylum Chordata, but in the class Aves.

A **species** is loosely limited to all organisms that can reproduce fertile offspring with one another. If two organisms can reproduce to create fertile offspring, they may be of the same species; if their gametes are incompatible, they are definitely not of the same species. Organisms may be prevented from producing fertile offspring for any number of reasons, such as living in different areas of the planet (*geographic isolation*), mating in different seasons (*temporal isolation*), and having gametes that do not form viable offspring (*genetic incompatibility*). An example of genetic incompatibility is a mule, the offspring of a female horse and a male donkey. When the pair mates, the mule receives 32 chromosomes from the horse and 31 chromosomes from the donkey. The mispairing of chromosomes prevents further reproduction. This definition of a species is by no means perfect or all-inclusive. Bacteria, for instance, do not reproduce sexually. Their classification system is entirely different and is beyond the scope of the MCAT®.

The evolutionary history of a single organism can be followed within its development from embryo to adult. That is, "ontogeny recapitulates phylogeny," or development retraces evolution. The human fetus has pharyngeal pouches, reflecting an ancestor with gills. Naming (taxonomy) makes sense when evolutionary or genetic history is considered.

As described by cell theory, the cell is the smallest basic unit of life and is shared by all living things. When microscopes were invented, biologists were able to observe the similar organization of small units – cells – across different forms of life. No matter how different two organisms are in their taxonomies and evolutionary histories, they are both made up of cells.

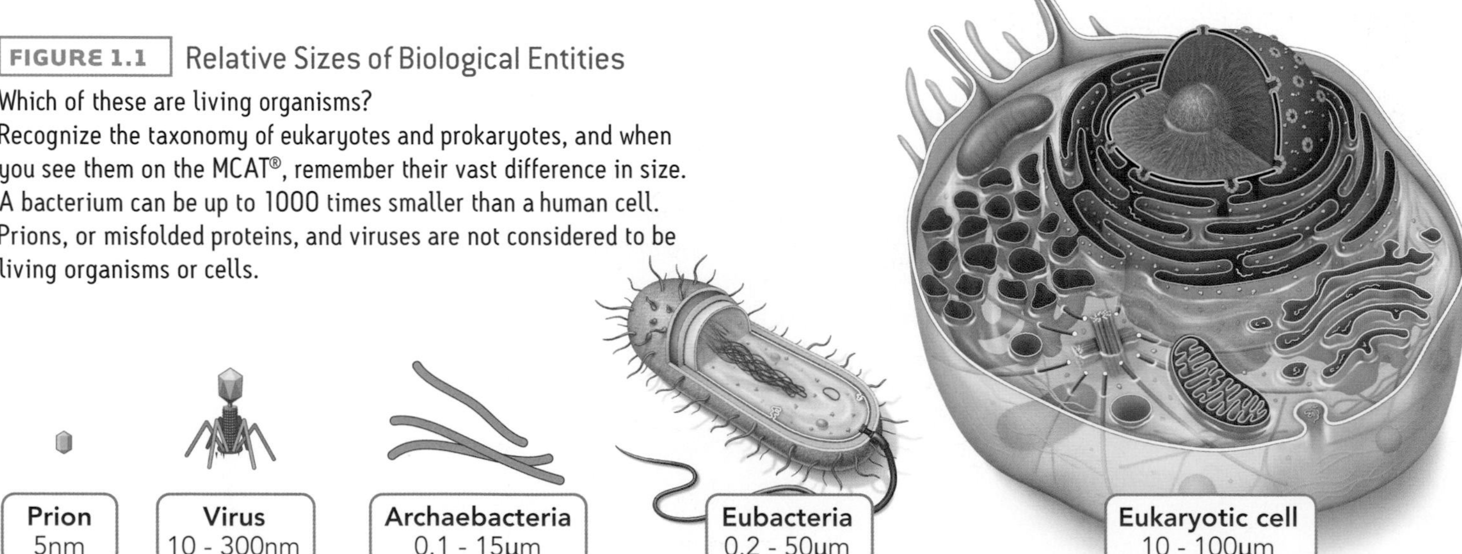

FIGURE 1.1 Relative Sizes of Biological Entities

Which of these are living organisms?
Recognize the taxonomy of eukaryotes and prokaryotes, and when you see them on the MCAT®, remember their vast difference in size. A bacterium can be up to 1000 times smaller than a human cell. Prions, or misfolded proteins, and viruses are not considered to be living organisms or cells.

1.3 Evolution: Genetics and Viruses

Evolution as Heritable Change in Genes

Our genome contains our entire evolutionary history. The genome is a patchwork of genetic material introduced through a variety of mechanisms including exchange with viruses. Evolution is the result of natural selection for traits that increase the fitness of a population.

An understanding of evolution requires terms used in genetics. Any particular trait has both a **genotype** and a **phenotype**. The genotype describes the chromosomes of an organism, while the phenotype describes the product of the genes that can be observed. Consider the flowers of a pea plant, which can be either purple or white. These colors comprise the phenotype and can be described as either dominant (purple) or recessive (white). On the chromosomal level, any dizygotic organism has two copies of each gene, each called an **allele**. For this flower, there are only two forms of the gene: P, which codes for purple, and p, which codes for white. P is the dominant allele so whenever it appears as one of the plant's genes, the flowers will be purple; thus either PP or Pp plants can be purple. Whenever P is absent, the flowers will be white. In the simple example of flower color, there are only two possible alleles, the dominant allele P and the recessive allele p. Not all phenotypes are so simple. Some, such as height, vary gradually within a species.

A gene that has multiple alleles, corresponding to distinct forms of a phenotype, is called a **polymorphism**. The existence of polymorphisms in the population makes evolution and the related process of speciation possible. The **gene pool** is the total of all alleles in a population and **evolution** is a change in a population's gene pool. Figure 1.2 shows the flower color alleles for a small population. There are 70% P alleles and 30% p alleles. Even if the ratio of purple (PP and Pp) to white (pp) flowers changes temporarily, as long as the ratio of alleles remains 70 to 30, the population has not evolved. If the ratio does change, or if a new allele is formed by a mutation (such as a mutation coding for red flowers), evolution has occurred.

FIGURE 1.2 Gene Pool of a Population

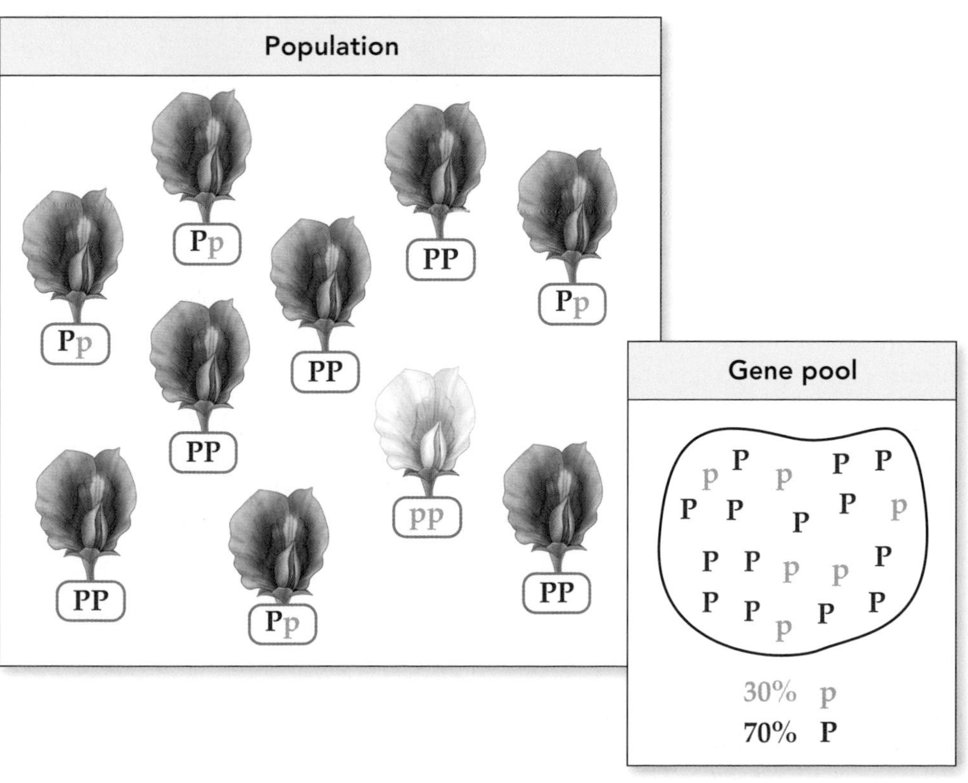

Cheetahs have a very little genetic variation, due to an evolutionary bottleneck event experienced by the species. As a result, they are more vulnerable to environmental threats, such as disease.

Evolution can sometimes result in **speciation**, the formation of a new species. It is possible for a species to evolve into one or more new species. Members of different species are unable to produce fertile offspring. Speciation is said to have occurred when members from a single species have evolved into different groups that can no longer produce fertile offspring.

The MCAT® requires an understanding of a variety of mechanisms that can contribute to speciation. Polymorphisms provide the raw material for speciation. **Inbreeding**, the mating of relatives, increases the number of homozygous individuals in a population without changing the allele frequency. Inbreeding can lead to speciation under certain circumstances, such as if a population is geographically divided into groups and a mutation arises in one group but not the other. **Outbreeding**, or mating of nonrelatives, maintains genetic flow between populations and so would not be expected to contribute to speciation.

Speciation can also occur due to random events. A species may face a crisis so severe that there are few survivors, whose allelic frequencies are not representative of the original population. The result is that the allelic frequencies of the population are shifted. This situation, called a **bottleneck**, can ultimately lead to speciation.

Finally, specialization and adaptation are processes that can contribute to speciation. **Specialization** is the process by which the members of a species tailor their behaviors to exploit their environment. They carve out a particular role in the environment, including the use of certain resources and advantageous behaviors. When distinct groups within a population specialize such that they differ substantially, such as in their habits or locations, speciation may ultimately result. Similarly, genetic or behavioral changes that are advantageous in the given environment, or **adaptations**, can lead to speciation.

The Five Mechanisms of Evolution

Evolution can be understood by considering the circumstances that are required to prevent it from happening. The scientists Hardy and Weinberg defined a theoretical population in which no evolution would occur. This population is said to be in **Hardy-Weinberg equilibrium** and has the following features:

1. mutational equilibrium,
2. large population,
3. random mating,
4. immigration or emigration must not change the gene pool, and
5. no selection for the fittest organism.

Hardy-Weinberg equilibrium is an idealized concept, and no real population ever possesses these characteristics completely. When one of the conditions is broken, evolution occurs. The implications of each condition will be examined below.

The first feature, mutational equilibrium, is the most intuitive. *Mutational equilibrium* means that the rate of forward mutations exactly equals the rate of backward mutations. Mutational equilibrium rarely occurs in real populations. Random mutations can be used to track the progress of evolution. In other words, **evolutionary time can be measured by gradual random changes in the genome.** Comparing the genomes of species that share a common ancestor can be used to determine how long ago they diverged.

The second feature, a large population, is also fairly intuitive. Small populations are subject to **genetic drift**, where one allele may be permanently lost due to the death of all members having that allele. Large populations are significantly less affected by loss of a few members because the lost alleles are likely contained in other members of the population. Genetic drift is not caused by selective pressure, so its results are random in evolutionary terms. A sudden decline in population size can also lead to changes in the gene pool, as in the example of bottlenecks.

The third feature, random mating, means that organisms are equally likely to mate with each other regardless of their phenotypes. Random mating can be described as the lack of selection for certain phenotypes in mating (*sexual selection*). In real populations, certain behavioral or physical phenotypes lead to increased mating success; as a result, the associated genotypes are preferentially passed on to the following generations, and evolution results.

The fourth feature, lack of immigration and emigration, is the requirement that new alleles are not introduced to the population by immigrating organisms, and that alleles are not decreased or eliminated by the emigration of organisms with those alleles. Either scenario could change the allelic makeup of a population.

In real populations, the fifth requirement, no selection of the fittest, is violated when certain members of a population exploit the environment more efficiently than others. According to the **fitness concept**, the "fittest" organism is the one that can best survive to reproduce offspring; these offspring will, in turn, reproduce offspring, and so on generation after generation.

From an evolutionary standpoint, the success of a gene can be measured as its **increase in percent representation in the gene pool of the next generation.** The driving principle of **natural selection** is that genes that are advantageous in a given environment are preferentially passed down from generation to generation. At the core of natural selection is not how long an organism lives, but rather how well it can reproduce to contribute its genes to the next generation. Such selection occurs by **differential reproduction**, which may include living beyond reproduction in order to give offspring a better chance to reproduce.

Viruses

Viruses are tiny infectious agents. They have the ability to transfer genetic material, which can have positive effects, as in gene therapy. Viruses have played a role in evolution by transferring genetic "information."

Viruses are generally comparable in size to large proteins, although some are larger and some are much smaller. In its most basic form, a virus consists of a protein coat, called a **capsid**, and from one to several hundred genes in the form of DNA or RNA inside the capsid. No virus contains both DNA and RNA, and viruses do not contain organelles or nuclei. Most animal viruses, some plant viruses, and very few bacterial viruses surround themselves with a lipid-rich envelope that is borrowed from the membrane of their host cell or synthesized in the host cell's cytoplasm. The envelope typically contains some virus-specific proteins. A mature virus outside the host cell is called a **viral particle** or virion. All organisms experience viral infections.

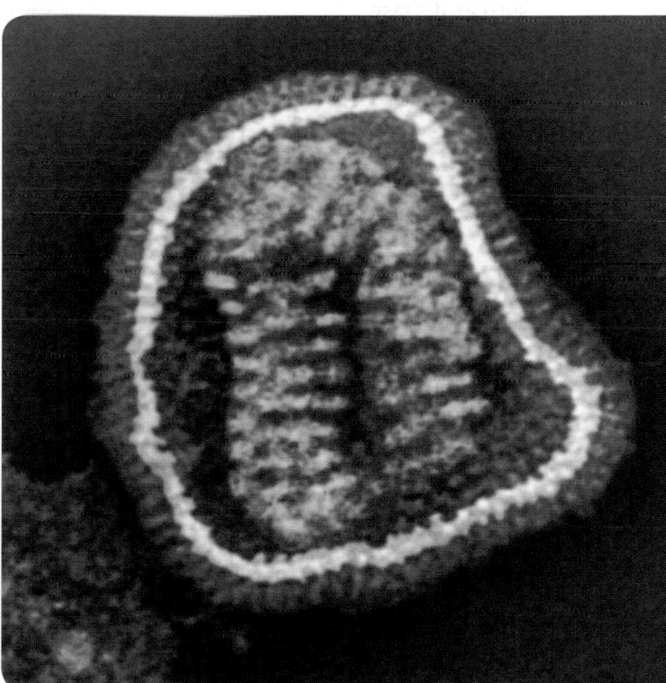

The transmission electron micrograph above shows the structure of an influenza virus particle. This virus has RNA as its genome and is an example of an enveloped virus.

Viruses are not currently classified as living organisms, so they do not belong to any of the taxonomical kingdoms of organisms. They are analogous to a portion of a cell, namely the eukaryotic nucleus. A nucleus does not metabolize nutrients and cannot make proteins to reproduce itself. Instead it uses the ATP made available in the cytosol and mitochondria, and it uses the ribosomes in the cytosol to make proteins and enzymes to reproduce. Viruses similarly exploit the cell's organelles to survive. Viruses and eukaryotic nuclei differ in that a virus can contain either DNA or RNA as its inherited genome (never both), while eukaryotic nuclei use only DNA. When inside a cell, a virus will remove its capsid and envelope to expose its genetic material in the cytosol. The nucleus remains membrane bound except during mitosis and meiosis.

Whenever you see a virus in a cell, think of it as an extra nucleus.

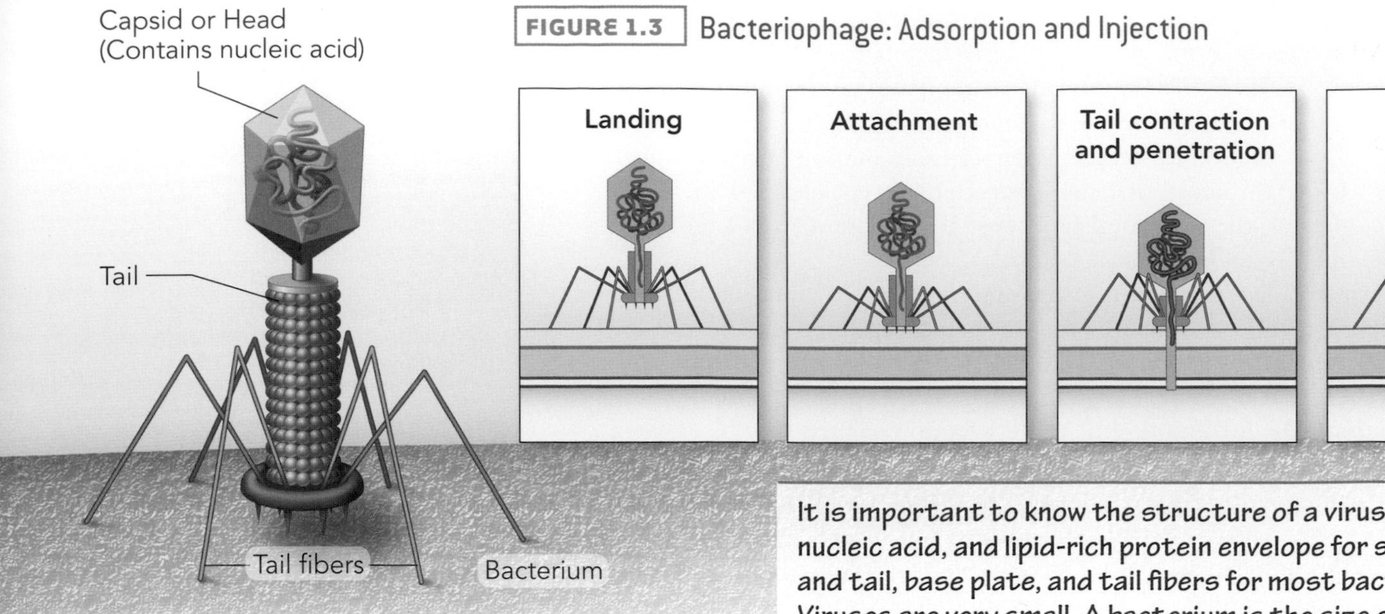

FIGURE 1.3 Bacteriophage: Adsorption and Injection

It is important to know the structure of a virus: capsid, nucleic acid, and lipid-rich protein envelope for some viruses; and tail, base plate, and tail fibers for most bacteriophages. Viruses are very small. A bacterium is the size of a mitochondrion, and hundreds of viruses may fit within a bacterium.

Most animal viruses do not leave capsids outside the cell, but rather enter the cell through receptor-mediated endocytosis.

A viral infection begins when a virus binds to a specific chemical receptor site on the **host cell**, the cell that is being infected. The chemical **receptor** is usually a specific glycoprotein on the host cell membrane. A virus cannot infect a cell that does not have the specific receptor for that virus. Next, the nucleic acid of the virus penetrates into the cell. A **bacteriophage** (Greek: phagein → to eat), a virus that infects bacteria, typically injects nucleic acids into the host cell through its **tail** after viral enzymes have digested a hole in the cell wall, as shown in Figure 1.3. (Notice that this indicates that some viruses also include enzymes within their capsids.) Most viruses that infect eukaryotes are instead engulfed by an endocytotic process, meaning that the cell membrane surrounds the virus and brings it into the cell. Once inside the host, there are two possible paths: a lysogenic infection or a lytic infection (Figure 1.4).

In a **lytic** (Greek: lysis → separation) infection, the virus commandeers the cell's **synthetic machinery**, which is normally used to produce proteins for the cell, to replicate viral components. The RNA (either directly from the virus or transcribed from viral DNA) is translated to form proteins. These proteins self-assemble to form a new virus. The cell may fill with new viruses until it lyses (bursts), or it may release new viruses one at a time in a reverse endocytotic process. The period from infection to lysis is called the *latent period*. A virus following a lytic cycle is a *virulent virus*, a virus that is capable of causing disease.

In a **lysogenic** infection, the viral DNA is incorporated into the host genome. When the host cell replicates its DNA, the viral DNA is replicated as well. A virus in a lysogenic cycle is called a *temperate virus*. A host cell infected with a temperate virus may show no symptoms of infection. While the viral DNA remains incorporated in the host DNA, the virus is said to be *dormant* or *latent*, and is called a *provirus* (a *prophage* [Greek: pro → before, phagein → eat] if the host cell is a bacterium). The dormant virus may become active when the host cell is under stress. When the virus becomes active, it takes on a lytic phenotype and becomes virulent.

As described previously, some viruses have **viral envelopes**, formed as they undergo exocytosis from the cell. These envelopes are pinched off pieces of host cellular membrane. To a certain extent, the envelope protects an **enveloped virus** from detection by the immune system. Even more importantly, the receptors on the envelope allow it to bind to a new host cell and start the process all over again. The original cell may not die right away, but due to degradation of its membrane

FIGURE 1.4 Viral Life Cycles

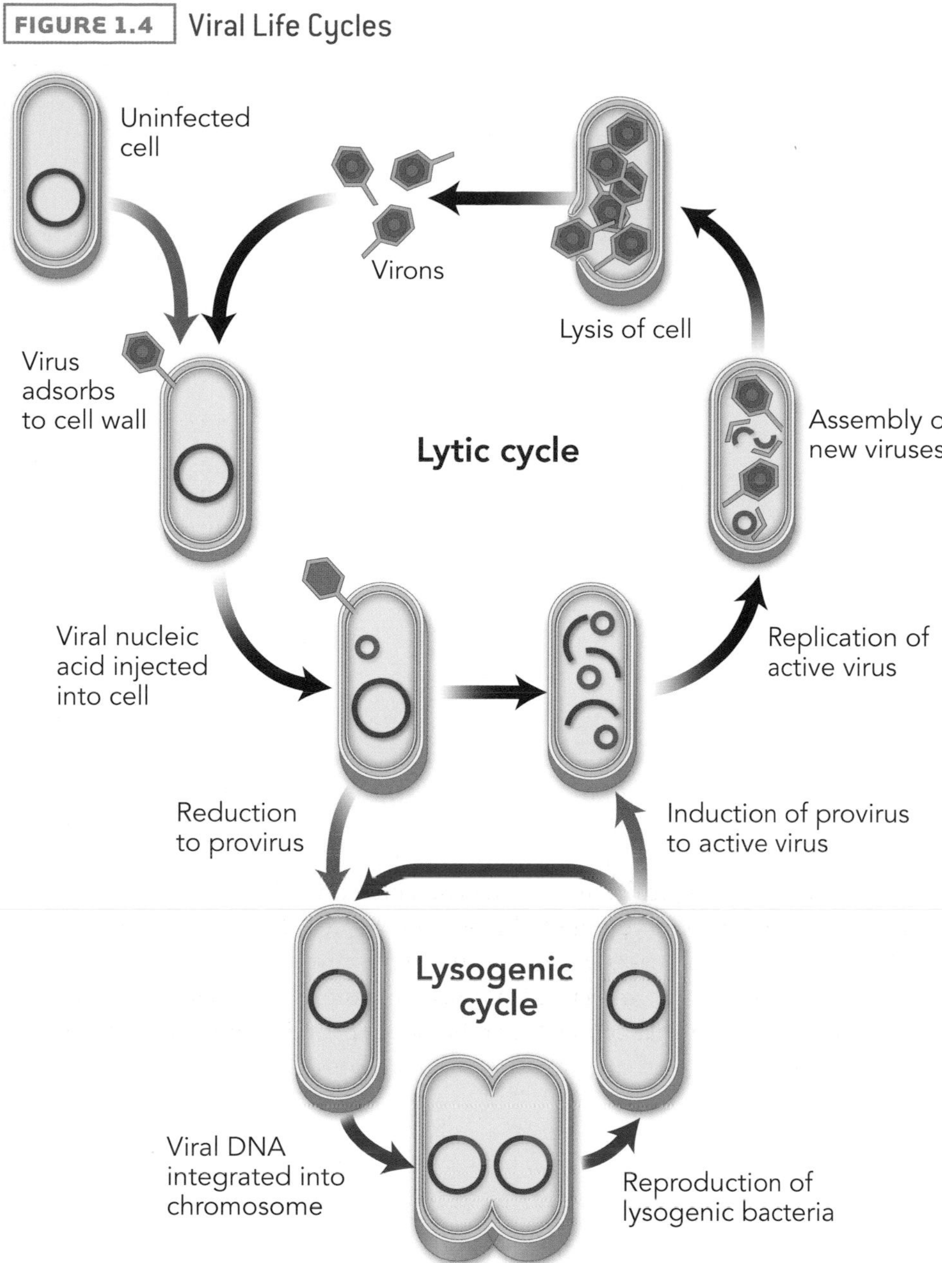

Be sure to know the differences between the two viral life cycles. In particular, remember how the viral genetic material is converted into proteins and also how it can be incorporated into the host cell genome. See the Genetics Lecture for information about transcription and translation.

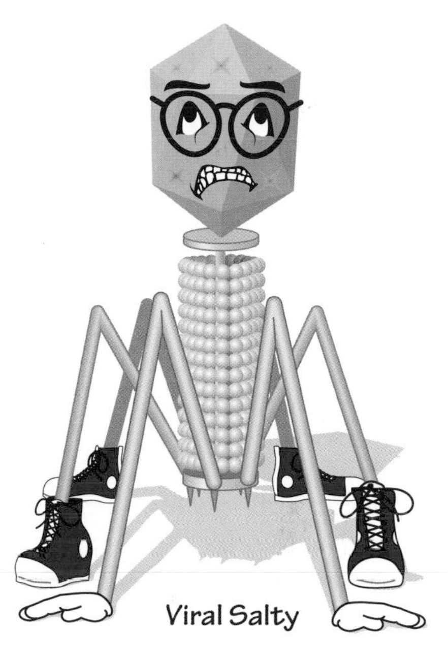

and usable cellular machinery, it usually does die. By contrast to enveloped viruses, **nonenveloped viruses** typically do lyse a cell and cause cell death on their release.

The replication and transcription of viral double stranded DNA (dsDNA) occurs as described in the Genetics Lecture in the *Biology 1: Molecules* manual. After injection into a cell, the viral DNA translocates to the nucleus, where it hijacks the cellular machinery in order to replicate and translate its genetic material.

Almost all RNA viruses replicate via an *RNA-dependent RNA polymerase* (*RdRP*). This enzyme can synthesize a new strand of RNA from a previously existing RNA strand. Some RNA strands (*+RNA*) actually code for proteins. However, when RdRP makes a copy of this strand, it will create a complimentary strand that cannot code for protein products. This is a –RNA strand, which must undergo another replication with RdRP in order to form a new +RNA strand. All single stranded RNA viruses specifically carry either +RNA or –RNA. If a virus carries –RNA, it must undergo one replication by RdRP to form +RNA in order to code for proteins (as shown in Figure 1.5) and then another one to be packaged into its progeny.

FIGURE 1.5 Viral RNA

A virus may have plus-strand or minus-strand RNA.

mRNA
UAGGCAUCUUUCGCA ⇓ (translation) protein

Plus-strand RNA
UAGGCAUCUUUCGCA ⇓ (translation) protein

Minus-strand RNA
AUCCGUAGAAAGCGU ⇓ (transcription) UAGGCAUCUUUCGCA ⇓ (translation) protein

Other genome types unique to viruses include single stranded DNA (ssDNA) and double stranded RNA (dsRNA). When a virus inserts ssDNA into a cell, it is translocated to the nucleus, where DNA polymerase transcribes a complimentary strand, thus forming dsDNA. The replication and transcription processes then proceed as they usually do for dsDNA. dsRNA in a cell simply separates to form a free +RNA and –RNA strand. These strands are then replicated by RdRP.

Immune System
BIOLOGY 2

Some single-stranded RNA viruses, called **retroviruses**, are able to transcribe their RNA into double stranded DNA, in contrast to the usual process of transcription of DNA to produce RNA. The transcription is carried out by an enzyme called **reverse transcriptase**, which is carried by the retrovirus. The DNA produced by retroviruses can potentially be integrated into host DNA. It is exceedingly difficult to eradicate these viruses. **HIV** (*human immunodeficiency virus*) is a retrovirus that attacks cells involved in the immune response.

You can remember that reverse transcriptase transcribes RNA into DNA because it *reverses* the usual process, where DNA is transcribed into RNA.

Subviral particles, infectious agents related to viruses, include viroids and prions. **Viroids** are small rings of naked RNA without capsids, which only infect plants. Naked proteins called **prions** cause infections in animals. Prions are capable of reproducing themselves, apparently without DNA or RNA.

1.4 Bacteria

Prokaryotes do not have membrane bound nuclei. They are split into two domains called Bacteria and Archaea. **Archaea** have as much in common with eukaryotes as they do with bacteria. They are typically found in extreme environments such as salty lakes and boiling hot springs. Unlike bacteria, the cell walls of archaea are not made from peptidoglycan. Most known prokaryotes are members of the domain **Bacteria** (Greek: bakterion: small rod).

Don't get caught up in all the minute differences between Archaea and Bacteria. Just know that there is a distinction and that, although both Archaea and Bacteria are prokaryotes, Archaea have similarities to eukaryotes.

The typical structure of prokaryotes (such as the bacterium shown in Figure 1.6) is simpler than that of eukaryotes. The most basic distinction between eukary-

FIGURE 1.6 Bacterium

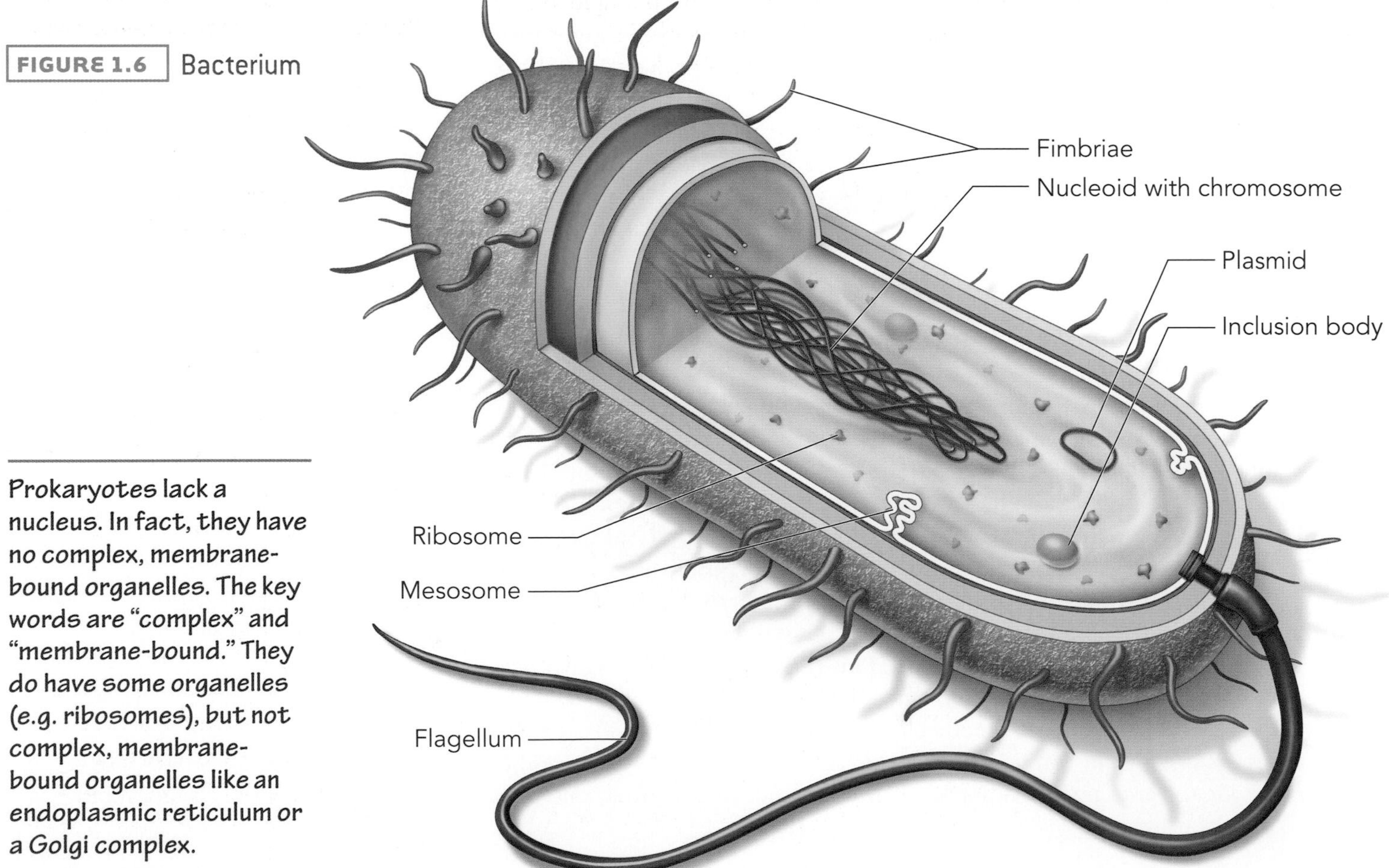

Prokaryotes lack a nucleus. In fact, they have no complex, membrane-bound organelles. The key words are "complex" and "membrane-bound." They do have some organelles (e.g. ribosomes), but not complex, membrane-bound organelles like an endoplasmic reticulum or a Golgi complex.

otes and prokaryotes is that prokaryotes do not have membrane-bound organelles, including a nucleus, while eukaryotes do. Instead of a nucleus, prokaryotes usually have a single circular double-stranded molecule of DNA. This molecule is twisted into *supercoils* and is associated with histones in Archaea and with proteins that are distinct from histones in Bacteria. The complex of DNA, RNA and proteins in prokaryotes forms a structure visible under the light microscope, called a *nucleoid*. Prokaryotes translate proteins and therefore have ribosomes. Prokaryotic ribosomes are smaller than eukaryotic ribosomes.

Bacteria come in three basic shapes: **cocci (spherical)**, **bacilli (rod-shaped)**, and **spirilla (spiral-shaped)**. Spiral, or *helical*, bacteria are called *spirochetes* if they are flexible.

The name of a bacterium often reveals the shape, like: *spiroplasma*, *staphylococcus*, or *streptococcus*.

E. coli bacteria, gram-negative bacilli (rod-shaped) bacteria, are normal inhabitants of the human intestine, and are usually harmless. However, under certain conditions their numbers may increase to such an extent that they cause infection. They cause 80% of all urinary tract infections, traveler's diarrhea (particularly in tropical countries), and gastroenteritis in children. They are widely used in genetic research.

Bacteria can co-exist with other organisms, including humans, with beneficial or harmful effects. The relationship can be **symbiotic**, or mutually beneficial for both: bacteria in the intestinal tract help humans digest food efficiently and benefit by receiving nutrients. Bacteria can also have a **parasitic** relationship with other organisms, where the relationship is beneficial to the bacteria, allowing them to grow and reproduce, but hurts the other organism.

Prokaryotes can be **anaerobic** (not dependent on oxygen for growth and survival) or **aerobic** (requiring oxygen). Early prokaryotes, which lived in an environment with no oxygen, were anaerobes. As atmospheric oxygen increased, aerobic prokaryotes evolved. A component of the eukaryotic cell, the mitochondrion, may have evolved from a symbiotic relationship between aerobic prokaryotes and eukaryotes.

Bacterial Envelopes: Protection and Movement

The cytosol of nearly all prokaryotes is surrounded by a phospholipid bilayer called the **plasma membrane**. Each phospholipid is composed of a **phosphate group**, two **fatty acid** chains, and a **glycerol backbone**. The bacterial plasma membrane, along with everything it contains, is called the *protoplast*. Surrounding the protoplast is the **bacterial envelope** (Figure 1.7). The component of the envelope adjacent to the plasma membrane is the cell wall.

The cell walls of bacteria are made of **peptidoglycan**. Peptidoglycan consists of a series of disaccharide polymer chains with amino acids, three of which are not found in proteins. These chains are connected by their amino acids, or crosslinked by an *interbridge* of more amino acids. The chains are continuous, forming a single molecular sac around the bacterium. The cell wall is also porous, so it allows large molecules to pass through. Many antibiotics such as *penicillin* attack the amino acid crosslinks of peptidoglycan. The cell wall is disrupted and the cell lyses, killing the bacterium. Some bacteria have been able to form new enzymes to attack penicillin, or even to change the molecular target so that penicillin can no longer recognize it. These bacteria gain penicillin resistance through transfer of plasmids via conjugation, an increasingly common problem in medicine.

Penicillin resistance is also a form of evolution. Bacteria that possess genes that neutralize penicillin are more likely to survive and reproduce. Their clones will also have penicillin resistance, resulting in a population that will be unaffected by certain antibiotics.

Bacteria can be classified according to the type of cell wall that they possess. *Gram staining* is a technique used to prepare bacteria for viewing under the light microscope, which stains two major cell wall types differently. Bacteria with one type of cell wall are called *gram-positive bacteria* because the thick peptidoglycan cell wall prevents the gram stain from leaking out. These cells appear purple when they undergo gram staining. In gram-positive bacteria, the cell wall is located just outside the cell membrane. The space between the plasma membrane and the cell wall is called the *periplasmic space*. The periplasmic space contains many proteins that help the bacteria acquire nutrition, such as hydrolytic enzymes.

In gram-positive bacteria, the peptidoglycan is copious and on the outside. In gram-negative bacteria, a smaller amount of peptidoglycan is located between two membranes.

FIGURE 1.7 Bacterial Envelope

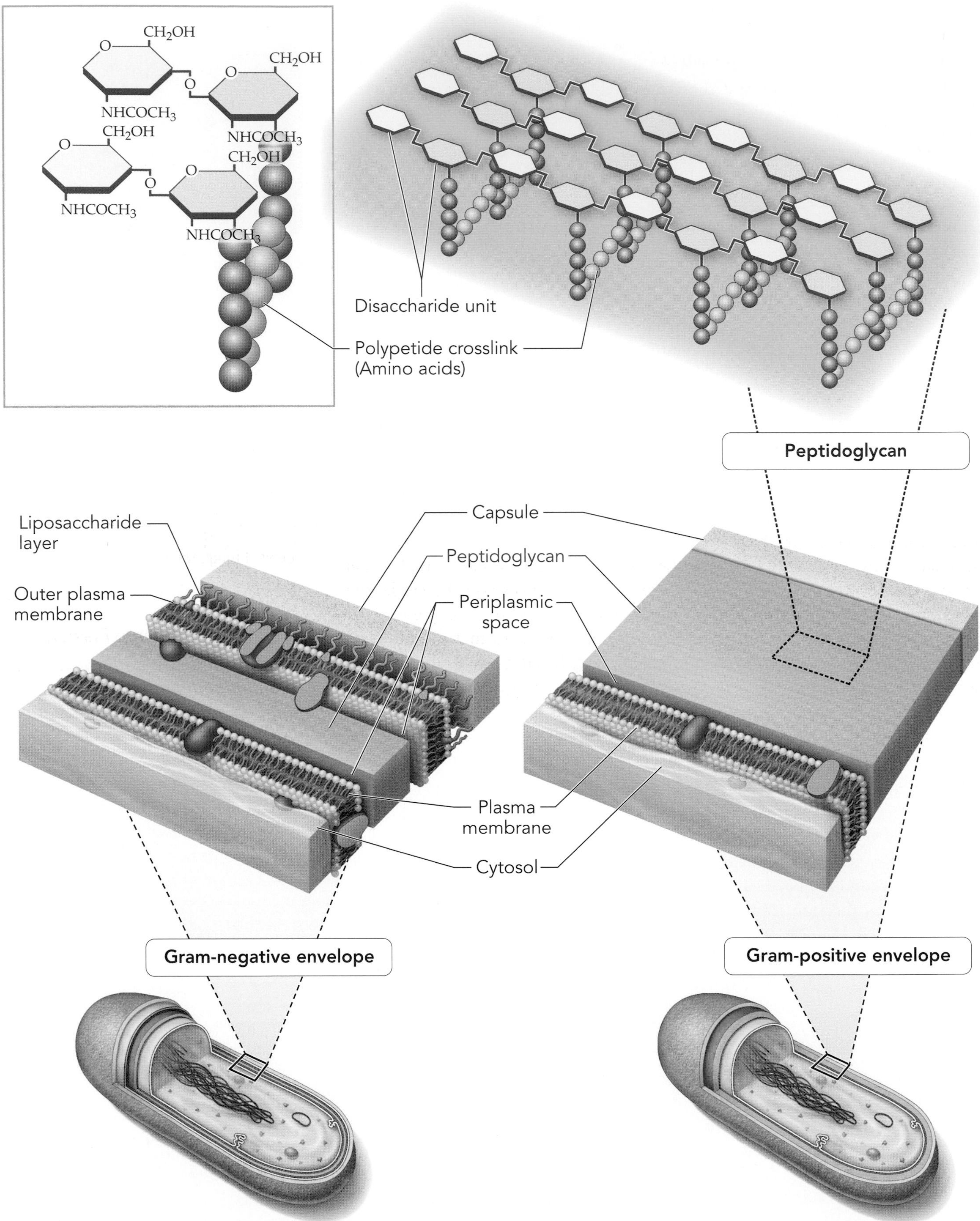

Gram-negative bacteria appear pink when gram stained. Their thin peptidoglycan cell walls allow most of the gram stain to be washed off. The small cell wall is located between two plasma membranes. The outer membrane is more permeable than the inner one, even allowing molecules the size of glucose to pass through easily. It also possesses *lipopolysaccharides*. The polysaccharide is a long chain of carbohydrates that protrudes outward from the cell. These polysaccharide chains can form a protective barrier from antibodies and many antibiotics. Some gram-negative bacteria possess *fimbriae*, or *pili*. Fimbriae are short tentacles that can attach a bacterium to a solid surface. They are not involved in cell movement.

Return to these details when you read about mitochondria in the Metabolism Lecture. Mitochondria are similar to gram-negative bacteria in that they also have two membranes and protons in their intermembrane space.

Bacterial **flagella** are long, hollow, rigid, helical cylinders made from a globular protein called **flagellin**. Bacterial flagella rotate counterclockwise (from the point of view of looking at the cell from the outside) to propel the bacterium in a single direction. When they rotate clockwise, the bacterium *tumbles*. The tumbling changes the orientation of the bacterium, allowing it to move in a new direction. The flagellum is propelled using the energy from a proton gradient rather than from ATP. **Flagellar propulsion** allows bacteria to move toward favorable products, such as food sources. This directed movement toward substances that will promote the survival and growth of the bacterium is called **chemotaxis**.

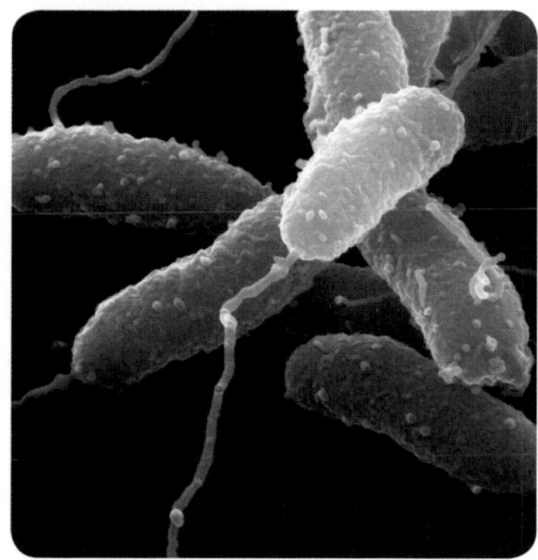

Vibrio cholerae, gram-negative rod-shaped bacteria, have a single polar flagellum (long, thin), which they use to propel themselves through water.

Reproduction and Genetic Recombination

Bacteria do not undergo meiosis or mitosis. Instead they undergo cell division called **binary fission**, a type of asexual reproduction. (Sexual reproduction, by contrast, is one method of recombining genetic information between individuals of the same species to produce a genetically distinct individual.) Three alternative forms of **genetic recombination** allow bacteria to trade DNA: conjugation, transformation, and transduction.

In binary fission, circular DNA is replicated in a process similar to replication in eukaryotes. Two DNA polymerases begin at the same point on the circle (the **origin of replication**) and move in opposite directions, making complementary single strands that combine with their template strands to form two complete DNA double-stranded circles. The cell then divides, leaving one circular chromosome in each daughter cell. The two daughter cells are genetically identical. Bacteria increase by **exponential growth**, meaning that each organism produces two offspring, which then each produce two offspring, and so on. They produce these organisms with little or no parental care until the essential nutrients of the environment are exhausted.

FIGURE 1.8 Binary Fission

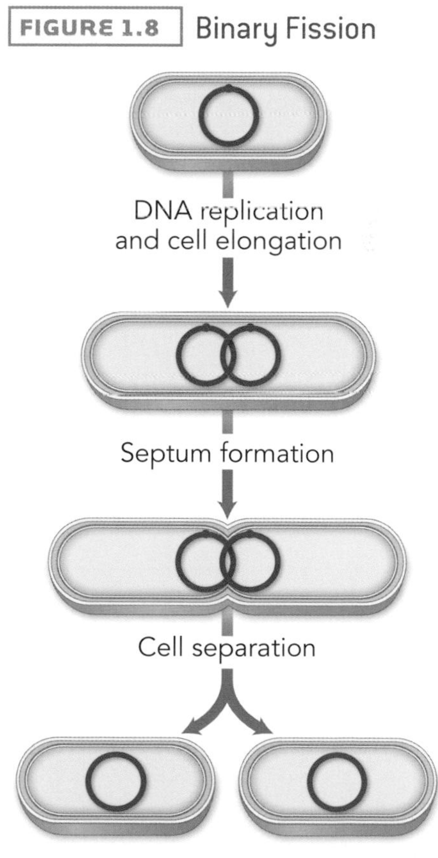

Binary fission results in two genetically identical daughter cells.

The first method of genetic recombination, **conjugation** (Figure 1.9), involves the transfer of a **plasmid**. Plasmids are small circles of **extragenomic DNA**, meaning that they exist and replicate independently from the bacterial chromosome. If the plasmid can integrate into the chromosome, it is also called an *episome*. Plasmids are not essential to the survival of the bacteria that carry them.

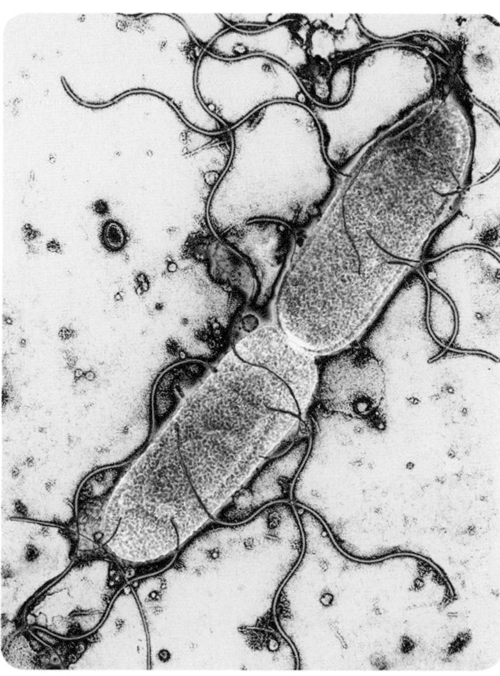

A single bacterium divides into identical daughter bacteria. Under optimal conditions, some bacteria can grow and divide extremely rapidly, and bacterial populations can double as quickly as every 10 minutes.

FIGURE 1.9 Conjugation

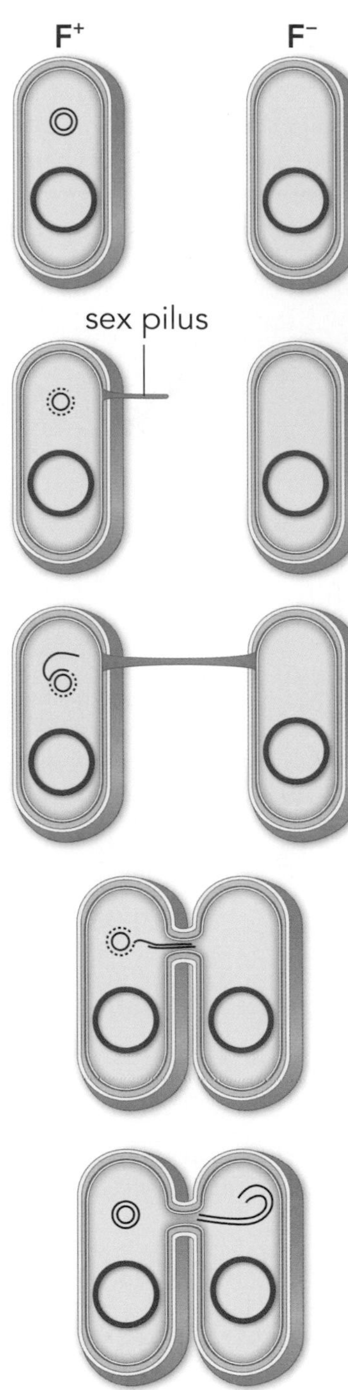

Plasmids are also used in laboratory research to express desired genes in a cell or animal model.

Not all bacteria with plasmids can conjugate. In order for a bacterium to initiate conjugation, it must contain a *conjugative plasmid*. A conjugative plasmid possesses the gene for the *sex pilus*. The sex pilus is a hollow protein tube that connects two bacteria to allow the passage of the plasmid from one to the other. The passage of DNA is always from the cell that contains the conjugative plasmid to the cell that does not.

Two types of plasmids may be mentioned on the MCAT®: the *F plasmid* and the *R plasmid*. The F plasmid is called the *fertility factor* or *F factor* because it codes for the sex pilus. A bacterium with the F factor is called F^+; one without the F factor is called F^-. The F plasmid can be in the form of an episome, and if the pilus is made while the F factor is integrated into the chromosome, some or all of the rest of the chromosome may be replicated and transferred.

The R plasmid donates resistance to certain antibiotics. It is also a conjugative plasmid. It was once common practice to prescribe multiple antibiotics for patients to take simultaneously. Such treatment promotes conjugation of multiple R plasmids, producing a super-bacterium that contains many antibiotic resistances on one or more R plasmids. Some R plasmids are readily transferred between different species of bacteria, further promoting resistance and causing health problems for humans.

The second method of genetic recombination is transformation (Figure 1.10). **Transformation** is the process by which bacteria incorporate DNA from the external environment into their genomes. DNA may be added to the external environment in the lab, or it may be released by lyses of other bacteria. Transformation can be demonstrated by mixing heat-killed virulent bacteria with harmless living bacteria. The living bacteria can receive the genes of the heat-killed bacteria through transformation and in this way become virulent.

The third type of genetic recombination, **transduction**, involves the transfer of genetic material by a virus (Figure 1.11). Transduction can occur when the capsid of a bacteriophage mistakenly encapsulates a DNA fragment of the host cell. When this virion infects a new bacterium, it injects harmless bacterial DNA fragments instead of virulent viral DNA fragments. The virus that mediates transduction is called a **vector**.

Within a bacterium, the genetic material of the chromosome is frequently rearranged. **Transposons** provide a way for nucleotides to move from one position to another. Transposons are pieces of DNA that are capable of "jumping" from one place to another along the genome. The enzyme *transposase* catalyzes the transposon's removal from and incorporation into the chromosome. Transposons are one way that genes from a plasmid can be incorporated into the genome.

FIGURE 1.10 Bacterial Transformation

1 When a bacterium dies, DNA fragments are released.

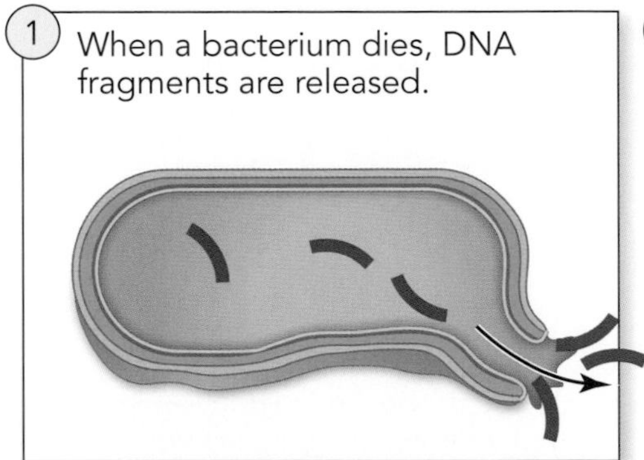

2 A live bacterium takes up the DNA fragment.

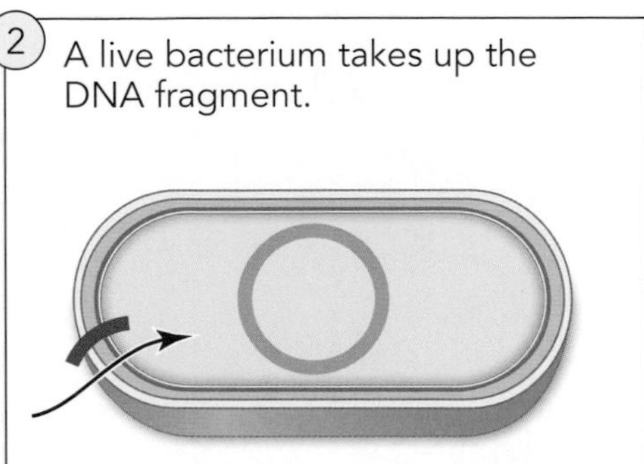

3 Through homologous recombination, DNA is incorporated into the genome of live bacterium.

FIGURE 1.11 Transduction

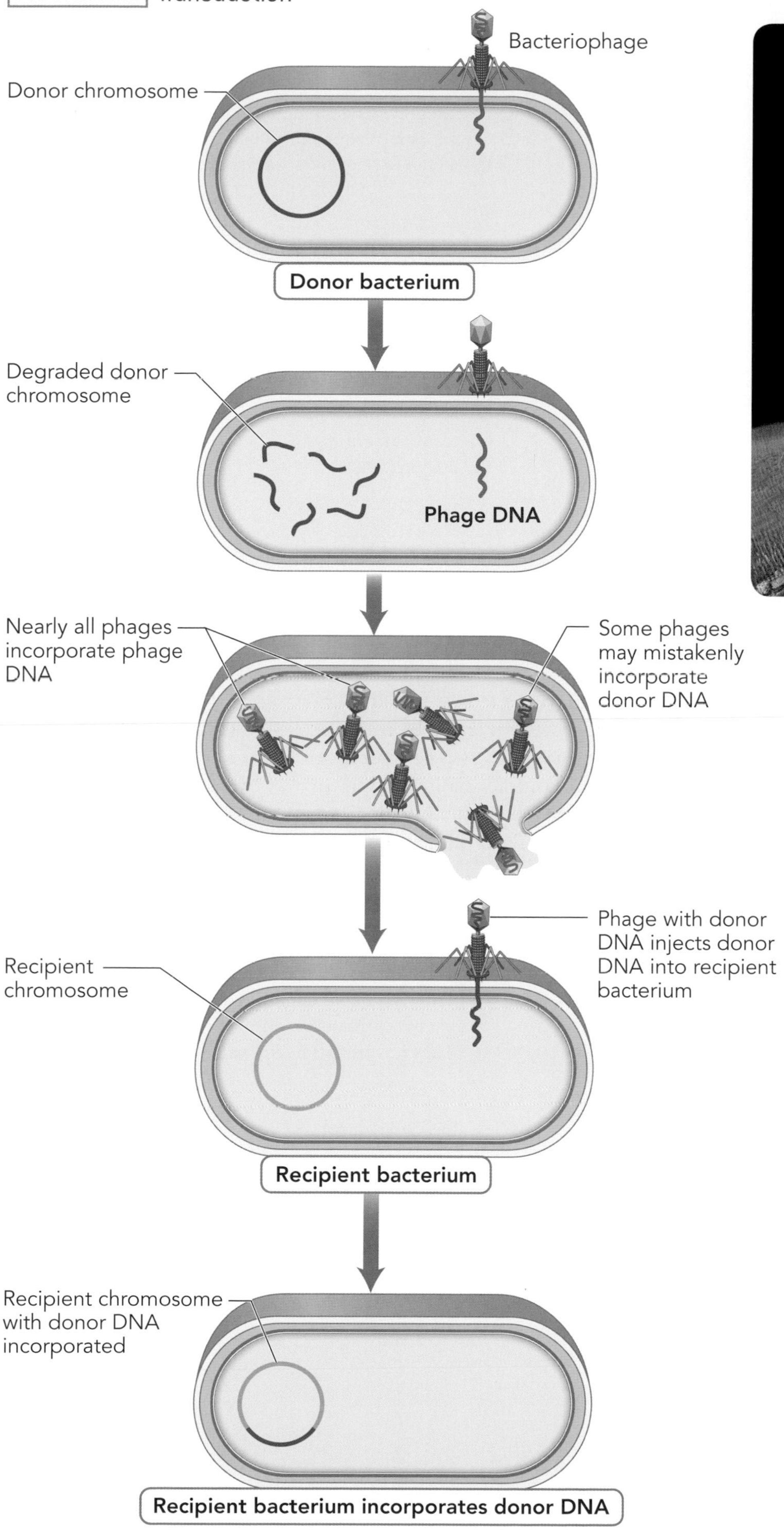

Bacteriophage

Donor chromosome

Donor bacterium

Degraded donor chromosome

Phage DNA

Nearly all phages incorporate phage DNA

Some phages may mistakenly incorporate donor DNA

Phage with donor DNA injects donor DNA into recipient bacterium

Recipient chromosome

Recipient bacterium

Recipient chromosome with donor DNA incorporated

Recipient bacterium incorporates donor DNA

A bacteriophage lands on the surface of a bacterium and injects its DNA into the cell.

These questions are NOT related to a passage.

Question 1

Which of the following events does NOT play a role in the life cycle of a typical retrovirus?

- O **A.** Viral DNA is injected into the host cell.
- O **B.** Viral DNA is integrated into the host genome.
- O **C.** The gene for reverse transcriptase is transcribed and the mRNA is translated inside the host cell.
- O **D.** Viral DNA incorporated into the host genome may be replicated along with the host DNA.

Question 2

Which of the following virus types require that an RNA-dependent RNA polymerase be carried within the virus for replication?

I. Retrovirus
II. Positive sense virus
III. Negative sense virus

- O **A.** I only
- O **B.** III only
- O **C.** I and II only
- O **D.** II and III only

Question 3

Most viruses that infect animals:

- O **A.** enter the host cell via endocytosis.
- O **B.** do not require a receptor protein to recognize the host cell.
- O **C.** leave their capsid outside the host cell.
- O **D.** can reproduce independently of a host cell.

Question 4

Which of the following structures is (are) found in bacteria?

I. A cell wall containing peptidoglycan.
II. A plasma membrane lacking cholesterol
III. Ribosomes

- O **A.** I only
- O **B.** II only
- O **C.** I and II only
- O **D.** I, II, and III

Question 5

Prior to infecting a bacterium, a bacteriophage must:

- O **A.** reproduce, making copies of the phage chromosome.
- O **B.** integrate its genome into the bacterial chromosome.
- O **C.** penetrate the bacterial cell wall completely.
- O **D.** attach to a receptor on the bacterial cell membrane.

Question 6

DNA from phage resistant bacteria is extracted and placed on agar with phage-sensitive E. coli. After incubation it is determined that these E. coli are now also resistant to phage attack. The most likely mechanism for their acquisition of resistance is:

- O **A.** transduction.
- O **B.** sexual reproduction.
- O **C.** transformation.
- O **D.** conjugation.

Question 7

Which of the following would least likely disrupt the Hardy-Weinberg equilibrium?

- O **A.** Emigration of part of a population
- O **B.** A predator that selectively targets the old and sick
- O **C.** A massive flood killing 15% of a large homogeneous population
- O **D.** Exposure of the entire population to intense radiation

Question 8

Methanogenic archaea are prokaryotes that produce methane and reside in the guts of human hosts. These microorganisms would be expected to possess which of the following characteristics?

- O **A.** They are smaller than bacteria that live in the gut.
- O **B.** They are resistant to the antibiotic penicillin.
- O **C.** They have abundant mitochondria to survive in harsh conditions.
- O **D.** They have small nuclei and rod-like cell shapes.

Questions 1-8 of 144

STOP

1.5 Eukaryotic Cells: Membrane-Bound Organelles

Organisms in the domain Eukarya are characterized by cells that each have a nucleus and other membrane bound organelles. The major differences between prokaryotes and eukaryotes are summarized in Table 1.1.

TABLE 1.1 > **Prokaryotes vs. Eukaryotes**

Prokaryotes	Eukaryotes
No membrane bound nucleus	True membrane bound nucleus
No membrane bound organelles	Membrane bound organelles
"Naked" DNA, without histone proteins	DNA is coiled with histone proteins
mRNA does not undergo post-translational modifications	mRNA undergoes splicing, addition of poly-A tail, and addition of 5′ cap
Ribosomes are smaller	Ribosomes are larger
Cell walls are composed of peptidoglycan	Cell walls, if present, are composed of chitin (fungi) or cellulose (plants)
Flagella are made of flagellin	Flagella are made of microtubules
Division by binary fission	Division by mitosis

The domain Eukarya encompasses four distinct kingdoms. The kingdom of *Protista* contains the oldest organisms of the domain Eukarya, as well as the evolutionary precursors of the other three kingdoms: animal-like, plant-like, and fungi-like protists. The kingdoms *Fungi*, *Plantae*, and *Animalia* all separated from the end of this evolutionary branch at approximately the same time. Fungi are known for their filaments called *hyphae* that form masses of mycelium. Hyphae are not truly multicellular, but are divided by thin walls called *septa* that separate the nuclei.

Plants are true multicellular eukaryotes. They are all *photosynthetic autotrophs*, meaning that they obtain energy from sunlight.

Animals exhibit tissue complexity and specialization as well as body symmetry. They also exhibit *cephalization*, a nervous system-like structure with a concentration of nerve tissue at one end. In addition, they have true digestive tracts.

The rest of this lecture will mostly apply to human cells, although some concepts (such as membranes) are applicable to all cells. As cellular organelles are discussed, their roles in the functioning of organs will be emphasized. All cells contain all of these organelles, but organelle functions can sometimes be understood more clearly using specific organs as analogies or examples. Also, notice that components of the cell are organized according to their physical properties. DNA, proteins, and lipids are all stored and processed in separate places within the cell.

Don't worry about memorizing the features of the Fungi, Plantae, and Animalia kingdoms. They are included here to help you see what makes animals unique.

Apply Key 1: The cell separates biological processes into compartments within the cell. This is especially true for competing processes such as fatty acid oxidation, which occurs in the mitochondria, and fatty acid synthesis, which occurs in the cytosol. Separating competing processes allows them to be reciprocally regulated.

The Nucleus as the "Brain"

The defining feature that distinguishes eukaryotic (Greek: eu → well, karyos → kernel) cells from prokaryotic cells is the **nucleus** (shown in Figure 1.12). The nucleus contains all of the DNA in an animal cell, other than a small amount in the mitochondria. DNA is transcribed to create proteins that carry out functions according to the needs of the cell. The nucleus directs the activities of the cell, much as the brain directs the activities of the body. Both the nucleus and the brain receive input from the environment and coordinate a response.

The aqueous fluid inside the nucleus is called the *nucleoplasm.* The nucleus is wrapped in a double phospholipid bilayer called the **nuclear envelope**, which is perforated with large holes called **nuclear pores**. RNA can exit the nucleus through the nuclear pores, but DNA cannot. The nucleus contains an area called the **nucleolus**, where ribosomal RNA (rRNA) is transcribed and the subunits of the ribosomes are assembled. The nucleolus is not separated from the rest of the nucleus by a membrane.

FIGURE 1.12 Eukaryotic Cell

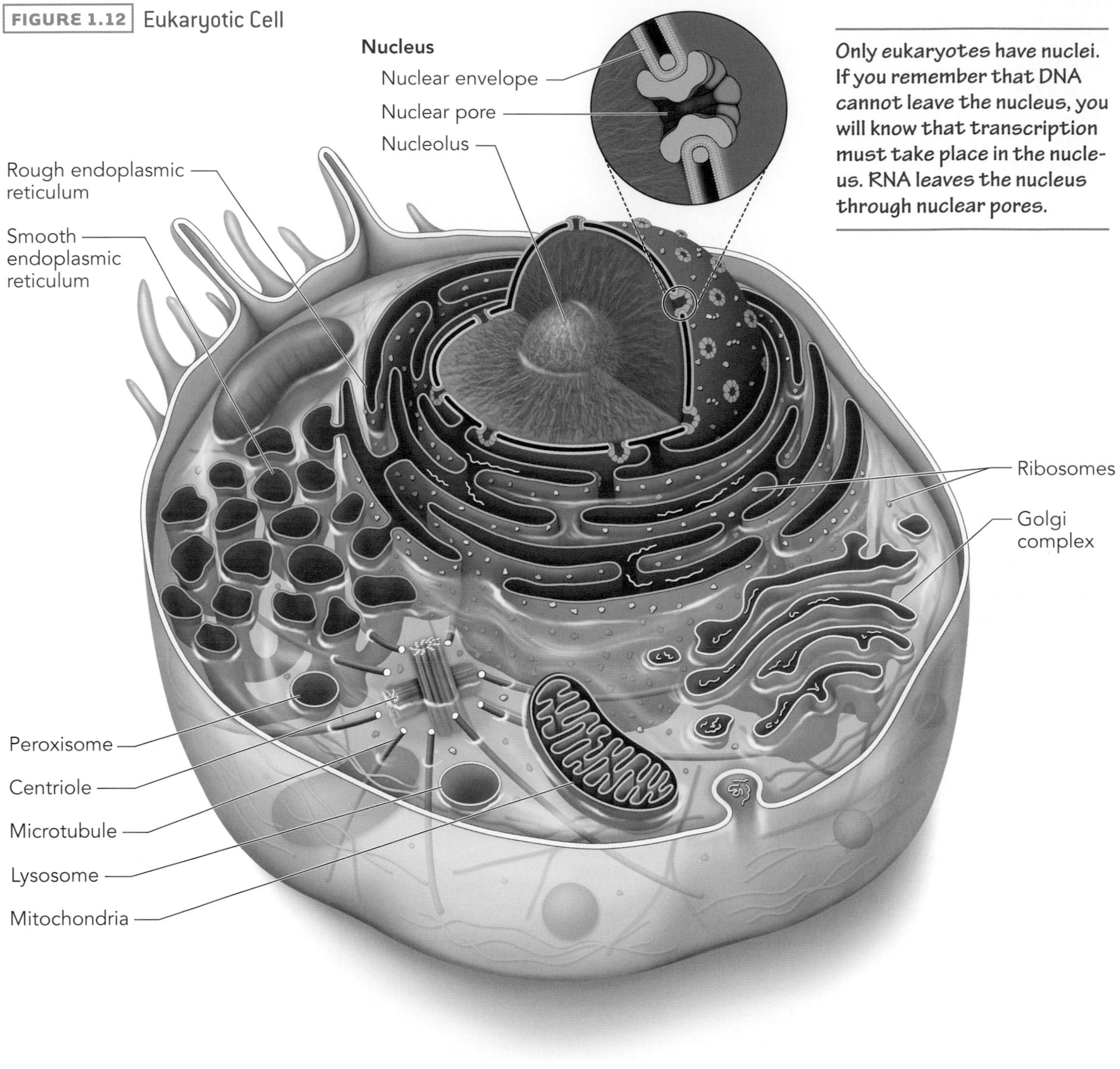

Nucleus
Nuclear envelope
Nuclear pore
Nucleolus
Rough endoplasmic reticulum
Smooth endoplasmic reticulum
Ribosomes
Golgi complex
Peroxisome
Centriole
Microtubule
Lysosome
Mitochondria

Only eukaryotes have nuclei. If you remember that DNA cannot leave the nucleus, you will know that transcription must take place in the nucleus. RNA leaves the nucleus through nuclear pores.

Proteins and Vesicles

Protein synthesis begins on a ribosome and uses mRNA as a template. As described in the Genetics Lecture, mRNA is transcribed from DNA in the nucleus. It then exits the nucleus so that protein synthesis can occur in the cytosol. The mRNA is translated on a ribosome, which feeds the newly synthesized protein into the cytosol or the endoplasmic reticulum, depending on the protein's final destination.

FIGURE 1.13 Trafficking of Secretory Proteins

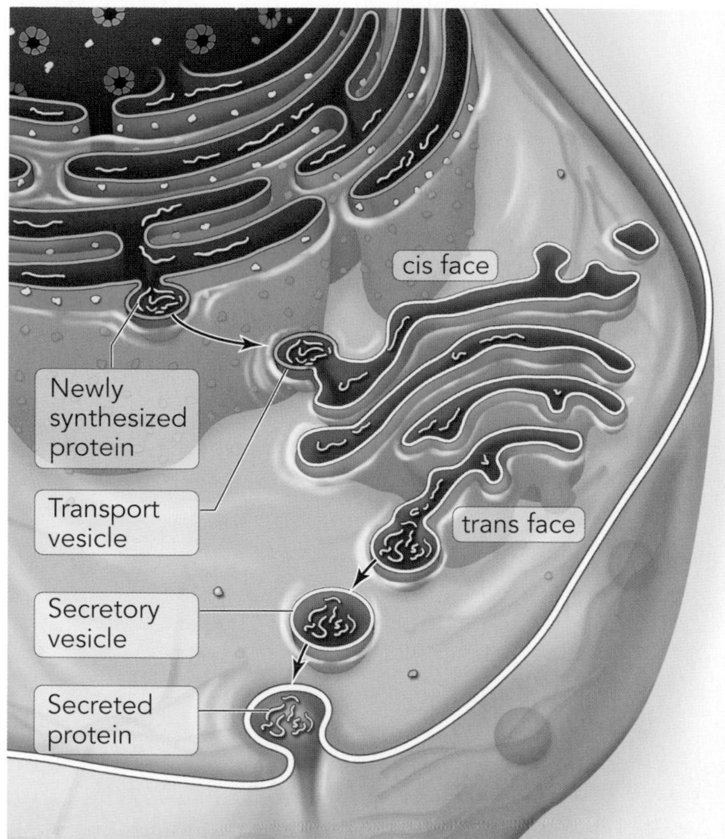

FIGURE 1.14 Transport and Fate of Lysosomal Proteins

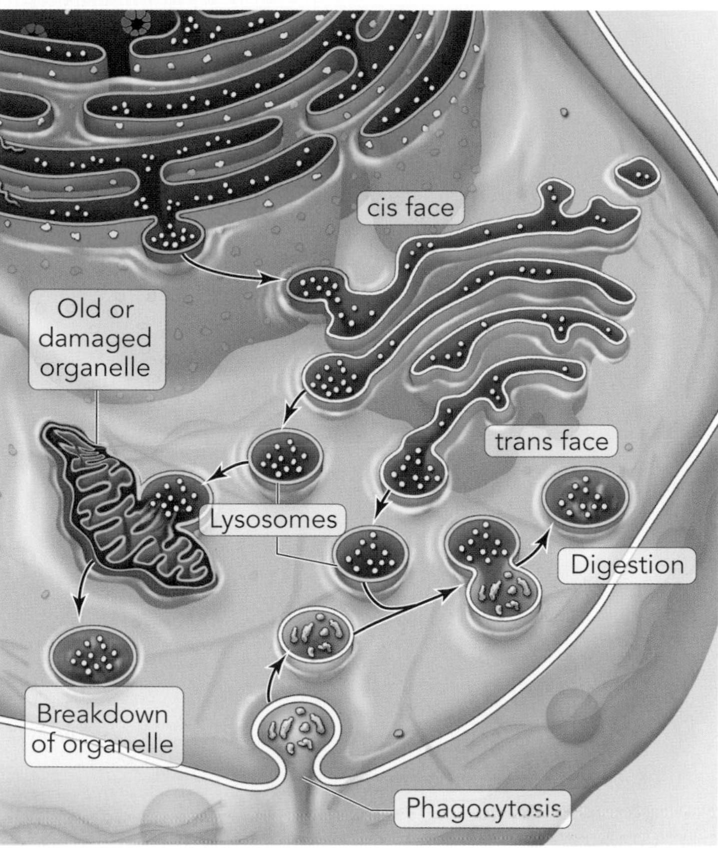

Proteins that will remain in the cytosol, the aqueous solution inside the cell, are translated in the cytosol. Examples of these proteins include the enzymes of glycolysis and actin filaments in muscle. Both are translated in the cytosol and remain there.

Proteins that will ultimately be exported from the cell or sequestered in a vesicle are translated on the **endoplasmic reticulum (ER)**. Proteins that are exported from the cell include protein hormones, such as ACTH, and bloodstream proteins, such as albumin. The ER is a thick maze of membranous walls in the eukaryotic cell separating the cytosol from the **ER lumen** or *cisternal space*. In some places the ER lumen is contiguous with the outer layer of the nuclear envelope. ER near the nucleus has many ribosomes attached to it on the cytosolic side, giving it a granular appearance, so it is called the **rough ER**. As mRNA is translated, a particular sequence of amino acids known as a *signal sequence*, if present, directs the protein to the ER membrane for the completion of translation. Proteins that are translated on the rough ER are propelled into the ER lumen as they are created.

The movement of a protein from the cytosol to the ER is just one example of the use of a signal sequence. Signal sequences direct molecules all around the cell – to lysosomes, the mitochondrion, and even to the nucleus.

The newly synthesized proteins are moved through the lumen toward the **Golgi apparatus** or *Golgi complex*. The Golgi apparatus is a series of flattened, membrane bound sacs whose major functions are packaging and secreting proteins. Small *transport vesicles* bud off from the ER and carry proteins across the cytosol to the Golgi. The Golgi then organizes and concentrates the proteins as they are shuttled by transport vesicles progressively outward from one compartment to the next. Proteins are distinguished by their signal sequences and carbohydrate chains. The Golgi may alter proteins chemically by *glycosylation*, the addition of a carbohydrate, or by removing amino acids. The end-products of the Golgi are vesicles full of proteins. The vesicles may be expelled from the cell as secretory vesicles, released from the Golgi to mature into lysosomes, or transported to other parts of the cell such as the mitochondria or even back to the ER, as shown in Figures 1.13 and 1.14. Various molecules on the surfaces of vesicles direct them to specific subcellular compartments.

The Golgi apparatus can be thought of as an assembly line. Proteins are passed from one compartment to the next and modified along the way.

Secretory vesicles may contain enzymes, growth factors, or extracellular matrix components. They release their contents through exocytosis. Since exocytosis incorporates vesicle membranes into the cell membrane, secretory vesicles can supply the cell membrane with its integral proteins and lipids. Secretory vesicles can also contribute to membrane expansion. In the reverse process, endocytotic vesicles made at the cell membrane are shuttled back to the Golgi for recycling of the cell membrane. Secretory vesicles are continuously released by most cells (*constitutive secretion*). Some specialized cells can release secretory vesicles in response to certain chemical or electrical stimuli (*regulated secretion*). Some proteins are activated within secretory vesicles. Proinsulin, for example, is cleaved to insulin only after the secretory vesicle buds off the Golgi.

Lysosomes are a type of vesicle that contain **hydrolytic enzymes**. This type of enzyme catalyzes the breakdown of macromolecules by hydrolysis. The hydrolytic enzymes contained in lysosomes are *acid hydrolases*, which function best in an acidic environment. Together, these enzymes are capable of breaking down every major type of macromolecule. Lysosomes usually have an interior pH of 5. They fuse with endocytotic vesicles (the vesicles formed by phagocytosis and pinocytosis, described later in this lecture) and digest their contents. Any material not degraded by the lysosome is ejected from the cell through exocytosis. Lysosomes also take up and degrade cytosolic proteins in an endocytotic process. Under certain conditions lysosomes rupture and release their contents into the cytosol, killing the cell.

Endocrine System
BIOLOGY 2

Sites of extracellular protein secretion include some endocrine glands and the liver. Endocrine glands such as the pituitary gland and the pancreas create protein hormones that follow the path of secretion described above. Albumin and blood clotting factors are also secreted from the liver by this route. Any cell that secretes proteins can be expected to have well-developed ER and Golgi bodies. However, lysosomes are not usually seen in high concentration in living cells. They exist in large concentrations in cells that are about to undergo **apoptosis**, also called **programmed cell death**. Apoptosis can contribute to development. At one point during the development of a human embryo, the hands are webbed. The skin between the fingers eventually undergoes apoptosis, which is mediated by an increased concentration of lysosomes.

There is a lot of background information here that the MCAT® will not directly test. For the MCAT®, know the bolded terms and focus on the following points:

1. There are many internal compartments in a cell (organelles) separated from the cytosol by membranes.
2. Rough ER has ribosomes attached to its cytosolic side, and it synthesizes virtually all proteins that do not belong in the cytosol. Proteins synthesized on the rough ER are pushed into the ER lumen and sent to the Golgi apparatus.
3. The Golgi apparatus modifies and packages proteins for use in other parts of the cell and outside the cell.
4. Lysosomes contain hydrolytic enzymes that digest substances taken into the cell by endocytosis. Lysosomes come from the Golgi apparatus.

Smooth ER and Peroxisomes: Fat Storage and Toxin Breakdown

The portion of the endoplasmic reticulum that lacks ribosomes is called the **smooth endoplasmic reticulum**. Smooth ER tends to be tubular, in contrast to rough ER, which tends to resemble flattened sacs. Smooth ER has a number of functions that differ according to the type of cell. In the liver and kidney, smooth ER contains *glucose 6-phosphatase*, the enzyme used in the liver, intestinal epithelial cells, and renal tubule epithelial cells to hydrolyze *glucose 6-phosphate* to glucose, an important step in the breakdown of glycogen to produce glucose. In muscle cells, the smooth ER is known as the sarcoplasmic reticulum, and it sequesters calcium away from actin and myosin.

When you see smooth ER on the MCAT®, think of lipid metabolism and storage as well as detoxification. Don't get carried away with the other details.

The smooth ER also plays a role in lipid metabolism. In the liver, triglycerides are produced in the smooth ER. These lipids are stored in **adipocytes**, cells containing predominantly fat droplets, again inside of smooth ER. Such cells contribute to energy storage and body temperature regulation. In some endocrine glands, such as the adrenal gland, the smooth ER and the cytosol share in the role of cholesterol formation and its subsequent conversion to various steroids. In addition, smooth ER oxidizes foreign substances, detoxifying drugs, pesticides, toxins, and pollutants.

Peroxisomes are vesicles in the cytosol that are involved in both lipid and protein storage. They grow by incorporating lipids and proteins from the cytosol. Rather than budding off membranes like lysosomes from the Golgi, peroxisomes self-replicate. They are involved in the production and breakdown of **hydrogen peroxide**, a byproduct that has the potential to harm the cell. Peroxisomes inactivate toxic substances such as alcohol, regulate oxygen concentration, play a role in the synthesis and breakdown of lipids, and are involved in the metabolism of nitrogenous bases and carbohydrates.

Mitochondria and Energy

Mitochondria (Figure 1.15) are the powerhouses of the eukaryotic cell. They are the site of ATP production, as described in the Metabolism Lecture. According to the *endosymbiotic theory*, mitochondria may have evolved from a symbiotic relationship between ancient prokaryotes and eukaryotes. Like prokaryotes, mitochondria contain circular DNA that replicates independently from the nuclear DNA. Mitochondrial DNA contains no histones or nucleosomes. Most animals have a few dozen to several hundred molecules of circular DNA in each mitochondrion.

The genes in mitochondrial DNA code for mitochondrial RNA that is distinct from the RNA in the rest of the cell. Mitochondria have their own ribosomes. However, most proteins used by mitochondria are coded for by nuclear DNA, not mitochondrial DNA. Antibiotics that block translation by prokaryotic ribosomes but not eukaryotic ribosomes also block translation by mitochondrial ribosomes. Some of the codons in mitochondria differ from the codons in the rest of the cell, an exception to the universal genetic code. Mitochondrial DNA is passed to offspring from the mother even in organisms in which male gametes contribute to the cytoplasm of the egg.

Like gram-negative bacteria, mitochondria are surrounded by two phospholipid bilayers. The **inner membrane** invaginates (folds in on itself) to form *cristae*. The inner membrane holds the electron transport chain of aerobic respiration. Between the inner and **outer membrane** is the **intermembrane space**.

High concentrations of mitochondria are seen in the cell wherever energy needs are high. Muscle cells, for example, contain a large number of mitochondria.

FIGURE 1.15 Mitochondria

The inner membrane of a mitochondrion folds inward to form cristae.

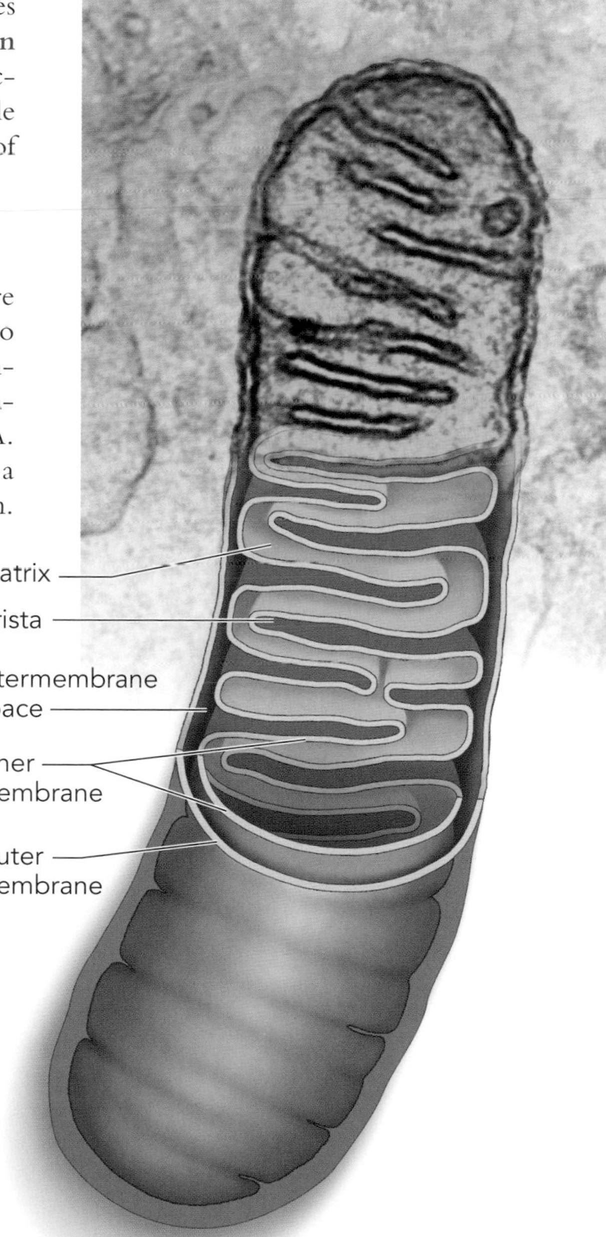

Consider Key 2: The organelles of a cell are similar to the organs of the body. Each organelle has a specific function in the life of a cell, whether that be providing energy, like the mitochondria, or storing genetic information, like the nucleus.

Don't confuse eukaryotic flagella with prokaryotic flagella. Eukaryotic flagella are made from a 9+2 microtubule configuration; a prokaryotic flagellum is a thin strand of a single protein called flagellin. Eukaryotic flagella undergo a whip-like action, while prokaryotic flagella rotate.

The Cytoskeleton: Organizing Structure and Substances

The structure and motility of a cell is determined by a network of filaments called the **cytoskeleton**. The cytoskeleton anchors some membrane proteins and other cellular components, moves components within the cell, and moves the cell itself. The three basic parts of the cytoskeleton are microtubules, microfilaments, and intermediate filaments. Microtubules are larger than intermediate filaments, which are larger than microfilaments.

Microtubules provide a platform for **transport** within cells. Molecular motors can use microtubules to taxi secretory vesicles throughout the cell. The mitotic spindle that moves chromosomes around the cell is made of microtubules. They also **support** the shape of the cell.

Structurally, microtubules are hollow tubes made from the protein *tubulin*. Tubulin is a globular protein that polymerizes into long straight filaments under certain conditions. Thirteen of these filaments lie alongside each other to form the tube. Microtubules have a + and – end. The – end attaches to a **microtubule-organizing center (MTOC)**. A microtubule grows away from an MTOC at its + end. The major MTOC in animal cells is the *centrosome*. The centrosome is composed of a pair of **centrioles**, which function in the production of flagella and cilia but are not necessary for microtubule production.

FIGURE 1.16 Structure of Flagella and Cilia

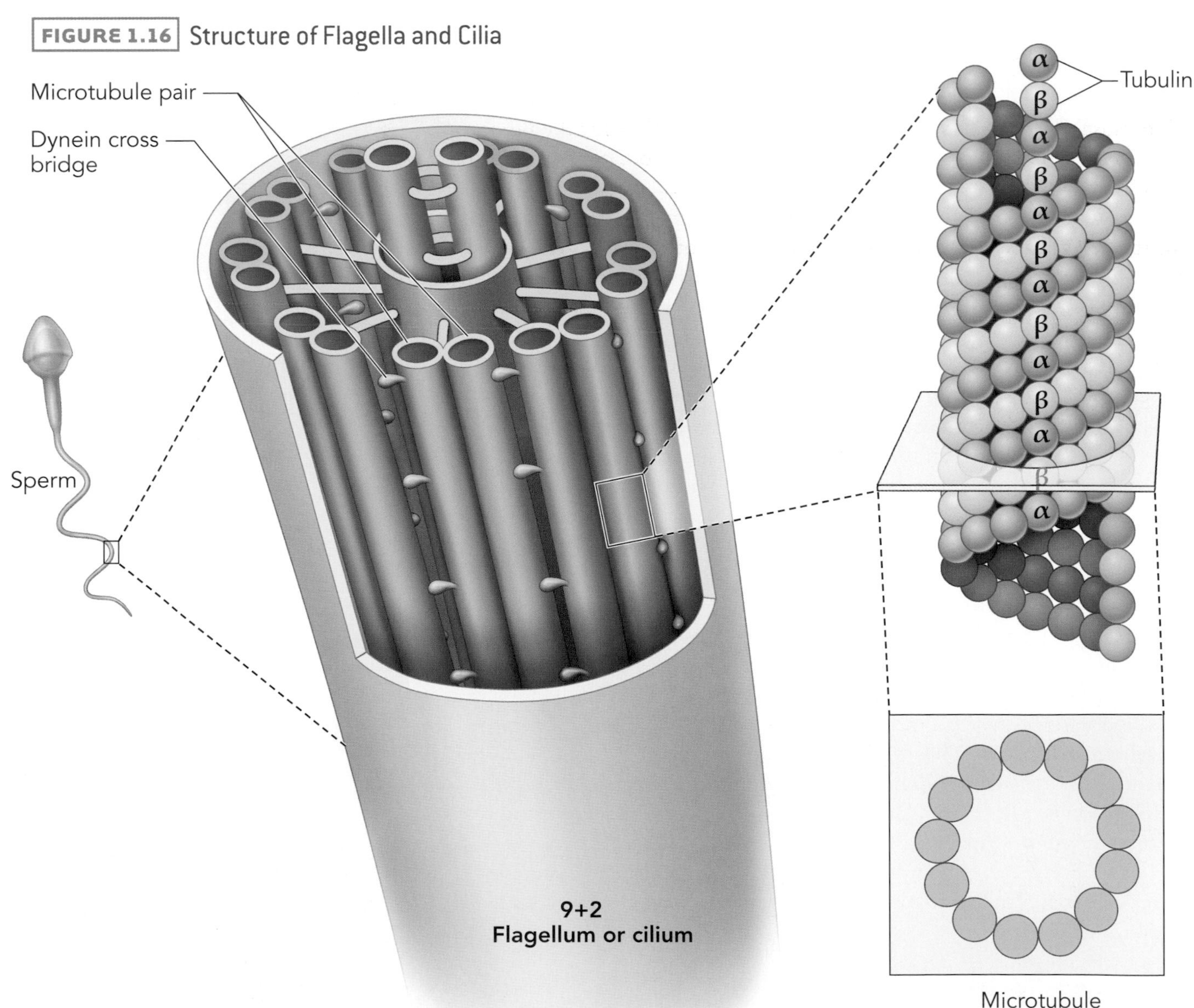

FIGURE 1.17 Motion of Flagella and Cilia

Lung and trachea cilia

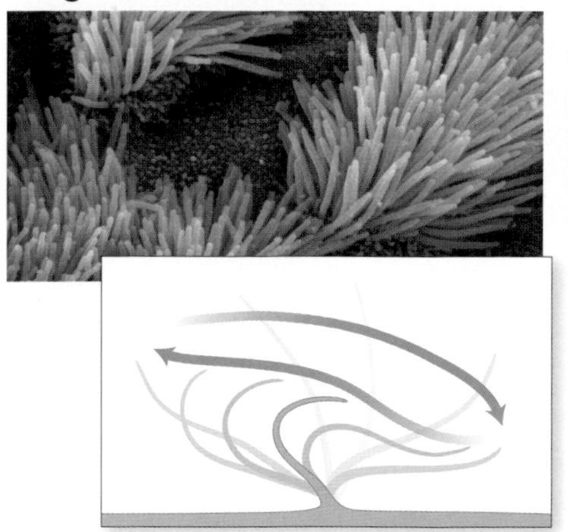

Spermatozoan flagellum

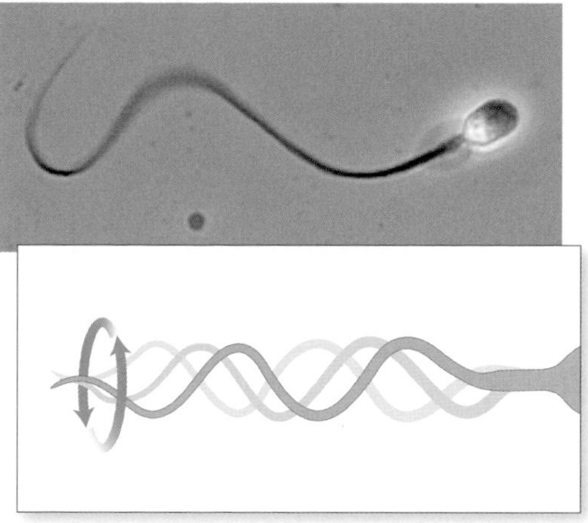

Why would it be useful for a cell to be able to shift the fluid around it, other than for movement? One example is in the respiratory tract, where cilia move fluid and foreign substances so that they can be removed from the body.

Flagella and **cilia** are specialized structures made from microtubules. They function to move fluid, causing the cell itself or nearby substances to move. The major portion of each flagellum and cilium contains nine pairs of microtubules that form a circle around two lone microtubules in a **9+2 arrangement**. Cross bridges made from the protein **dynein** connect each outer pair of microtubules to its neighbor. The cross bridges cause microtubule pairs to slide along their neighbors, creating a whip action in cilia that causes fluid to move laterally, or a wiggle action in flagella that causes fluid to move directly away from the cell.

Microfilaments are another name for actin filaments. They interact with myosin to cause muscle contraction. They are also responsible for the pinching of the cytoplasm during cytokinesis (**cleavage**). Notice that both of these actions reshape the cell membrane.

Intermediate filaments are the third portion of the cytoskeleton, which maintain the cell's shape. Intermediate filaments are not nearly as dynamic as microtubules or microfilaments and primarily serve to impart structural rigidity to the cell. Keratin, a type of intermediate filament found in epithelial cells, is associated with hair and skin.

FIGURE 1.18 Cellular Filaments

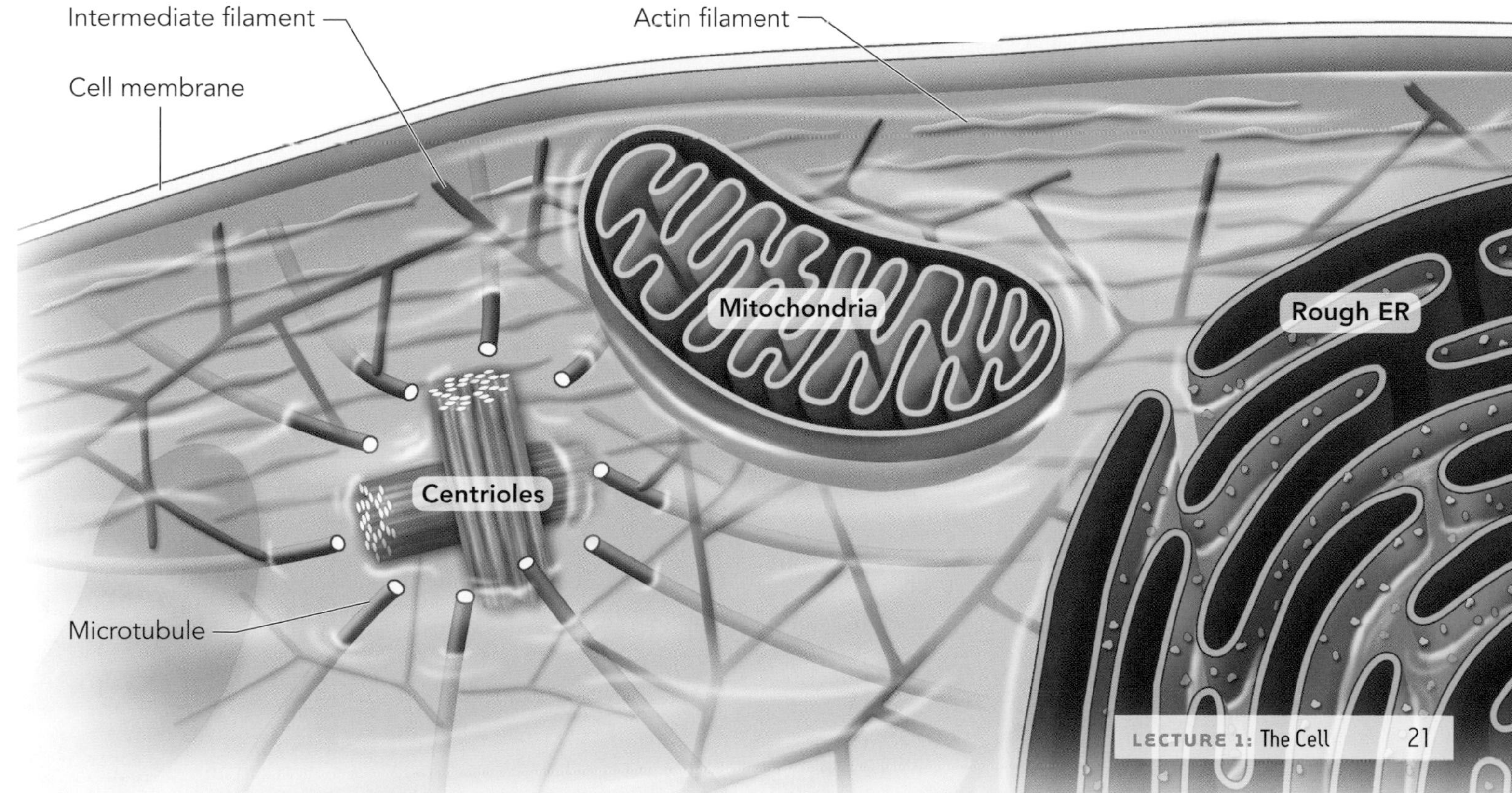

These questions are NOT related to a passage.

Question 9

All of the following are composed of microtubules EXCEPT:

- O **A.** the tail of a sperm cell.
- O **B.** the spindle apparatus.
- O **C.** the cilia of the fallopian tubes.
- O **D.** the flagella of bacteria.

Question 10

Which of the following is true concerning the nucleolus?

- O **A.** It is bound by a phospholipid membrane.
- O **B.** It disappears during prophase.
- O **C.** It is the site of translation of ribosomal RNA.
- O **D.** It is found in most bacteria.

Question 11

Researchers discover a toxin that prevents the activity of all ribosomes. Which of the following statements best describes this compound?

- O **A.** The toxin will only affect prokaryotes.
- O **B.** The toxin will only affect eukaryotes.
- O **C.** The toxin will bind most strongly to the smooth ER and enzymes inside the peroxisome.
- O **D.** The toxin will bind most strongly to the rough ER and enzymes inside the mitochondria.

Question 12

Which of the following cells would be expected to contain the greatest concentration of smooth endoplasmic reticulum?

- O **A.** A liver cell
- O **B.** An islet cell from the pancreas
- O **C.** A mature sperm
- O **D.** A zygote

Question 13

Researchers studying the impact of alcohol on the body inject alcohol into the bloodstream of a mouse. Which organelle would most likely show increased activity in the hours that follow?

- O **A.** Rough ER
- O **B.** Smooth ER
- O **C.** Nucleolus
- O **D.** Lysosome

Question 14

One function of the liver is to detoxify alcohol taken into the body. The organelle within the liver cell that most directly affects this process is:

- O **A.** the smooth endoplasmic reticulum.
- O **B.** the nucleus.
- O **C.** the Golgi apparatus.
- O **D.** the rough endoplasmic reticulum.

Question 15

When a primary lysosome fuses with a food vesicle to become a secondary lysosome:

- O **A.** its pH drops via active pumping of protons into its interior.
- O **B.** its pH drops via active pumping of protons out of its interior.
- O **C.** its pH rises via active pumping of protons into its interior.
- O **D.** its pH rises via active pumping of protons out of its interior.

Question 16

Which of the following cells would be expected to have a well-developed rough endoplasmic reticulum?

- O **A.** Pancreatic cells
- O **B.** Adipocytes
- O **C.** Muscle cells
- O **D.** Neurons

1.6 Membrane Transport

The eukaryotic cell, like the prokaryotic cell, has a phospholipid bilayer membrane surrounding the cytosol. A major function of the cell membrane is to regulate the substances that enter and leave the cell. Cell membranes also facilitate communication between cells.

A phospholipid (Figure 1.19) is made up of a glycerol backbone with a phosphate group and two fatty acid chains attached. The phospholipid is often drawn as a balloon with two strings. The balloon portion represents the phosphate group, and the strings represent the fatty acids. The phosphate group is polar, while the fatty acid chains are nonpolar, making the molecule **amphipathic** (having both a polar and a nonpolar portion). When placed in aqueous solution, amphipathic molecules spontaneously aggregate, turning their polar ends toward the solution and their nonpolar ends toward each other. The resulting spherical structure is called a **micelle** (Figure 1.20). If enough phospholipids are present in a solution that is subjected to ultrasonic vibrations, *liposomes* may form. A liposome is a vesicle surrounded by and filled with aqueous solution. It contains a lipid bilayer like that of a plasma membrane.

The inner and outer layers of a membrane are called leaflets. In addition to phospholipids, the plasma membrane contains other types of lipids such as glycolipids. Eukaryotic membranes, unlike prokaryotic membranes, contain steroids such as cholesterol. Cholesterol helps maintain membrane fluidity and stability across a wide range of temperatures. Lipid types are arranged asymmetrically between the leaflets. For instance, glycolipids are found on the outer leaflet only.

Biological Molecules and Enzymes
BIOLOGY 1

Make sure the chemical properties of a phospholipid bilayer are clear in your mind: at the edges it is hydrophilic, and at the center it is hydrophobic. Most of the depth of a membrane consists of the hydrophobic fatty acid chains.

FIGURE 1.19 Phospholipid Structure

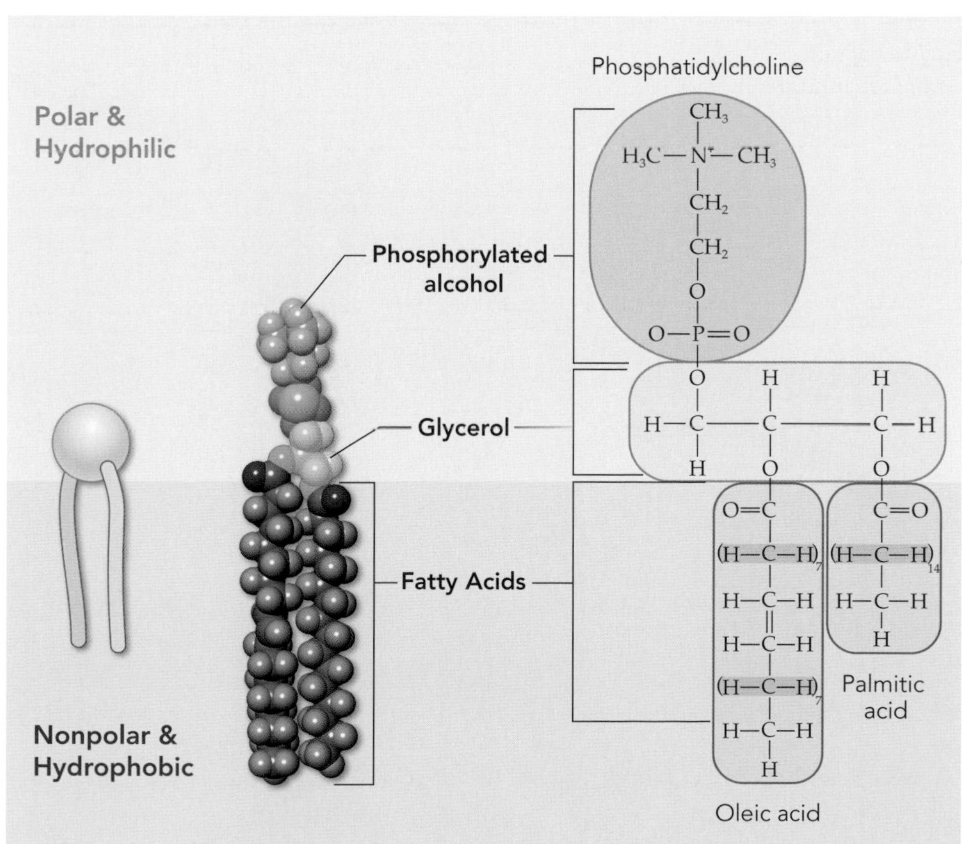

FIGURE 1.20 Micelle

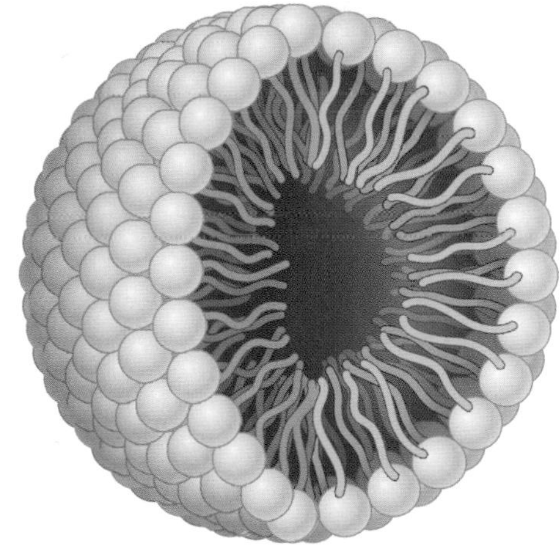

Micelles form spontaneously, whereas membranes must be assembled. If phospholipids are dumped into an aqueous solution, a micelle will form because it is the most thermodynamically stable arrangement.

FIGURE 1.21 Functions of Membrane Proteins

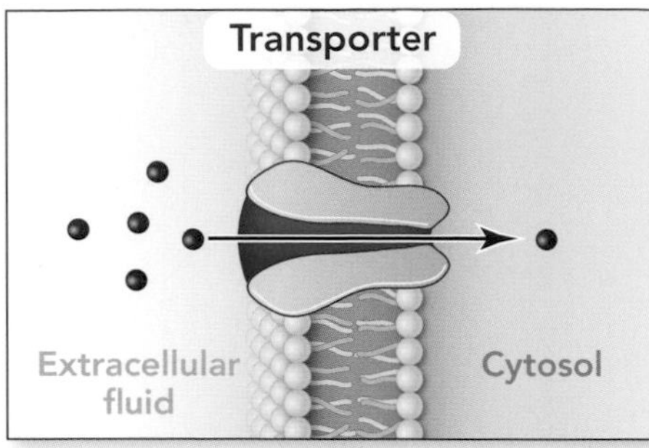

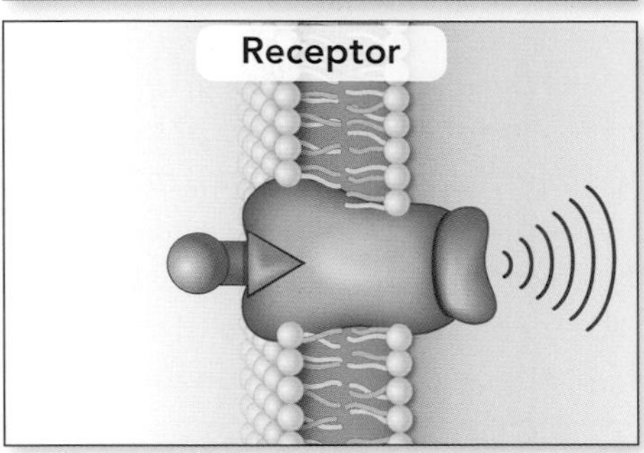

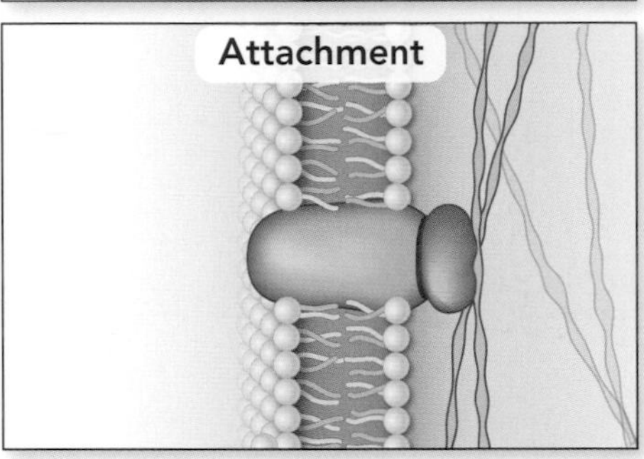

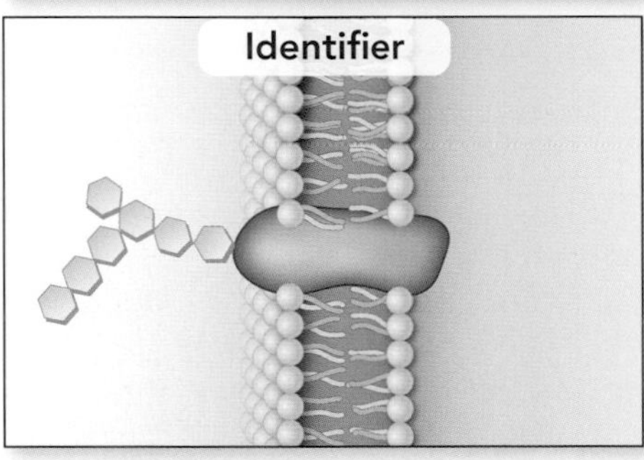

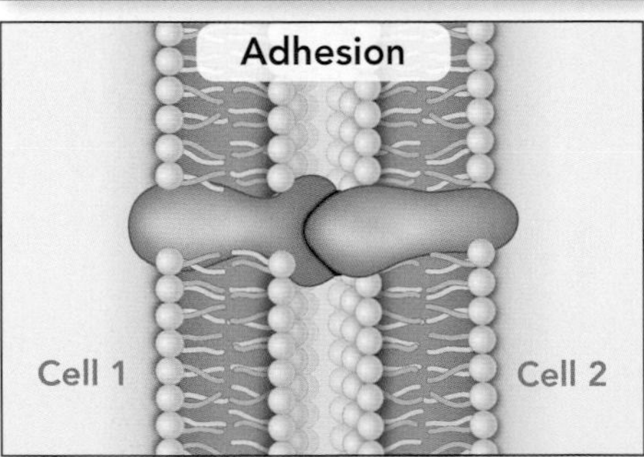

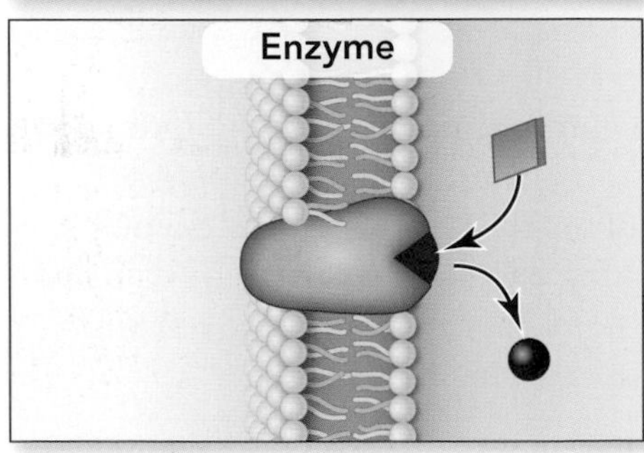

Proteins are also embedded within the plasma membrane. Amphipathic proteins can cross the membrane from the inside of the cell to the outside and are called **integral** or **intrinsic proteins**. By contrast, **peripheral** or **extrinsic proteins** are located on the surfaces of the membrane and are generally hydrophilic. They are ionically bonded to integral proteins or the polar group of a lipid. Both integral and peripheral proteins may contain carbohydrate chains, making them *glycoproteins*. The carbohydrate portion of a membrane glycoprotein always protrudes toward the outside of the cell. Membrane proteins are distributed asymmetrically throughout the membrane and between the leaflets. Neither proteins nor lipids flip easily from one leaflet to the other.

The model of the membrane just described is called the **fluid mosaic model** (Figure 1.22). The fluid part refers to the phospholipids and proteins, which can slide past each other. Since the forces holding the entire membrane together are intermolecular, the membrane is fluid; its parts can move laterally but cannot separate. The mosaic part refers to the asymmetrical layout of a membrane's lipids and proteins.

Most of the functions of membranes can be attributed to their proteins. Membrane proteins act as transporters, receptors, attachment sites, identifiers, adhesive proteins, and enzymes (Figure 1.21). As transporters, membrane proteins select which solutes enter and leave the cell. Other membrane proteins act as receptors by receiving chemical signals from the cellular environment. Some membrane proteins are attachment sites that anchor to the cytoskeleton. Membrane proteins can act as identifiers that other cells recognize. Adhesion by one cell to another is accomplished by membrane proteins. Many of the chemical reactions that occur within a cell are governed by membrane proteins on the inner surface.

FIGURE 1.22 Fluid Mosaic Model

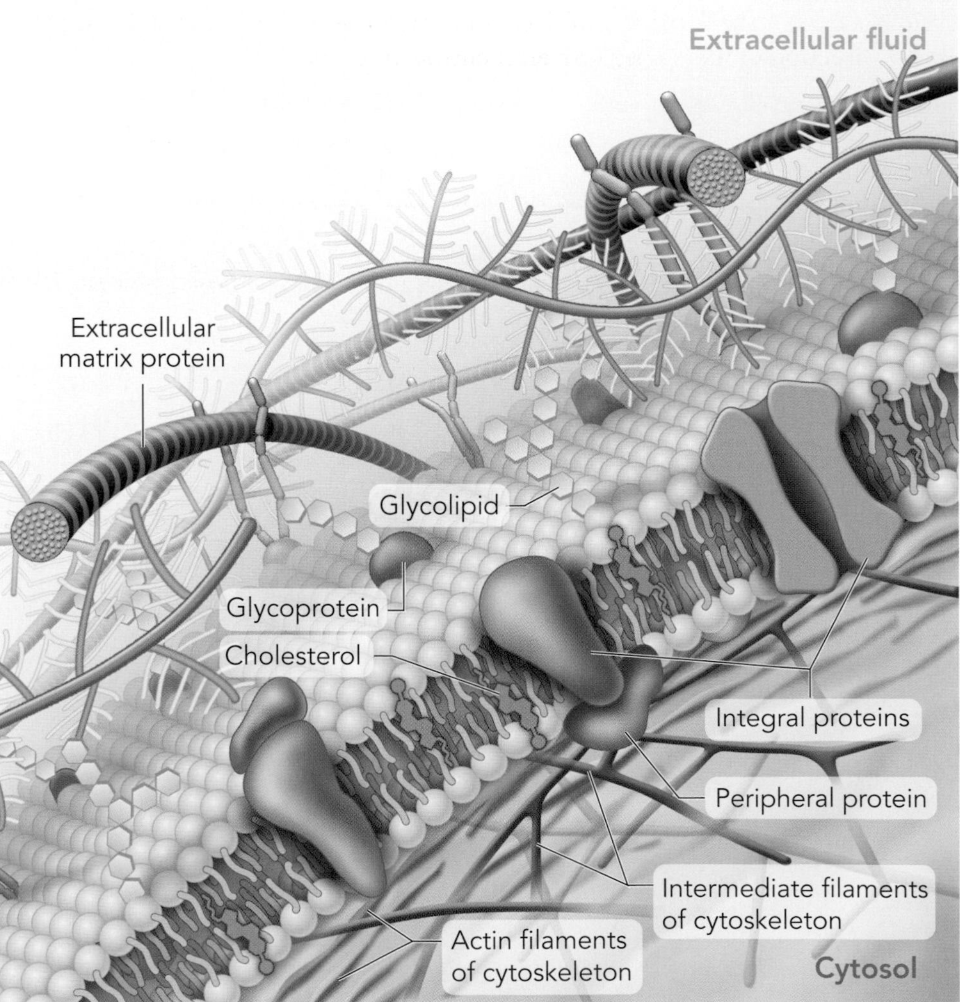

A mosaic is a picture made by placing many small, colorful pieces side by side.

Membrane Transport and Solution Chemistry

The movement of a molecule across a membrane depends on two key factors: the electrochemical gradient experienced by the molecule and the chemical properties of the molecule.

A membrane is not just a barrier between two aqueous solutions of different compositions; it actually creates the difference in the compositions of the solutions. At normal temperatures for living organisms, all molecules move rapidly in random directions, frequently colliding with one another. This random movement is called *Brownian motion*. Brownian motion leads to the tendency of solutions to mix completely with each other over time. If solutions of two solutes, X and Y, are placed on opposite sides of the same container, the net movement of X will be toward Y. This movement is called **diffusion**. For solutes without an electric charge, diffusion occurs in the direction of lower concentration. In Figure 1.23, X diffuses toward the lower concentration of X, and Y diffuses toward the lower concentration of Y. A gradual change in the concentration of a compound over a distance is called a chemical concentration gradient.

The **chemical concentration gradient** is a series of vectors pointing in the direction of lower concentration. For charged solutes, there is also an **electrical gradient** pointing in the direction that a positively charged particle will tend to move. The two gradients can be added to form a single **electrochemical gradient** for a specific solute. The electrochemical gradient for compound X points in the direction that particle X will tend to move. (Note that other factors, including heat and pressure, also affect the direction of diffusion. This manual will assume that these factors are included in the electrochemical gradient. In strict terms, diffusion occurs in the direction of decreasing free energy, or in the strictest terms, in the direction of increasing universal entropy.)

FIGURE 1.23 Diffusion

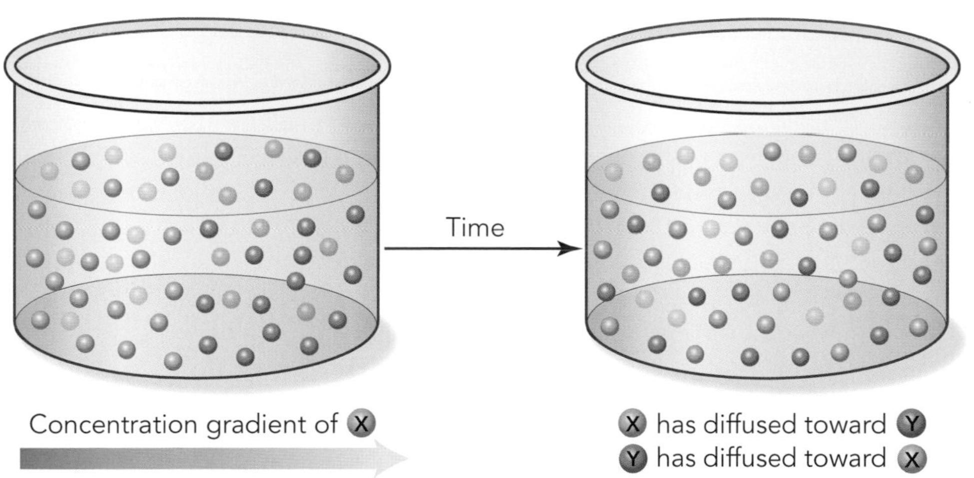

Blue food coloring begins to diffuse in a beaker of water. Diffusion is caused by the tiny, random movements of solutes in a solution. Over a period of time, these movements cause the dissolved substance to become evenly dispersed throughout the solvent.

If compounds X and Y are separated by an impermeable membrane, diffusion stops. However, if the molecules of X can squeeze through the membrane, diffusion of X is only slowed. Since the membrane slows the diffusion of X but does not stop it, the membrane is **semipermeable** to compound X.

Figure 1.24 shows examples of membrane transport. Natural membranes are semipermeable to most compounds, but to varying degrees. Two characteristics of a compound, **size** and **polarity**, largely determine the permeability of the membrane to that compound. The bigger the molecule, the less permeable the mem-

FIGURE 1.24 Types of Membrane Transport

Interstitial fluid

Simple diffusion

Simple diffusion

Passive diffusion

Facilitated diffusion

Passive diffusion

Active transport

Cytosol

Water | Glucose | Steroid | Na^+ | K^+ | ATP

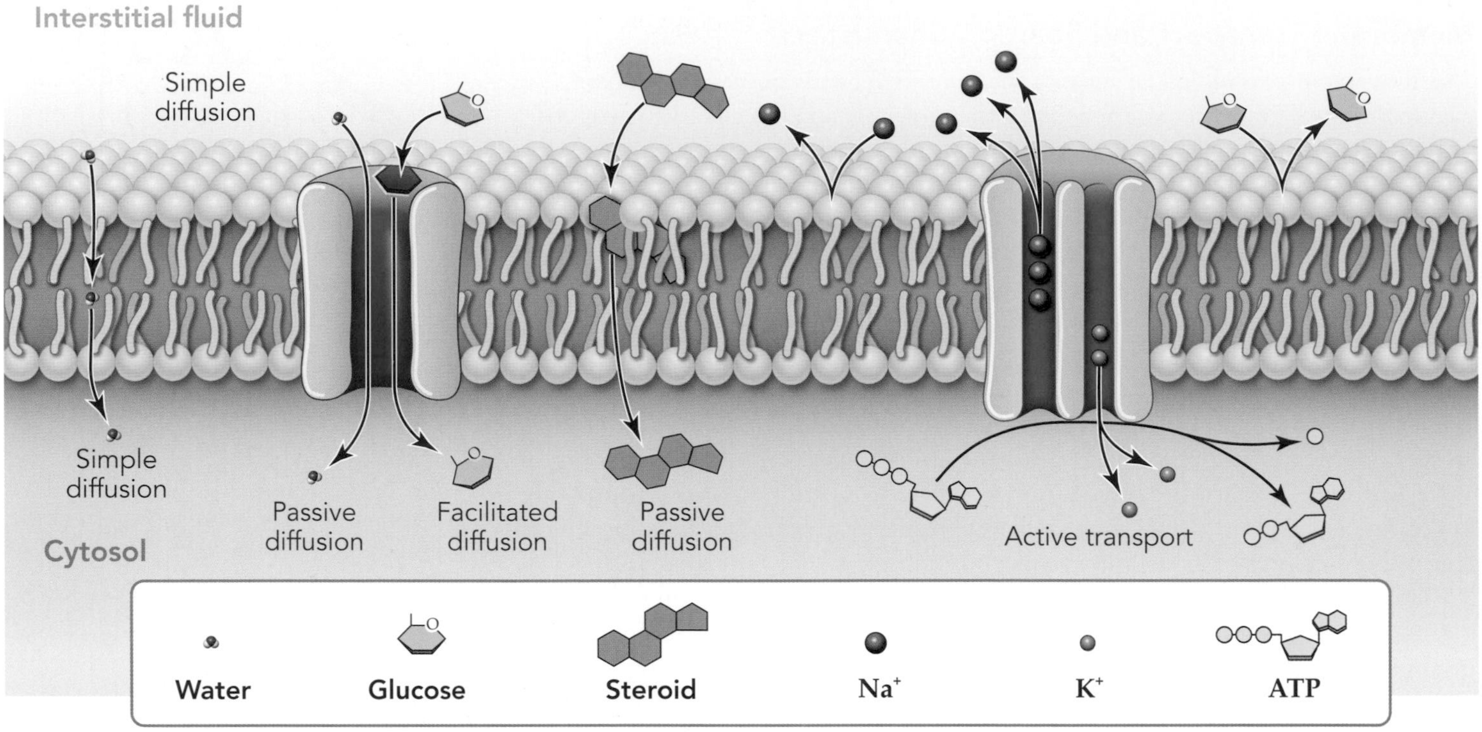

There are some important concepts here that require understanding and memorization. The membrane stuff is worth a second read through. For any molecule or ion, first determine whether it is moving against a gradient or with a gradient. If it is moving against a gradient, it requires active transport, regardless of size and charge. If it is moving with a chemical gradient, consider its chemical properties. Anything that is lipid soluble (nonpolar enough to slide right through the phospholipid bilayer) and small enough to fit around the cracks in the integral proteins can pass through the membrane without the aid of a protein. If it meets these criteria, the molecule or ion crosses the membrane by passive diffusion. Otherwise it requires a protein to cross the plasma membrane. This type of transport is called facilitated diffusion.

brane is to that molecule. A natural membrane is generally impermeable to polar molecules that have a molecular weight greater than 100 g/mol without some type of assistance. The greater the polarity of a substance (or if it has a full charge), the less permeable the membrane is to that substance. Very large lipid soluble (i.e., nonpolar) molecules, such as steroid hormones, can easily diffuse through the membrane. The various types of membrane transport are discussed in more detail below.

When predicting the permeability of the membrane to a substance, both size and polarity must be considered. Water is larger than a sodium ion, but water is polar, while the sodium ion possesses a complete charge. Therefore, a natural membrane is more permeable to water than it is to sodium. In this case, the charge difference outweighs the size difference. A cell membrane is, in fact, highly permeable to water. Water crosses the plasma membrane by simple diffusion. However, if the membrane were made only of a phospholipid bilayer, and did not contain proteins, the rate of diffusion for water would be very slow. Most of the diffusion of polar or charged molecules across a membrane takes place through incidental holes (sometimes called *leakage channels*) created by the irregular shapes of integral proteins. The function of these proteins is not to aid in diffusion. This is just an incidental contribution.

The diffusion described above, where molecules move through leakage channels across the membrane through random motion, is called **passive diffusion**. As mentioned previously, some molecules are too large or too polar to passively diffuse, yet they are needed for the survival of the cell. Proteins embedded into the cell membrane assist these molecules in moving across the membrane. These proteins, called **carrier proteins** or **membrane channels**, facilitate the diffusion of specific molecules or ions across the membrane. Several mechanisms are used by transport proteins in **facilitated diffusion**, but in order for the passage to be considered facilitated diffusion, the diffusion must occur down the electrochemical gradient of all species involved. Most, but not all, human cells rely on facilitated diffusion for their glucose supply. Facilitated diffusion contributes to the selective permeability of the membrane by selecting between molecules of similar size and charge.

A living organism must be able to transport some substances against their electrochemical gradients. Of course, such transport cannot be accomplished by diffusion. Movement of a compound against its electrochemical gradient occurs by **active transport**. Active transport requires expenditure of energy. It can be accomplished by the direct expenditure of ATP to acquire or expel a molecule against its electrochemical gradient. Active transport can also be powered indirectly by using ATP to create an electrochemical gradient, which is then used to acquire or expel a second molecule down its electrochemical gradient. The latter method is called *secondary active transport*.

Remember Key 3: Transport of molecules in and out of the cell can occur by many processes, depending on the size and charge of the molecule. Small or nonpolar molecules can diffuse down their concentration gradients, while large and charged molecules need to be moved via facilitated diffusion or active transport.

The Na^+/K^+ pump is an example of secondary active transport that maintains the negative membrane potential across the lipid bilayer.

MCAT® THINK

Which of the following could cross through the cell membrane via passive diffusion?

Answer on page 36.

A.

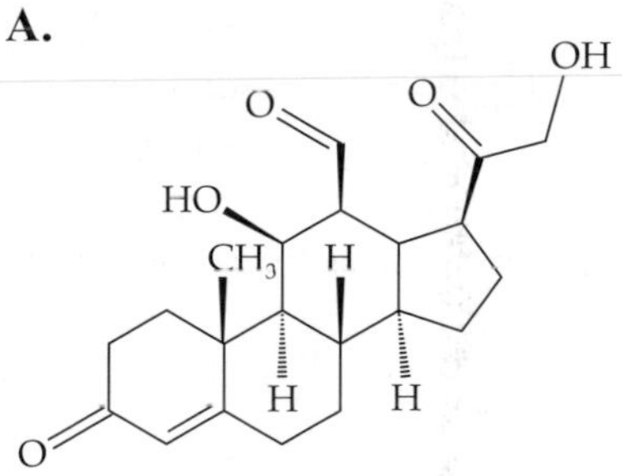

B.

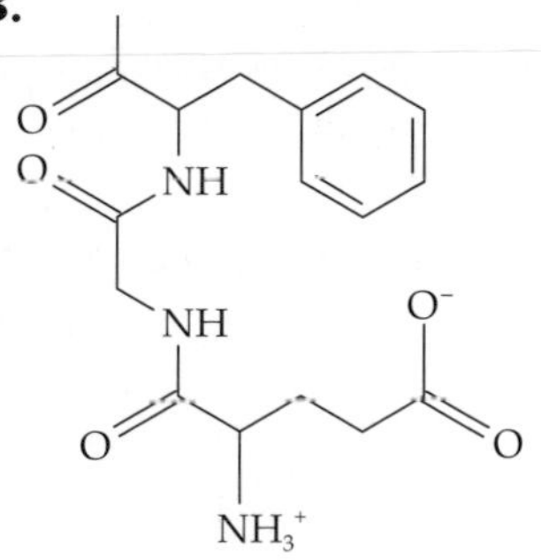

C.

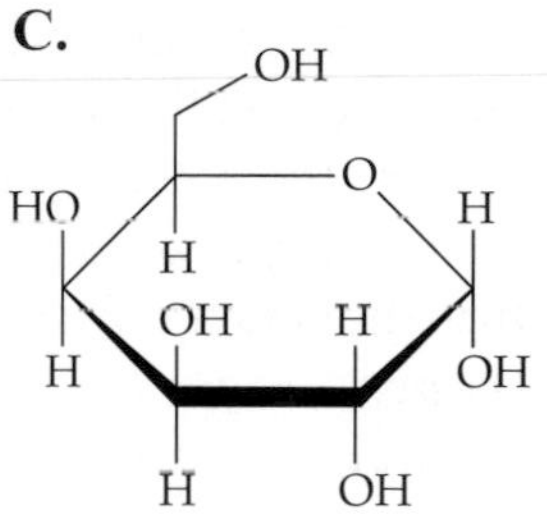

D. H_2O

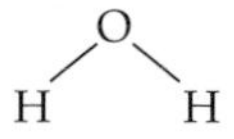

E. O_2

F. CO_2

G. Na^+

Most cells in the human body are roughly **isotonic** to their environment, meaning that the aqueous solution of their cytosol contains approximately the same concentration of particles as the aqueous solution surrounding them. While molecules may randomly move into or out of the cell, the number moving in each direction will be roughly the same. **Hypertonic** cells are more concentrated than their environment, and **hypotonic** cells are less concentrated than their environment. When the overall concentrations of molecules on either side of a membrane are not equal, osmotic pressure is generated. **Osmotic pressure** is the "pulling" pressure generated by a concentration gradient, which encourages osmosis. It is an example of a **colligative property**, a property that is based on the number of particles present rather than the type of particle.

When you see tonic, think solute concentration. Hypertonic means more concentrated. Hypotonic means less concentrated.

The body works to maintain an extracellular environment that is isotonic to its cells. If the extracellular environment suddenly became hypotonic with respect to the cell, a concentration gradient would be established. Recall that when there is a higher concentration of particles on one side of the cell membrane, the particles tend to move down their concentration gradient to the other side of the barrier. If

The opposite process can happen if a cell is hypotonic to its environment. Water will exit the cell, and the cell will shrivel.

the particles are prevented from crossing the barrier, perhaps because they are too large or charged, water will cross in the opposite direction, down its own electrochemical gradient. Since human cells have no rigid cell wall, the cell will fill until the osmotic pressures are equal on the inside and outside of the cell. Water continues to move in and out of the cell very rapidly, but an equilibrium is reached. If the plasma membrane cannot withstand the pressure, the cell will burst.

Immune System
BIOLOGY 2

Besides transport across the membrane by diffusion, cells can acquire substances from the extracellular environment through **endocytosis** (shown in Figure 1.25). There are several types of endocytosis: phagocytosis (Greek: phagein → to eat), pinocytosis (Greek: pinein → to drink), and receptor mediated endocytosis. In **phagocytosis**, the cell membrane protrudes outward to envelop and engulf particulate matter. Only a few specialized cells are capable of phagocytosis. Phagocytosis is triggered by the binding of particulate matter to protein receptors on the phagocytotic cell. In humans, antibodies or complement proteins bind to particles and stimulate receptor proteins on macrophages and neutrophils to initiate phagocytosis. Once the particulate matter is engulfed, the membrane bound body is called a *phagosome*.

FIGURE 1.25 Types of Endocytosis

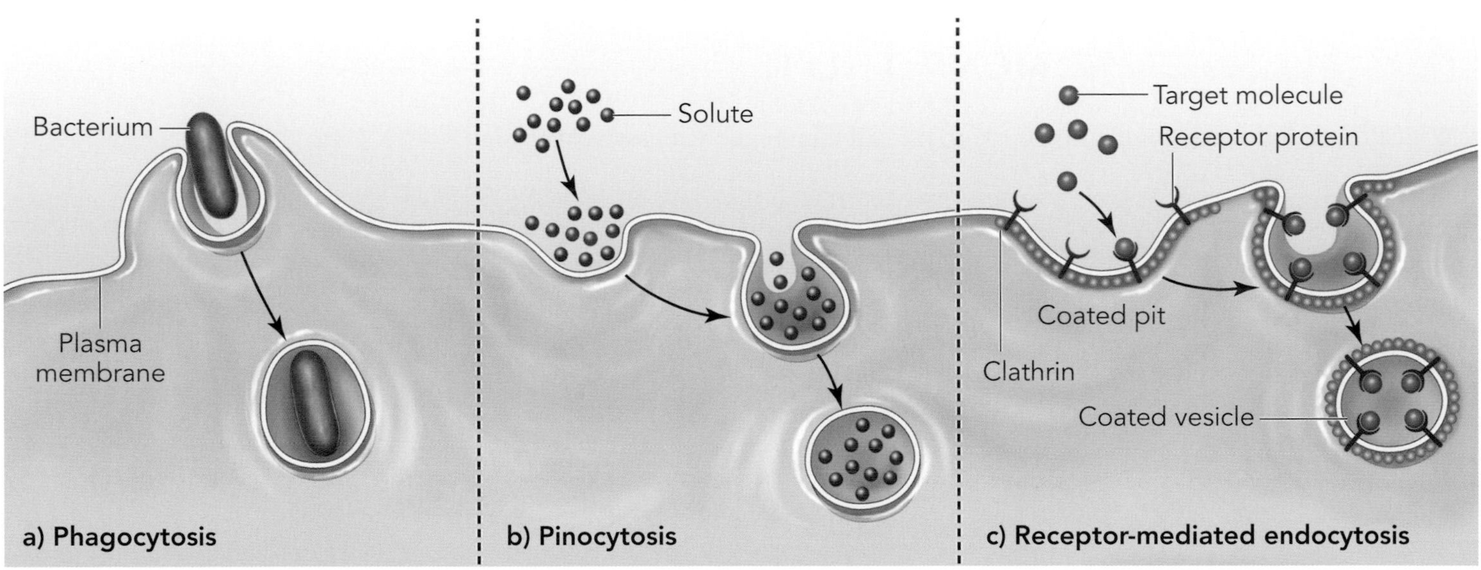

The MCAT® probably will not ask you to distinguish between the types of endocytosis, but understand the basic concept and be aware that there are multiple methods for particles to gain access to the interior of the cell.

In *pinocytosis*, extracellular fluid is engulfed by small invaginations of the cell membrane. This process is performed by most cells, and in a random, nonselective fashion.

Receptor-mediated endocytosis refers to specific uptake of macromolecules such as hormones and nutrients. In this process, the ligand binds to a receptor protein on the cell membrane and is then moved to a *clathrin-coated pit*. Clathrin is a protein that forms a polymer, adding structure to the underside of the coated pit. The coated pit invaginates to form a *coated vesicle*. This process differs from phagocytosis in that its purpose is to absorb ligands, whereas ligands are only involved in phagocytosis as signals that initiate phagocytosis of other particles.

Endo sounds like "in door," so molecules are entering the cell. "Exo" sounds like "exit," so molecules are leaving the cell.

After endocytosis, the substance that was engulfed remains sequestered in a secretory vesicle. It moves backward along the route of proteins discussed earlier in this lecture. It can fuse back into the Golgi apparatus, or it can be targeted for destruction by a lysosome.

Exocytosis is simply the reverse of endocytosis. It provides a way for substances to leave the cell.

FIGURE 1.26 Exocytosis

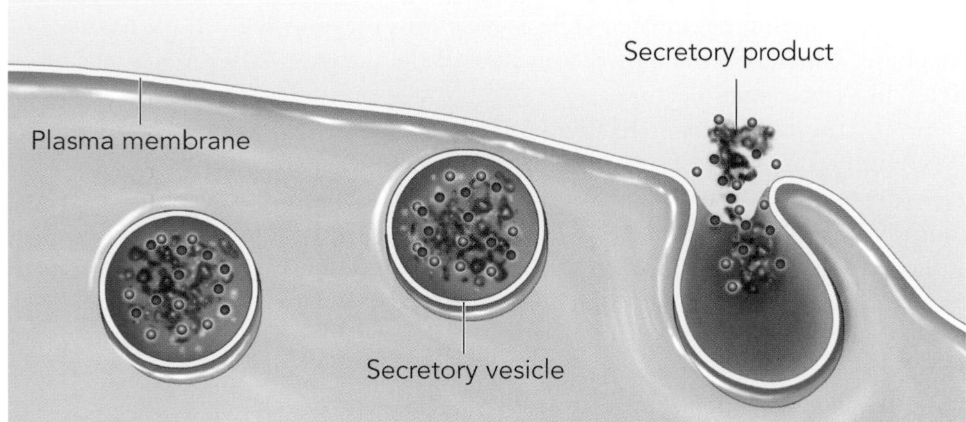

The Cell Cycle: Cell Growth and Division

Every cell has a life cycle (Figure 1.27) that begins with the birth of the cell and ends with the death or division of the cell. The life cycle of a typical somatic cell in a multicellular organism can be divided into four stages: the first growth phase (G_1), synthesis (S), the second growth phase (G_2), and mitosis or meiosis (M). G_1, S, and G_2 collectively are called **interphase**.

FIGURE 1.27 The Cell Life Cycle

In G_1, the cell has just divided, and begins to grow in size, producing new organelles and proteins. RNA synthesis and protein synthesis are highly active. The cell must reach a certain size and synthesize sufficient proteins in order to continue to the next stage. As it grows, the endoplasmic reticulum creates phospholipids and new portions of cell membrane as part of secretory vesicles. As these vesicles fuse with the membrane, the membrane grows. Cell growth is assessed at the G_1 checkpoint near the end of G_1. If conditions are favorable for division, the cell enters the S phase; otherwise the cell enters the G_0 phase. The main factor in triggering the beginning of S is cell size, based upon the ratio of cytoplasm to DNA. G_1 is normally, but not always, the longest stage.

G_0 is a non-growing state distinct from interphase. The G_0 phase is responsible for variations in length of the cell cycle between different types of cells. In humans, enterocytes of the intestine divide more than twice per day, while liver cells spend a great deal of time in G_0, dividing less than once per year. Mature neurons and muscle cells remain in G_0 permanently.

In S, the cell devotes most of its energy to DNA replication. Organelles and proteins are produced more slowly than in G_1. In this stage, an exact duplicate of each chromosome is created. The S cycle thus prepares the cell for **mitosis**, or **M phase**, which is defined by the division of the nucleus. (Mitosis is discussed in detail in the Genetics Lecture.)

Cancer genes are like a car that is trying to get past a cop while speeding. If the cop is blinded, he can no longer check for the car. This is like the loss of a tumor repressor. If you paint the car a new color so that the cop can no longer recognize it, the car can also get by the cop. This is like the gain of an oncogene.

In G_2, the cell prepares to divide. Cellular organelles continue to duplicate. RNA and proteins (especially tubulin for microtubules) are actively produced. G_2 typically occupies 10-20% of the cell life cycle. Near the end of G_2 is the G_2 checkpoint, which checks for *mitosis promoting factor* (*MPF*). When the level of MPF is sufficiently high, mitosis is triggered. An M checkpoint towards the end of mitosis ensures that the chromosomes are aligned correctly before the cell divides.

As described, there are a number of checkpoints that regulate the cell cycle, step by step. Sometimes a cell acquires a mutation that allows the cell to bypass these checkpoints. Such cells can grow unchecked and develop into **cancer**. There are two types of mutations that can cause cancer: the deactivation of a checkpoint protein (a *tumor repressor*) and the activation of a gene that causes the proliferation of the cell (an *oncogene*).

1.7 Tissues, Organ Systems, Organisms...

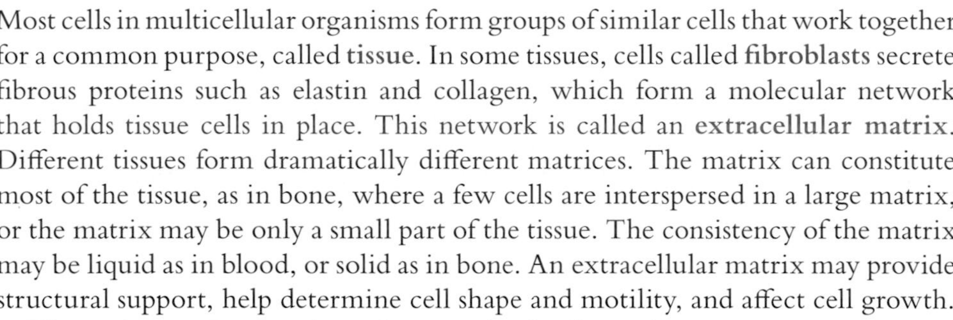

Most cells in multicellular organisms form groups of similar cells that work together for a common purpose, called **tissue**. In some tissues, cells called **fibroblasts** secrete fibrous proteins such as elastin and collagen, which form a molecular network that holds tissue cells in place. This network is called an **extracellular matrix**. Different tissues form dramatically different matrices. The matrix can constitute most of the tissue, as in bone, where a few cells are interspersed in a large matrix, or the matrix may be only a small part of the tissue. The consistency of the matrix may be liquid as in blood, or solid as in bone. An extracellular matrix may provide structural support, help determine cell shape and motility, and affect cell growth.

Three classes of molecules make up animal cell matrices:

1. *glycosaminoglycans* and *proteoglycans*,
2. structural proteins, and
3. adhesive proteins.

Glycosaminoglycans are polysaccharides that typically have proteoglycans attached. They make up over 90% of the matrix by mass. This first class of molecules provides pliability to the matrix. Structural proteins provide the matrix with strength. The most common extracellular matrix structural protein in the body is collagen. Collagen is particularly important to the strength of cartilage and bone. Adhesive proteins help individual cells within a tissue stick together.

The basal lamina (which, along with the reticular lamina, forms the basement membrane) could appear in an MCAT® passage. The *basal lamina* is a thin sheet of matrix material that separates epithelial cells from *support tissue*. (Epithelial cells separate the outside environment from the inside of the body. Support tissue is composed of the cells adjacent to epithelial cells on the inside of the body.) Basal lamina is also found around nerves, and muscle and fat cells. Basal lamina usually acts as a sieve-like type barrier, selectively allowing the passage of some molecules but not others.

Don't worry about memorizing details about the extracellular matrix. Just know that is the material that surrounds the cell and that it is formed by the cell itself.

Many animal cells contain a carbohydrate region analogous to the plant cell wall or bacterial cell wall, called the *glycocalyx*. The glycocalyx separates the cell membrane from the extracellular matrix. However, a part of the glycocalyx is made from the same material as the matrix, so the glycocalyx is often difficult to identify. The glycocalyx can be involved in cell-cell recognition, adhesion, cell surface protection, and permeability.

FIGURE 1.28 The Extracellular Matrix

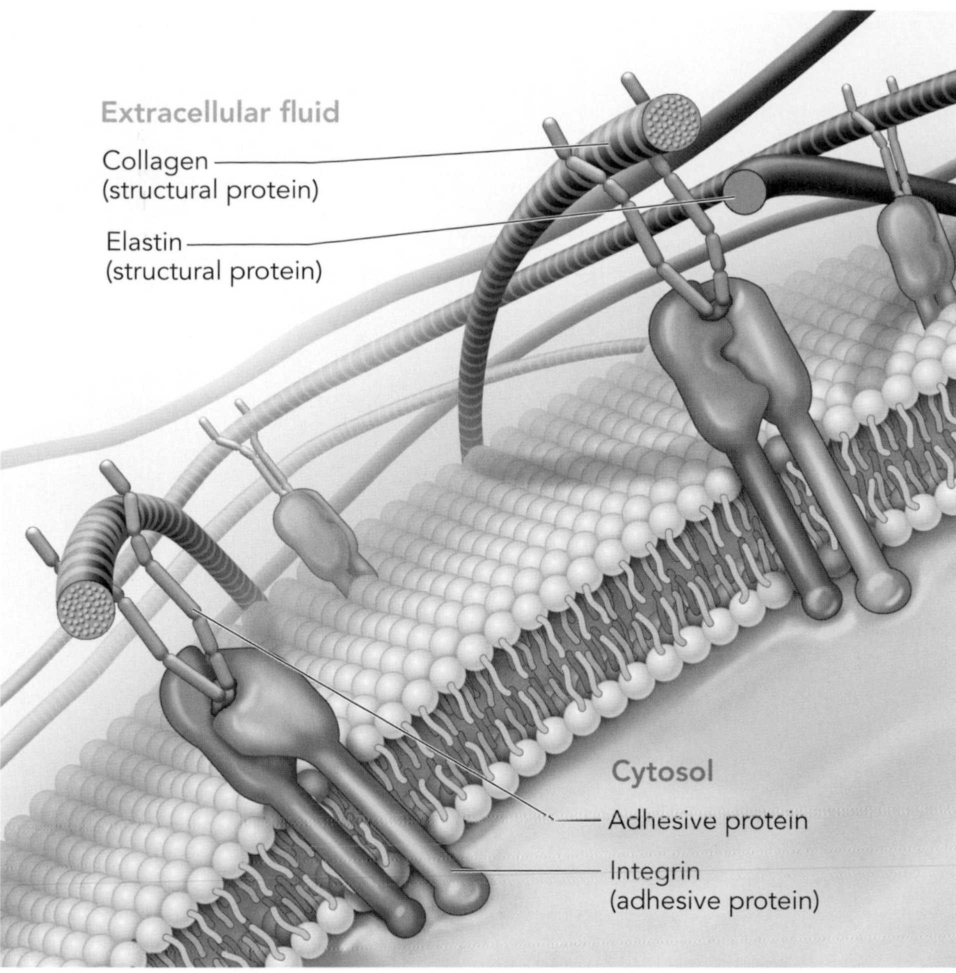

Connections Between Cells

Three types of **intercellular junctions** connect animal cells: tight junctions, desmosomes, and gap junctions (Figure 1.29). Each type of junction performs a unique function. **Tight junctions** form a watertight seal from cell to cell that can block water, ions, and other molecules from moving around and past cells. Tissue held together by tight junctions can act as a complete fluid barrier. Epithelial tissue cells in organs like the bladder, the intestines, and the kidney are held together by tight junctions in order to prevent waste materials from seeping around the cells and into the body. Since proteins have some freedom to move laterally about the cell membrane, tight junctions also act as a barrier to protein movement between the *apical* and the *basolateral* surface of a cell. (The part of a cell facing the lumen of a cavity is called the apical surface. The opposite side of a cell is called the basolateral surface.)

Tight junctions are like the plastic rings around a six pack of soda. The soda cans are the cells. The plastic rings hold the cans together and provide a watertight barrier around them. Although soda cans are impermeable to water, real cells may or may not be impermeable to water.

Desmosomes join two cells at a single point. They attach directly to the cytoskeleton of each cell. Desmosomes do not prevent fluid from circulating around the sides of a cell. They are found in tissues that normally experience a lot of stress due to sliding, like skin or intestinal epithelium. Desmosomes often accompany tight junctions.

Gap junctions are small tunnels that connect cells, facilitating the movement of small molecules and ions between the cells. Gap junctions in cardiac muscle allow the spread of the action potential from cell to cell. They also exist in some kinds of smooth muscle, allowing fibers to contract as a unit.

FIGURE 1.29 Types of Cellular Junctions

Cell 1 cytosol

Cell 1 membrane

Cell 2 membrane

Cell 2 cytosol

Tight junctions
Found in epithelial tissues like bladder, intestines, kidneys

Gap junctions
Found in cardiac muscle

Desmosomes
Found in stressed epithelial tissues like skin or intestinal epithelium

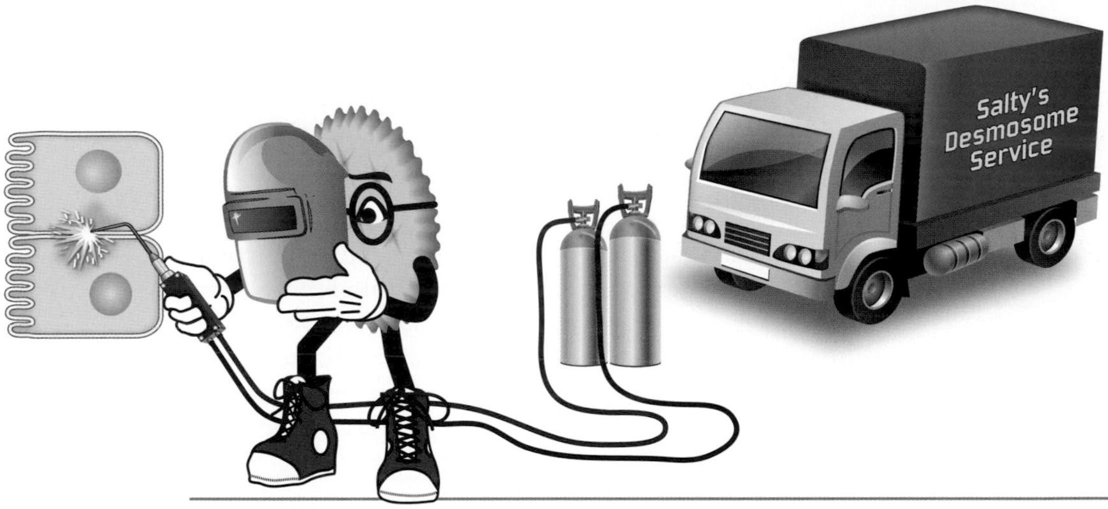

Salty's Desmosome Service

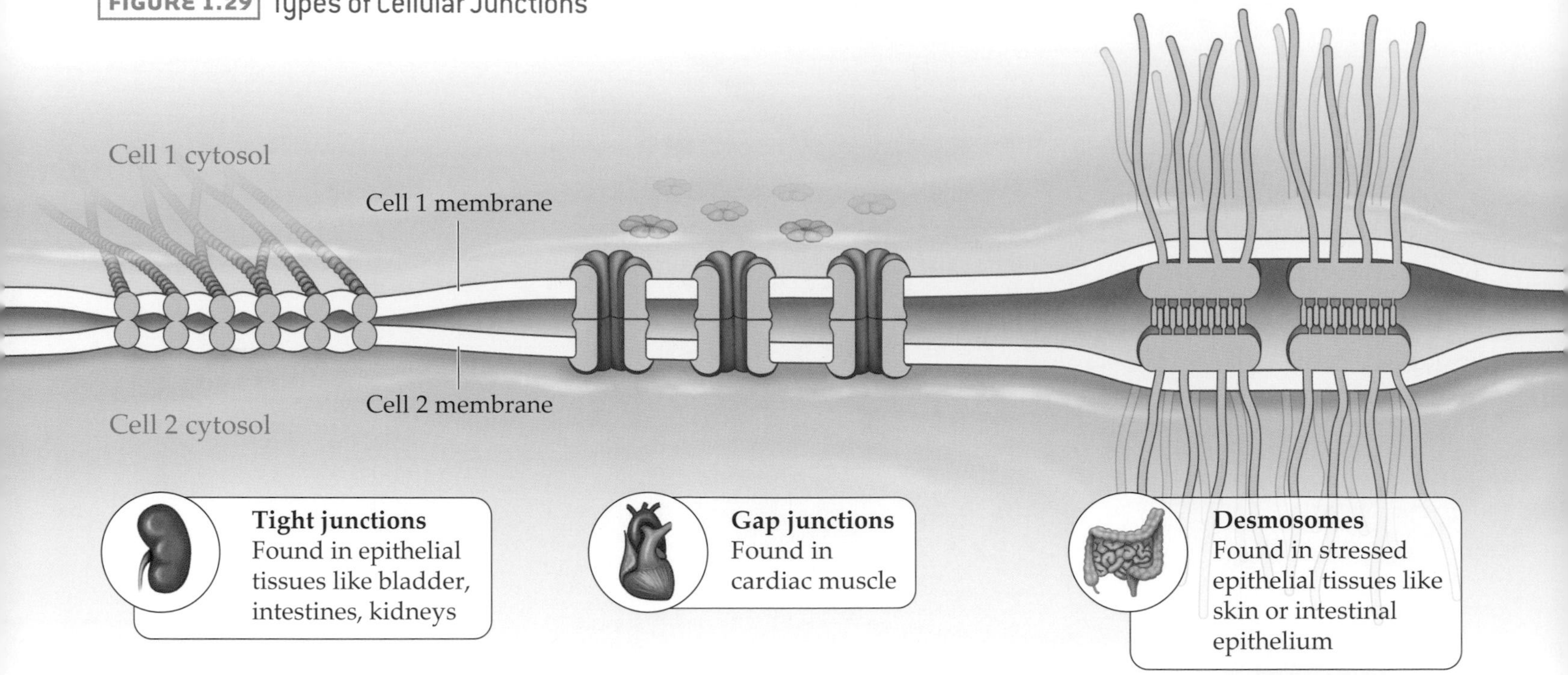

Remember the three types of cellular junctions. Tight junctions act as a fluid barrier around cells. Desmosomes are like spot-welds holding cells together. Gap junctions are tunnels between cells that allow for the exchange of small molecules and ions.

Cell Communication Through Receptors

It is often the case that cells and tissues that are not directly connected must coordinate their activities. Gap junctions can only allow communication between adjacent cells. In the more common case where a cell has to communicate with a cell that is further away, the cell sends out a chemical message, or **hormone**, to be picked up by a receptor on another cell.

The location of the hormone receptor depends on the nature of the hormone. If a molecule is small and lipophilic, such as a steroid (cholesterol derivative), it can cross the cell membrane and bind to receptors inside of a cell. These receptors are located in the cytosol or the nucleus and act at the level of transcription. This means that the typical effect of a steroid hormone is to increase certain membrane or cellular proteins within the target cell.

These receptors are like an ear for the nucleus. They receive input from the outside world and communicate it to the nucleus. The nucleus can then process this information and coordinate cell activities.

If a molecule is large and lipophobic, such as a protein, it cannot cross the cell membrane and instead binds to a receptor on the cell membrane. The receptor may itself act as an ion channel and increase membrane permeability to a specific ion in response to the hormone, or the receptor may activate or deactivate other intrinsic membrane proteins that act as ion channels. Another effect of the hormone binding to the receptor may be to activate an **intracellular second messenger** such as cAMP, cGMP, or calmodulin. These chemicals are called second messengers because the hormone is the first messenger to the cell. The second messenger activates or deactivates enzymes and/or ion channels and often creates a 'cascade' of chemical reactions that amplifies the effect of the hormone. Through a second messenger cascade, a small concentration of hormone can have a significant effect.

Types of hormones and their effects will be discussed in much greater detail in the Endocrine System Lecture. For now, focus on the fact that receptors allow cells to communicate with each other from a distance.

Cells in Context

The existence of tissues, where groups of cells work together with each type of cell performing a unique function that contributes to the specialized function of the group, is a defining feature of multicellular eukaryotes. There are four basic types of tissue in animals: *epithelial tissue*, *muscle tissue*, *connective tissue*, and *nervous tissue*. Epithelial tissue separates free body surfaces from their surroundings. Epithelium includes *endothelium*, which lines the vessels of the body, including the heart. Connective tissue is characterized by an extensive matrix. Examples include blood, lymph, bone, cartilage, and connective tissue proper, which makes up tendons and ligaments.

Various tissue types work together to form *organs*. The stomach, for example, is an organ with an outer layer made from epithelial tissue and connective tissue, a second layer of muscle tissue, and an innermost layer of epithelial tissue.

Organs that work together to perform a common function are called *systems*. This manual will discuss many significant organ systems, such as the respiratory system. Systems work together to create *organisms*.

Expanding outward, organisms live in social groups, as described in the *Psychology & Sociology* manual. Biology includes the study of such groups. Humans live in families, which exist in communities. Just as the cells of a multicellular organism do not exist in isolation, humans and other organisms do not live independent of one another. The key to biology is not just understanding biological entities, but also understanding how these entities interact with their environments.

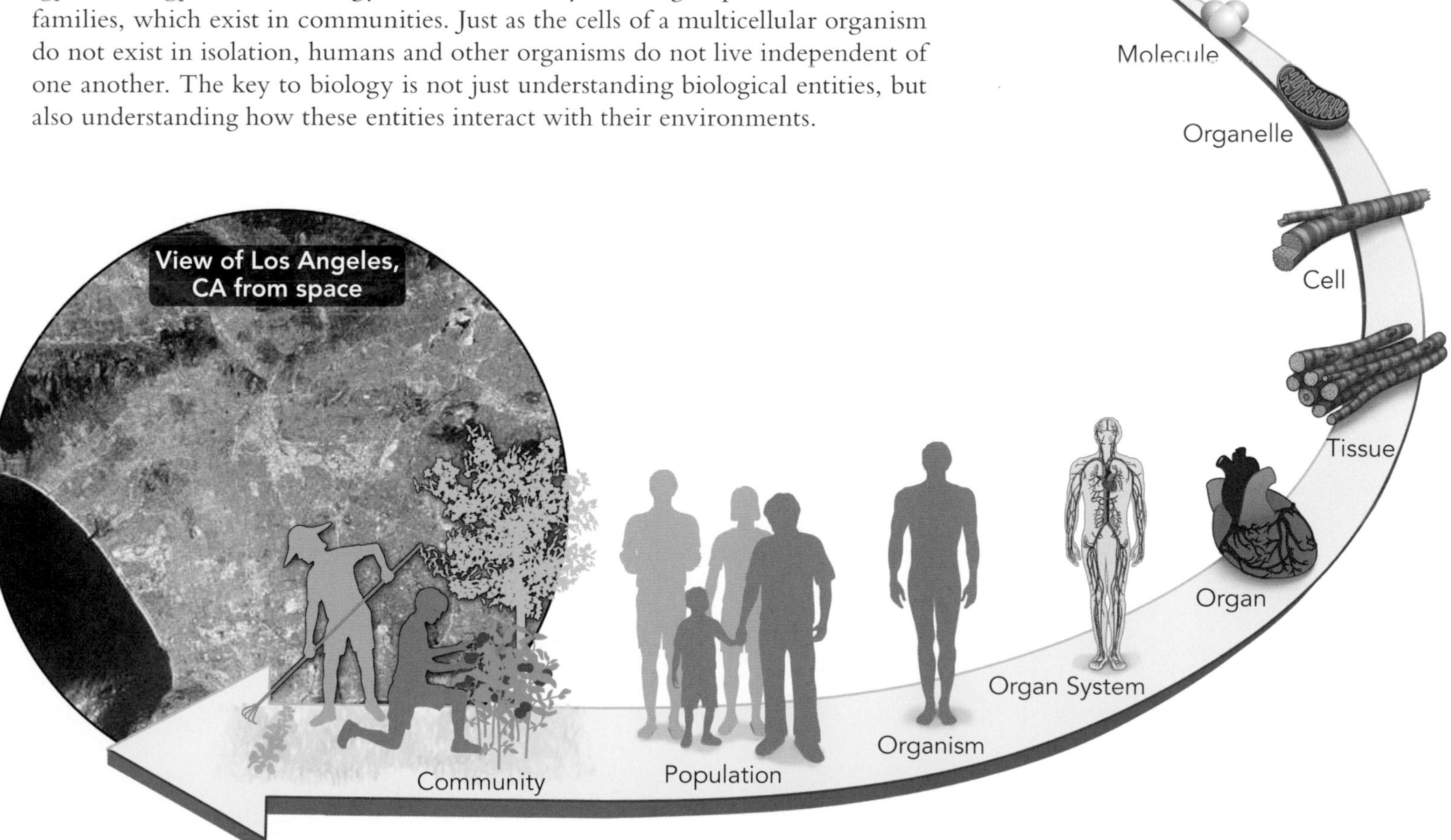

These questions are NOT related to a passage.

Question 17

The sodium-potassium ATPase hydrolyzes ATP to pump three sodium outside of the cell and two potassium into the cell. Which of the following is true about the sodium-potassium ATPase?

I. It is an example of secondary active transport.
II. Sodium is pumped against its concentration gradient.
III. Potassium is pumped against its concentration gradient.

- O **A.** I only
- O **B.** II only
- O **C.** II and III only
- O **D.** I, II, and III

Question 18

Intestinal epithelium uses the diffusion of sodium down its concentration gradient to power the translocation of glucose inside of a cell. This is an example of:

- O **A.** primary active transport.
- O **B.** secondary active transport.
- O **C.** passive diffusion.
- O **D.** facilitated diffusion.

Question 19

Which of the following is least likely to require a protein to be transported into a cell?

- O **A.** Protein
- O **B.** Glucose
- O **C.** Amino acids
- O **D.** Carbon dioxide

Question 20

A protein hormone is least likely to act by which of the following mechanisms?

- O **A.** Binding to a receptor in the nucleus
- O **B.** Opening a membrane bound channel
- O **C.** Activating a membrane bound enzyme
- O **D.** Activating a G protein

Question 21

Which of the following statements is (are) true concerning tight junctions?

I. They connect adjacent cells.
II. They may form a barrier to extracellular fluids.
III. They have the greatest strength of all cellular adhesions.

- O **A.** I only
- O **B.** I and II only
- O **C.** II and III only
- O **D.** I, II, and III

Question 22

What would be the likely consequence of adding pure water to a petri dish of human red blood cells?

- O **A.** The cells would stay the same size.
- O **B.** The cells would expand.
- O **C.** The cells would burst.
- O **D.** The cells should shrink.

Question 23

A neuron in an adult human is in which stage of the cell cycle?

- O **A.** G_0
- O **B.** G_1
- O **C.** S
- O **D.** G_2

Question 24

Which of the following is not involved in cell-to-cell communication?

- O **A.** Gap junctions
- O **B.** Desmosomes
- O **C.** Membrane receptors
- O **D.** Nuclear receptors

Questions 17-24 of 144

STOP

TERMS YOU NEED TO KNOW

9 + 2 arrangement
Active transport
Adaptations
Adipocytes
Aerobic
Allele
Amphipathic
Anaerobic
Apoptosis
Archaea
Bacilli (rod-shaped)
Bacteria
Bacterial envelope
Bacteriophage
Binary fission
Bottleneck
Cancer
Capsid
Carrier protein
Centrioles
Chemical concentration gradient
Chemotaxis
Cilia
Cleavage
Cocci (spherical)
Colligative property
Conjugation
Cytoskeleton
Desmosomes
Differential Reproduction
Diffusion
Domain
Dynein
Electrochemical gradient
Endocytosis
Endoplasmic reticulum (ER)
Enveloped viruses
ER lumen
Evolution
Evolutionary time
Exocytosis
Exponential growth
Extracellular matrix
Extragenomic DNA
Facilitated diffusion
Fatty acid
Fibroblasts
Fitness concept
Flagella
Flagellar propulsion
Flagellin
Fluid mosaic model
G_1, G_2, S, and G_0 phases
Gap junctions
Gene pool
Genetic drift
Genetic recombination
Genotype
Glycerol backbone
Golgi apparatus
Hardy-Weinberg Equilibrium
HIV
Hormone
Host cell
Hydrogen peroxide
Hydrolytic enzymes
Hypertonic cell
Hypotonic cell
Inbreeding
Increase in percent representation
Inner membrane
Integral proteins (intrinsic proteins)
Intercellular junctions
Intermediate filaments
Intermembrane space
Interphase
Intracellular second messenger
Isotonic
Lysogenic
Lysosomes
Lytic
Membrane channels
Micelle
Microfilaments
Microtubules: transport, support
Microtubule-organizing center (MTOC)
Mitochondria
Mitosis (M phase)
Natural selection
Nonenveloped viruses
Nucleolus
Nucleus
Nuclear envelope
Nuclear pores
Origin of replication
Osmotic pressure
Outbreeding
Outer membrane
Parasitic
Passive diffusion
Peptidoglycan
Peroxisomes
Peripheral proteins (extrinsic proteins)
Phagocytosis
Phenotype
Phosphate group
Plasma membrane
Plasmid
Polarity
Polymorphism
Prions
Programmed cell death
Prokaryotes
Receptor
Retroviruses
Reverse transcriptase
Rough ER
Size
Semipermeable membrane
Smooth endoplasmic reticulum
Speciation

TERMS YOU NEED TO KNOW (CONTINUED)

Species
Specialization
Spirilla (spiral-shaped)
Subviral particles
Symbiotic
Synthetic machinery
Tail
Tight junctions
Tissue
Transduction
Transformation
Transposons
Vector
Viral envelopes
Viral particle
Viroids
Viruses

MCAT® THINK ANSWER

Pg. 27: D, E, and F

Small, non-charged molecules such as O_2 and CO_2 can easily pass through the cell membrane via passive diffusion. Although H_2O is polar, it is small and does not have a full charge, so it can also pass through the membrane by passive diffusion. Ions that have a full charge, such as Na^+, cannot diffuse passively through the membrane. Choices A, B, and C can all be eliminated because of their size; also note that the molecule shown in choice B has multiple full charges.

DON'T FORGET YOUR KEYS

1. The cell has compartments to separate chemical properties and processes.
2. A cell is like an organism – understand the function of organelles by treating them as organs.
3. Transport in and out of a cell occurs through diffusion down the concentration gradient, by pumps against the diffusion gradient, or via endocytosis/exocytosis.

The Nervous System

LECTURE 2

2.1 Introduction

The nervous system is responsible for an organism's ability to sense stimuli, process information, and respond to changes in its environment. This lecture will review the nervous system and its functions on the molecular, structural, organ, and systemic levels. Following an introduction to the organization and essential functions of the nervous system, including the contrast with other communication systems in the body, the lecture will describe the smallest unit of the nervous system: the neuron. A review of the chemistry and physics of the neuron will elucidate the biological and physical principles involved in the transmission of a signal down the length of the cell. Next, the site of interaction between neurons will be considered, including the conversion of information between a chemical and electrical state. These early sections provide the basis for understanding the higher-level organization and functions of the nervous system. Following a consideration of the major divisions of the nervous system and their unique characteristics, the lecture concludes by describing how each of the sensory systems of the body senses environmental stimuli and conveys this information to the central nervous system for processing and the generation of a response.

THE 3 KEYS

1. The nervous system senses the environment, processes information, and responds.
2. Neural signals are fast, fleeting, and specific; endocrine signals are slow, sustained, and general.
3. The type and number of channels and receptors are the hidden determinants of many cellular and systemic phenomena.

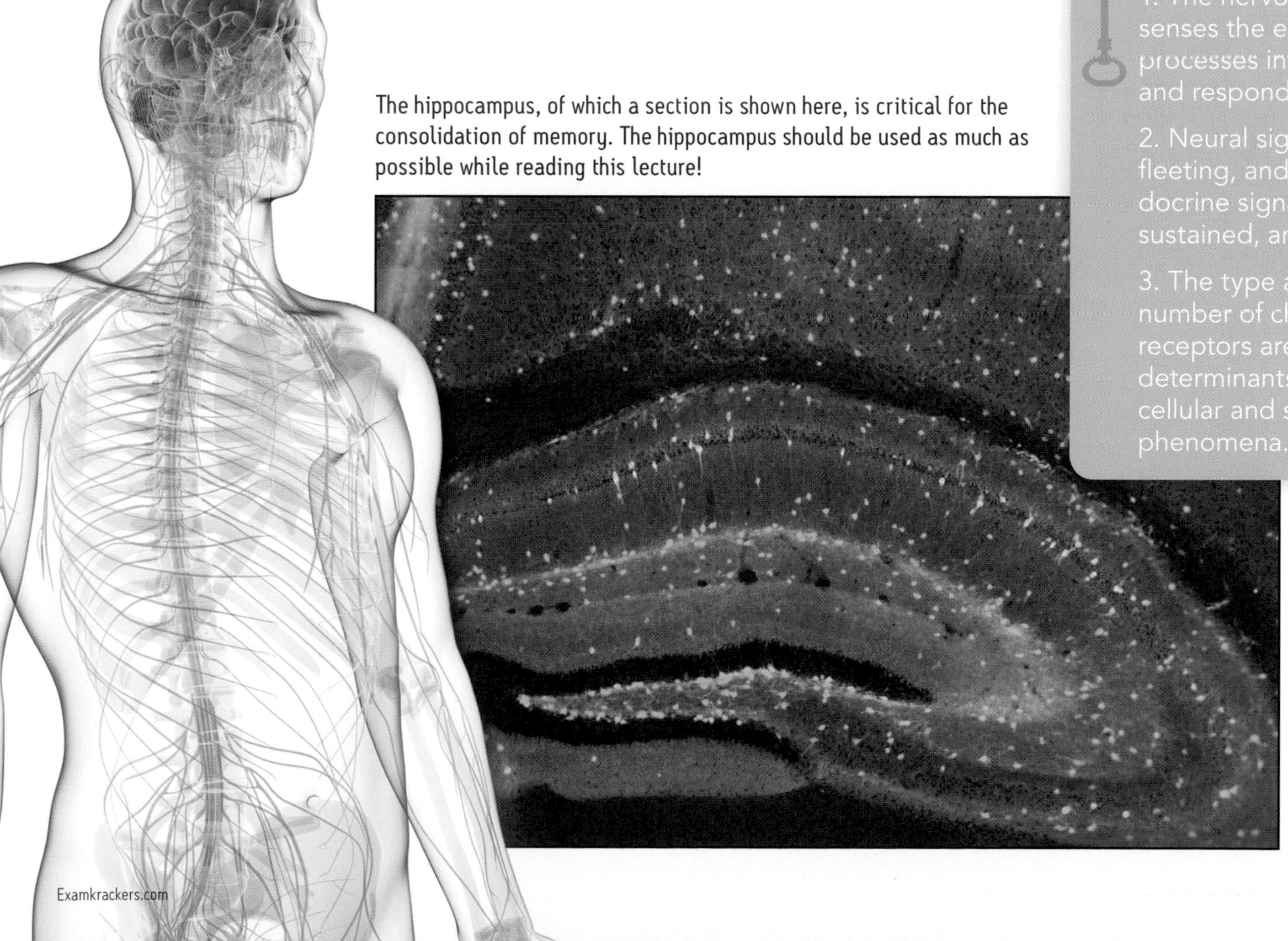

The hippocampus, of which a section is shown here, is critical for the consolidation of memory. The hippocampus should be used as much as possible while reading this lecture!

2.2 Communication within the Body

A consideration of the body's communication systems provides a basis for understanding the features of the nervous system. In multicellular organisms, cells must be able to communicate with each other so that the organism can function as a single unit. Communication is accomplished chemically via three types of molecules: 1. neurotransmitters, 2. local mediators, and 3. hormones. These methods of communication are governed by the nervous system, the paracrine system, and the endocrine system, respectively. These communication systems can be understood by looking at a few key characteristics: a signal can be fast or slow, specific or generalized, and fleeting or sustained in its effects. The specificity of a signal refers to the part of the body upon which the signal acts. A signal can be directed at a very specific site or can be more generalized, affecting multiple organs and systems. The category of fleeting or sustained has to do with how long the effect of the signal continues to influence the target. Usually, a signal that is fast is also fleeting, whereas a signal that is slow is usually sustained. Once a sustained signal reaches its target, it stays around for a while and continues to exert an effect. The nervous, paracrine, and endocrine systems each have a unique position along these continuums, as shown in Figure 2.1.

Remember Key 1: The nervous system is responsible for integrating information it obtains from the environment and making judgement calls about how to respond. Some actions of the nervous system are automatic, like reflexes, while others can be controlled, such as movement of skeletal muscles during exercise.

The nervous system involves rapid and direct communication between specific parts of the body, allowing quick but generally short-lived changes in muscular contractions, organ functions, or glandular secretions in response to incoming stimuli. Signals in the nervous system are generally specific, fast, and fleeting. By contrast, the endocrine system acts by releasing hormones into the bloodstream. The effects of the endocrine system are more generalized, slow, and sustained when compared to communication through the nervous system. The paracrine system lies somewhere between the extremes of the nervous and endocrine systems; for example, clotting factors to help in the coagulation of blood.

Understand Key 2: The signaling of the nervous system is fast, fleeting, and specific. The controlled movement of ions across neuron membranes in specific locations in the body allows for rapid, precise control. Endocrine signals are slow, sustained, and affect many parts of the body. Hormones that travel through the blood have the opportunity to interact with diverse cell types over a longer period of time until they are cleared from circulation.

Another major distinction between the body's methods of communication is the distance traveled by the molecules used in each system. Neurotransmitters travel over very short intercellular gaps, local mediators function in the immediate area around the cell from which they were released, and hormones travel throughout the entire organism via the bloodstream. Thus neurotransmitters are released right at the target site, producing a quick and direct signal; hormones act on many areas throughout the body and take longer to arrive; and local mediators fall somewhere between these extremes.

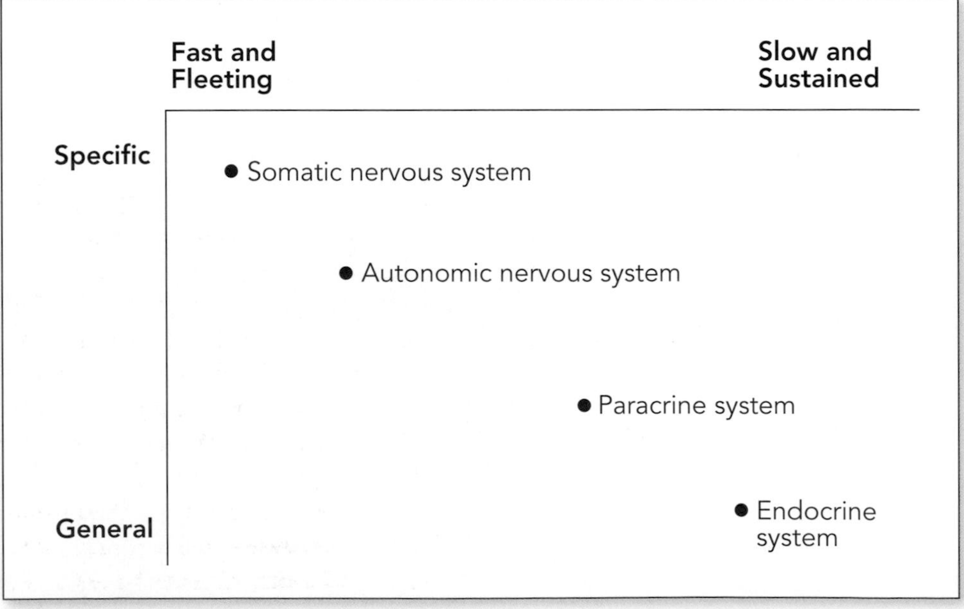

FIGURE 2.1 Comparison of Communication Systems: Somatic and Autonomic NS

The motor and sensory neurons of the somatic nervous system have specific targets, allowing for fine control of movement and detailed mapping of sensory stimuli. The autonomic nervous system has more general actions involving multiple targets and longer-lasting effects. Signals in the endocrine system are much more general, slow, and sustained than those of the nervous system.

You can think about different types of signals in terms of modern communication systems. A slow and sustained signal is like a letter. It takes a while to arrive and then has a lasting effect, since the recipient can read it multiple times. A fast and fleeting signal is more like a short text. It arrives instantly and then can be read and forgotten just as quickly.

2.3 Functions and Features of the Nervous System

The characteristics that define the nervous system as a communication system allow it to carry out its most crucial functions in the body: sensing the environment, selecting and processing the most significant information, communicating it to different parts of the body, and coordinating a response. These actions allow the nervous system to play a key role in homeostasis. It is important to be aware that the nervous and endocrine systems often work together to sense the changing needs of various systems in the body and respond accordingly. The particular mechanisms by which the nervous system regulates functions in the organism are described in this lecture and throughout this manual, as well as in the *Psychology & Sociology* manual.

Organisms perceive a huge amount of stimuli in their surroundings, but only the information that is most relevant is actually preserved by the conscious and unconscious mind for use. The nervous system selects the most important information based on past experience, preconceived ideas, and evolutionary instinct. While conscious perception of stimuli is a significant part of the function of the nervous system, homeostatic regulation that occurs outside of conscious awareness is also crucial to the functioning of the organism. Unconscious regulation involves simpler processing mechanisms than those of the conscious mind, but it still involves the gathering and integration of information, as well as the prioritization of the most important stimuli and functions of the body.

Sensory information is converted into different forms as it travels throughout the nervous system. First, input from the environment, including the body's internal environment, reaches sensory receptors. Physical information stimulates sensory neurons and is then translated into an electrical signal that travels along sensory neurons. Electrical signals travel the length of neurons, while a chemical signal is required for information to pass between adjacent neurons. The synapse, discussed later in this lecture, is the junction between two neurons that allows the transfer of the chemical signal. The conversion of signals from one form to another will be described throughout this lecture.

The nervous system includes the brain, spinal cord, nerves, and neural support cells, as well as sense organs such as the eye and the ear. Several sets of nerves in the nervous system carry out different actions. The brain and spinal cord make up the central nervous system (CNS), which allows for the highest level of integration of input and output. The CNS transmits information to the rest of the body by way of the peripheral nervous system (PNS), which is divided into two parts: one that signals voluntary movement and one that regulates unconscious and involuntary bodily functions. The overall organization and division of the nervous system will be described in greater detail later in this lecture.

Rather than being like a camera that records exactly what it sees, the nervous system is like a computer program that processes the image, enhancing certain parts and deleting others.

FIGURE 2.2 Nervous System

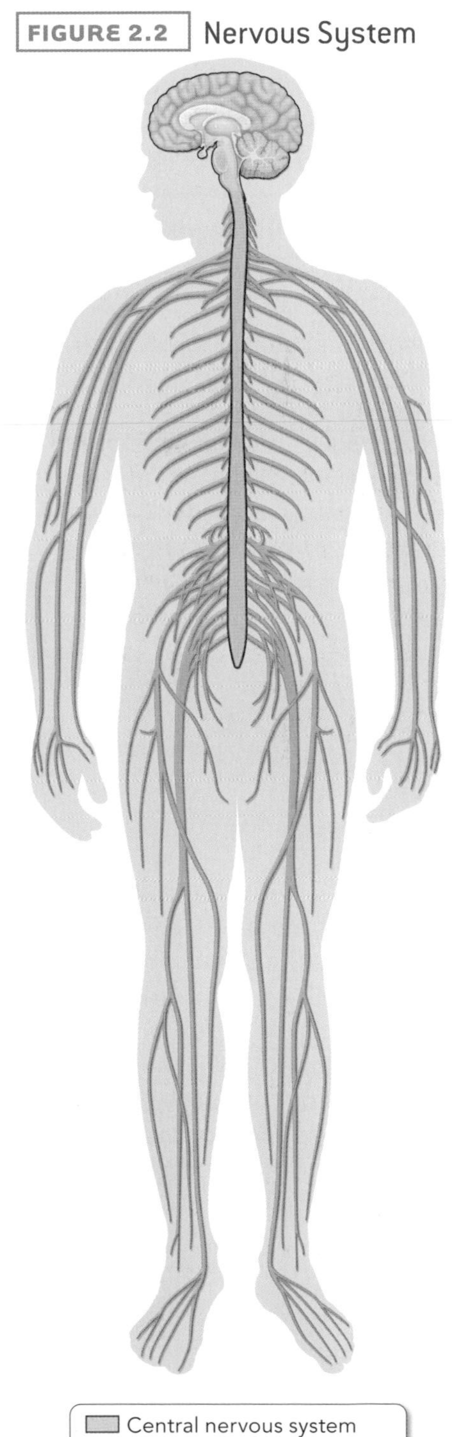

2.4 Electrochemistry and the Neuron

It is crucial to understand that the nervous system acts at different levels of organization. For the purposes of the MCAT®, the most important levels to be aware of are the cellular level (the neuron), the level of the organ (i.e. the synapse between a neuron and organ), and the level of the system (the different systems are described in section 2.7). As you read this lecture and proceed from the neuron to the larger levels of organization, review earlier sections as necessary to understand how the different pieces fit together to determine the function of the nervous system as a whole.

The smallest functional unit of the nervous system is the **neuron**. A neuron is a highly specialized cell capable of transmitting a signal from one cell to another through a combination of electrical and chemical processes. Neurons join to form the higher-level divisions of the nervous system. The physical structure of the neuron and the mechanism by which it passes on a signal create the characteristic features of communication in the nervous system: fast, fleeting, and specific. The long axon of a neuron allows for contact with a specific site, the signal travels down the axon through a fast electrical mechanism, and the slowest (chemical) portion of the signal only has to cross a tiny space. Meanwhile, the effect of the signal is fleeting because the neurotransmitter is taken back up by the pre-synaptic neuron to prevent a sustained effect.

The neuron is so highly specialized that it has lost the ability to divide. In addition, it depends almost entirely upon glucose for its chemical energy. Although the neuron uses facilitated transport to move glucose from the blood into its cytosol, the neuron is not dependent upon insulin for this transport, unlike most other cells. The neuron depends heavily on the efficiency of aerobic respiration. However, it is not able to store significant amounts of glycogen and oxygen, and so must rely on the blood to supply sufficient metabolic resources.

Neurons in different parts of the body have different physical appearances corresponding with their particular functions, but all neurons have a basic anatomy consisting of many **dendrites**, a single **cell body**, and usually one axon with many small branches (see Figure 2.3). The physical characteristics of each of these parts of the neuron correspond to its function, as described below.

Think of the dendrites as a voting machine. The structure of the dendrites allows them to collect and sum multiple signals in order to select the most important pieces of information that the neuron will transfer to other neurons.

The dendrites are stubby structures that receive a signal to be transmitted. Because a single neuron has many dendrites, each of which has multiple branches, inputs from many other neurons can be received simultaneously. **Summation** provides a way for the nervous system to screen for the most important stimuli. Spatial summation occurs when multiple dendrites receive signals at the same time, whereas temporal summation adds up the effects of signals that are received by a single dendrite in quick succession. The utility of summation is particularly evident in the gathering of sensory information by the central nervous system. Since the intensity of a stimulus can be coded by the **frequency of firing** of the sensory neuron or the number and type of receptors that respond, a stimulus of higher intensity will be more likely to trigger an action potential than a less intense stimulus.

The small size of the dendrites facilitates the quick transfer of an electrical signal to the cell body, to which the dendrites are attached. The cell body contains the organelles of the neuron, including the nucleus. Typically, the cytosol of the cell body is highly conductive and any electrical stimulus creates a disturbance in the electric field that is transferred immediately to the axon hillock. If the signals received by the dendrites sum to produce a large enough change in voltage across the membrane, the **axon hillock** (the site of connection between the cell body and the axon) generates an action potential in all directions, including down the **axon**. The membrane of the cell body usually does not contain enough ion channels to sustain an action potential. The axon, however, carries the action potential to a synapse, which passes the signal to another cell. The single axon is much longer than the dendrites, allowing the neuron to project to distant neurons or organs.

Some axons are extremely long (they can measure as much as a meter or even more!) Why have one long cell instead of several shorter ones? The answer is that having a single axon minimizes error and maximizes efficiency. Multiple synapses would provide opportunities for information to be transferred incorrectly and would also slow down the signal.

Receptors and **ion channels** are two types of molecules that are crucial to the function of neurons. Both are proteins that are embedded in the cellular membrane of neurons and determine what processes are possible at a given time or place along the membrane. Receptors bind substances (ligands) such as neurotransmitters and hormones, and respond by triggering processes within the cell. Different types of receptors are specialized to bind to particular chemicals. Some types of sensory receptors (cells that detect environmental stimuli, as described later in this lecture) have receptors in their membranes that bind chemicals in the environment. By contrast, ion channels open to allow ions to travel from one side of the membrane to the other, facilitating the transmission of signals. Different ion channels can specialize in terms of which ions they allow to pass through the membrane. In addition, ion channels vary in how quickly they open in response to changes in electrical potential. Some receptors are themselves ion channels and open in response to the binding of a ligand. These properties are crucial to the transfer of information between neurons. The role of receptors and ion channels in the nervous system will be discussed throughout this lecture.

Apply Key 3: The types and numbers of various receptors and channels determine how a cell will react to a given stimulus. Channels can open and close to control the movement of ions and molecules across the cell membrane. Receptors respond to specific hormones and can facilitate signaling cascades in the cell that affect transcription, translation, and cellular phenotypes.

The Neuron as an Electrochemical Cell

The neuron is a biological example of an electrochemical **concentration cell** where differences in concentration initiate the movement of charge, creating a voltage. This property of neurons, along with features of the surrounding environment, allows neurons to carry out their function of transferring information throughout the nervous system. The mechanism for the transfer of signals between neurons will be considered later in this lecture, but it is first necessary to understand how neurons generate voltage by acting like concentration cells.

The neuron is similar to an electric circuit. Circuits need batteries to provide an electromotive force. In the neuron, this driving force comes from the electric potential, or voltage, established when charges are separated across the plasma membrane. Current flows through the axon via the movement of cations, namely Na+ and K+. As in a wire, current moves in the direction of flow of positive charge. In the neuron, current is conducted as sodium ions flow across the plasma membrane into the cell. The neuron holds separated charges across its membrane, much like a capacitor. As sodium enters, it dissipates the negative membrane potential. This phenomenon is similar to the action of discharging a capacitor. Permeability in the neuron can be compared to electrical resistance. At rest, the neuron is largely impermeable to sodium ions and has a high resistance. As voltage gated sodium channels open during an action potential, the membrane's permeability increases along with its conductance, which is the reciprocal of resistance.

FIGURE 2.3 Myelinated Neuron

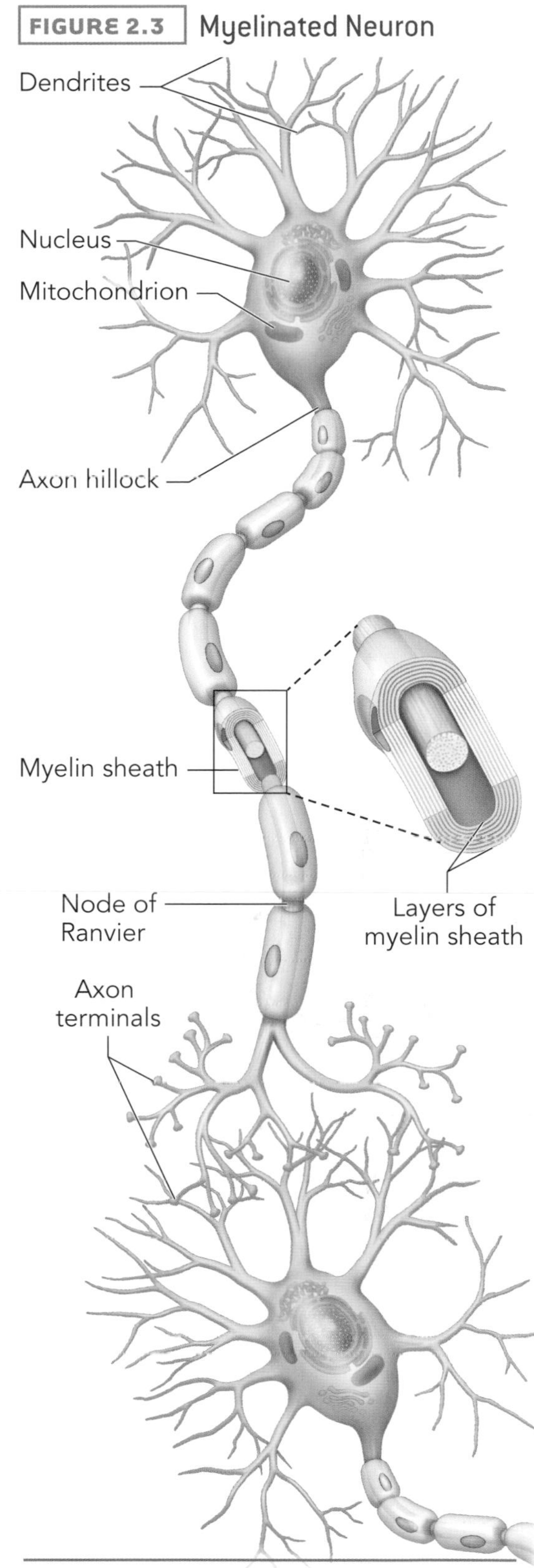

It is important to know the basic anatomy of a neuron. Picture this basic structure to help you remember how neurons receive and convey information. Multiple branched dendrites receive multiple signals, the signals travel from the dendrites to the adjacent axon hillock, and an action potential is generated at the axon hillock and moves down the axon to the synapse.

FIGURE 2.4 Concentration Cell

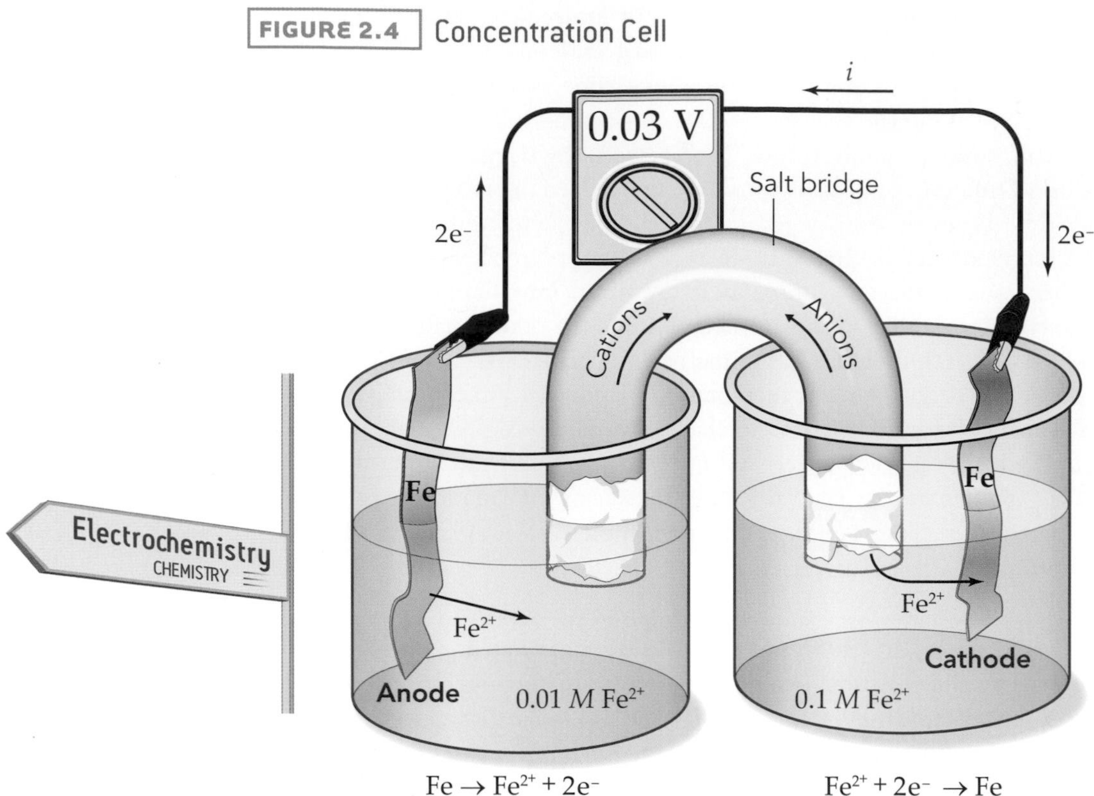

As in all concentration cells, neurons use an insulating membrane to separate two conducting solutions. In the case of the neuron, the two solutions are the intracellular and extracellular fluids, and the insulating membrane is the cellular membrane. Concentration cells separate solutions that contain different concentrations of the same ion, generating an electrical potential. For the purpose of considering the neuron as a concentration cell, the two most important ions in solution are Na^+ and K^+. The uneven distribution of these ions across the membrane, with a much higher extracellular concentration of Na^+ than the intracellular concentration and the reverse for K^+, is responsible for the resting potential difference across the membrane, which is approximately -70 mV. There is a buildup of negative charge just inside the cell membrane, and a buildup of positive charge just outside the membrane. The voltage generated by the separation of different concentrations of an ion in solution can be calculated using the Nernst equation, as will be described.

Remember the uneven distribution of Na^+ and K^+ across the membrane of a neuron: more Na^+ outside the cell than in, and more K^+ inside the cell than out.

MCAT® THINK

The neuron can be thought of as a biological concentration cell because of the differences between the intracellular and extracellular concentrations of K^+ and Na^+. A concentration cell is comprised of two half cells, each containing the same ion but in different concentrations. The half cells are separated by a salt bridge and connected by a conducting wire. In the context of the neuron, the two half cells are the inside and outside of the neuron membrane. The concentration of sodium is high outside and low inside, while potassium is high inside and low outside.

The salt bridge in a chemical cell allows for the transfer of electrons without the buildup of a significant charge difference. Remember that separating charges is energetically unfavorable. In the neuron, the very large negatively charged proteins on the inside of the plasma membrane are anions behaving like the chemical salt bridge. Ions such as Na^+ and K^+ often associate with and cluster around the negatively charged R groups of proteins – if they did not, the proteins would be reactive. When the chemical gradient is strong enough to force the movement of these small cations across the membrane, the very large proteins around which they are clustered cannot fit through the channels and remain on the original side. The proteins and ions continue to align across the plasma membrane because of their sustained attraction. The separation of the proteins and cations results in an accumulation of negative charge. When K^+ diffuses out of the cell, the negative proteins that cluster on the inner surface of the membrane create a resting membrane potential of around -70 mV within the cell. When Na^+ diffuses into the cell, the outward-facing protein anions with which they were associated remain on the external side of the membrane, creating a positive action potential of around +60 mV within the cell.

In chemistry, current flows through the wire between the half cells. In biology, the movement of cations through ion channels generates current. Remember that the concentration of sodium is high outside the cell and low inside the cell. During an action potential, Na^+ rushes down its electrochemical gradient, through ion channels, and into the cytosol of the neuron, creating an action potential. This movement of charge is equivalent to current moving between half cells of a concentration cell.

The Resting Potential

The **resting potential** is the electrical potential, or voltage, across the neuronal membrane at rest. It is the same as the voltage that is discussed in the context of electric charge and circuits in the *Physics* manual; it is related to potential energy. It can be thought of as a stored potential that is used by the neuron to transmit signals. A sufficient disruption in the balance of forces that maintain the resting potential leads to a cascade of events that constitute an action potential, as described in the next section. If a neuron loses the ability to establish the resting potential, it is no longer able to transmit action potentials.

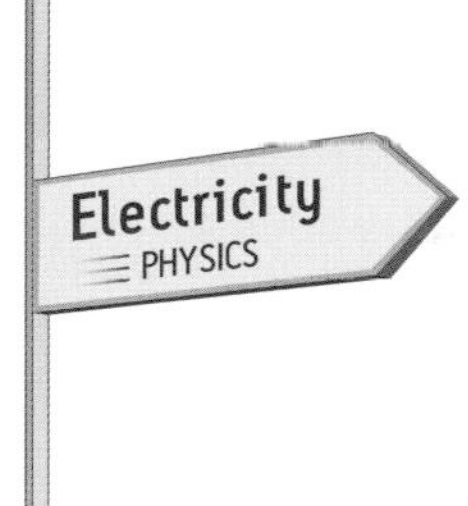

The resting potential is established by the diffusion of potassium out of the cell through channels in the membrane. Because the concentration of K^+ is much greater inside the cell, a chemical gradient is established that pushes potassium ions out of the cell. The high intracellular K^+ concentration is maintained by the Na^+/K^+ ATPase, which pumps three Na^+ out for every two K^+ brought in. When the K^+ ion approaches its chemical equilibrium, meaning that the concentrations of the ion on the inner and outer surfaces of the membrane start to equilibrate, there is an electric voltage created by the difference in charge between the ions that have crossed the membrane and the protein anions left behind. In the neuron at rest, the membrane is highly permeable to K^+ but almost completely impermeable to Na^+. In other words, many more channels that allow the diffusion of K^+ are open compared to the number of channels specific to the passage of Na^+. As a result, when the neuron is at rest, the diffusion of K^+ across the membrane is the major determinant of the membrane potential across the membrane.

As with all chemical gradients, as K^+ diffuses across the membrane, the chemical gradient starts to equilibrate and the diffusional force decreases. As the chemical gradient starts to break down, an electrical gradient starts to build up. The two factors that cause this electrical gradient are the diffusion of positively charged K^+ ions into the extracellular environment adjacent to the membrane and the buildup

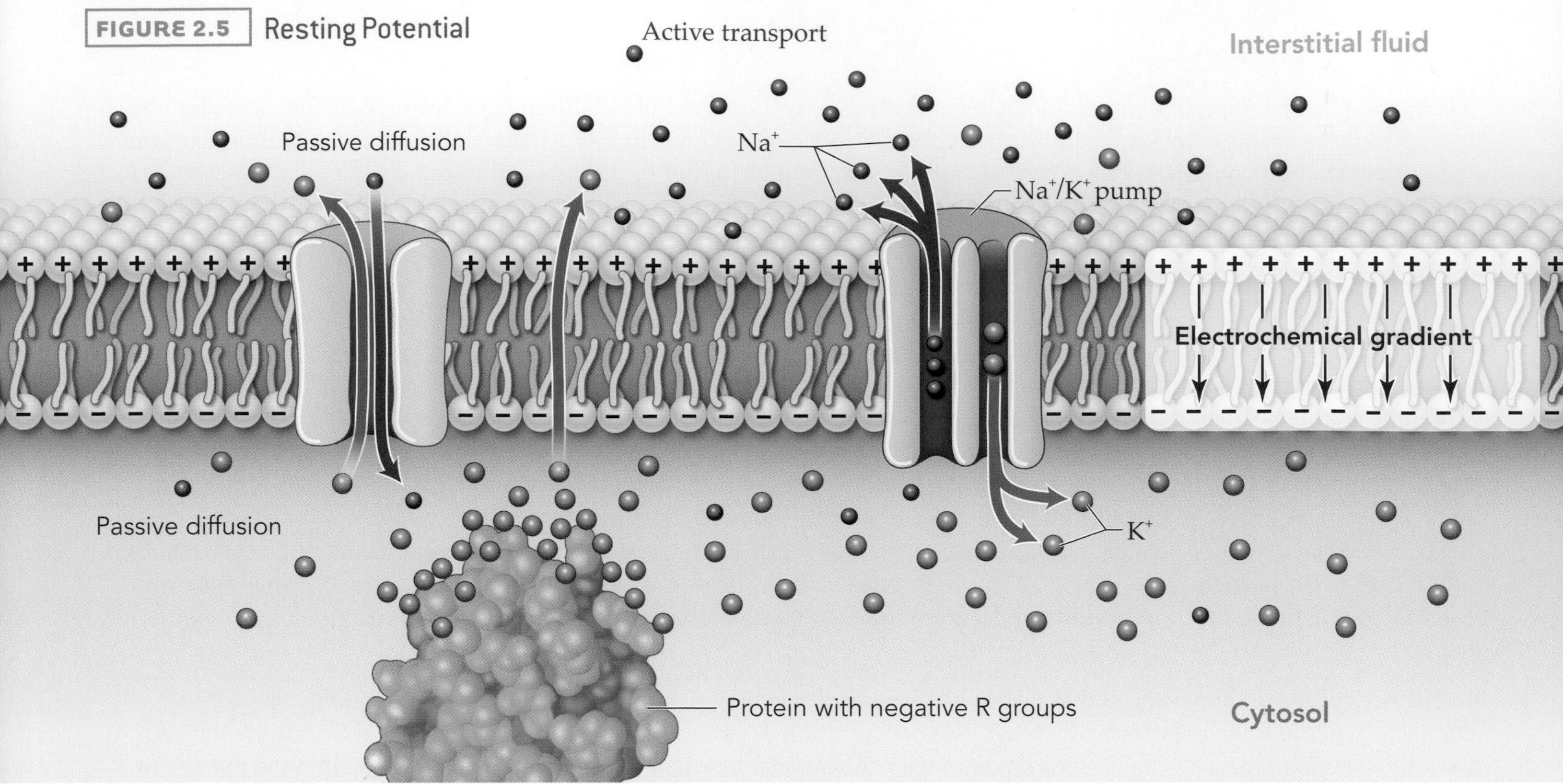

FIGURE 2.5 Resting Potential

You may be confused about how the resting membrane potential can be set up without dramatically altering the chemical environment of the cell. Students often ask whether the concentrations used in the Nernst equation refer to the intracellular and extracellular concentrations before or after the diffusion of K^+ to achieve its equilibrium potential. The answer is both! These processes occur locally, in the vicinity of the membrane. Only a small number of ions have to cross the membrane to set up the voltage. The overall concentration of ions on either side of the membrane is unaffected. Similarly, the negative and positive charge buildup also occurs right next to the membrane; the inside of the neuron does NOT have an overall negative charge.

of negatively charged substances along the inner side of the membrane. The direction of the electrical gradient is therefore opposite that of the chemical gradient. The buildup of positive charge in the extracellular environment discourages the movement of more K^+ ions out of the cell down the chemical gradient, while the negative charge on the intracellular side of the membrane exerts an attractive force on the positive ions. The combination of these opposing gradients creates the **electrochemical gradient** of potassium.

The buildup of negative charge along the inside of the membrane occurs as a side effect of the diffusion of K^+. The K^+ ions in solution associate themselves with the negatively charged R groups of proteins within the cell. Thus, as they start to move towards the membrane, they drag along the protein. If the proteins could move through the membrane along with the ions, they would. But of course proteins are much larger than the ion channels that are just barely large enough to allow the passage of K^+ ions, so they get left behind. The positive charge building up along the outside of the membrane as K^+ ions flow out of the cell attracts the negatively charged proteins, causing them to stay in their position near the membrane. Other anionic substances, particularly Cl^-, also contribute to the buildup of negative charge.

The membrane contains a pump called the **sodium/potassium** or **Na^+/K^+ pump** that functions to maintain or reestablish the chemical gradient that is lost by diffusion. The pump moves ions in the direction opposite to the flow of diffusion, so it uses an active transport mechanism and requires energy input from ATP. The Na^+/K^+ pump moves three positively charged sodium ions out of the cell while bringing two positively charged potassium ions back into the cell, preventing equilibrium and replenishing the concentration gradients of these ions. As a result, the resting potential is maintained at a constant -70 mV when the neuron is at rest.

The value of the resting membrane potential is determined by the **equilibrium potential** of potassium. When different concentrations of an ion are separated across a membrane that is permeable to that ion, as in the case of K^+ across the neuronal cell membrane, an electrical gradient develops that precisely balances the chemical gradient caused by the unequal concentrations. As with any concentration cell, the value of the resulting voltage can be calculated by the **Nernst equation**:

$$E = E^{\circ} - \frac{RT}{nF}\ln(Q)$$

When temperature is constant, the whole expression RT/F is a constant. The charge of the ion is represented by *n*. E° is the standard cell potential, meaning the voltage that exists when ion concentrations are equal in both parts of the cell. In any concentration cell, where the same reaction is taking place but proceeding in opposite directions, voltage can only be generated by unequal concentrations, so E° is equal to zero. Q stands for the ratio of products to reactants; in the case of the neuron, it can be represented by the ratio of the extracellular and intracellular concentrations of potassium. For the purposes of calculating the resting potential of the neuron, the Nernst equation can be rewritten as:

$$E_{K^+} = -\frac{RT}{nF}\ln\left(\frac{[K^+]_{intracellular}}{[K^+]_{extracellular}}\right)$$

Because the intracellular concentration of potassium is greater than the extracellular concentration, the ratio will be more than one, and the natural logarithm of the ratio will be positive. However, the negative sign results in a negative potential. As expected, the inner side of the membrane is found to be negatively charged with respect to the outer side. Also notice that the equation shows the proportionality between the concentration difference across the membrane and the potential that is generated. In other words, the more the concentrations differ, the more negative the electrical potential will be. The MCAT® requires an understanding of the variables in the Nernst equation, plus the sign and direction of the change in electric potential.

The equilibrium potential of potassium can be used to approximate the resting potential of the neuron because the membrane of the neuron at rest is much more permeable to potassium than to any other ion. This is because many potassium channels are open compared to the number of open channels that allow the passage of other types of ions, such as sodium. However, the resting potential is slightly more positive than it would be if only potassium were involved because there is some leakage of sodium ions across the membrane and into the cell. Put another way, the equilibrium potential of Na^+ is positive because the extracellular concentration is greater than the intracellular concentration (in contrast to potassium), so the leakage of sodium slightly offsets the negative equilibrium potential of potassium. Other ions also have concentration gradients, but they do not significantly affect the resting potential due to the low permeability of the membrane for these ions.

The take home messages about the resting potential are: 1) Most importantly, the resting potential is set up by the diffusion of K^+. 2) K^+ diffuses out of the cell, dragging along negatively charged proteins that get stuck along the inner side of the membrane. 3) When the chemical gradient of K^+ equilibrates, the inner side of the membrane is negatively charged compared to the outer side. The resulting potential across the membrane is very close to the equilibrium potential of K^+. Another thing: understand the Nernst equation and its variables.

MCAT® THINK

It may seem strange that neurons use such a complex process to transmit information. Why wouldn't electricity just flow down a neuron, as it does in a wire? The answer is that unlike wires, the membranes of axons are not conductive of electricity because they are primarily composed of lipids. Instead, the solution of the cell is made up of ions, also called electrolytes, which are conductors of electrical signals. The action potential is the evolved solution to the problem of how to convey an electrical signal. The signal can be transmitted down the axon exactly as fast as ions can travel and channels can open. Think of an axon as a revised wire that depends on the travel of ions rather than electrons. Note that ions are much larger than electrons. Drag slows the transmission of the signal if the cross-sectional area of the axon is too small.

The Action Potential

The concentration gradients across the membrane, and the membrane potential that develops as those gradients equilibrate, are crucial for the generation of the action potential. The **action potential** is the mechanism by which a signal travels down the length of a neuron and ultimately is transferred to the next cell. It is generated by a change in voltage across the neuronal membrane. The entire mechanism of the action potential can be conceptualized as a flip between permeability to potassium and permeability to sodium. At first there are many more open potassium channels than sodium channels. The action potential begins when many sodium channels suddenly open up. The potential across the membrane moves towards the equilibrium potential of the ion to which the membrane is most permeable at any given moment, due to the unequal intracellular and extracellular concentrations of the ions. In the most common type of synapse, the end result of the action potential is the release of neurotransmitter from the end of the axon.

FIGURE 2.6 Steps of the Action Potential

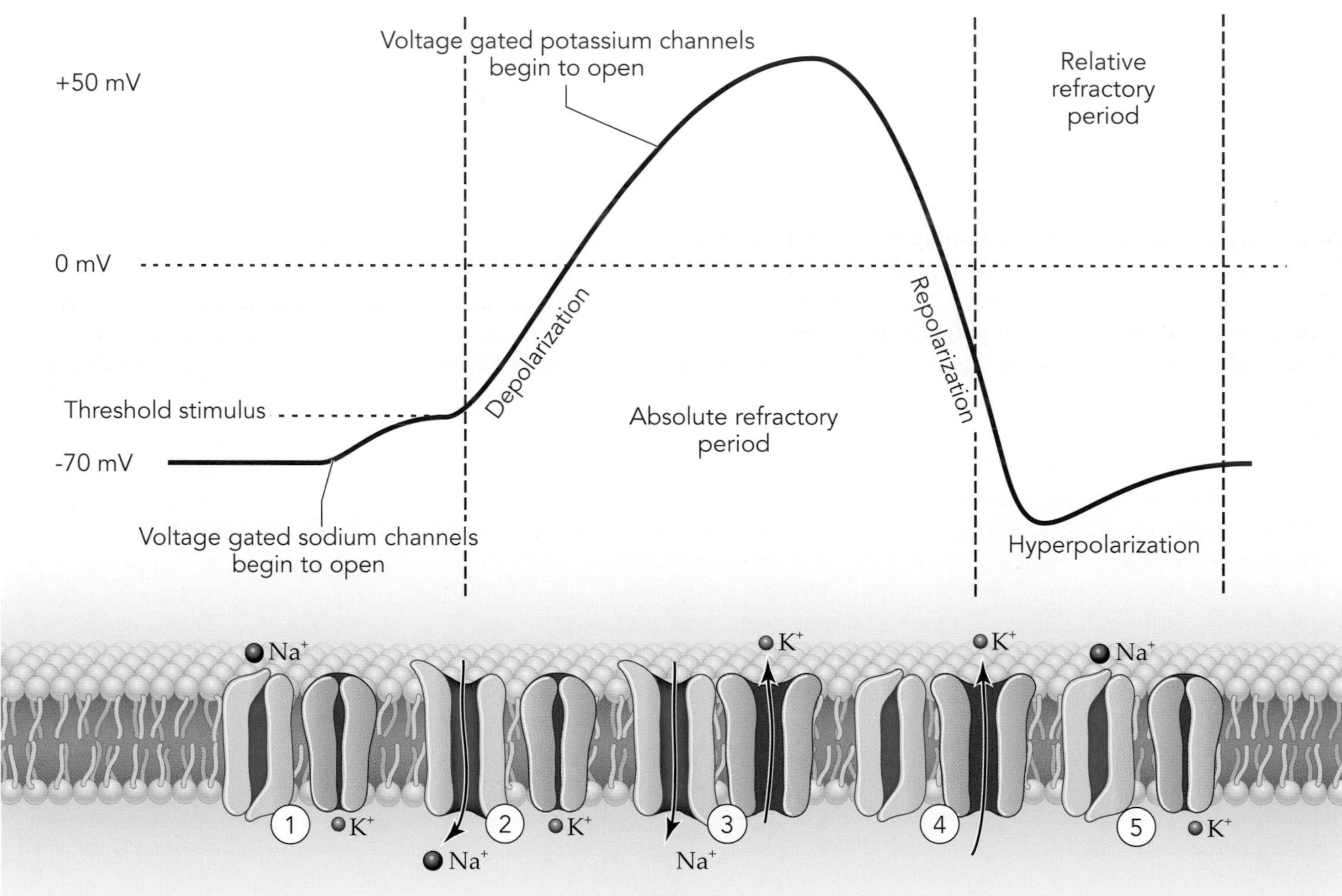

(1) Membrane is at rest. Voltage gated sodium and potassium channels are closed.

(2) Voltage gated sodium channels open and the cell depolarizes.

(3) Voltage gated potassium channels open as sodium channels begin to inactivate.

(4) Voltage gated sodium channels are inactivated. Open potassium channels repolarize the membrane.

(5) Voltage gated potassium channels close and the membrane equilibrates to its resting potential.

The membrane of a neuron contains integral membrane proteins called **voltage gated sodium channels**. These proteins change configuration when the voltage across the membrane is disturbed, allowing Na^+ to flow through the membrane for a fraction of a second. As Na^+ flows into the cell, the voltage changes further, causing more sodium channels to change configuration, still allowing more sodium to flow into the cell in a positive feedback mechanism. Since the Na^+ concentration moves toward equilibrium (i.e. towards the positive equilibrium potential of sodium) while the K^+ concentration remains higher inside the cell, the membrane potential reverses polarity so that it is positive on the inside and negative on the outside. This process is called **depolarization**.

The neuronal membrane also contains **voltage gated potassium channels**. The potassium channels are less sensitive to voltage change than the sodium channels, so they take longer to open. By the time they begin to open, most of the sodium channels are closing, diminishing the membrane's permeability to sodium. At this point K^+ flows out of the cell, moving the potential back towards the negative equilibrium potential of potassium. This process is called **repolarization**. The potassium channels are also so slow to close that, for a fraction of a second, the potential across the membrane becomes even more negative than the resting potential. This portion of the process is called **hyperpolarization**. Passive diffusion returns the membrane to its resting potential. The entire process just described is called the action potential (Figure 2.6). Throughout the action potential, the Na^+/K^+ pump keeps working, helping to maintain the unequal concentrations of potassium and sodium across the membrane that allow the resting potential to be reset.

The action potential occurs at a point on the membrane and propagates by depolarizing the section of membrane immediately adjacent to it, which then causes the depolarization of the next adjacent section and so on down the membrane. Figure 2.7 shows an action potential propagating from right to left down the axon. The voltage at any given point on the membrane is indicated. The entire action potential as measured at one point on the membrane of a neuron takes place in a fraction of a millisecond.

An action potential is **all-or-none**, meaning that unless the membrane completely depolarizes, no action potential is generated. In order to create an action potential, the stimulus to the membrane must be greater than the **threshold stimulus**. Any stimulus greater than the threshold stimulus creates the same size action potential. The threshold stimulus is usually reached through the combination of many stimuli, each of which causes a small change in the membrane potential. Some have an **excitatory** (depolarizing) effect, while some have an **inhibitory** (hyperpolarizing) effect. An action potential is generated when the total effect of all inputs reaches the level of the threshold stimulus. The dendrites carry out both temporal and spatial summation of inputs, as described previously.

If the threshold stimulus is reached, but is reached very slowly, an action potential still may not occur. This is called *accommodation*. Once an action potential has begun, there is a short period of time called the *absolute refractory period* (Figure 2.6) in which no stimulus will create another action potential. The

FIGURE 2.7 Unidirectional Propagation of an Action Potential

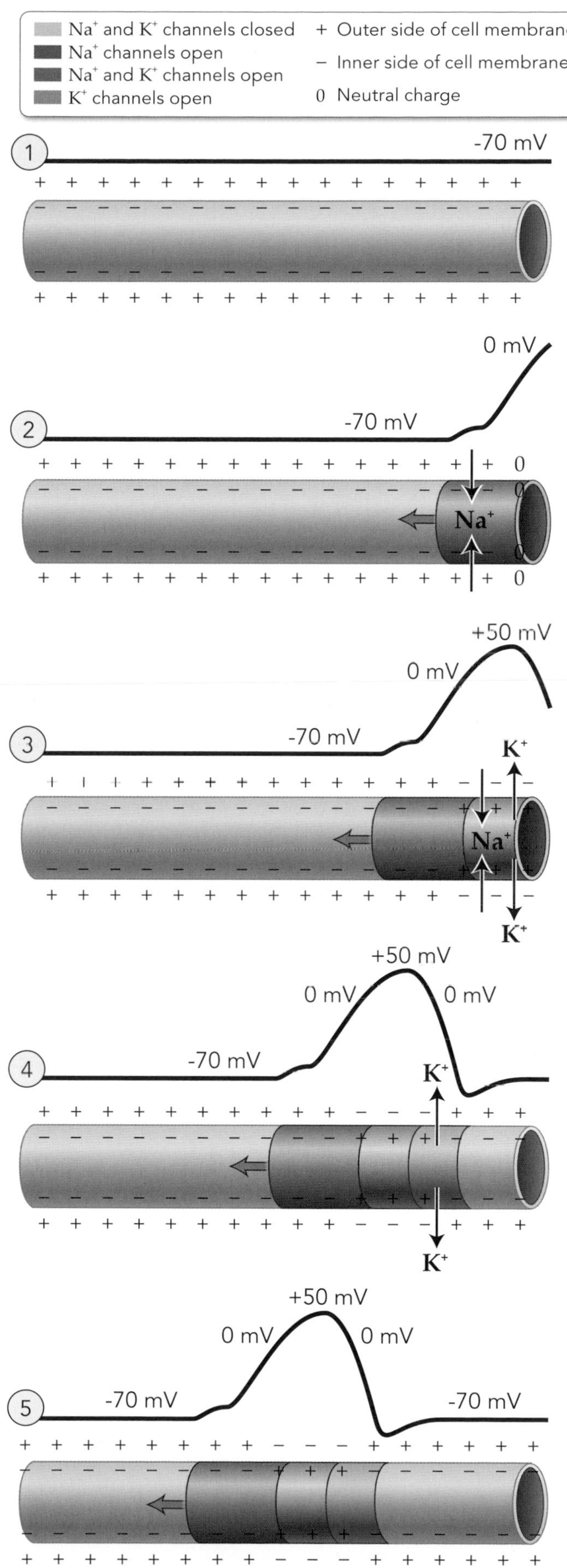

absolute refractory period occurs because the membrane potential is already more positive than the resting potential, so the driving force of the action potential is absent. The *relative refractory period* is the time during which only an abnormally large stimulus will create an action potential because the membrane is hyperpolarized, requiring a greater threshold stimulus. A neuron cannot generate an action potential during the absolute refractory period but can do so if provided with a large enough stimulus during the relative refractory period. Other cells, such as skeletal and cardiac muscle cells, also conduct action potentials. Although these action potentials are slightly different in duration, timing of stages, and even the types of ions involved, they work on the same principles.

Understand the dynamics of the membrane potential. Use the Na^+/K^+ pump to help you remember that the inside of the membrane is negative with respect to the outside. The action potentials of different types of cells have different characteristics, but if you understand the basic principle, you can understand any action potential. Remember that an action potential originates at the axon hillock. This section is important for the MCAT®. If you don't thoroughly understand it, reread it.

MCAT® THINK

Consider a mutation in the voltage-gated sodium channel that allows a small amount Na^+ ions through, even when the channel is closed. What effect is this likely to have on the resting membrane potential of the neuron? Would this make it easier or more difficult for an action potential to occur?

Answer on page 75.

These questions are NOT related to a passage.

Question 25

Which of the following gives the normal direction of signal transmission in a neuron?

- O **A.** From the axon to the cell body to the dendrites
- O **B.** From the dendrites to the cell body to the axon
- O **C.** From the cell body to the axon and dendrites
- O **D.** From the dendrites to the axon to the cell body

Question 26

Novocaine is a local anesthetic used by many dentists. Novocaine most likely inhibits the action potential of a neuron by:

- O **A.** stimulating calcium voltage gated channels at the synapse.
- O **B.** increasing chloride ion efflux during an action potential.
- O **C.** uncoiling Schwann cells wrapped around an axon.
- O **D.** blocking sodium voltage gated channels.

Question 27

If a neuronal membrane were to become suddenly impermeable to potassium ions but retain an active Na^+/K^+-ATPase, the neuron's resting potential would:

- O **A.** become more positive because potassium ion concentration would increase inside the neuron.
- O **B.** become more positive because potassium ion concentration would increase outside the neuron.
- O **C.** become more negative because potassium ion concentration would increase inside the neuron.
- O **D.** become more negative because potassium ion concentration would increase outside the neuron.

Question 28

If only Na^+/K^+ pumps were allowed to function in the membrane of a cell, which of the following would be expected to occur?

- O **A.** The membrane potential would drop, then rise.
- O **B.** A net influx of sodium would occur.
- O **C.** The membrane potential would continuously drop.
- O **D.** The membrane potential would continuously rise.

Question 29

Compared to the endocrine system, the nervous system is:

- O **A.** faster and more specific.
- O **B.** faster and more generalized.
- O **C.** slower and more specific.
- O **D.** slower and more generalized.

Question 30

The Nernst equation, when used to determine the resting membrane potential of a neuron, indicates that, as intracellular K^+ increases, the membrane potential:

- O **A.** stays the same.
- O **B.** changes, but not in a predictable way.
- O **C.** becomes more positive.
- O **D.** becomes more negative.

Question 31

Which of the following changes to the neuron at rest would be LEAST likely to prevent the formation of the normal resting membrane potential?

- O **A.** Disruption of the electrochemical gradient of chlorine
- O **B.** Disruption of the electrochemical gradient of sodium
- O **C.** Disruption of the electrochemical gradient of potassium
- O **D.** A change to the membrane resulting in impermeability to potassium

Question 32

A defect in a particular neuron blocks all voltage gated potassium channels. Which of the following is true of this neuron?

- O **A.** No action potentials could occur
- O **B.** No return to resting potential after firing
- O **C.** Inability to maintain resting potential
- O **D.** Inability to undergo saltatory conduction

2.5 Communication Between Neurons: The Synapse

Neural impulses are transmitted from one cell to another chemically or electrically via a synapse. The transmission of the signal from one cell to another is the slowest part of the process of nervous system cellular communication, yet it occurs in a fraction of a second.

Electrical synapses are uncommon. They are composed of gap junctions between cells and are often used when coordinated action is required from a group of cells. Cardiac muscle, visceral smooth muscle, and a very few neurons in the central nervous system contain electrical synapses. Since electrical synapses do not involve diffusion of chemicals, they transmit signals much more quickly than chemical synapses. In addition, the signal can propagate bidirectionally, meaning that both cells involved in the synapse can send and receive signals.

A more common type of synapse, the **chemical synapse** (Figure 2.8), consists of a space between two neurons that is crossed by neurotransmitters. Notice that unlike cells involved in electrical synapses, cells that are separated by a chemical synapse do not actually touch each other. Chemical synapses also differ from electrical synapses in that they are **unidirectional**; in a given chemical synapse, one cell always releases neurotransmitter and the other always receives it. An example of a chemical synapse is a **motor end plate**, which is the connection between a neuron and a muscle.

In a chemical synapse, small vesicles filled with neurotransmitters rest just inside the presynaptic membrane. **Neurotransmitters**, chemicals that are often derived from amino acids, are released into the synapse by a presynaptic neuron and attach to receptors on the post-synaptic cell. They are required because the electrical signal of the action potential cannot be transferred across the synapse between neurons. The membrane of the presynaptic neuron near the synapse contains a large number of Ca^{2+} voltage gated channels. When an action potential arrives at a synapse, these channels are activated, allowing Ca^{2+} to flow into the cell. In a mechanism that is not completely understood, the sudden influx of calcium ions causes some of the neurotransmitter vesicles to be released into the synaptic cleft through exocytosis. The neurotransmitter then diffuses across the synaptic cleft.

The postsynaptic membrane contains neurotransmitter receptor proteins. When the neurotransmitter attaches to the receptor proteins, the postsynaptic membrane becomes more permeable to ions. Ions move across the postsynaptic membrane through ion channels, completing the transfer of the neural impulse. In this way, the impulse is not attenuated by electrical resistance as it moves from one cell to the next. However, if a presynaptic cell is fired too often it will not be able to replenish its supply of neurotransmitter vesicles, resulting in *fatigue* (the impulse will not pass to the postsynaptic neuron).

FIGURE 2.8 Chemical Synapse

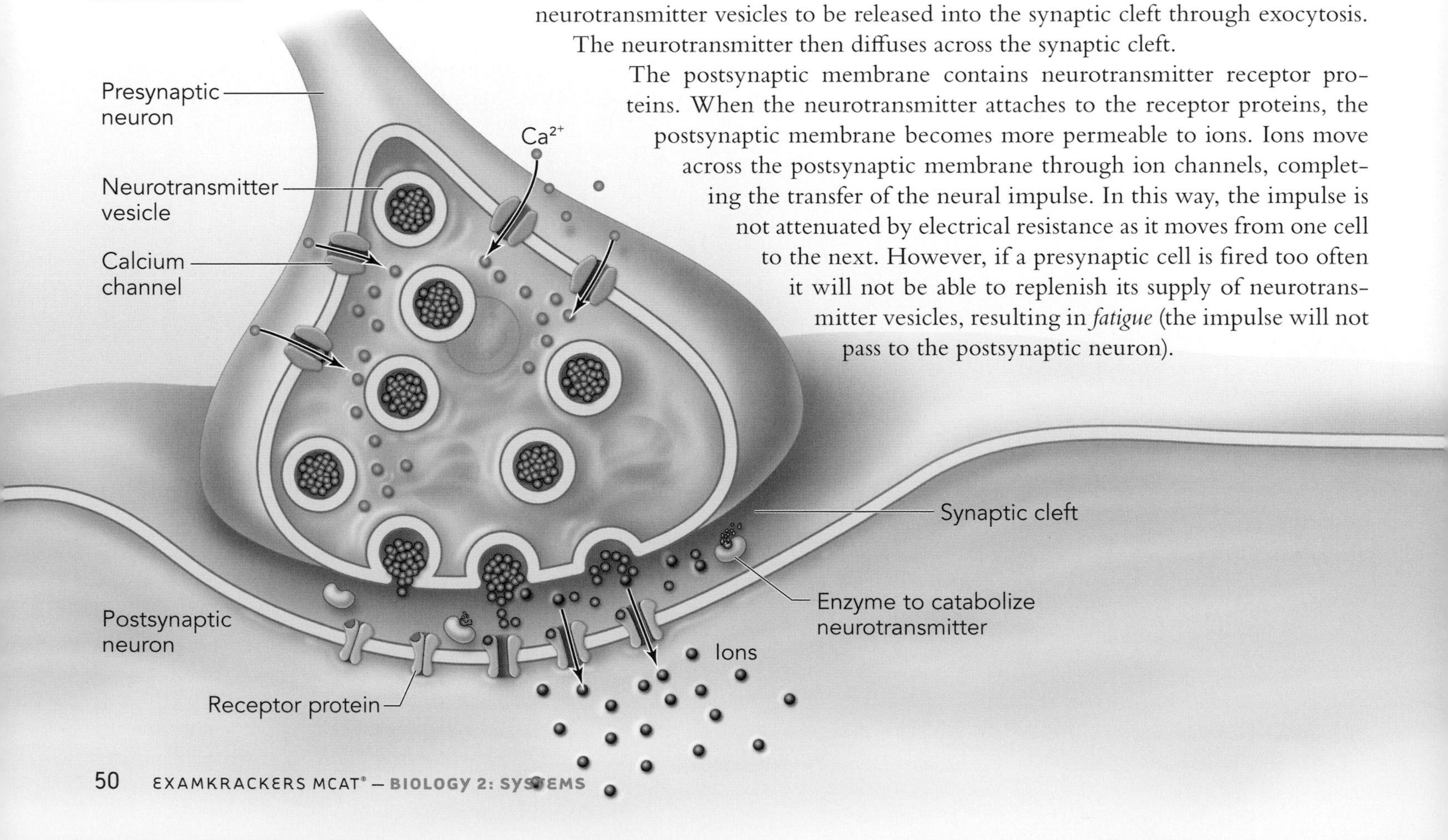

FIGURE 2.9 G-protein Second Messenger System

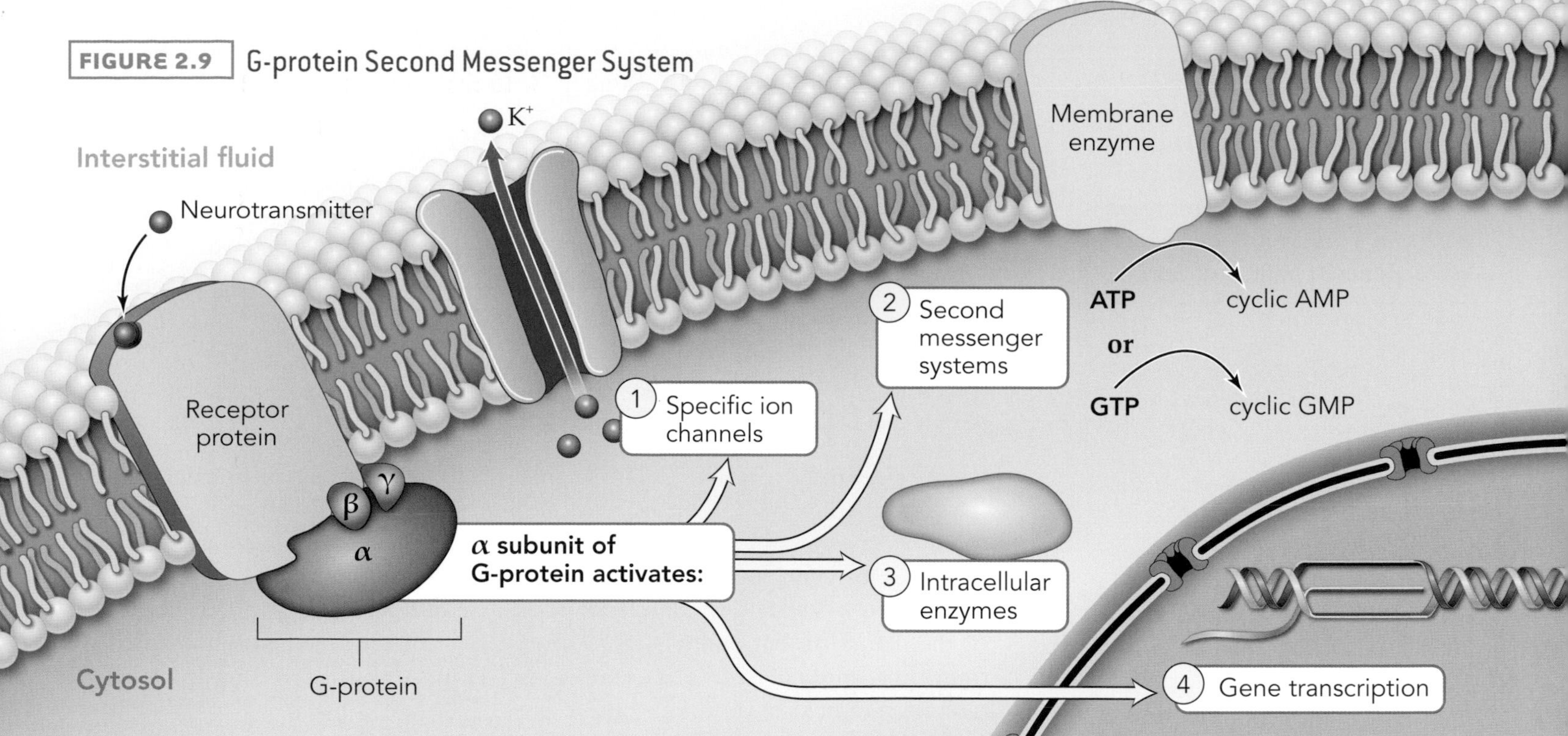

The neurotransmitter attaches to its receptor for only a fraction of a second and is then released back into the synaptic cleft. If the neurotransmitter remains in the synaptic cleft, the postsynaptic cell may be stimulated repeatedly. The cell uses several mechanisms to deal with this problem. The neurotransmitter may be destroyed by an enzyme in the matrix of the synaptic cleft and its parts recycled by the presynaptic cell; it may be directly absorbed by the presynaptic cell via active transport; or the neurotransmitter may diffuse out of the synaptic cleft.

Over 50 types of neurotransmitters have been identified. Different neurotransmitters are characteristic of different parts of the nervous system. A single neuron usually secretes only one type of neurotransmitter. Rather than having to select different neurotransmitters for different purposes, a neuron can simply manufacture a single neurotransmitter for storage and release it when stimulated by an action potential. (However, a neuron may be able to respond to multiple types of neurotransmitter if its dendrites have the corresponding receptors.)

Any given synapse is designed either to inhibit or to excite, but not both. A single synapse cannot change from inhibitory to excitatory, or vice versa. On the other hand, some neurotransmitters can produce an inhibitory or excitatory effect depending on the type of receptor in the postsynaptic membrane. Acetylcholine, a common neurotransmitter, has an inhibitory effect on the heart, but an excitatory effect on the visceral smooth muscle of the intestines.

Receptors may be ion channels themselves, which open when their respective neurotransmitters attach, or they may act via a **second messenger system**, meaning that they activate another molecule inside the cell to make changes. For prolonged changes, such as those involved in memory formation, the second messenger system is preferred. *G-proteins* commonly initiate second messenger systems, as shown in Figure 2.9. A G-protein is attached to the receptor protein along the inside of the postsynaptic membrane. When the receptor is stimulated by a neurotransmitter, part of the G-protein, called the α *subunit*, breaks free. The α subunit may:

1. activate separate specific ion channels,
2. activate a second messenger (i.e. cyclic AMP or cyclic GMP),
3. activate intracellular enzymes, and
4. activate gene transcription.

Notice that the second messenger system allows nervous signals to have an effect on the expression of genes in target neurons.

When a man rubs his stubbed toe to decrease pain, he is taking advantage of IPSP. The rubbing stimulates neurons that also have inhibitory synapses on pain pathways, decreasing the overall perception of pain.

The formation of an action potential in a single neuron is influenced by information from many synapses. In fact, the number of synapses made by a neuron can range from a few to as many as 200,000 synapses. Most synapses contact dendrites, but some may directly contact other cell bodies, other axons, or even other synapses. The firing of one or more of these synapses creates a change in the neuron cell potential. This change in the cell potential is called either the *excitatory postsynaptic potential (EPSP)* or the *inhibitory postsynaptic potential (IPSP).* Typically, 40–80 synapses must fire simultaneously on the same neuron in order for an EPSP to create an action potential within that neuron.

2.6 Supporting Cells: Glia

Besides neurons, nervous tissue contains many support cells called **glial cells** or **neuroglia** (Figure 2.10). In fact, in the human brain, glial cells can outnumber neurons by as much as 10 to 1. Unlike neurons, glial cells do not convey electrical signals. However, neuroglia act in many ways to support neuronal function. They are capable of cellular division, and, in the case of traumatic injury to the brain, it is the neuroglia that multiply to fill any space created in the central nervous system. The ability of neuroglia to divide, unlike most neurons, also means that they are often responsible for brain cancer.

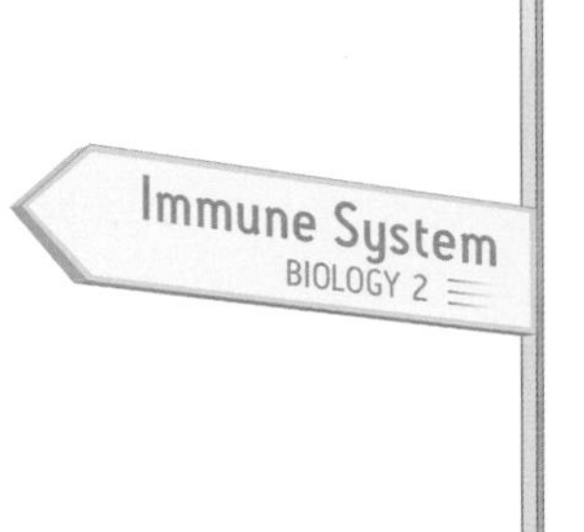

There are six types of glial cells: microglia, ependymal cells, satellite cells, astrocytes, oligodendrocytes, and Schwann cells. *Microglia* are the central nervous system's macrophages. Like other macrophages, they arise from white blood cells called monocytes. They carry out an immune function, phagocytizing microbes and cellular debris. *Ependymal cells* are epithelial cells that line the space containing the cerebrospinal fluid. Ependymal cells use cilia to circulate the cerebrospinal fluid. *Satellite cells* support ganglia, which are groups of cell bodies in the peripheral nervous system. *Astrocytes* are star-shaped neuroglia in the central nervous system that give physical support to neurons and help maintain the mineral and nutrient balance in the interstitial space.

FIGURE 2.10 Glial Cells

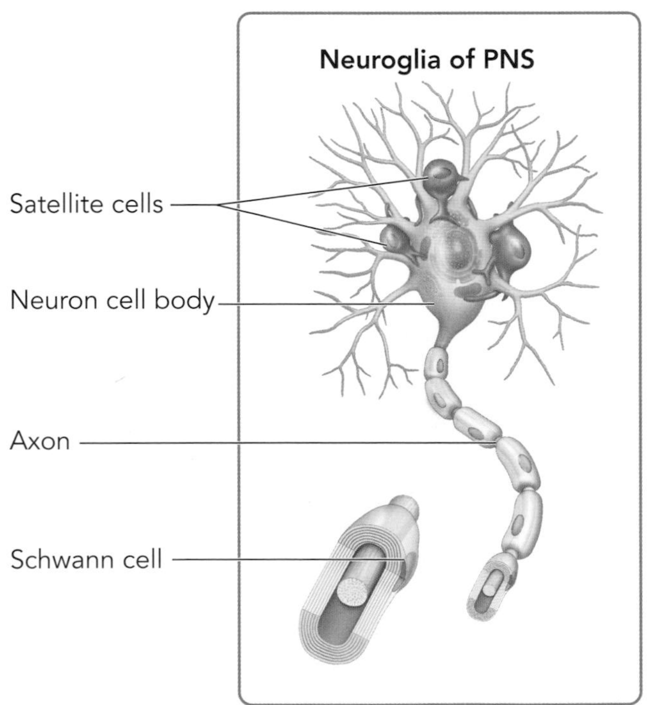

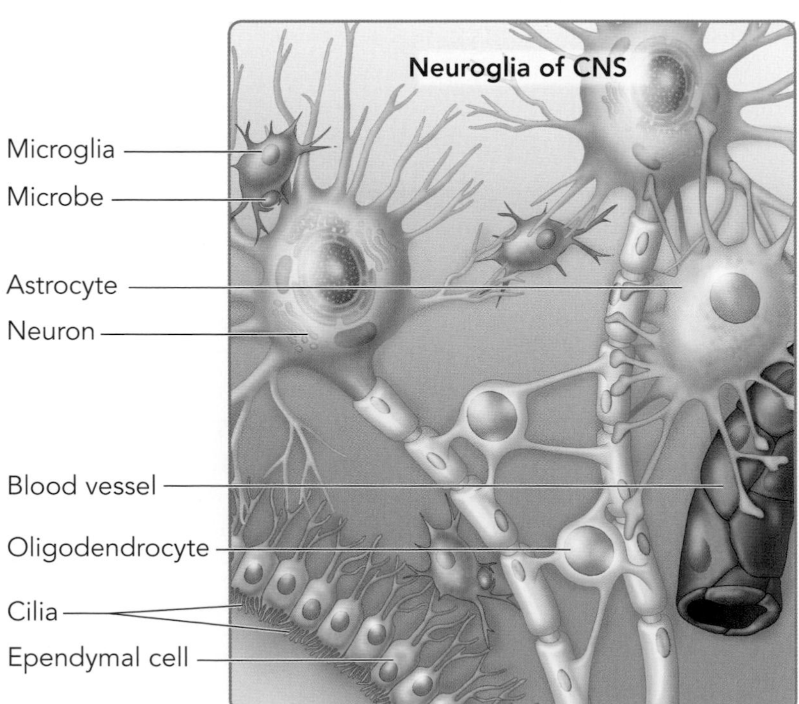

Oligodendrocytes have processes that wrap many times around axons in the central nervous system, creating electrically insulating sheaths called **myelin**. In the peripheral nervous system, myelin is produced by **Schwann cells**, which wrap their entire cell bodies around axons. To the naked eye, myelinated axons appear white, while neuronal cell bodies appear gray. For this reason, **white matter** refers to areas of the nervous system that are composed of myelinated axons of neurons, while **grey matter** refers to bundles of the cell bodies of neurons.

Understand how myelin increases the speed with which the action potential moves down the axon. Only vertebrates have myelinated axons. Memorizing the names of each type of glial cell is not necessary, but know the general functions of support cells.

Myelin increases the rate at which signals can travel down an axon. There are tiny gaps between myelinated areas called **nodes of Ranvier**. When an action potential is generated down a myelinated axon, the myelin acts as an insulator around the axon, increasing resistance to the passage of ions through the membrane. As a result, the action potential jumps from one node of Ranvier to the next as quickly as the disturbance moves through the electric field between them. This phenomenon is called **saltatory conduction** (Latin *saltus*: a jump). In the absence of myelin, the action potential travels much more slowly because each tiny adjacent portion of the membrane must be depolarized in sequence.

MCAT® THINK

The effect of myelin can be understood by visualizing the membrane as a capacitor. As discussed in the Electricity Lecture of the *Physics* manual, capacitance is defined as the amount of charge that a capacitor can store in a given area. Myelin is a non-conductive insulator, so it reduces the capacitance of the axon, which encourages ion flow just as the increased resistance does. Lowering capacitance also decreases the time required to charge a capacitor. In this biological example, the effect is that the speed of depolarization (and so the generation of the action potential) is increased at the nodes of Ranvier. This connection could appear in a physics passage on the MCAT®.

The nodes of Ranvier are the only places along the myelinated axon that ions cross the neuron's membrane. This propagates the action potential between regions of the axon that are insulated by myelin.

2.7 The Structures of the Nervous System

Neurons may perform one of three functions:

1. **Sensory** (also called **sensor or afferent**) **neurons** receive signals from a receptor cell that interacts with the environment. The sensory neuron then transfers this signal to other neurons. The brain filters for the most relevant information and discards 99% of sensory input.

2. **Interneurons** transfer signals from neuron to neuron. 90% of neurons in the human body are interneurons.

3. **Motor** (also called **effector or efferent**) **neurons** carry signals to a muscle or gland called the **effector**. Sensory neurons are located dorsally (toward the back) on the spinal cord (Figure 2.11), while motor neurons are located ventrally (toward the front or abdomen). The neurons of the sensory and motor systems constitute physically separated systems, much like the parasympathetic and sympathetic divisions of the autonomic nervous system.

You can remember that EFFerent neurons are motor neurons because they cause the body to make an EFFort to respond to sensory information.

Neurons with these differing functions work in a coordinated fashion to sense incoming stimuli, communicate information to other neurons, and produce a response. Neuron processes (axons and dendrites) are typically bundled together to form **nerves** (called tracts in the CNS).

Figure 2.11 shows a simple **reflex arc** using all three types of neurons. A **reflex** is a quick response to a stimulus, such as one with the potential to cause tissue damage, that occurs without direction from the central nervous system. This adaptive response is what causes an organism to jerk away from a hot stimulus before having a conscious experience of pain. Some reflex arcs do not require an interneuron. For example, the *stretch reflex* involves direct communication between a motor and sensory neuron. In response to the stretching of a muscle, a sensory neuron signals a motor neuron in the spinal cord. The motor neurons respond such that the stretched muscle contracts to regain its resting length. The stretch reflex demonstrates how a reflex arc can act according to a **negative feedback loop**. The sensory neuron senses a change and stimulates a corresponding adjustment by the motor neuron that acts in opposition to the change, ensuring a return to the resting muscle length without the need for conscious control.

Even though reflexes do not require a higher-level command, they do not occur independently from the CNS. Information about the stimulus is sent to the CNS as the reflex is generated, allowing the perception of the stimulus and coordination of a more complex response. In addition, the CNS can send signals that dampen or enhance reflex responses according to the needs of the organism. In the example of the stretch reflex given above, inhibitory input from **supraspinal circuits** descending from the CNS can decrease the sensitivity and speed of the reflex so that it proceeds smoothly and only when needed.

For the MCAT®, think of the CNS as the brain and spinal cord, and the PNS as everything else.

As discussed previously, the nervous system has two major divisions: the **central nervous system** (CNS) and the **peripheral nervous system** (PNS). The CNS consists of the neurons and support tissue within the brain and the spinal cord. A major function of the CNS is to integrate nervous signals between sensory and motor neurons.

The CNS is connected to the peripheral parts of the body by the PNS. Parts of the PNS, such as the *cranial nerves* and the *spinal nerves*, project into the brain and spinal cord. The PNS handles the sensory and motor functions of the nervous system. The PNS is further divided into the **somatic nervous system** and **autonomic nervous system** (ANS). The somatic nervous system contains sensory and motor functions and primarily functions to respond to the external

FIGURE 2.11 Simple Reflex Arc of the Somatic Nervous System

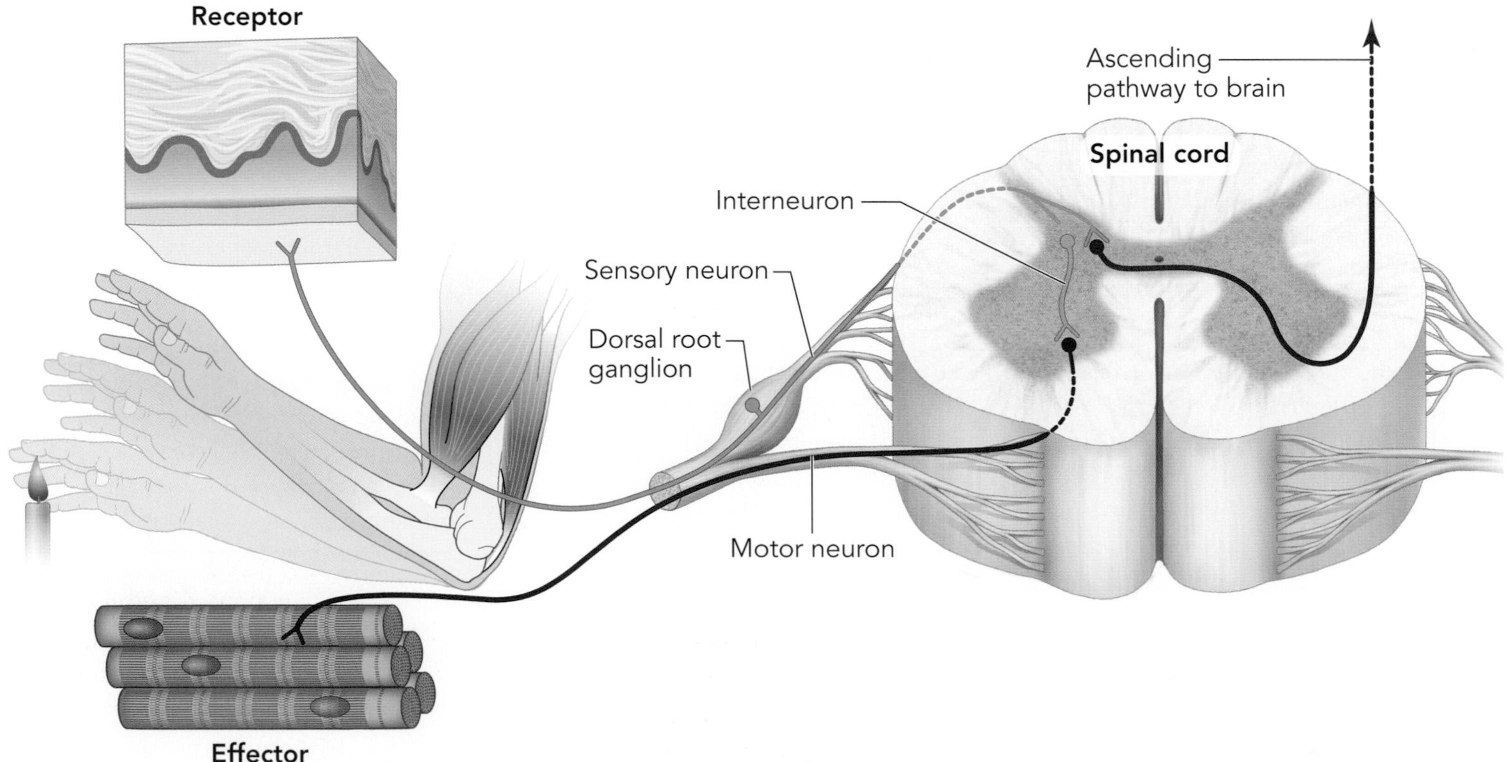

environment. The sensory neuron cell bodies are located in the *dorsal root ganglion*. The motor neurons innervate only skeletal muscle, which is involved in voluntary movement. The cell bodies of somatic motor neurons are located in the ventral horns of the spinal cord. These neurons synapse directly on their effectors and release the neurotransmitter acetylcholine. The motor functions of the somatic nervous system can be consciously controlled and are considered voluntary.

The ANS coordinates an involuntary response to environmental stimuli, altering processes within the body to produce the most adaptive physiological state and behavior. The sensory portion of the ANS receives signals primarily from the viscera (the organs inside the ventral body cavity). The motor portion of the ANS then

Associate the somatic nervous system with voluntary movement and conscious control. By contrast, when you think of the autonomic nervous system, think of involuntary processes that occur outside of conscious awareness.

FIGURE 2.12 The Autonomic Nervous System

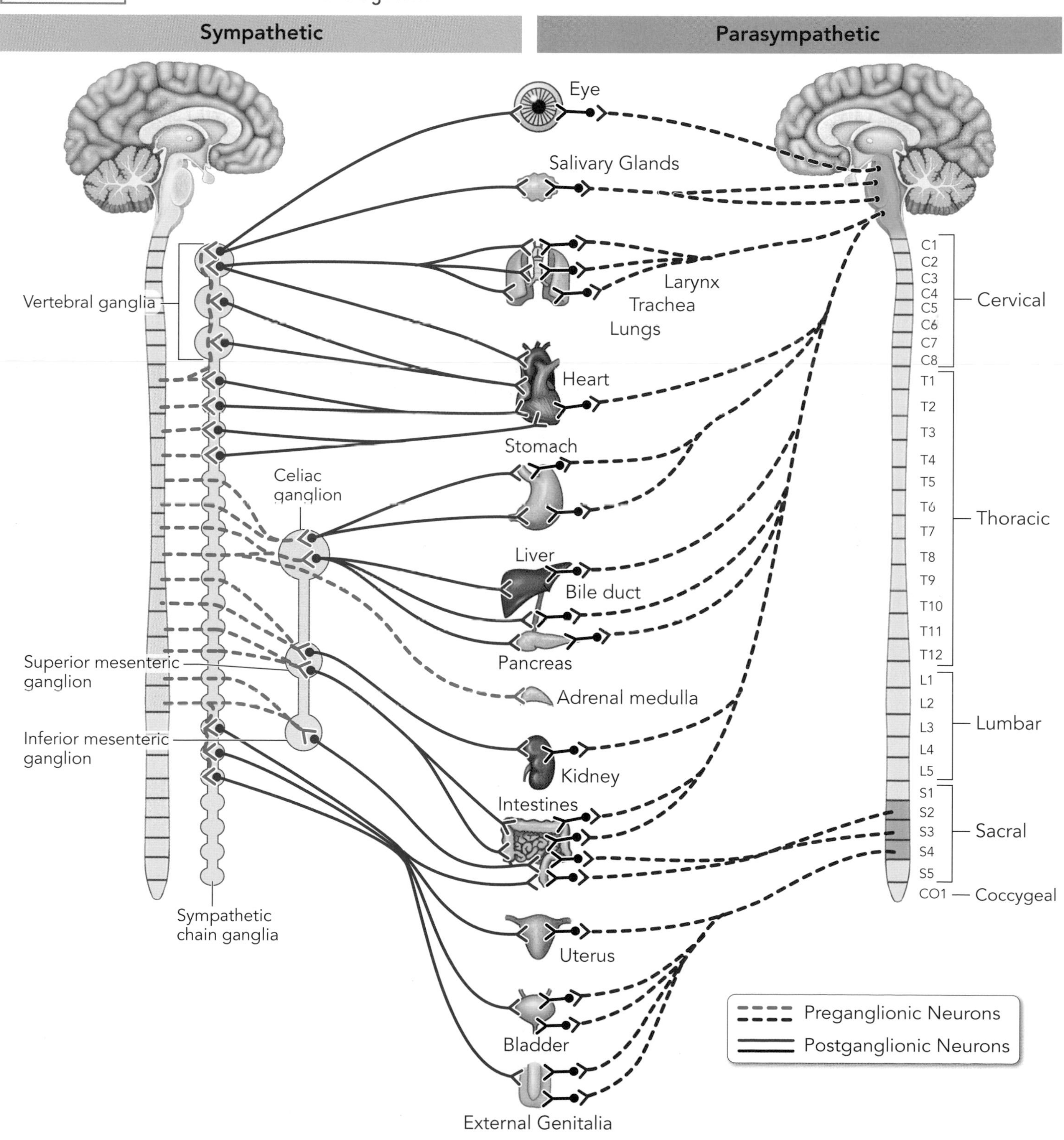

Don't make the mistake of thinking that sympathetic stimulates and parasympathetic inhibits; this is INCORRECT. Their actions depend on the organs that they are innervating. For instance, parasympathetic neurons have an inhibitory effect on the heart but an excitatory effect on the digestive system. Use "fight or flight" and "rest or digest" to make predictions.

conducts these signals to smooth muscle, cardiac muscle, and glands. The motor portion of the ANS is divided into two systems: **sympathetic** and **parasympathetic** (see Figure 2.12). Most internal organs are innervated by both systems in a competitive manner. In other words, the two systems have opposing influences on the same organs and thus exert **antagonistic control**. The parasympathetic and sympathetic nervous systems are two physically separable sets of nerves, which are able to independently exert opposite influences on the same organs by each secreting their own neurotransmitter, as will be described.

The sympathetic ANS deals with "fight or flight" responses. For instance, it acts on the heart to increase heart rate and stroke volume. It also causes the constriction of blood vessels around organs of the digestive and excretory systems in order to divert more blood flow to skeletal muscles. Both of these effects help prepare the organism for fight or flight. Parasympathetic action, on the other hand, generally works toward the opposite goal, to "rest and digest." Parasympathetic activity slows the heart rate and increases digestive and excretory activity.

Sympathetic signals originate in neurons whose cell bodies are found in the spinal cord, while parasympathetic signals originate in neurons whose cell bodies can be found in both the brain and spinal cord. (A group of cell bodies located in the CNS is called a *nucleus*; if located outside the CNS, it is called a *ganglion*.) These neurons extend out from the spinal cord to synapse with neurons whose cell bodies are located outside the CNS. As shown in Figure 2.12, the former neurons are called preganglionic neurons, and the latter are called postganglionic neurons. The cell bodies of sympathetic postganglionic neurons lie far from their effectors, generally within the paravertebral ganglion, which runs parallel to the spinal cord, or within the prevertebral ganglia in the abdomen. The gathering of signals in large ganglia far from the effectors allows for a strong, coordinated signal, important for the sympathetic nervous system's "fight or flight" function. The parasympathetic nervous system's "rest and digest" functions do not require the careful coordination of signaling found in the Sympathetic Nervous System (SNS), so the cell bodies of the parasympathetic postganglionic neurons lie in ganglia inside or near their effectors.

With few exceptions, the neurotransmitter used by all preganglionic neurons in the ANS and by postganglionic neurons in the parasympathetic branch is **acetylcholine**. In contrast, the postganglionic neurons of the sympathetic nervous system use either **epinephrine** or **norepinephrine** (also called **adrenaline** and

The autonomic nervous system is involuntary and innervates cardiac and smooth muscle, as well as some glands; the somatic nervous system is voluntary and innervates skeletal muscle. Autonomic pathways are controlled mainly by the hypothalamus (described in the next section).

Familiarize yourself with the following neurotransmitters: acetylcholine, epinephrine, and norepinephrine. Relate acetylcholine to both the somatic and parasympathetic nervous systems, and epinephrine and norepinephrine to the sympathetic nervous system only.

FIGURE 2.13 Types of Motor Neurons

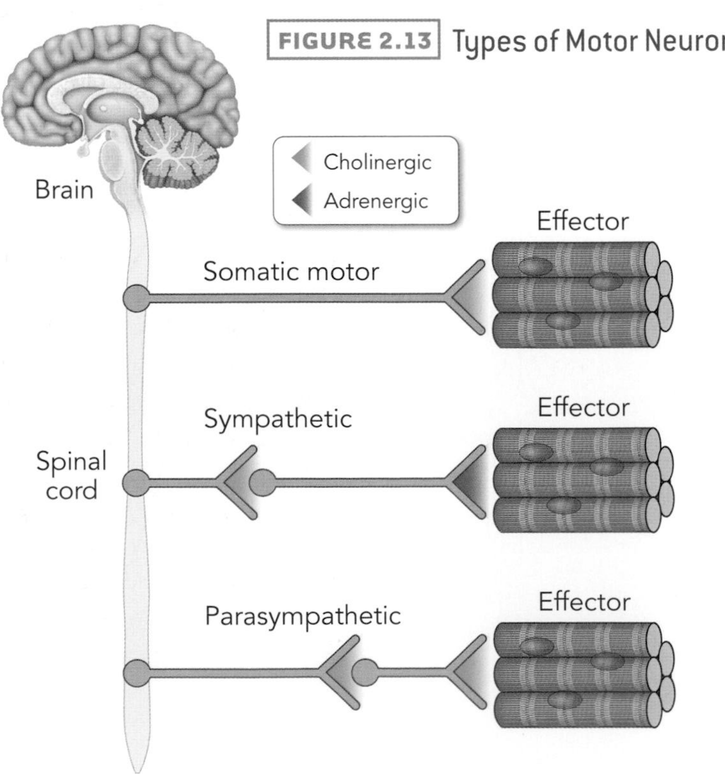

FIGURE 2.14 Sensory and Motor Neurons

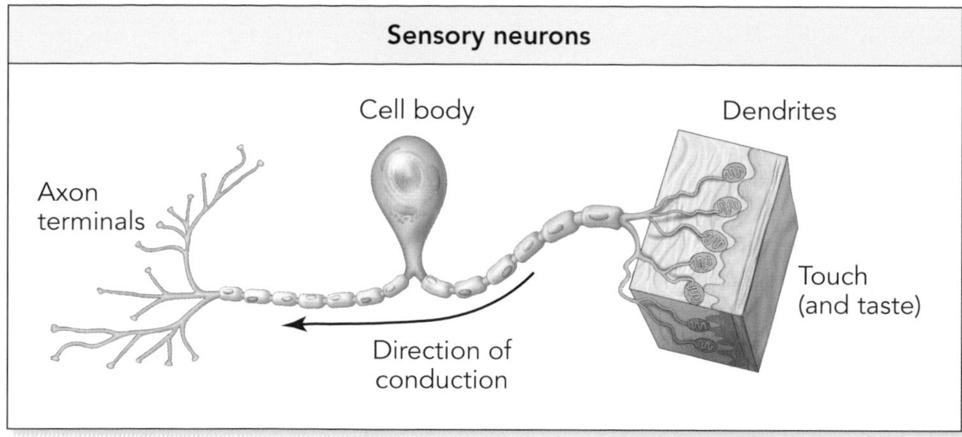

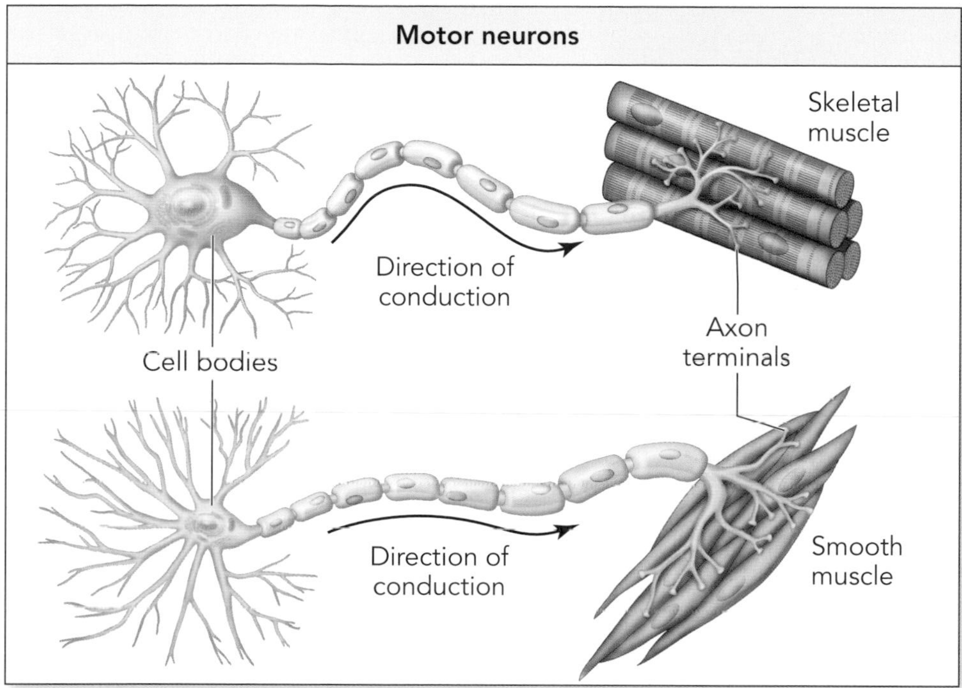

Think of GANGlia as gangs of cell bodies in the PNS. Recall that the cell bodies of neurons receive the electrical disturbances generated in dendrites at the synapse. As the interface between the input and output of the neuron, the cell body is the ideal location to gather together and integrate multiple signals. The axons of these neurons can then gather to form nerves.

Also, remember that groups of cell bodies that lie within the central nervous system are referred to as nuclei. You will see examples in the sections regarding processing of sensory information in the brain.

MCAT® THINK

The sympathetic and parasympathetic nervous systems usually function in opposition to one another, creating a balance within the human body. As an example, when the heart receives neuronal stimulation from the parasympathetic nervous system, the heart slows down. Alternatively, when the heart receives neuronal stimulation from the neurons of the sympathetic nervous system, the heart will speed up. This also occurs in terms of sexual arousal. Which of the two systems do you believe is acting during the onset of an erection and which is active during the period of ejaculation? Why do you believe this is?

Answer on page 75.

noradrenaline). Receptors for acetylcholine are called *cholinergic receptors*. There are two types of cholinergic receptors: *nicotinic* and *muscarinic*. Generally, nicotinic receptors are found on the postsynaptic cells of the synapses between ANS preganglionic and postganglionic neurons and on skeletal muscle membranes at the neuromuscular junction. Muscarinic receptors are found on the effectors of the parasympathetic nervous system. The receptors for epinephrine and norepinephrine are called *adrenergic* receptors.

The Central Nervous System

The central nervous system consists of the lower brain, the higher brain, and some of the spinal cord. The spinal cord acts as a bridge between the peripheral nervous system and the brain, conveying sensory and motor signals. (Although the spinal cord acts mainly as a conduit for nerves to reach the brain, it does possess limited integrating functions such as walking reflexes, leg stiffening, and limb withdrawal from pain, as described previously.) The brain is the site of integration for all of the information that travels through the nervous system. The most important functions of the brain are the processing of sensory information, responding to stimuli, and regulation of the body's internal environment. The connection between the nervous system and psychological processes will be considered further in the last lecture of the *Psychology & Sociology*.

The areas of the CNS that originated in earlier stages of evolution control vital functions and convey information to more recently developed structures that carry out higher-level thought and control motor responses to sensory input. For this reason, the brain can be considered in two sections known as the lower and higher brains, each of which contains areas that are specialized for certain functions. The lower brain represents an earlier stage of evolutionary development and consists of the brainstem, cerebellum, and diencephalon. It integrates unconscious activities such as the respiratory system, arterial pressure, salivation, emotions, and reaction to pain and pleasure.

The **brainstem** controls the basic involuntary functions necessary for survival. It includes the **medulla**, pons, and midbrain. The medulla plays an important role in the regulation of the cardiovascular and respiratory systems. Chemoreceptors (described in the following section) in the medulla monitor levels of carbon dioxide in the bloodstream and trigger changes in respiration rate accordingly. In addition, the medulla receives information about blood pressure and can respond by altering relative levels of sympathetic and parasympathetic innervation of the heart. The **pons** coordinates communication between the motor cortex and the cerebellum, facilitating the transfer of motor commands. Finally, the midbrain is a relay station for auditory and visual signals.

The **cerebellum** is heavily involved in the coordination and planning of movement. It is a highly integrative structure in that it receives and processes sensory, motor, and vestibular input.

The thalamus and hypothalamus are collectively called the **diencephalon**. The **thalamus** is often described as a control center or waystation because it processes almost all sensory information before it reaches higher cortical centers, as well as receiving motor commands from these cortical areas on their way to the spinal cord. The **hypothalamus** regulates many of the body's basic physiological needs by maintaining homeostasis in multiple systems such as temperature and water balance. The hypothalamus and pituitary gland form a major site of interaction between the nervous and endocrine systems, allowing **feedback control** of hormone levels in the body. As discussed in the Biological Molecules and Enzymes Lecture of *Biology 1: Molecules*, negative feedback within an enzyme pathway leads to down-regulation of earlier steps. In an analogous process, rising levels of hormones released by the pituitary gland in response to signals from the hypothalamus feed back to these structures to inhibit further hormone release. The next lecture in this manual will describe the hormones that are released by the anterior pituitary and which participate in this mechanism of feedback control.

FIGURE 2.15 Major Structures of the Lower Brain

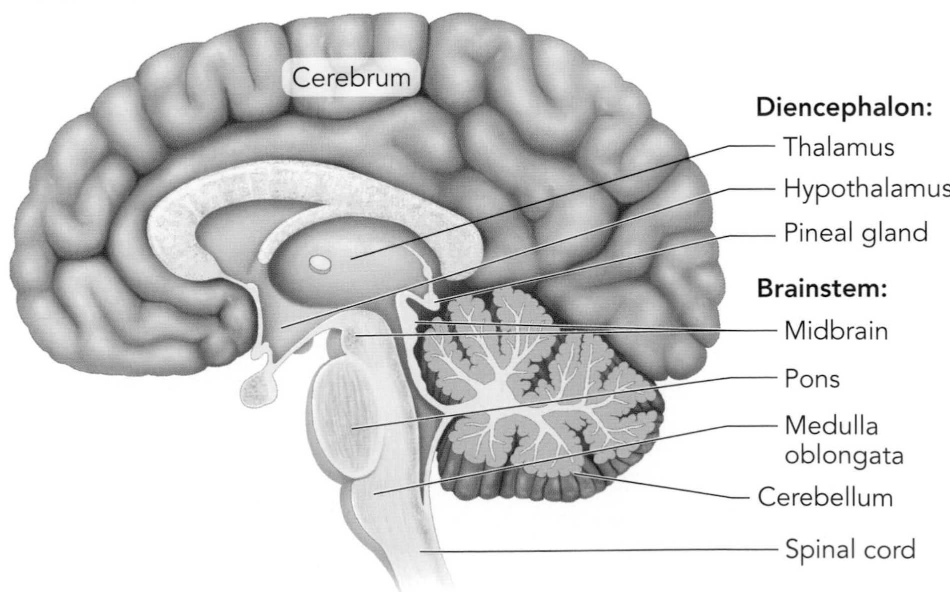

FIGURE 2.16 Lobes of the Cerebral Cortex

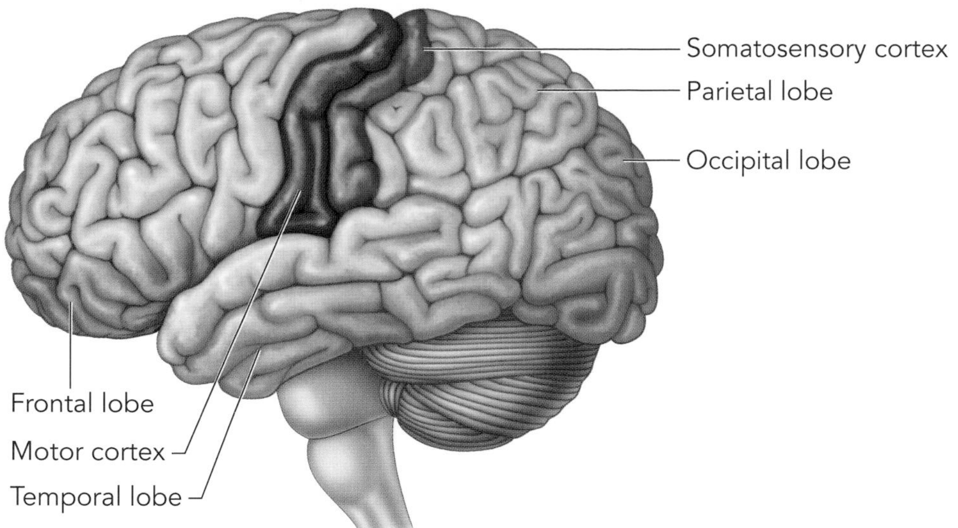

The higher brain, or cortical brain, contains the **cerebrum** or **cerebral cortex**. The cerebral cortex is the most recently evolved portion of the brain and is incapable of functioning without the lower brain. It has a folded, grooved structure, which was the evolutionary response to the problem of fitting a large new part of the brain into the restricted space of the skull. The cerebral cortex is the location of many higher-level functions of the nervous system that influence the experiences that most make us human, such as consciousness, memory, cognition, planning, and emotion.

The cerebral cortex is the main endpoint for the majority of neuronal input and can be divided into areas that receive and integrate specific types of information from the environment to achieve the highest level of perception. The **frontal lobe** is the location of higher-level (executive) functions such as planning and impulse inhibition. It includes the motor cortex, which sends signals to control voluntary movements. The **motor cortex** creates a map of the parts of the body, such that specific sets of neurons control certain body parts. The amount of the motor cortex that is devoted to each part of the body depends on the complexity of movement required. The **parietal lobe** contains the **somatosensory cortex**, which maps the body's sensation of touch. The somatosensory cortex is part of the **somatosensory system**, which involves the detection of physical stimuli such as touch, temperature, and pain. Like the motor cortex, the somatosensory cortex creates a map of the body. The devotion of comparatively large portions of the somatosensory cortex to particular parts of the body, such as the hands, makes these regions particularly sensitive to tactile stimuli.

The **occipital lobe** is the site where visual information is processed. Finally, the **temporal lobe** is primarily concerned with auditory and olfactory information. The cortical brain also contains the hippocampus and the amygdala. These structures are central to the function of the **limbic system**, which is primarily concerned with memory and emotion.

It is important to understand that the cerebrum does not simply receive information from the sensory system and coordinate a response (although this is an important part of its function). It also influences our experience of the world in two important ways. First, the cerebrum processes and interprets information received from the senses, integrating information and imposing higher-level meaning and interpretation based on previous experience. (The field of cognitive psychology describes the cerebrum's interpretation of stimuli using an **information processing model**, as discussed in Lecture 4 of the *Psychology & Sociology* manual.) Second, it can exert a top-down influence on the sensory system, influencing what is considered most important and thereby altering the perception of environmental

Both the limbic system and the interplay between the senses and higher-level interpretation will be considered further in the *Psychology & Sociology* manual. It is likely that you would see questions about the different lobes of the cerebral cortex in the Psychosocial section of the MCAT®. Understand that the brain contains specialized areas for different sensory and motor functions, but realize that the brain also has plasticity. The brain can often partly compensate for damage to one part of the brain by recruiting neurons from adjacent areas to take over the functions of that area.

stimuli. For instance, in addition to carrying signals from sensory receptors to the cerebrum, the thalamus is also a waystation for signals in the opposite direction. In some cases the cerebrum can even influence sensory receptors, such as in the auditory system, where signals can travel all the way from the cerebral cortex and through the thalamus to produce an effect on hair cells.

Thought and Emotion
PSYCH & SOC

The cerebrum is physically divided into two halves (cerebral hemispheres), connected by the *corpus callosum*. For the most part the two hemispheres of the brain carry out the same functions. However, the brain does exhibit some **lateralization of cortical functions**. Lateralized functions take place primarily in one hemisphere or the other. The production and comprehension of language are localized to two areas in the left hemisphere. Emotional experience seems to be roughly divided between the two hemispheres such that the left hemisphere is more involved in positive emotions while the right hemisphere is associated with negative emotions.

Many **methods of studying the brain** have been used to characterize the specific neural functions of certain areas. These methods can be placed into three categories: lesion studies, imaging, and recording of electrical activity. Many known areas of localized function, including the language centers mentioned above, were discovered by observation of people who had damage to those areas. It is now possible to carry out more direct observation of the brain in living organisms. Various methods of imaging, relying on the increased blood flow to areas of the brain that are active and therefore have increased metabolic needs, can be used to examine the brain while the subject carries out tasks. In addition, it is possible to measure the firing of individual neurons (using electrodes in animal studies) or overall patterns of electrical activity at the scalp. However, none of these methods perfectly capture neurological activity. There is constant communication between different parts of the brain such that no one structure carries out its functions in isolation.

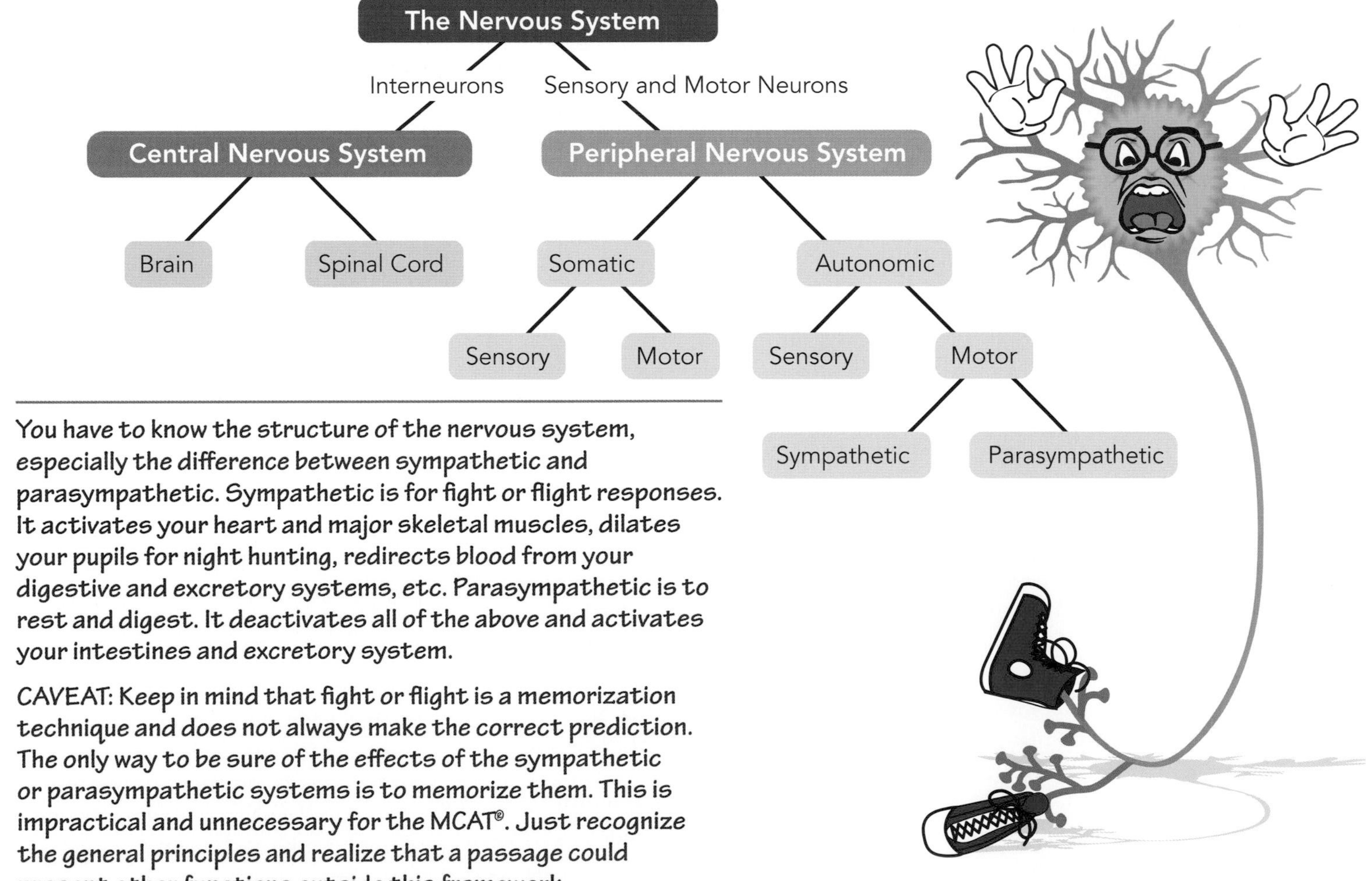

You have to know the structure of the nervous system, especially the difference between sympathetic and parasympathetic. Sympathetic is for fight or flight responses. It activates your heart and major skeletal muscles, dilates your pupils for night hunting, redirects blood from your digestive and excretory systems, etc. Parasympathetic is to rest and digest. It deactivates all of the above and activates your intestines and excretory system.

CAVEAT: Keep in mind that fight or flight is a memorization technique and does not always make the correct prediction. The only way to be sure of the effects of the sympathetic or parasympathetic systems is to memorize them. This is impractical and unnecessary for the MCAT®. Just recognize the general principles and realize that a passage could present other functions outside this framework.

These questions are NOT related to a passage.

Question 33

If an acetylcholinesterase inhibitor were administered into a cholinergic synapse, what would happen to the activity of the postsynaptic neuron?

- O **A.** It would decrease, because acetylcholine would be degraded more rapidly than normal.
- O **B.** It would decrease, because acetylcholinesterase would bind to postsynaptic membrane receptors less strongly.
- O **C.** It would increase, because acetylcholine would be produced more rapidly than normal.
- O **D.** It would increase, because acetylcholine would be degraded more slowly than normal.

Question 34

A small stroke localized to the occipital lobe would impact which of the following senses?

I. Sight
II. Smell
III. Hearing

- O **A.** I only
- O **B.** II only
- O **C.** I and II only
- O **D.** II and III only

Question 35

There are several different types of neurons in the body. Most of these neurons:

- O **A.** contain multiple somas.
- O **B.** are entirely white matter.
- O **C.** are entirely grey matter.
- O **D.** transfer signals from one neuron to another.

Question 36

Which of the following activities is controlled by the cerebellum?

- O **A.** Involuntary breathing movements
- O **B.** Fine muscular movements during a dance routine
- O **C.** Contraction of the thigh muscles during the knee-jerk reflex
- O **D.** Absorption of nutrients across the microvilli of the small intestine

Question 37

If an acetylcholine antagonist were administered generally into a person, all of the following would be affected EXCEPT:

- O **A.** the neuroeffector synapse in the sympathetic nervous system.
- O **B.** the neuroeffector synapse in the parasympathetic nervous system.
- O **C.** the neuromuscular junction in the somatic nervous system.
- O **D.** the ganglionic synapse in the sympathetic nervous system.

Question 38

Which of the following occurs as a result of parasympathetic stimulation?

- O **A.** Vasodilation of the arteries leading to the kidneys
- O **B.** Increased rate of heart contraction
- O **C.** Piloerection of the hair cells of the skin
- O **D.** Contraction of the abdominal muscles during exercise

Question 39

Reflex arcs:

- O **A.** involve motor neurons exiting the spinal cord dorsally.
- O **B.** require fine control by the cerebral cortex.
- O **C.** always occur independently of the central nervous system.
- O **D.** often involve inhibition as well as excitation of muscle groups.

Question 40

Which of the following structures is NOT part of the central nervous system?

- O **A.** A parasympathetic effector
- O **B.** The medulla
- O **C.** The hypothalamus
- O **D.** The cerebral cortex

Questions 33-40 of 144

STOP

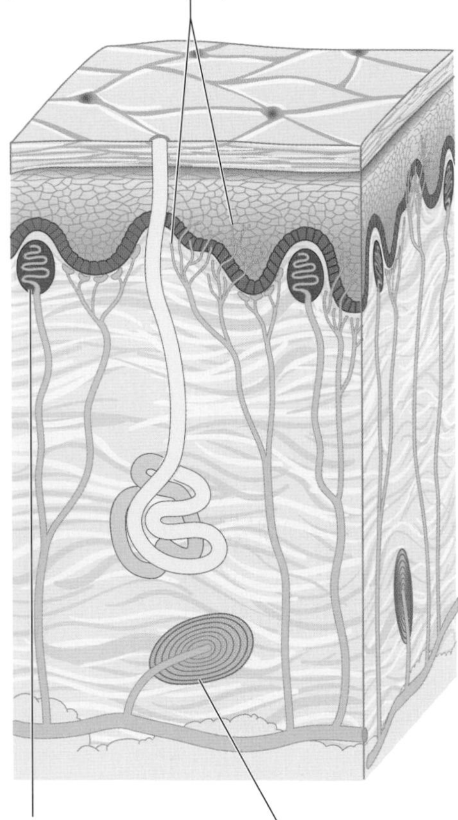

FIGURE 2.17 Sensory Receptors of the Somatosensory System

2.8 Sensing and Sensory Receptors

Sensing is bringing information into the nervous system from the outside. By contrast, motor neurons carry information out. The body uses cells called **sensory receptors** to detect internal and external stimuli. The sensory receptors communicate information by transducing physical stimuli to electrical signals that can then be conveyed to the central nervous system via sensory nerves for processing, integration, and interpretation. There are five types of sensory receptors:

1. **mechanoreceptors** for touch;
2. **thermoreceptors** for temperature change;
3. **nociceptors** for pain;
4. **electromagnetic receptors** for light; and
5. **chemoreceptors** for taste, smell, and blood chemistry.

Some types of sensory receptors are actually part of sensory neurons rather than being separate cells. The receptors associated with the somatosensory system are extensions of sensory neurons, as are the sensory receptors of the smell system. Sensory neurons that are themselves sensory receptors differ from other types of neurons in that they respond to environmental stimuli rather than to signals from other cells.

Since the somatosensory system processes the sensations of touch, temperature, and pain, the sensory receptors of the somatosensory system are mechanoreceptors, nociceptors, and thermoreceptors. Mechanoreceptors respond to the physical stimuli of touch and pressure change. (Mechanoreceptors are not only involved in the somatosensory system. As described later in this lecture, the hair cells of the organ of Corti are mechanoreceptors that send signals in response to the increased and decreased pressure of sound waves.) Nociceptors detect some of the same stimuli as the other types of receptors, such as temperature, pressure, and chemicals, but they are particularly sensitive to the extremes of these stimuli. Nociceptors specialize in the detection of stimuli that generate the experience of pain.

Photoreceptors are a type of electromagnetic receptor. They detect the physical stimulus of photons that enter the eye. Unlike other types of sensory receptors, they do not generate action potentials. Instead, in a similar but distinct process, the relative level of light in the environment affects the rate of neurotransmitter release by photoreceptors into the synapses that they share with sensory neurons. This unique setup will be described in the discussion of the visual system.

In contrast to the other types of receptors that detect and interpret non-chemical stimuli, chemoreceptors bind chemicals, acting as detectors of chemical levels in the external and internal environment. Chemoreceptors are extremely important for the maintenance of homeostasis throughout the body; they are involved in the regulation of the cardiovascular and respiratory systems. Chemoreceptors are the type of sensory receptor used by the taste and smell systems, as described at the end of this lecture.

Each receptor responds strongly to its own type of stimulus and weakly or not at all to other types of stimuli. The sensory receptors transduce physical and chemical stimuli to neural signals that can then be conveyed by sensory neurons. Each type of receptor has its own neural pathway and termination point in the central nervous system, resulting in the various sensations.

Of course, it would not be particularly convenient if every little stimulus caused a response by sensory receptors. Just as small voltage disturbances in the neuronal resting potential caused by individual stimuli are summed to produce an overall excitatory or inhibitory effect, receptor potentials produced by outside stimuli each increase or decrease the likelihood that an action potential will be produced by the sensory receptor to transfer information to the associated sensory neurons. In addition, the sensory system codes information about the intensity of stimuli through activation of different numbers of sensory receptors, activation of different types of receptors (e.g. nociceptors for pain and extreme stimuli), and the frequency of signaling by sensory neurons.

Just as the nervous system has mechanisms for distinguishing between stimuli of different intensity levels, it is able to tone down the response to a repeating stimulus. In a process known as **sensory adaptation**, a stimulus that occurs repeatedly at the same intensity level evokes fewer and fewer action potentials in the sensory receptors. Sensory adaptation is part of how the nervous system achieves its function of filtering out less important information, allowing the perception of changes in the environment. Some types of sensory receptors, called *phasic receptors*, adapt very quickly and specialize in the perception of changes in stimuli. By contrast, *tonic receptors* adapt more slowly. The length of time required for tonic receptors to stop producing action potentials provides information to the nervous system about the intensity of the stimulus.

Why would the nervous system want to pay less attention to repeated stimuli? Imagine what would happen if I kept noticing these wool socks all day just as much as I did when I first put them on. I wouldn't be able to notice more important new information in my surroundings, like this hungry tiger.

The rest of this lecture is concerned with sensory systems that specialize in particular types of environmental input. Each of these systems can be characterized by the type(s) of receptors involved, the process by which the physical stimulus is converted into electrical information that can be transferred by the nervous system, and the interpretation of this information by higher brain centers. Within most of these systems, as will be described, the sensory receptors create a map of the surroundings that is preserved as the information travels throughout the nervous system.

2.9 The Eye and Vision

The eye detects patterns of light and transmits this information to the nervous system, ultimately resulting in the perception of vision. The eye can be thought of as a biological example of an optical instrument. Much like binoculars or a microscope, the eye can adjust to focus on objects of different sizes and distances. For the MCAT®, the test taker must know the basic anatomy of the eye (as shown in Figure 2.19) and understand the function of a few of its parts. A good way to remember the basics of vision is to follow the path of light as it enters the eye.

Waves: Sound and Light
PHYSICS

Light reflects off an object in the external environment and first strikes the eye on the **cornea**. (Technically the light first encounters a very thin, protective layer known as the corneal epithelium.) The cornea is nonvascular and made largely from collagen. It is clear with a refractive index of about 1.4, which means that the incoming light is actually bent further at the interface of the air and the cornea rather than at the lens.

From the cornea, the light enters the *anterior chamber*, which is filled with *aqueous humor*. Aqueous humor is formed by the *ciliary bodies* and leaks out of the *canal of Schlemm*. Blockage of the canal of Schlemm increases intraocular pressure, resulting in one form of *glaucoma* and possibly blindness.

From the anterior chamber, light enters the **lens**. The lens naturally has a spherical shape, but stiff suspensory ligaments tug on it and tend to flatten it. This allows the shape of the lens to be adjusted according to the focal length needed to ensure that the image produced by a given object will be focused precisely on the retina, rather than behind or in front of it. The suspensory ligaments are connected to the **ciliary muscle** circling the lens. When the ciliary muscle contracts, the opening of the circle decreases, allowing the lens to become more like a sphere and bringing its focal point closer to the lens; when the muscle

FIGURE 2.18 Image Formation on the Retina

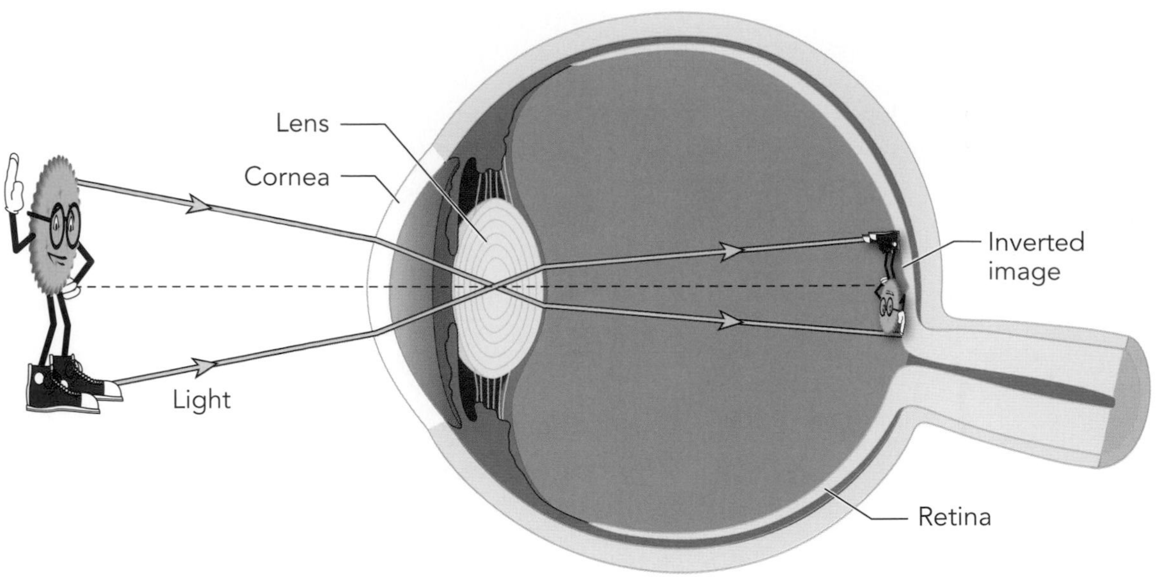

relaxes, the lens flattens, increasing the focal distance. The elasticity of the lens declines with age, making it more difficult to decrease the focal length of the lens, with the result that it becomes harder to focus on nearby objects as one gets older.

When light passes through the structures described above, it is focused through the gel-like *vitreous humor* and onto the retina. Since the eye acts as a converging lens, and the object is outside the focal distance, the image on the retina is real and inverted (Figure 2.18).

The **retina** covers the inside of the back (distal portion) of the eye. It contains **rods** and **cones**, two types of photoreceptors named for their characteristic shapes. The tips of these cells contain light sensitive photochemicals called *pigments* that undergo a chemical change when one of their electrons is struck by a single photon. Vitamin A is a precursor to all the pigments in rods and cones. (As discussed in the Biological Molecules and Enzymes Lecture of *Biology 1: Molecules*, vitamin A is an example of a terpene.) The restricted range of wavelengths of light to which the pigments respond determines the spectrum of light visible to humans.

The pigment in rod cells is called *rhodopsin*. Rhodopsin is made of a protein bound to a prosthetic group called *retinal*, which is derived from vitamin A. The photon isomerizes retinal, causing the membrane of the rod cell to become less permeable to sodium ions, which in turn causes it to hyperpolarize. Rods respond to all photons with wavelengths in the visible spectrum (390 nm to 700 nm), so they cannot distinguish colors.

Remember that cones distinguish colors and rods don't.

There are three types of cones, each with a different pigment that is stimulated by a slightly different spectrum of wavelengths. Cones, unlike rods, are able to distinguish colors.

The *fovea* is a small point on the retina containing mostly cones. This is the point on the retina where vision is most acute.

Another important feature of the eye is the **iris**. The iris is the colored portion of the eye that creates the opening called the **pupil**. The iris contains both circular and radial muscles. In a dark environment, the sympathetic nervous system contracts the iris, dilating the pupil and allowing more light to enter the eye. In a bright environment, the parasympathetic nervous system contracts the circular muscles of the iris, constricting the pupil and screening out light.

FIGURE 2.19 Anatomy of the Eye

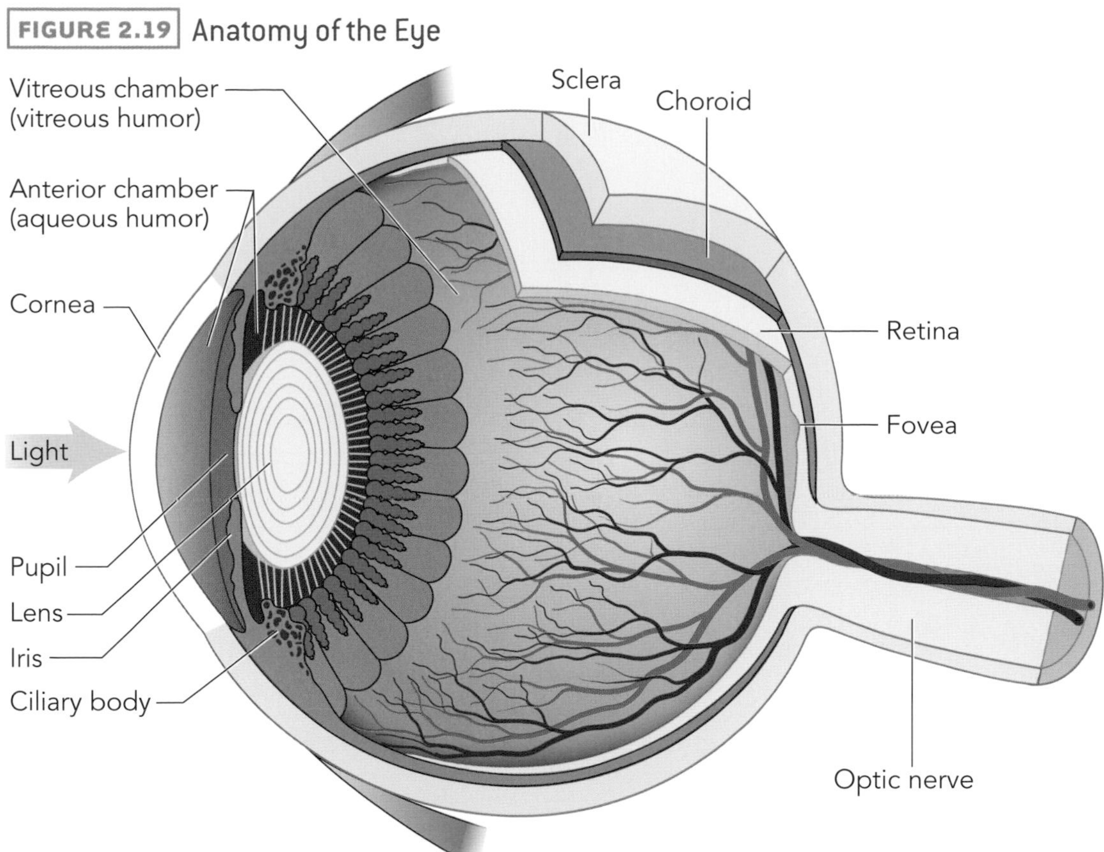

Processing of Visual Information

Once light travels through the eye and is detected by the photoreceptors, this information must be conveyed to higher levels of the nervous system. As described previously, light striking the photopigments contained in cone and rod cells triggers a series of events that lead to the hyperpolarization of the membrane of the photoreceptor. Photoreceptors do not generate action potentials, but hyperpolarization has an inhibitory effect by reducing the rate of neurotransmitter release. All photoreceptors release glutamate, which is usually an excitatory neurotransmitter. However, the cells that receive glutamate from photoreceptors respond to it differently.

The next step in the visual pathway after the photoreceptors is the passage of visual information to *bipolar cells*. Bipolar cells receive signals from their associated photoreceptors ("vertical" information). Depending on the type of glutamate receptor, a given bipolar cell may be inhibited or excited by changes in the amount of glutamate released by photoreceptors. Bipolar cells are also affected by signals from *horizontal cells*, which provide "horizontal" information from photoreceptors at the edge of the bipolar cell's *receptive field*, the distinct area of visual information to which the cell responds. The integration of horizontal and vertical information gives the eye the ability to focus on changes and edges in the visual field.

If the bipolar cells experience an overall excitatory effect from both the vertical and horizontal inputs, they release neurotransmitter at an increased rate (much like photoreceptors), producing an excitatory effect on the *ganglion cells*. These are the sensory neurons that finally produce action potentials.

The axons of the ganglion cells gather to form the **optic nerve**, which leaves the eye to convey visual information to the brain. Some of the axons from each eye cross to the opposite side of the brain. The optic nerve travels to the **lateral geniculate nucleus (LGN)** of the thalamus, which preserves the visual map created by the ganglion cells and projects this information to the **primary visual cortex**, located in the occipital lobe. Groups of neurons within the primary visual

Think about what would happen if all bipolar cells responded equally to stimuli in all areas of the visual field. This would be a very inefficient and metabolically expensive system! Just as the parasympathetic and sympathetic branches are physically separate systems that are able to complete separate tasks, different bipolar cells respond to different receptive fields. Subsequent levels of visual processing retain the use of these receptive fields.

FIGURE 2.20 Anatomy of the Retina

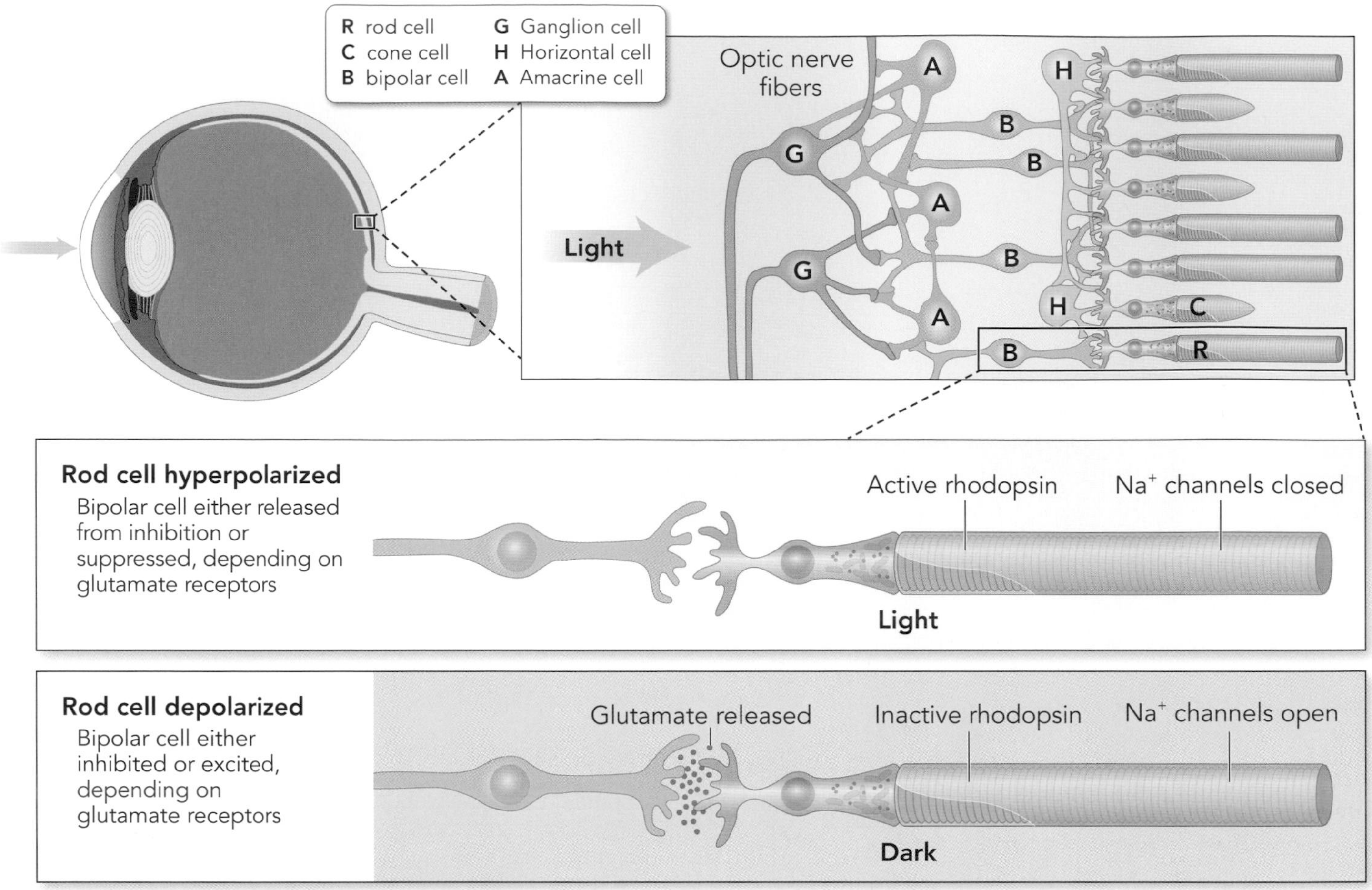

FIGURE 2.21 Visual Pathways in the Brain

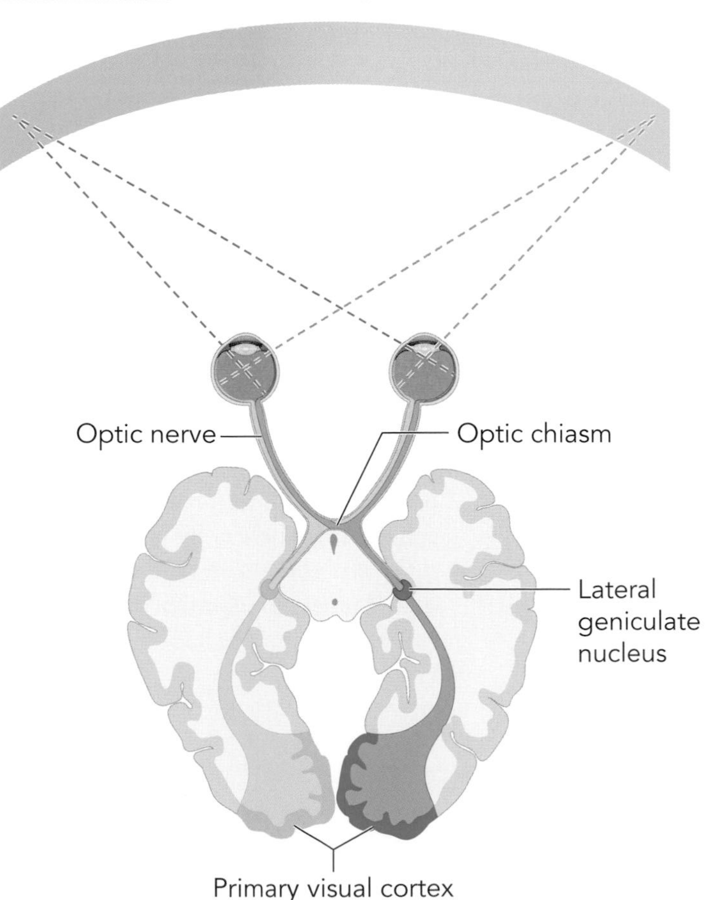

cortex are specialized for the detection of specific aspects of visual stimuli, particularly lines and edges of different orientations.

Neurons project from the primary visual cortex along two visual pathways, which detect different features of visual stimuli. The **ventral ("what") pathway** travels to the temporal lobe towards the base of the brain and is involved in object recognition, while the **dorsal ("where") pathway** projects to the parietal cortex and is more involved in perceiving the location of objects. However, the two pathways communicate with each other, and information from both is ultimately integrated in other areas of the brain.

You probably won't be asked about the nitty gritty details of the visual pathways in the brain, but you should definitely know that visual information passes from the optic nerve to the thalamus and finally to the primary visual cortex. Also remember the "what" and "where" pathways. More will be said about specific psychological theories of visual processing in the *Psychology & Sociology* manual.

2.10 The Ear and Hearing

The ear can be thought of as an instrument that condenses sound into a tiny space. The ear transforms sound from waves of pressure in the air (a gas) into physical vibrations, first in solid bone and then in fluid, before finally transforming the physical phenomenon into an electrical signal. As with the eye, it is important to know the basic parts (see Figure 2.22) and the functions of these parts. The ear is divided into three sections:

You can remember the bones of the middle ear with the mnemonic, "I MISheard you" for *malleus, incus,* and *stapes.*

1. the **outer ear**,
2. the **middle ear**, and
3. the **inner ear**.

Following the path of a sound wave through the ear can be helpful in remembering the structures.

The *auricle* or *pinna* is the flap of skin and cartilage that is commonly called the ear. The auricle directs the sound wave into the *external auditory canal*, which then carries the wave to the *tympanic membrane* or *eardrum*. The tympanic membrane begins the middle ear.

The middle ear contains three small bones: the *malleus*, the *incus*, and the *stapes*. These three bones act as a lever system translating the wave into a physical vibration that is conveyed to the *oval window*. Like any lever system, these bones change the combination of force and displacement from the inforce to the outforce. In this case the displacement is lessened, which causes an increase in force. In addition, the oval window is smaller than the tympanic membrane, acting to increase the pressure. (See the Energy and Equilibrium Lecture in the *Physics* manual for more on machines and mechanical advantage.) This increase in force is necessary because the wave is being transferred from the air in the outer ear to a more resistant fluid within the inner ear.

The ear could show up on a physics passage concerning waves or the mechanics of the middle ear. Make sure you understand how structures in the auditory system transduce sound waves to electrical signals. Also know that the cochlea detects sound, while the semicircular canals detect orientation and movement of the head.

The wave in the inner ear moves through the *scala vestibuli* of the **cochlea** to the center of the spiral, and then spirals back out along the *scala tympani* to the *round window*. As the wave moves through the cochlea, the alternating increase and decrease in pressure causes vibration of the **organ of Corti**, which is located on the cochlea's basilar membrane. This, in turn, causes movement of the specialized microvilli, called *stereocilia,* on the **hair cells** of the organ of Corti. The movement of these hair cells is transduced into neural signals that are sent to the brain, as will be described.

Also in the inner ear are the **semicircular canals** and the adjacent **otolith organs**, which are involved in the **vestibular system**. This system is responsible for detecting changes in position and signaling the body to make necessary adjustments to maintain balance. The semicircular canals are primarily responsible for detecting twisting of the head, while the otolith organs detect tilting and linear acceleration. Both structures contain fluid and hair cells. When the body moves or the head position changes with respect to gravity, the momentum of the fluid impacting on the hair cells is changed, and the body senses motion. The canals are oriented at right angles to each other, allowing the detection of movement in all directions. Damage to the vestibular system can be devastating, as the vestibular sense is crucial for everyday functioning.

The vestibular sense works together with the **kinesthetic sense**, which contributes to the awareness of the body's location and movement, to give the individual an overall sensation of the body.

FIGURE 2.22 Anatomy of the Ear

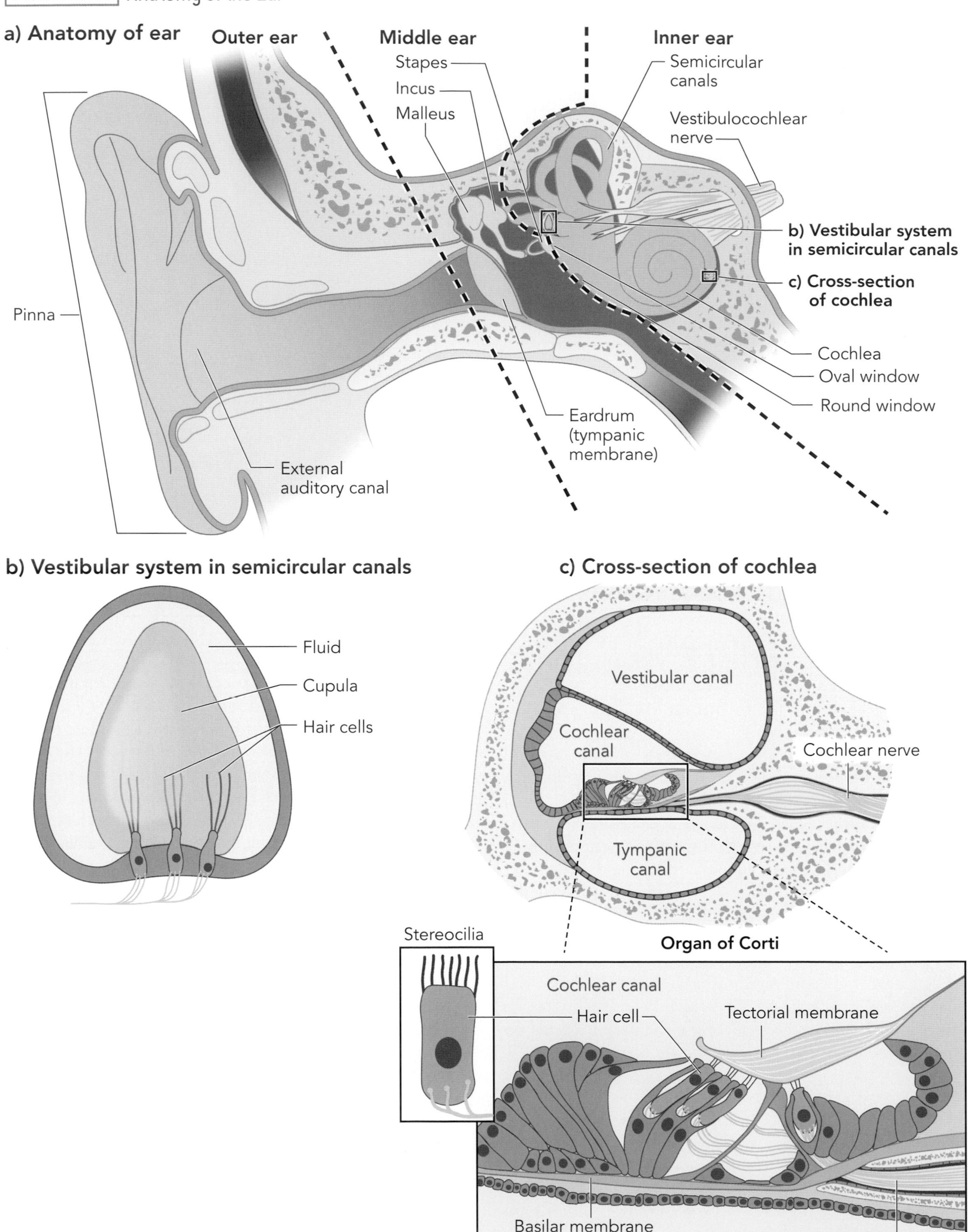

Processing of Auditory Information

As with all of the sensory systems, the next step after exploring the physical structures that receive stimuli in the auditory system is the consideration of how these stimuli are converted into neural signals. Recall that the hair cells of the organ of Corti are mechanoreceptors. As described above, they sense the bending of microvilli that results when the organ of Corti vibrates under the influence of sound waves in the fluid. Hyperpolarization or depolarization occurs depending on the direction in which the microvilli bend. When hair cells experience depolarization, they release a neurotransmitter into their synapses with neurons of the cochlear nerve (also called the auditory nerve), which can then convey an electrical signal to higher levels of the nervous system.

As auditory information travels through the nervous system, it undergoes processing and integration at each stage. Neural signals encoding sound information primarily travel along a pathway that starts with the cochlear nerve and then proceeds to the *cochlear nuclei* within the medulla. Next, axons of the cochlear nuclei synapse with neurons within the *inferior colliculus*. (On the way to the inferior colliculus, some crossover occurs such that auditory information from each ear is sent to the opposite hemisphere.) From there, the electrical signals are passed on to the **medial geniculate nucleus** of the thalamus. (Recall that visual information similarly passes through an area of the thalamus, the lateral geniculate nucleus, on its way to a higher cortical area.) Finally, the **auditory cortex** in the temporal lobe receives information from the thalamus. The auditory cortex is where the detection of complex features of auditory information, such as patterns, takes place.

Senses like vision and hearing depend just as much on the brain as they do on the sensory organs involved. Damage to an area of the brain involved in sensory processing can cause a deficit in the associated sense, even if the sensory organ is intact.

FIGURE 2.23 Auditory Pathways in the Brain

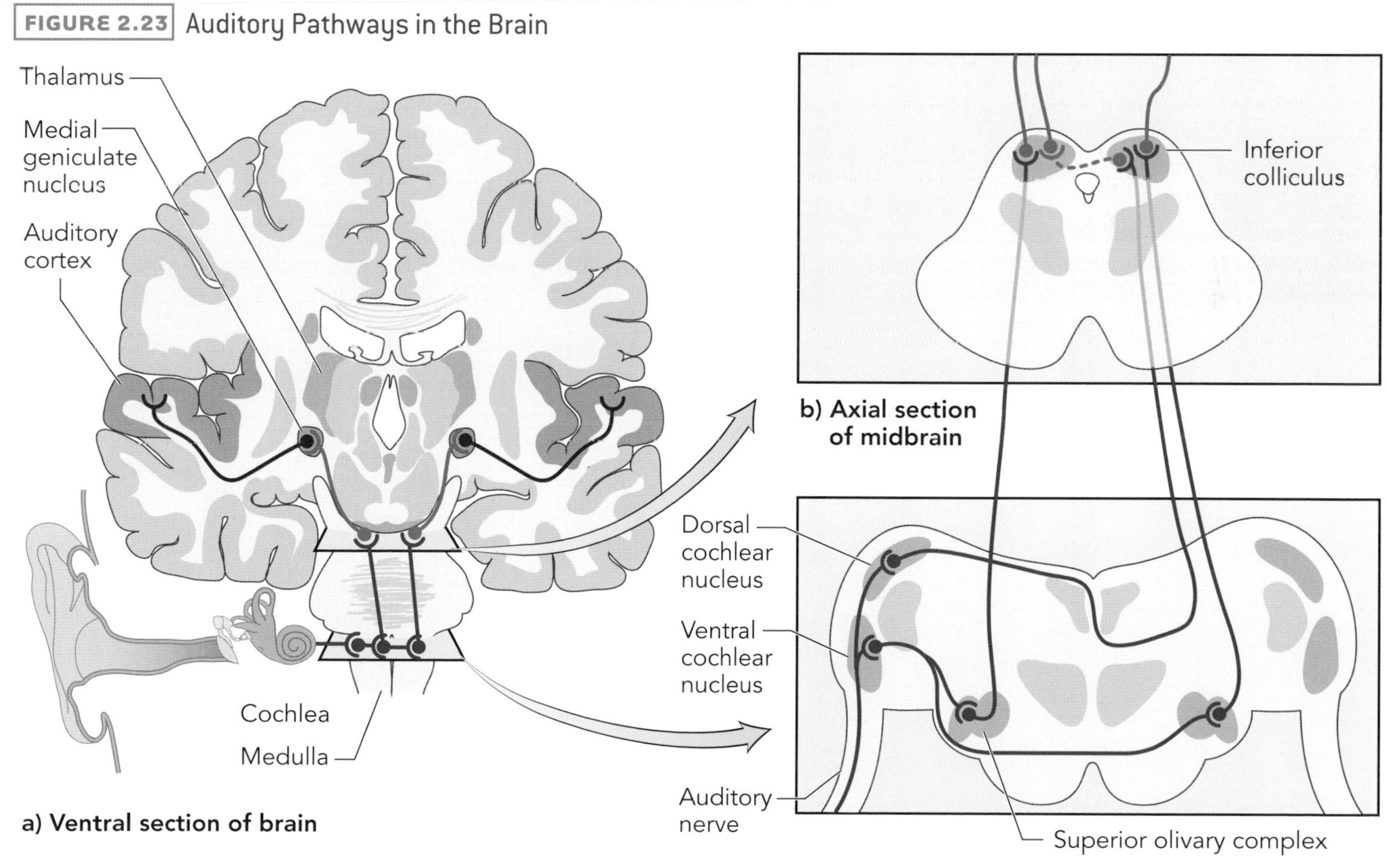

Hair cells (in yellow) in the organ of Corti, shown in an electron micrograph.

Consider the most important features of the experience of sound. Loudness, which is the psychological perception of the intensity of the sound wave, is one. Another is the frequency of the sound wave, experienced as pitch. How are these pieces of information transferred from the level of sensory receptors to the areas of the brain that process and interpret auditory information?

The answer for frequency lies with the spatial organization of hair cells along the organ of Corti. Hair cells in the part of the basilar membrane that is closest to the stapes and therefore the most stiff, respond best to higher frequencies. (It is difficult to move these hair cells embedded in a stiff substance, so only higher frequencies will do the trick!) The opposite is true for hair cells that are further from the entrance to the inner ear, in the more flexible portion of the basilar membrane. In this way, a spatial map of frequencies is created that is preserved at each level of auditory processing, all the way to the auditory cortex.

As for intensity, think about how the intensity of physical stimuli is usually communicated by sensory receptors. Just like in other areas of the sensory system, a stimulus of higher intensity causes higher firing rates of the associated sensory neurons. In addition, some sensory neurons of the auditory nerve only respond to sounds that pass a certain threshold of intensity, much as nociceptors only respond to extreme stimuli.

2.11 Smell, Taste, and Pain

Think about how food doesn't taste as good when you have a cold. The taste and smell systems work together, so when your sense of smell is blocked, your overall experience of food is affected.

The highly linked senses of smell and taste are called olfactory and *gustatory*, respectively. The mechanisms of both of these systems are not as well understood as those of the auditory and visual systems. Both senses involve chemoreceptors. Different chemoreceptors sense different chemicals.

The taste system evolved for the function of alerting the organism to substances that are harmful or nutritionally beneficial. There are five primary taste sensations:

1. bitter,
2. sour,
3. salty,
4. sweet, and
5. umami (savory).

All taste sensations are combinations of these five, and the receptor cells of the taste system are able to detect all five taste sensations. The binding of chemicals to chemoreceptors in the taste buds causes these sensory receptors to trigger action potentials in associated sensory neurons. Sensory information from the binding of many chemicals to different chemoreceptors is combined in higher-level processing to produce the psychological experience of taste.

Similarly to taste, the olfactory sense evolved for the protection of the organism. Chemoreceptors located in the nasal cavity bind certain gaseous chemicals (*odorants*) that enter during inspiration. Each type of olfactory receptor has selectivity for a certain set of chemicals, rather than just one type. Furthermore, there are a multitude of types of olfactory chemoreceptors (many more than in the taste system) and, as in the taste system, processing in the brain combines input from the binding of many different receptors. This allows for the perceived experience of a wide variety of complex smells.

FIGURE 2.24 Taste Structures

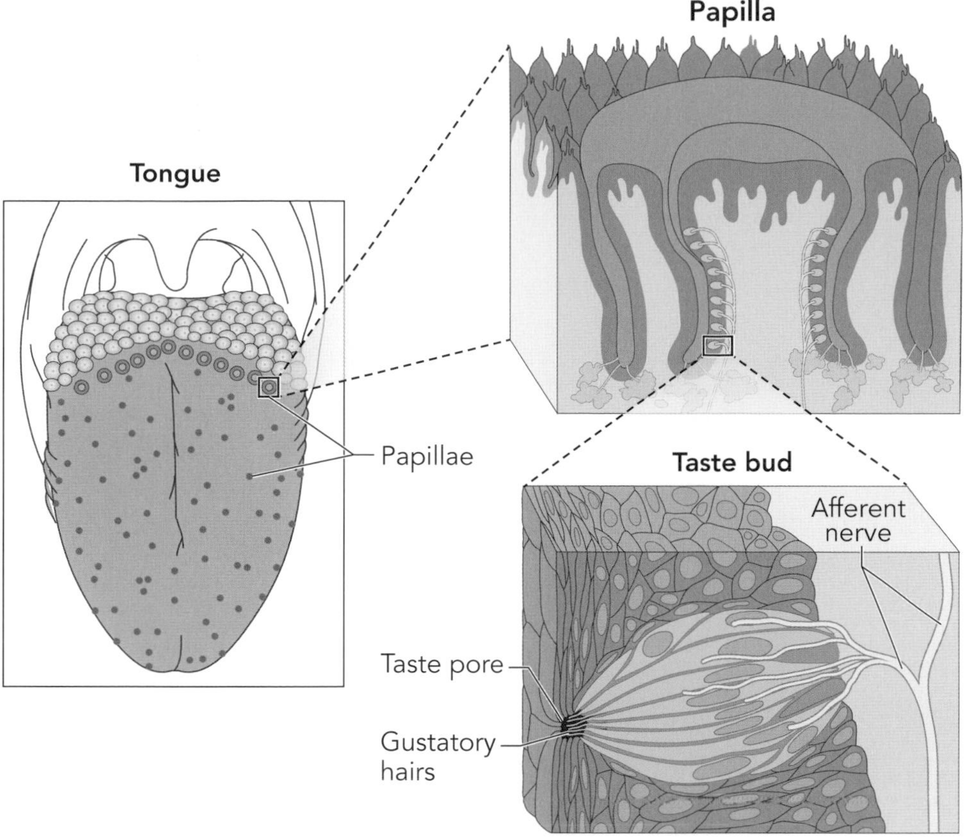

Unlike chemoreceptors in the gustatory system, olfactory chemoreceptors are themselves sensory neurons. The axons of the olfactory chemoreceptors form the **olfactory nerve**. The binding of chemicals to G-protein coupled receptors on the chemoreceptors triggers an action potential that is transmitted by the olfactory nerve to neurons of the **olfactory bulb**. An important feature of the **olfactory pathways** is that unlike in the other sensory systems described in this lecture, neural signals do not pass through the thalamus before reaching higher cortical levels in the brain. The olfactory bulb projects directly to the **pyriform (olfactory) cortex** in the temporal lobe. The pyriform cortex in turn conveys olfactory information to the orbitofrontal cortex, where it is integrated with other sensory information. The olfactory bulb also projects to the amygdala and hippocampus in the temporal lobe, providing a quick and direct pathway for smell to influence processes of memory and emotional experience. Like the pyriform cortex, the amygdala projects to the orbitofrontal cortex, allowing information about the emotional significance of smells to reach consciousness. It also projects to the hypothalamus; for this reason, olfactory stimuli (and the emotional interpretation of these stimuli) can influence the many functions that are linked with the hypothalamus.

Some species have a related but separate set of sensory receptors that detect a specific type of odorant called **pheromones**. The key feature that makes pheromones different from other types of smell stimuli is that they are not perceived consciously. Pheromones play a much greater (and better understood) role in the behavior of non-primate species. Information about pheromones travels to the amygdala and hypothalamus and is thought to exert a subconscious influence on behaviors related to these structures, particularly aggression and sexual behavior. It is unclear whether pheromones affect human behavior, but they may have some influence on social interaction and the timing of menstrual cycles.

I might not be able to see, hear, or feel a predator until it's too late to escape, but I can smell that it's been here recently and run away!

It makes sense that the olfactory pathway bypasses the thalamus and goes straight to the amygdala (among other structures) when you think about how certain smells can quickly lead to an emotional reaction before you even remember what they are! The olfactory system is also unusual in that smell information cannot be mapped the way visual, auditory, and tactile stimuli can be.

FIGURE 2.25 Olfactory Pathways

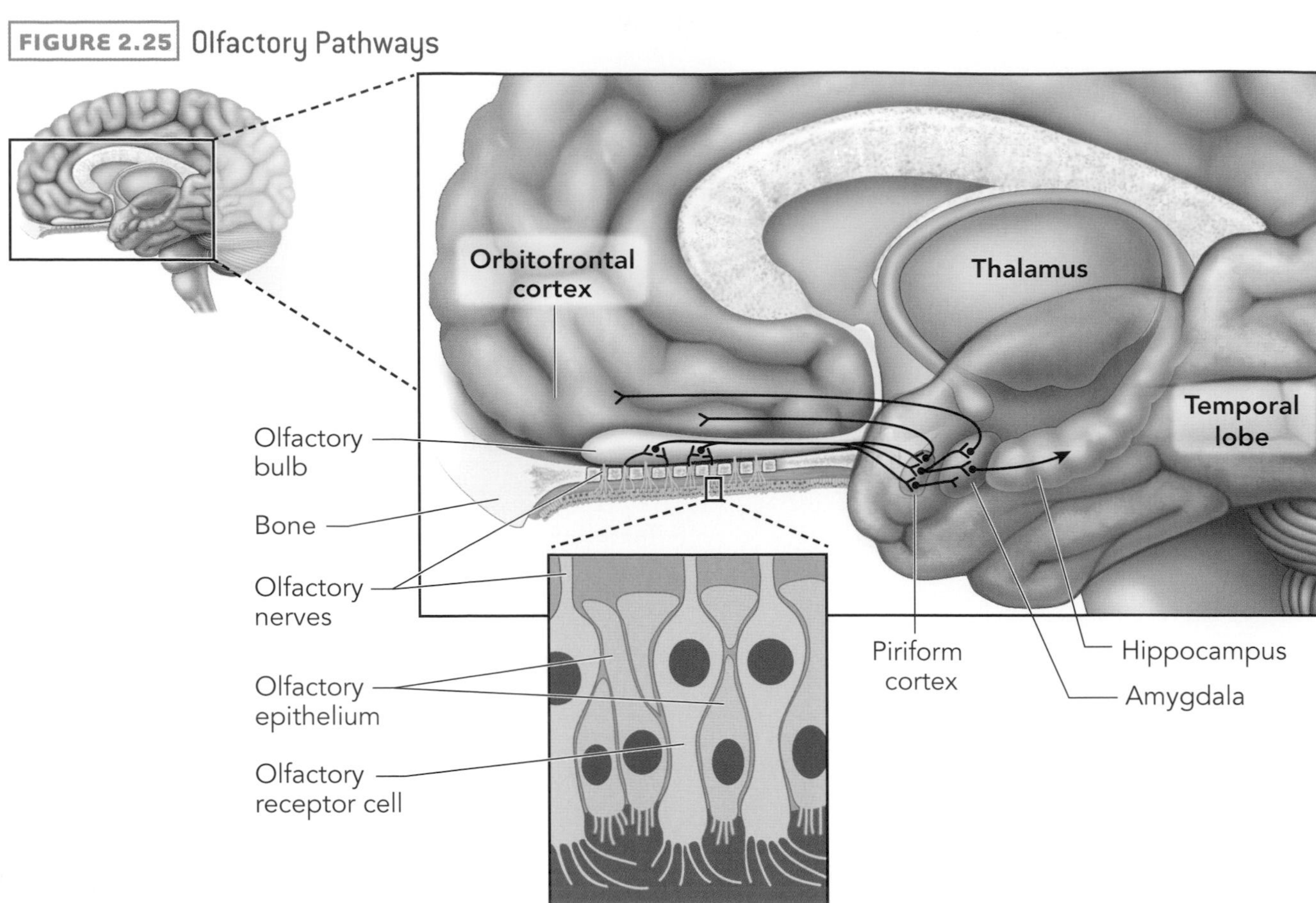

Finally, the way that the experience of **pain** is produced by the somatosensory system is representative of the type of complex processing that allows higher areas of the central nervous system to both interpret and influence the organism's experience of the environment. The perception of potentially damaging stimuli conforms to the general pattern seen in most of the other sensory systems: nociceptors transduce the physical stimulus to an electrical signal that is conveyed through the spinal cord, ultimately reaching the thalamus and then a higher cortical area for processing (in this case, the somatosensory cortex). However, the perception of pain is far from being just a bottom-up process. The brain can exert top-down influence to enhance or diminish the experience of pain. Both types of influence have evolutionary advantages, depending on the situation: pain is adaptive in that it alerts the organism to a dangerous stimulus, but it can also be maladaptive if it presents a distraction when attention to other factors is necessary for survival.

MCAT® THINK

If you have any doubts that pain and other sensory processing can be a top-down process, consider the phenomenon of phantom limb pain. This occurs when people experience the feeling of pain in a limb that they no longer have, such as after an amputation. They feel pain as though signals were being sent from sensory receptors that no longer exist! Phantom limb pain and other sensations associated with the missing limb occur because the neurons that were previously devoted to sensing that limb reassign themselves to structures that are adjacent in the map of the somatosensory cortex. However, the brain still interprets these neural representations as coming from the missing limb, resulting in the subjective experience of phantom limb sensations.

These questions are NOT related to a passage.

Question 41

Pressure waves in the air are converted to neural signals at the:

- O **A.** retina.
- O **B.** tympanic membrane.
- O **C.** cochlea.
- O **D.** semicircular canals.

Question 42

Which of the following physical stimuli cannot be converted to electrical signals by the chemoreceptors of the human body?

- O **A.** Light
- O **B.** Taste
- O **C.** Smell
- O **D.** Blood chemistry

Question 43

As light enters the eye, what is the correct order of the layers through which it passes before being converted into an electrical signal?

- O **A.** Lens → retina → cornea → aqueous humor → vitreous humor
- O **B.** Cornea → aqueous humor → lens → vitreous humor → retina
- O **C.** Cornea → vitreous humor → lens → retina → aqueous humor
- O **D.** Lens → aqueous humor → cornea → vitreous humor → retina

Question 44

Which of the following is NOT true of the olfactory system?

- O **A.** The olfactory system makes use of chemoreceptors that bind odorants.
- O **B.** Olfactory information does not reach the amygdala and hippocampus until after it undergoes processing in the cortex.
- O **C.** Unlike other sensory systems, the olfactory system does not create a "map" of olfactory information in the environment.
- O **D.** Olfactory information does not pass through the thalamus before reaching cortical areas for higher-level processing.

Question 45

Which type of sensory receptor is primarily associated with the sensation of pain?

- O **A.** Chemoreceptors
- O **B.** Mechanoreceptors
- O **C.** Photoreceptors
- O **D.** Nociceptors

Question 46

The vestibular system carries out which of the following functions?

- O **A.** Converting sound waves into auditory signals
- O **B.** Maintaining balance and detecting changes in body position
- O **C.** Detecting and interpreting touch sensations
- O **D.** Conveying information about pheromones in the environment to the amygdala

Question 47

Which of the following correctly describes a component of the visual system and its function?

- O **A.** The optic nerve carries out "what" and "where" processing of visual stimuli.
- O **B.** The lateral geniculate nucleus conveys visual information to the brain for processing.
- O **C.** Photoreceptors respond to the stimulus of photons by triggering signaling in the eye that ultimately causes visual information to be conveyed to the brain.
- O **D.** Hair cells distinguish between colors by responding differently to different wavelengths of light.

Question 48

The process of sensory adaptation can be described as:

- O **A.** the adaptation of sensory receptors to become most sensitive to a single type of stimulus.
- O **B.** the temporal and spatial combination of multiple inputs.
- O **C.** the experience of a decreased response to a repetitive stimulus.
- O **D.** the adaptation of physical stimuli into electrical stimuli.

STOP

TERMS YOU NEED TO KNOW

Acetylcholine
Action potential
All-or-none
Antagonistic control
Auditory cortex
Autonomic nervous system (ANS)
Axon
Axon hillock
Brainstem
Cell body
Central nervous system (CNS)
Cerebellum
Cerebral cortex
Cerebrum
Chemical synapse
Chemoreceptors
Ciliary muscle
Cochlea
Concentration cell
Cones
Cornea
Dendrites
Depolarization
Diencephalon
Dorsal ("where") pathway
Ear
Effector
Electrical synapse
Electrochemical gradient
Electromagnetic receptors
Epinephrine (adrenaline)
Equilibrium potential
Eye
Excitatory
Feedback control
Frequency of firing
Frontal lobe
Glial cells
Grey matter
Hair cells
Hyperpolarization
Hypothalamus
Information processing model
Inhibitory
Inner ear
Interneuron
Ion channels
Iris
Kinesthetic sense
Lateral geniculate nucleus (LGN)
Lateralization of cortical functions
Lens
Limbic system
Mechanoreceptors
Medial geniculate nucleus
Medulla
Methods of studying the brain
Middle ear
Motor cortex
Motor end plate
Motor (effector/efferent) neuron
Myelin
Negative feedback loop
Nernst equation
Nerves
Neuroglia
Neuron
Neurotransmitter
Nociceptors
Nodes of Ranvier
Norepinephrine (noradrenaline)
Occipital lobe
Olfactory
Olfactory bulb
Olfactory nerve
Olfactory pathways
Optic nerve
Organ of Corti
Otolith organs
Outer ear
Pain
Parasympathetic nervous system
Parietal lobe
Peripheral nervous system (PNS)
Pheromones
Photoreceptors
Pons
Primary visual cortex
Pupil
Pyriform (olfactory) cortex
Receptors
Reflex
Reflex arc
Repolarization
Resting potential
Retina
Rods
Saltatory conduction
Schwann cells
Second messenger system
Semicircular canals
Sensory adaptation
Sensory (sensor/afferent) neuron
Sensory receptors
Smell
Sodium potassium pump (Na^+/K^+ pump)
Somatic nervous system
Somatosensory cortex
Somatosensory system
Summation
Supraspinal circuit
Sympathetic nervous system
Synapse
Taste
Temporal lobe
Thalamus
Thermoreceptors
Threshold stimulus
Unidirectional
Ventral ("what") pathway
Vestibular system
Visual processing
Voltage gated potassium channels
Voltage gated sodium channels
White matter

MCAT® THINK ANSWERS

Pg. 48: If some sodium ions were able to enter the neuron, the resting membrane potential would become more positive because positively charged ions are flowing into the neuron. Remember than an action potential occurs when the cell acquires enough positive charge for the voltage-gated sodium channels to open, fully depolarizing the membrane. If the membrane were less negative due to the leakiness of the mutated channel, an action potential would occur more easily.

Pg. 57: Erection is mediated by the parasympathetic nervous system. During erection, the parasympathetic nervous system allows blood to flow into the penis by dilating blood vessels. However, strong muscle contractions are needed during ejaculation. The sympathetic nervous system mediates these muscular contractions.

DON'T FORGET YOUR KEYS

1. The nervous system senses the environment, processes information, and responds.
2. Neural signals are fast, fleeting, and specific; endocrine signals are slow, sustained, and general.
3. The type and number of channels and receptors are the hidden determinants of many cellular and systemic phenomena.

The Endocrine System

THE 3 KEYS

1. See a hormone, think of its gland, function, polarity, and target(s).
2. Hormones respond to imbalance in the body to restore homeostasis. Normally functioning hormones do not create imbalance.
3. Reproductive hormones and cells in males and females are similar in both structure and process but act along different timelines.

3.1 Introduction

The endocrine system, like the nervous system, allows communication within the organism and regulation in response to stimuli in the internal and external environment. While the nervous system has fast, fleeting, and specific signals, the endocrine system acts more slowly but exerts effects that are more general and sustained. The neurotransmitters and local mediators discussed in the Nervous System Lecture are often referred to as local hormones. They are secreted by neurons and cause action potentials. By contrast, the hormones released by the endocrine system are called general hormones. They are referred to as "general" because they are released into the body fluids, often the blood, and may affect many cell types in a tissue, as well as multiple tissues in the body. Although this lecture concentrates on general hormones, the chemistry described is accurate for local hormones as well.

Hormones can be placed into one of three major categories according to how they are synthesized. Each type has different chemical and physical characteristics that determine the mechanisms by which the hormones act and the effects that they can produce, as will be discussed.

FIGURE 3.1 Communication via the Endocrine and Nervous Systems

Motor neurons (fast, specific) | **Autonomic neurons (fast, multiple targets)** | **Hormones (slow, diffuse)**

Hormones are released by endocrine glands, which differ from exocrine glands in the following manner. **Exocrine glands** release enzymes to the external environment through ducts. Types of exocrine glands include sudoriferous (sweat), sebaceous (oil), mucous, and digestive glands. **Endocrine glands** release hormones directly into the bloodstream. For instance, the pancreas acts as both an exocrine gland, releasing digestive enzymes through the pancreatic duct, and an endocrine gland releasing insulin and glucagon (described further in this lecture) directly into the blood.

After their release from endocrine glands, hormones may take anywhere from seconds to days to produce their effects. They do not move directly to their target tissue, but are released into the general circulation. All hormones act by binding to protein **receptors**. Each receptor is highly specific to a particular hormone. One method of hormone regulation is the reduction or increase of a receptor type in the presence of high or low concentrations of its hormone. Some hormones have receptors on virtually all cells, while other hormones have receptors only on specific tissues. Very low concentrations of hormones in the blood can have significant effects on the body.

Remember that hormones need a receptor, either on the membrane or inside the cell. Also, when comparing the endocrine system to the nervous system, remember that the endocrine system is slower, less direct, and longer lasting.

FIGURE 3.2 Exocrine and Endocrine Glands

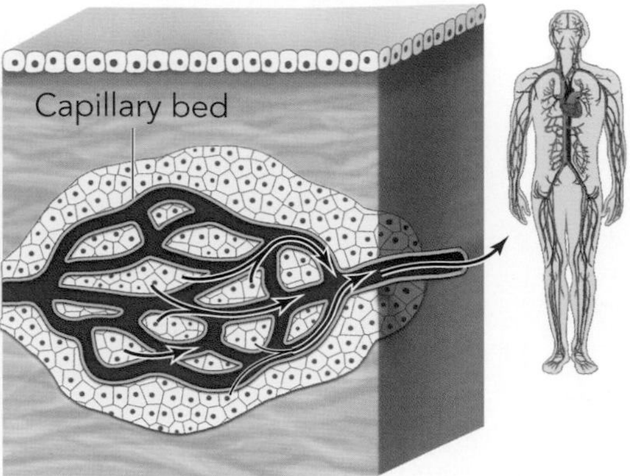

Exocrine gland
Releases enzymes to external environments through ducts

Endocrine gland
Releases hormones into fluids that circulate throughout the body

In general, the effects of the endocrine system are to alter metabolic activities, regulate growth and development, and guide reproduction. The endocrine system works in conjunction with the nervous system. Many endocrine glands are stimulated by neurons to secrete their hormones, so even though the endocrine system acts via generalized communication with multiple parts of the body, it can be tightly regulated by specific neural control.

Communication via the endocrine system can seem disorganized and nonspecific when compared to the nervous system. The brain has fine control of the body's musculature via the nervous system. What evolutionary advantage would there be in the brain not having precise control over all aspects of physiology?

The endocrine system controls body states. For example, after eating a meal, the body absorbs the consumed food over approximately two hours, causing a temporary elevation in blood glucose. Various hormones coordinate the metabolic activity of the liver, the musculature, and the adipose tissue to match the influx of glucose at that moment. If the nervous system were to assume these functions, separate nerves communicating with each organ would have to fire action potentials in accordance with the state of the body at that moment. Action potentials are metabolically expensive, so this setup would waste valuable energy.

3.2 The Chemistry of Hormones

Understand Key 1: When you see a hormone, think about the gland it comes from and the tissues it is traveling to. How a hormone binds its cognate receptor is dependent upon its polarity. Polar hormones tend to bind extracellular receptors while nonpolar steroid hormones typically bind cytoplasmic or nuclear receptors.

Hormones exist in three basic chemistry types:

1. peptide hormones,
2. steroid hormones, and
3. tyrosine derivatives.

The hormones of each type can be categorized according to solubility, which has implications for their physical characteristics, mechanisms of action, and ultimate effects on their targets.

Peptide hormones are derived from peptides. As discussed in the Cell Lecture, since peptide hormones are water soluble, they can move freely through the blood but have difficulty diffusing through the cell membrane of the effector. (The

FIGURE 3.3 Transport of Hormones in the Blood

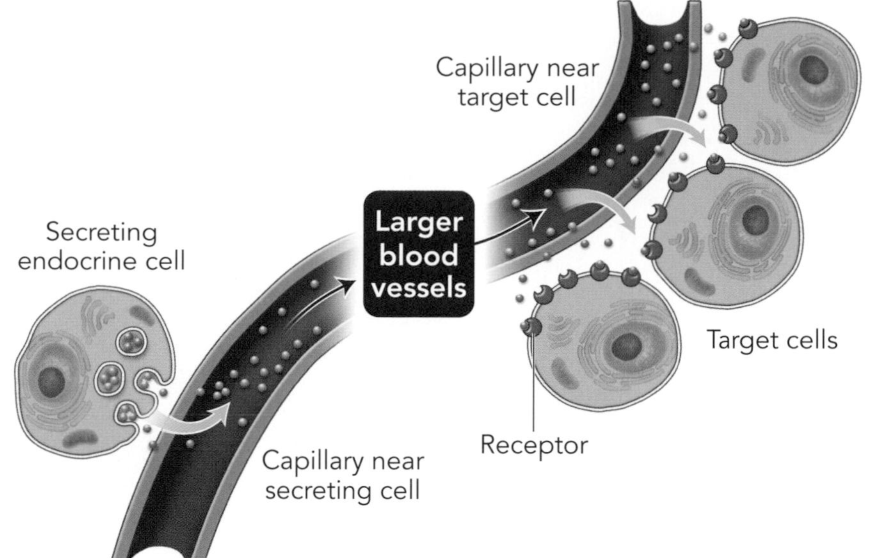

The endocrine system delivers hormones to target cells throughout the body via the circulatory system.

effector is the target cell of the hormone. This cell contains the receptor for the hormone.) Instead of diffusing through the membrane, peptide hormones attach to a membrane-bound receptor. Once bound by a hormone, the common mechanisms of the receptor include activating an enzyme and activating a second messenger system such as cAMP, cGMP, or calmodulin.

Peptide hormones may be large or small, and often include carbohydrate portions. All peptide hormones are manufactured in the rough ER, typically as *preprohormones* that are larger than the active hormones. The preprohormone is cleaved in the ER lumen to become a *prohormone* and then transported to the Golgi apparatus. In the Golgi, the prohormone is cleaved and sometimes modified with carbohydrates to produce the hormone's final form. The Golgi packages the hormone into secretory vesicles, and, upon stimulation by another hormone or a nervous signal, the cell releases the vesicles via exocytosis.

The peptide hormones that must be known for the MCAT® are:

1. the anterior pituitary hormones: FSH, LH, ACTH, HGH, TSH, and prolactin;
2. the posterior pituitary hormones: ADH and oxytocin;
3. the parathyroid hormone PTH;
4. the pancreatic hormones: glucagon and insulin; and
5. the thyroid C cell hormone: calcitonin.

The specifics of these hormones will be discussed later in this lecture.

Steroid hormones are derived from and are often chemically similar to cholesterol. They are formed in a series of steps that take place mainly in the smooth endoplasmic reticulum and the mitochondria. Since they are lipids and therefore hydrophobic, steroids typically require a protein transport molecule (carrier protein) in order to dissolve into the bloodstream. (Usually, a fraction of the steroid concentration is bound to a transport molecule and a fraction is free in the blood.) Steroids are able to diffuse through the cell membrane of their effectors. Once inside the cell, they combine with a receptor in the cytosol or the nucleus and

MCAT® THINK

cAMP stands for cyclic adenosine monophosphate. It is very similar to ATP, adenosine triphosphate, except that it only has one phosphate group, which attaches to the ribose sugar in two places to make a ring.

The anterior pituitary hormones can be remembered with the mnemonic "FLAT PiG." The G stands for Growth Hormone.

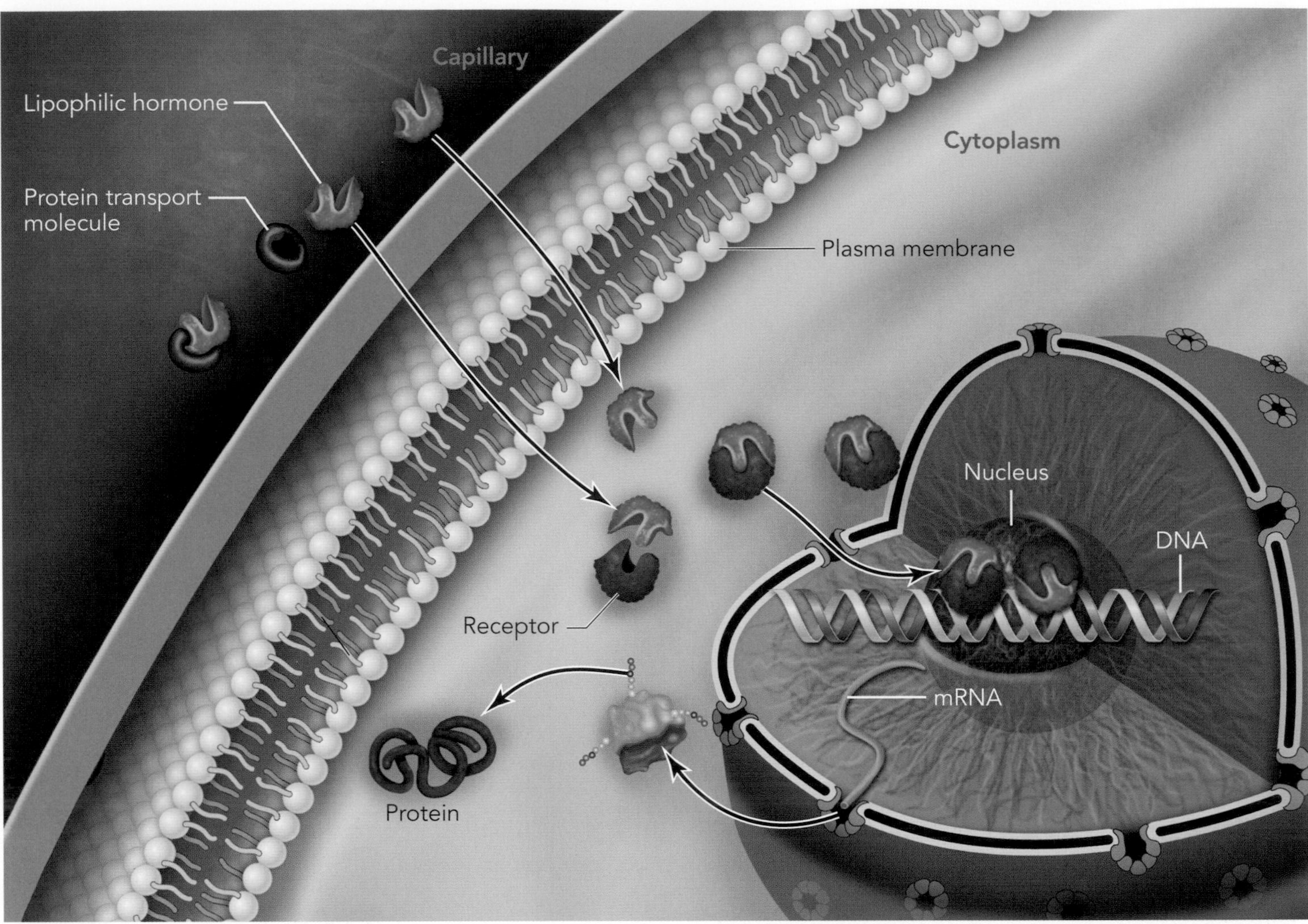

FIGURE 3.4 Mechanism of Nonpolar Hormones

Nonpolar hormones such as steroids and some tyrosine derivatives are transported in the blood by carrier proteins. Once they reach their target cell, they freely diffuse across the membrane and attach to receptors. The receptor-hormone complex then moves to the nucleus, where it regulates transcription of certain genes or gene families.

act at the level of transcription. Because of this, the typical effect of a steroid hormone is to increase certain membrane or cellular proteins within the effector. The important steroid hormones for the MCAT® are:

1. the glucocorticoids and mineral corticoids of the adrenal cortex: cortisol and aldosterone; and
2. the gonadal hormones: estrogen, progesterone, testosterone. (Estrogen and progesterone are also produced by the placenta.)

The specifics of these steroids will be discussed later in this lecture.

The tyrosine derivatives include the thyroid hormones: T_3 (triiodothyronine, containing 3 iodine atoms) and T_4 (thyroxine, containing 4 iodine atoms), as well as the catecholamines formed in the adrenal medulla: epinephrine and norepinephrine. Tyrosine derivative hormones are formed by enzymes in the cytosol or on the rough ER. Unlike the peptide and steroid hormones, tyrosine derivatives do not all have either water or lipid solubility in common.

Thyroid hormones are lipid-soluble, so they must be carried in the blood by plasma protein carriers. They are slowly released to their target tissues and bind to receptors inside the nucleus. Thyroid hormones have a strong affinity to their binding proteins in the plasma and in the nucleus, creating a latent period in the response produced and increasing the duration of the effects. Thyroid hormones increase the transcription of large numbers of genes in nearly all cells of the body.

Epinephrine and norepinephrine are water-soluble, so they dissolve in the blood. They bind to receptors on the target tissue and act mainly through the second messenger cAMP.

The specifics of the tyrosine derivative hormones will be discussed later in this lecture.

3.3 Regulation of Hormones

Endocrine glands tend to over-secrete their hormones. Typically, some aspect of their effect on the target tissue will inhibit this secretion. This is an example of negative feedback. A critical aspect of negative feedback in endocrine glands is that hormones act to bring the body back to normal, not to cause abnormalities. For instance, insulin, a hormone that lowers blood glucose, would be expected to be released when blood glucose is high. If the body had low blood glucose, insulin would be expected to be low. A second example involves the hypothalamus, anterior pituitary, and adrenal cortex, the so-called HPA axis. Corticotrophin releasing hormone (CRH) secretion from the hypothalamus causes the release of adrenocorticotropic hormone (ACTH) from the anterior pituitary. ACTH, in turn, causes release of cortisol from the adrenal cortex. Cortisol can negatively feedback to inhibit both the release of CRH from the hypothalamus and ACTH from the anterior pituitary. Notice that the control point of the feedback is the behavior of the effector, not the concentration of hormone. In other words, the gland lags behind the effector. So if an MCAT® question indicates that a person has high blood sugar and asks whether insulin levels would be high or low, the correct response is that insulin will try to correct this abnormality, so a high concentration will be present.

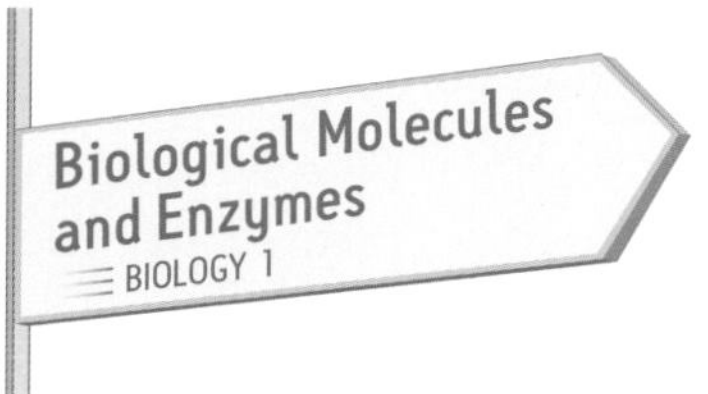

FIGURE 3.5 Negative Feedback Mechanism

High glucose levels result in the release of insulin from the pancreas. Insulin acts on the liver and other cells throughout the body to promote glucose utilization and storage of glucose as glycogen and fatty acids. If glucose levels become too low, glucagon is released from the pancreas, resulting in fatty acid mobilization, protein breakdown into amino acids, and gluconeogenesis in the liver. The result is a relatively steady concentration of glucose in the bloodstream.

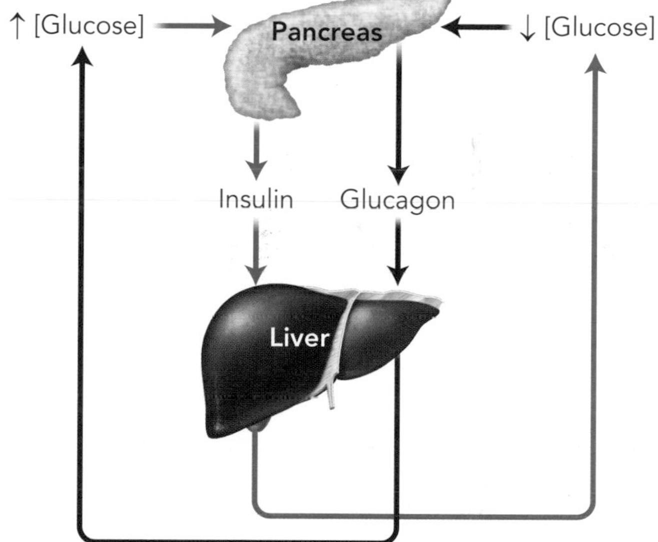

There will be a negative feedback question on the MCAT®. An air conditioner provides a good analogy. You might set your air conditioner to 78 °F during the summer. If your house were 65 °F, would you expect your air conditioner to be on or off? You would expect it to be off, because it should only come on when your house warms up to 78 °F. If your air conditioner had simply stayed on and caused your house to get so cold, you would say it is broken. The MCAT® does not test broken physiology.

See if you get the idea. Calcitonin is a hormone that decreases blood calcium levels. If blood calcium levels are low, would that person have a high or a low level of calcitonin?

If you said high, then you reasoned that the calcitonin created the low calcium level. WRONG! On the MCAT®, hormones act to correct problems, not to cause problems. If you said low, then you reasoned that the body decreased output to respond to the low calcium state. CORRECT!

How about another? One effect of aldosterone is to increase blood pressure. Would expected aldosterone levels be high or low in a person with low blood pressure?

The answer: Since aldosterone increases blood pressure, and hormones act to bring the body back to its normal state, the adrenal cortex should release more aldosterone into the blood to increase blood pressure. Expected aldosterone levels would be higher than normal.

3.4 Organizing the Endocrine System

Because of their different physical properties, hydrophilic and hydrophobic hormones necessarily play different roles in the body. Just as the nervous system has a faster, more specific and a less fast, more diffuse portion, the endocrine system has faster and slower acting hormones. This spectrum is illustrated in Figure 3.6.

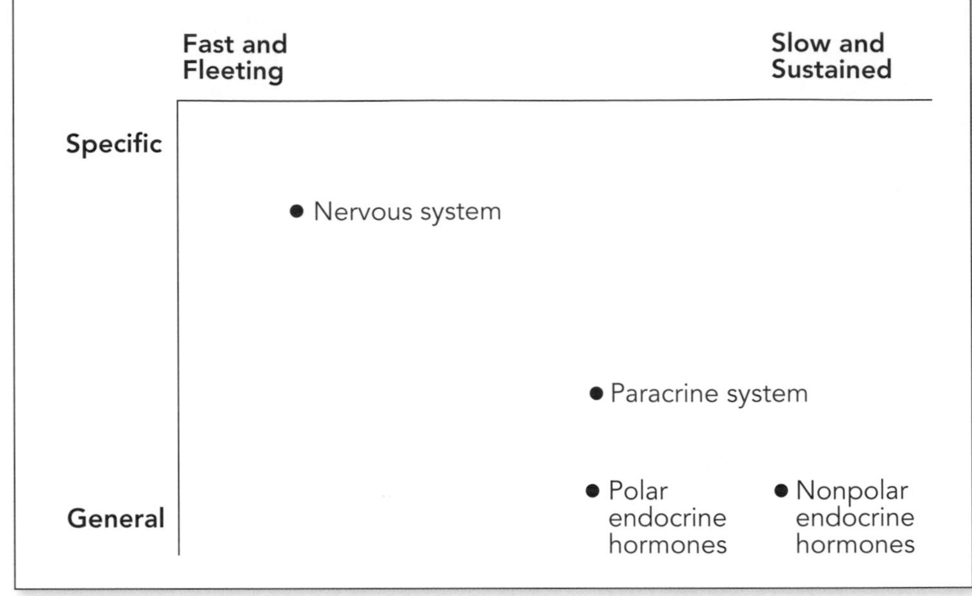

FIGURE 3.6 Comparison of Communication Systems: Polar and Nonpolar Hormones

Overall, the effects of the endocrine system are far more general, slow, and sustained than those of the nervous system. However, each type of endocrine hormone has unique properties: the effects of nonpolar hormones are more slow and sustained than those of polar hormones.

Hydrophilic hormones are the fast and fleeting players of the endocrine system. They are fast because they are readily soluble in water, so they do not need a carrier protein to travel in the plasma. Because they are free in the plasma, they can also be cleared quickly from the blood and have a relatively short half-life. As a result, the concentration of hydrophilic hormones can be adjusted on a minute to minute basis. The nature of their receptors also allows them to enact changes within their target cells quickly. Activating an enzyme or opening a channel is a relatively fast cellular event.

Hydrophobic hormones have effects that are slower and more sustained. They are slower to arrive at their targets because they are not readily soluble in water; in fact, only a small amount of these hormones are free-floating in the blood at any given time. Carrier proteins in the blood both transport hydrophobic hormones around the body and protect them from breakdown. Hormones bound to their carrier proteins are not active until they dissociate and diffuse into their target cells. Because they are hydrophobic, these hormones also have a tendency to dissolve in the fatty tissues of the body, where they can hide indefinitely from enzymes that would degrade them. These hormones therefore have a relatively long half-life, and their concentrations can only be adjusted hourly. Because nonpolar hormones operate at the level of transcription, the response by target cells is slow. However, the effect that these hormones produce is sustained because it involves the production of new proteins for the cell's use. Unlike polar hormones, which cause quick but short-lasting effects, nonpolar hormones stimulate long-lasting changes in the infrastructure and functioning of their target cells, typically by affecting transcription of target genes.

The endocrine system can be understood by organizing glands in terms of the solubility of the hormones that they secrete. The rest of this section will introduce this framework, as demonstrated in Table 3.1. Each major hormone will be briefly introduced in terms of its secreting gland, target, and major function. This intro-

TABLE 3.1 > Polar Hormones vs. Nonpolar Hormones

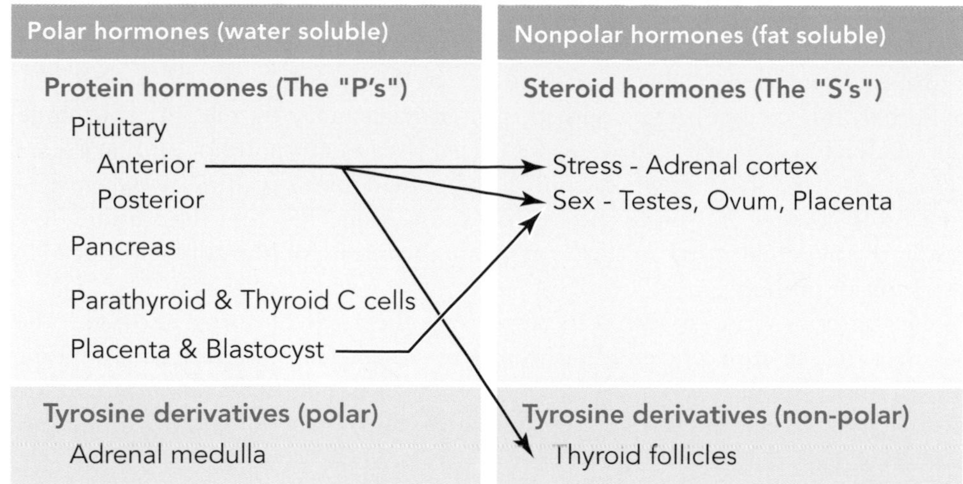

Polar hormones (water soluble)	Nonpolar hormones (fat soluble)
Protein hormones (The "P's")	**Steroid hormones (The "S's")**
Pituitary	
Anterior	Stress - Adrenal cortex
Posterior	Sex - Testes, Ovum, Placenta
Pancreas	
Parathyroid & Thyroid C cells	
Placenta & Blastocyst	
Tyrosine derivatives (polar)	**Tyrosine derivatives (non-polar)**
Adrenal medulla	Thyroid follicles

FIGURE 3.7 Tyrosine

The amino acid tyrosine is the precursor to both polar and nonpolar hormones. The hydroxyl functional group allows formation of ether linkages to create nonpolar hormones such as T_3. The addition of a second hydroxyl to the benzene ring allows for the formation of polar hormones such as epinephrine.

ductory section provides a basis from which to understand the rest of the lecture, which will describe the characteristics of glands and hormones in greater detail.

Almost all glands only secrete either hydrophilic or hydrophobic hormones. The glands can be organized along a spectrum according to the polarity (and solubility) of the hormones that they secrete. Tyrosine derivative hormones, which can be either polar or nonpolar, lie between the polar peptides and nonpolar steroids. There are two major types of tyrosine derivative hormones: those secreted by the thyroid and those secreted by the adrenal medulla. The thyroid secretes tyrosine derivatives that are nonpolar, like the steroid hormones, while the catecholamines of the adrenal medulla are polar, like the peptide hormones.

Notice that regulatory hormones, also called tropic hormones, indicated by arrows extending to the glands that they influence in Table 3.1, are all peptides. Their targets are glands that secrete hydrophobic hormones. In fact, every hydrophobic hormone has a correlating hydrophilic control hormone. This makes sense because the effects of hydrophilic hormones are faster and more fleeting than those of hydrophobic hormones, making them well-suited for a regulatory function.

Almost all of the glands that secrete Peptide hormones start with P. The exception is the adrenal medulla. Think of it as the P-Med (rhymes with premed!) to remember that the adrenal medulla also secretes Peptide hormones. Also, the glands that secrete Steroid hormones can be grouped into functions that start with S. Use these memory tools to keep your hormones and glands straight!

Overview of Endocrine Glands

The pituitary gland is an important site of interaction between the nervous and endocrine systems, as discussed in the Nervous System Lecture. The connection between the hypothalamus and pituitary gland allows fine nervous control over the levels of hormones released by the pituitary. Because many of these hormones have a regulatory effect on other endocrine glands, the interface between the hypothalamus and pituitary gland has a widespread impact on the functioning of the endocrine system.

The posterior pituitary gland is essentially a continuation of the nervous system. In fact, it is a bundle of axons of neurons whose cell bodies lie in the hypothalamus. Rather than producing hormones itself, it is a storage site for hormones that are synthesized by the hypothalamus. Two major types of hormones are released by the posterior pituitary: anti-diuretic hormone (ADH) and oxytocin. ADH plays a critical role in the regulation of plasma volume by stimulating receptors on cells of the kidneys' collecting ducts to facilitate reabsorption of water. The direct nervous control of ADH afforded by the fact that it is released from the posterior pituitary allows changes in ADH levels to occur rapidly in response to the body's needs, as described later in this lecture. Oxytocin is involved in the stimulation of labor and milk ejection for nursing, so receptors for oxytocin are present in the uterus and the milk ducts of the breasts. In the case of milk ejection, the benefit of close nervous control is that oxytocin can be quickly released by the posterior pituitary in response to suckling or other stimuli that indicate the infant's need for milk.

"Head, Shoulders, Knees, and Toes!" The endocrine system is organized from top to bottom, with the glands at the top regulating glands further downstream. For example, ADH released from the posterior pituitary regulates water reabsorption in the kidney.

Since the peptide hormones are fast and fleeting (more similar to nervous system signals) compared to the steroid hormones, it makes sense that the hormones released by the pituitary gland under nervous system control are peptides.

To remember that the posterior pituitary is the part of the pituitary that is really just an extension of the hypothalamus used for storage, think of the name: the posterior pituitary is like any posterior—it just sits there!

The anterior pituitary gland is the higher regulatory gland of the endocrine system, releasing hormones that in turn affect the release of hormones by other glands. Like the posterior pituitary, it is subject to nervous control; however, it is a group of endocrine cells rather than an extension of the nervous system. The hypothalamus communicates with the anterior pituitary by releasing hormones into shared blood vessels, called a *portal system*. Because the anterior pituitary gland is located such a short distance from the hypothalamus, signaling between these structures is faster and more direct than most endocrine communication. As a result, the hypothalamus can closely regulate the release of hormones produced by the anterior pituitary.

The anterior pituitary exerts its regulatory effects by releasing hormones that regulate the secretion of steroid hormones by other glands, as illustrated in Table 3.1. In this way, polar hormones control the slower nonpolar hormones, under the direction of the nervous system. The individual hormones secreted by the anterior pituitary gland will be considered later in this lecture.

The parathyroid glands release parathyroid hormone (PTH) in response to lowered levels of calcium in the bloodstream. Although PTH receptors are present on osteoblasts rather than osteoclasts, intercellular signaling ultimately causes increased osteoclast activity and therefore increased bone breakdown. The breakdown of bone releases calcium, causing the level of blood calcium to rise.

The thyroid gland is largely responsible for basal metabolic rate through secretion of the thyroid hormones T_3 and T_4, and for regulating blood calcium through calcitonin. T_3 and T_4 are nonpolar and are the other type of tyrosine derivative. While the polar hormone calcitonin is secreted by the C cells of the thyroid, the thyroid hormones are instead secreted by follicular cells. The release of thyroid hormones is regulated by one of the hormones released by the anterior pituitary. Thyroid hormones have a widespread effect due to the presence of receptors for these hormones on cells in almost all parts of the body. Heightened levels of thyroid hormones cause the basal metabolic rate to rise, while the opposite is true of low levels.

For now, think of osteoclasts as cells that eat up bone and osteoblasts as cells that bring bone its lunch. The Muscle, Bone and Skin Lecture will discuss the different types of cells that build up and break down bone. When you reach that lecture, review the hormones involved in calcium regulation and make sure you understand how they interact with bone.

Calcitonin is released by the C cells of the thyroid in response to rising calcium levels. Like parathyroid hormone, calcitonin functions by affecting the action of osteoclasts. While PTH causes osteoclasts to increase their activity, calcitonin has an inhibitory effect. As a result, the breakdown of bone decreases while the rate of bone formation is unchanged, and the level of calcium in the bloodstream decreases.

The parathyroid glands and the thyroid are physically adjacent glands that have complementary roles in the regulation of calcium levels in the bloodstream and bone. Calcium must be closely regulated because it is both necessary for the functioning of the body and potentially harmful at high levels. Although calcium must be readily available for muscles and neurons to function, it has low water solubility and must be prevented from precipitating in the plasma. Both the parathyroid and the C cells of the thyroid secrete polar hormones that act in opposition to each other to regulate calcium.

The pancreas secretes two hormones, insulin and glucagon, that act antagonistically to regulate cellular metabolism. One of the most significant effects of these hormones is the tight regulation of glucose. (See the Metabolism Lecture in *Biology 1: Molecules* for a more in-depth consideration of the multiple hormones that contribute to glucose regulation). Glucose must be maintained at a certain level for use by cells, but abnormally high glucose levels in the bloodstream can have significant negative effects.

Insulin is released when the level of glucose in the bloodstream increases (such as after a meal) and is higher than the amount needed by the body's cells. Insulin reduces the amount of glucose in the blood by inducing cells to take it up for storage; most cells in the body have receptors for insulin. The net effect of rising insulin levels is that excess nutrients are stored for later use.

It is crucial to know that the pairs of hormones that act antagonistically to regulate calcium and glucose levels are all peptides. This is easy to remember if you keep in mind that polar hormones are faster than non-polar ones. Compared to steroid hormones, peptide hormones are better able to respond quickly to changing bloodstream levels of calcium and glucose.

In contrast to insulin, glucagon release by the pancreas is triggered by a drop in the level of blood glucose, indicating a deficit of nutrients. The purpose of glucagon is opposite to that of insulin: mobilizing the release of nutrients from storage. Glucagon acts on the liver to stimulate the release of glucose into the bloodstream.

The adrenal medulla, one of the glands contained in the adrenal gland, is involved in the stress response. It secretes the polar tyrosine-derived hormones epinephrine and norepinephrine. They are also called adrenaline and noradrenaline, after the adrenal gland, and are catecholamines (amine derivatives). Note that these are the same chemicals that are used by the sympathetic nervous system to facilitate a fight-or-flight response. Rather than being secreted into the synapse as part of a nervous system response, adrenaline from the adrenal medulla is secreted into the bloodstream in response to SNS input, supplementing nervous system activity by sending signals that are slower but more sustained and generalized.

The adrenal cortex is the other gland that secretes steroid hormones. It is involved in blood pressure regulation and the production of the body's stress response. The adrenal cortex secretes two categories of hormones: glucocorticoids and mineralocorticoids. For the purposes of the MCAT®, it is necessary to know the actions of cortisol (the major glucocorticoid) and aldosterone (the major mineralocorticoid).

The adrenal cortex influences stress response, much like the adrenal medulla, in this case via the secretion of cortisol. However, cortisol produces a more long-term stress response such as when the organism is under some type of chronic stress.

The adrenal cortex also secretes the mineralocorticoid aldosterone, which affects blood pressure by altering processes in the kidney. Aldosterone stimulates increased reabsorption of the mineral Na^+, which in turn causes increased water reabsorption and plasma volume. That process will be described in this and future lectures.

The testes and ovaries release hormones that stimulate reproductive development and functions. These glands secrete their hormones in response to regulation by the anterior pituitary, another example of a peptide hormone regulating a steroid hormone. Testosterone, secreted by the testes, facilitates male puberty and spermatogenesis. The ovaries secrete estrogen and progesterone, which determine female pubertal development and the progression of menstruation and pregnancy. As discussed earlier in this section, the secretion of hormones by the ovaries is affected by HCG when pregnancy occurs. Later in pregnancy, the production of these hormones occurs in the placenta.

Much like the anterior pituitary gland, the placenta secretes a peptide hormone that regulates the secretion of steroid hormones. In this case, the result is that the uterus is prepared to support a pregnancy. The developing blastocyst has a layer of cells that ultimately contributes to the formation of the placenta. These cells start secreting human chorionic gonadotropin (HCG) after implantation, and the placenta then takes over the task of HCG secretion. HCG travels to the ovaries, which exert control over the uterus. In response to HCG, the ovaries signal the continuing synthesis and secretion of the steroid hormones progesterone and estrogen. These hormones suppress menstruation (as described below), allowing the buildup of the uterine wall in preparation for implantation. The placenta continues to secrete HCG at a low level throughout pregnancy.

The placenta is a specialized organ that exists only in pregnancy, and it is unusual in that it secretes both a polar and nonpolar hormone (although the secretion of HCG is greatly diminished when the secretion of steroid hormones begins). The only other glands that likewise secrete polar and nonpolar hormones are the thyroid gland and adrenal gland. One of the hormones secreted by each of these glands is a tyrosine; unlike the placenta, neither gland secretes both a peptide and a steroid. The adrenal gland is actually two glands, while the thyroid is truly a single gland that secretes a polar and nonpolar hormone.

The major hormones secreted by the Adrenal Cortex are Aldosterone and Cortisol. Also, know that the GLUCOcorticoids regulate glucose mobilization (and other aspects of the stress response), while MINERALocorticoids regulate levels of minerals in the plasma.

Epinephrine explains why I still feel nervous and jittery a few minutes after I'm done running from a bear. If I have to run from bears several times a day for a month, my cortisol level will rise and I will have an even more prolonged stress response. You can remember that cortisol contributes to a long-term stress response because it is a non-polar steroid hormone, whereas epinephrine is a polar tyrosine hormone and has a more short-term effect.

The Steroid hormones are involved in Stress and Sex (reproduction).

HCG is the developing blastocyst's way of asking the mother for room and board! You can think of the ovaries as the bosses of the uterus. They give the uterus commands based on information from both the blastocyst (if present) and the anterior pituitary gland.

The endocrine system may seem overwhelming at first, with all of its glands, hormones, targets, and functions. Don't make the mistake of focusing so much on memorizing that you forget to learn the bigger picture! Using the framework of solubility to organize the properties of different hormone types will help you keep things straight. As you read the rest of the lecture, refer back to this section as necessary to put all the pieces in place.

3.5 Endocrine Glands and Their Hormones

Memorization of several major hormones, their glands, and their target tissues is required for the MCAT®. As a memory aid, group hormones according to the gland that secretes them and remember that a given gland produces either steroids, peptides, or tyrosine derivatives, but not two categories of hormones. (The adrenal glands are really two glands. The cortex produces steroids; the medulla produces catecholamines. The thyroid is a true exception. The thyroid secretes T_3 and T_4, which are tyrosine derivatives, and calcitonin, which is a peptide. Calcitonin and the thyroid hormones are produced by different types of cells within the thyroid.) This section expands upon the overview presented in the previous section and will start by discussing the hormones of the posterior pituitary.

Posterior Pituitary

The **posterior pituitary** is also called the *neurohypophysis* because it is composed mainly of support tissue for nerve endings extending from the hypothalamus. The hormones oxytocin and ADH are synthesized in the neural cell bodies of the hypothalamus, and transported down axons to the posterior pituitary, where they are released into the blood. Both oxytocin and ADH are small polypeptides.

Antidiuretic hormone (ADH) (also called **vasopressin**) is a small peptide hormone which causes the collecting ducts of the kidney to become permeable to water, causing reabsorption of water from the collecting tubule. Because fluid is reabsorbed, ADH also increases blood pressure. Throughout the evolution of mammals, ADH secretion has always been under tight neural control. As a peptide hormone, it has a fast action and can quickly respond to blood loss.

Beer is an ADH blocker that increases urine volume.

Oxytocin is a small peptide hormone that increases uterine contractions during pregnancy and causes milk to be ejected from the breasts. Notice that both of these effects are associated with the periodic contraction of smooth muscle, although they are on different scales.

Anterior Pituitary

The **anterior pituitary** (Figure 3.8) (also called the *adenohypophysis*), which functions to regulate much of the endocrine system, is located in the brain beneath the **hypothalamus**. The hypothalamus controls the release of the anterior pituitary hormones with *releasing* and *inhibitory hormones* of its own. These releasing and inhibitory hormones are carried to the capillary bed of the anterior pituitary by small blood vessels. The release of the releasing and inhibitory hormones is, in turn, controlled by nervous signals.

FIGURE 3.8 Anterior Pituitary

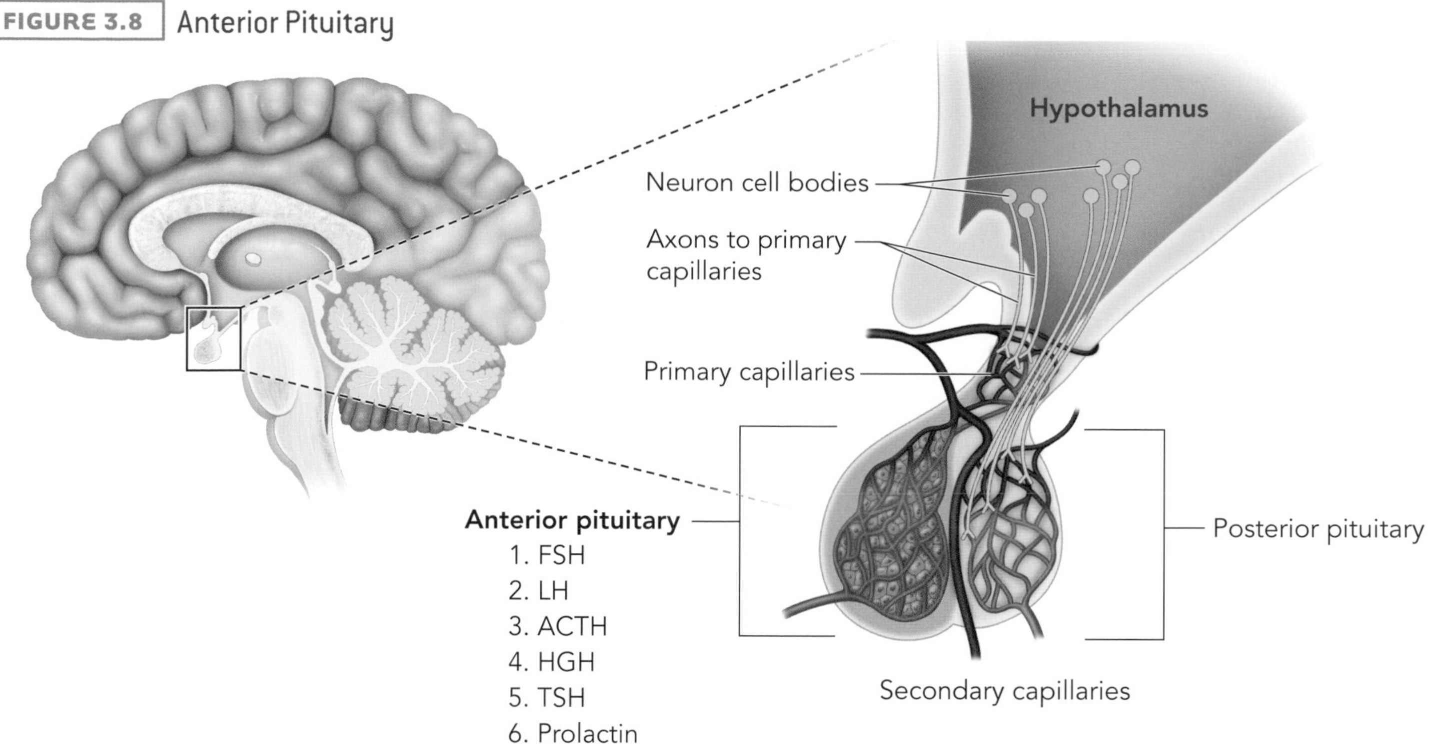

The anterior pituitary synthesizes and releases six major hormones and several minor hormones. Once released into the bloodstream, these hormones travel to their target tissues to affect cellular processes. All of these are peptide hormones. The MCAT® requires familiarity with the six major hormones, their target tissues, and their functions. The hormones are:

1. Thyroid-stimulating hormone (TSH),
2. Adrenocorticotropin (ACTH),
3. Follicle-stimulating hormone (FSH),
4. Luteinizing hormone (LH),
5. Human growth hormone (HGH), and
6. Prolactin.

Thyroid-stimulating hormone (TSH) (also called thyrotropin) is a peptide that stimulates the thyroid to release T_3 and T_4 via the second messenger system using cAMP. Among other effects on the thyroid, TSH increases thyroid cell size, number, and the rate of secretion of T_3 and T_4. It is important to note that high T_3 and T_4 concentrations have a negative feedback effect on TSH release, both at the anterior pituitary and the hypothalamus. (See below for effects of T_3 and T_4.)

Adrenocorticotropic hormone (ACTH) is a peptide that stimulates the adrenal cortex to release glucocorticoids via the second messenger system using cAMP. Release of ACTH is stimulated by many types of stress. Glucocorticoids are stress hormones. (See below for the effects of the adrenal cortical hormones.)

Human growth hormone (HGH) (also called *somatotropin*) is a peptide that stimulates growth in almost all cells of the body. All other hormones of the anterior pituitary have specific target tissues. HGH stimulates growth by increasing the rate of mitosis, increasing cell size, increasing the rate of protein synthesis, mobilizing fat stores, increasing the use of fatty acids for energy, and decreasing the use of glucose. HGH also increases transcription and translation and decreases the breakdown of protein and amino acids.

Prolactin is a peptide that promotes lactation (milk production) by the breasts. Milk is not normally produced before birth due to the inhibitory effects of progesterone and estrogen on milk production. Although the hypothalamus has

The hormones released by the anterior pituitary can be remembered by the acronym HALF PiT:

H- Human growth hormone

A- Adrenocorticotropin

L- Luteinizing hormone

F- Follicle-stimulating hormone

P- Prolactin

i- ignore

T- Thyroid stimulating hormone

PROlactin is involved in the PROduction of milk.

a stimulatory effect on the release of all other anterior pituitary hormones, it mainly inhibits the release of prolactin. The act of suckling, which stimulates the hypothalamus to stimulate the anterior pituitary to release prolactin, inhibits the menstrual cycle. It is not known whether this is directly due to prolactin. The milk production effect of prolactin should be distinguished from the milk ejection effect of oxytocin, which is secreted by the posterior pituitary.

FSH and LH are discussed later in this lecture in the context of reproduction.

FIGURE 3.9 Parathyroid

Parathyroid

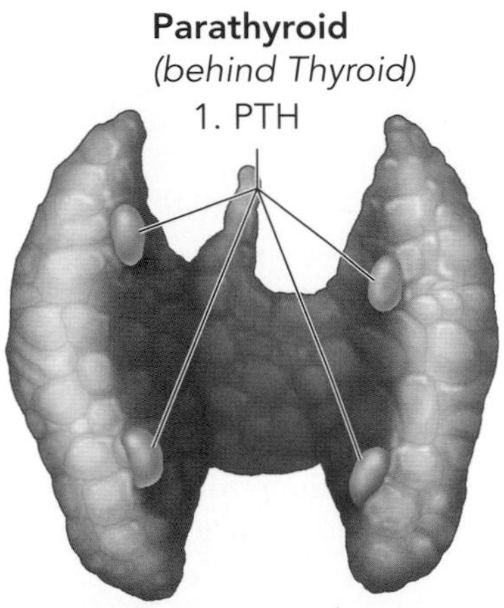

Parathyroid hormone (PTH) is a peptide that increases blood calcium. It has the opposite effect of calcitonin. In the bone, PTH increases osteocyte absorption of calcium and stimulates proliferation of osteoclasts, the cells that break down calcified bone. In the kidney, PTH increases renal calcium reabsorption and renal phosphate excretion. Excretion of phosphate allows the free concentration of calcium in the blood to remain high as calcium and phosphate react to form insoluble calcium phosphate. Finally, PTH increases calcium uptake from the gut by increasing renal production of a steroid derived from vitamin D. PTH secretion from the parathyroid glands is regulated by the concentration of calcium ions in the blood.

As with other hormones, PTH has multiple effectors, and its effects may vary between different effectors.

FIGURE 3.10 PTH Control of Blood Calcium Level

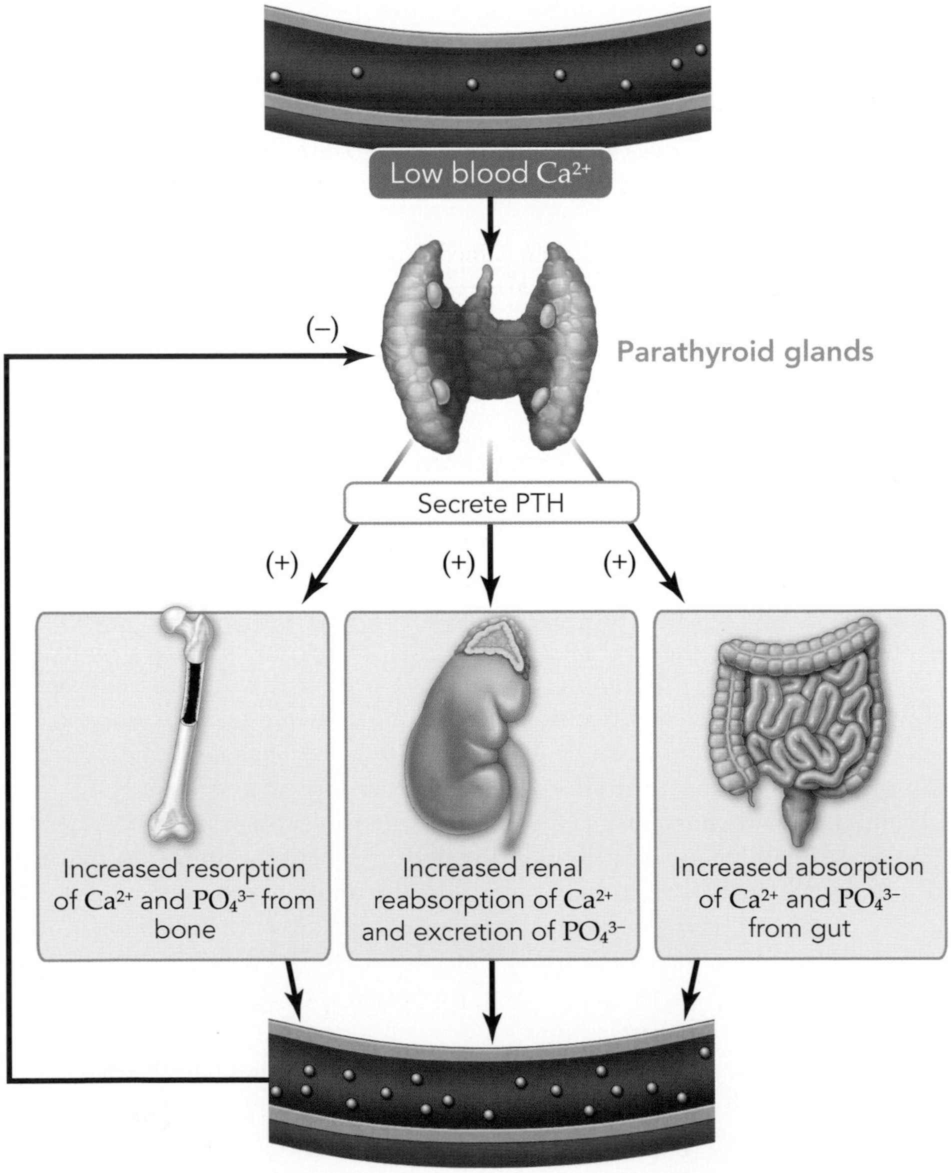

Thyroid

The thyroid hormones are *triiodothyronine* (T_3), thyroxine (T_4), and calcitonin. The thyroid (Figure 3.11) is located along the trachea (windpipe) just in front of the (vocal cords).

FIGURE 3.11 Thyroid

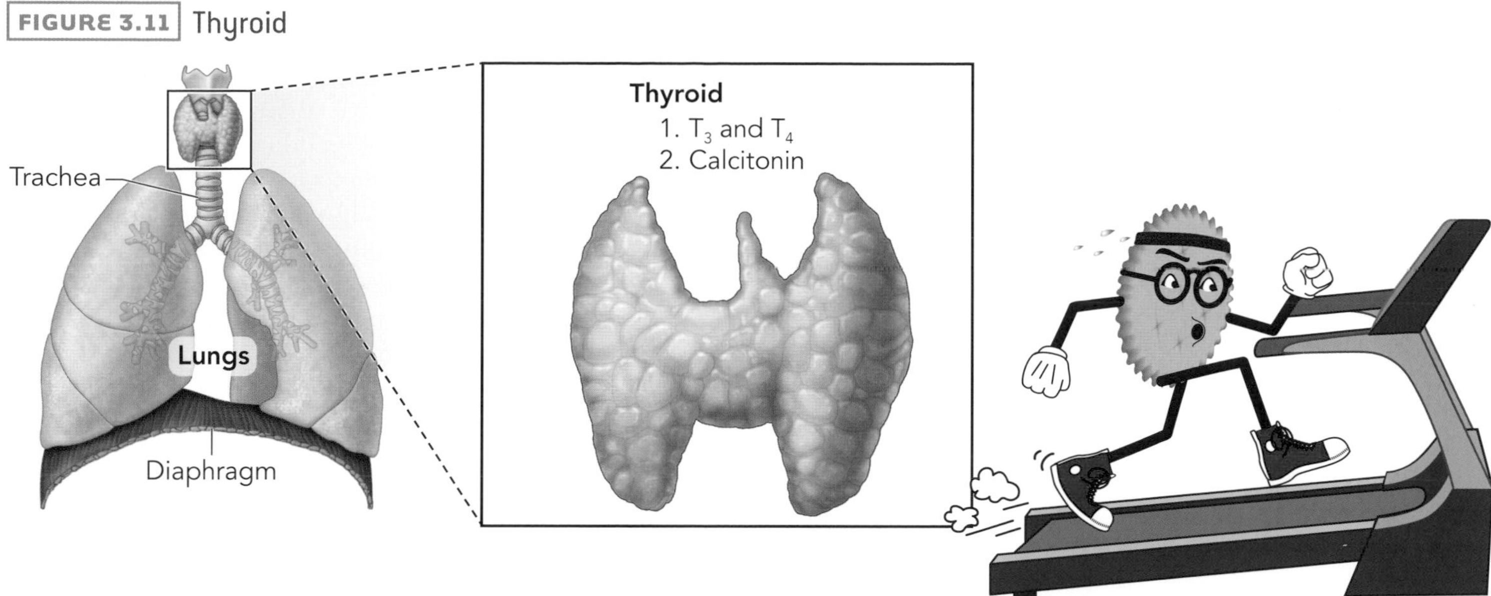

T_3 and T_4 help determine the basal metabolic rate but are not the sole determinants of it. Why some people have higher or lower basal metabolic rates is largely unknown.

The general effect of the thyroid hormones is to increase the basal metabolic rate (the resting metabolic rate). The basal metabolic rate is the amount of energy expended at rest in one day and is commonly measured as rate of oxygen consumption. The determinants of basal metabolic rate include heart rate and muscle mass.

T_3 and T_4 are very similar in their effects, and no distinction will be made by an MCAT® question unless it is thoroughly explained in a passage. T_3 contains three iodine atoms, and T_4 contains four. Both hormones are lipid soluble tyrosine derivatives that diffuse through the lipid bilayer and act in the nuclei of the cells of their effectors. Thyroid hormone secretion is regulated by TSH from the anterior pituitary.

Calcitonin is a large peptide hormone released by the thyroid gland. Calcitonin slightly decreases blood calcium by decreasing osteoclast activity and number. Calcium levels can be effectively controlled in humans in the absence of calcitonin.

Remember that calcitonIN makes calcium go IN the bones!

Some people say calcitonin "tones down calcium" in the bloodstream. Other people say calcitonin moves "calcium into bones." Both are correct and are good memory aids.

Adrenal Glands

The adrenal glands (Figure 3.12) are located on top of the kidneys. They are separated into the adrenal cortex and the adrenal medulla. The adrenal cortex is the outside portion of the gland. The cortex secretes only steroid hormones. There are two types of steroids secreted by the cortex: mineralocorticoids and glucocorticoids. (The cortex also secretes a small amount of sex hormones, significant in the female but not the male.) Mineralocorticoids affect the electrolyte balance in the bloodstream; glucocorticoids increase blood glucose concentration and have an even greater effect on fat and protein metabolism. Approximately 30 corticoids have been isolated from the cortex, but the major mineralocorticoids is aldosterone, and the major glucocorticoid is cortisol.

FIGURE 3.12 Adrenal Glands

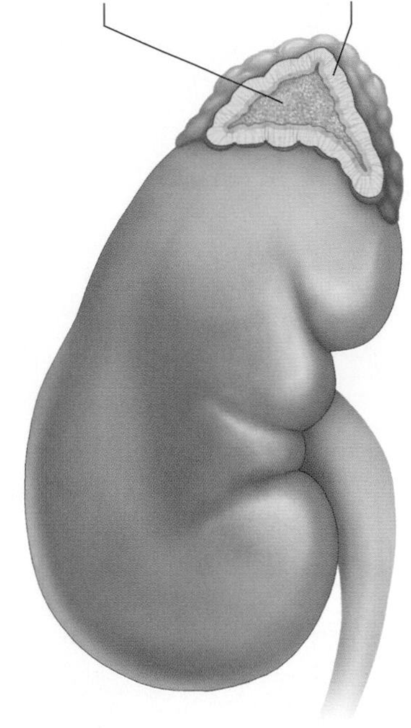

Aldosterone, a steroid, is a mineralocorticoids that acts in the distal convoluted tubule and the collecting duct to increase Na^+ and Cl^- reabsorption and K^+ and H^+ secretion. Na^+ reabsorption drives increased water reabsorption through water channels called aquaporins in the distal convoluted tubule. The end result is that the concentration of salt in the bloodstream remains constant while the blood volume rises. This reabsorption results in an eventual increase in blood pressure. Aldosterone has the same effect, but to a lesser extent, on the sweat glands, salivary glands, and intestines.

Cortisol, a steroid, is a glucocorticoid that increases blood glucose levels by stimulating **gluconeogenesis** in the liver. (See the Metabolism Lecture in *Biology 1: Molecules* for more on gluconeogenesis.) Cortisol mobilizes fatty acids from fat cells to be used for cellular energy and causes a moderate decrease in the use of glucose by the cells. Cortisol also stimulates protein degradation in nonhepatic cells, resulting in an increase in the amino acid concentration in the liver. These amino acids can serve as carbon sources for gluconeogenesis. However, high levels of cortisol also diminish the capacity of the immune system to fight infection.

The *catecholamines* **epinephrine** and **norepinephrine** (also called **adrenaline** and **noradrenaline**) are tyrosine derivatives synthesized in the adrenal medulla. The effects of epinephrine and norepinephrine on the target tissues are similar to their effects in the sympathetic nervous system, but the effects last much longer. Epinephrine and norepinephrine constrict the blood vessels of most internal organs and skin, but they dilate the blood vessels of skeletal muscle. This is consistent with the 'fight-or-flight' response of these hormones by increasing the amount of blood flow and oxygen delivered to the muscles.

Pancreas (Islets of Langerhans)

The pancreas (Figure 3.13) acts as both an endocrine and an exocrine gland. For the MCAT®, the two important endocrine hormones released into the blood by the pancreas are the peptide hormones insulin and glucagon. *Somatostatin*, not likely to be seen on the MCAT®, is released by the δ cells of the pancreas. Somatostatin inhibits both insulin and glucagon. The role of somatostatin may be to extend the period of time over which nutrients are absorbed.

FIGURE 3.13 Pancreas and Islets of Langerhans

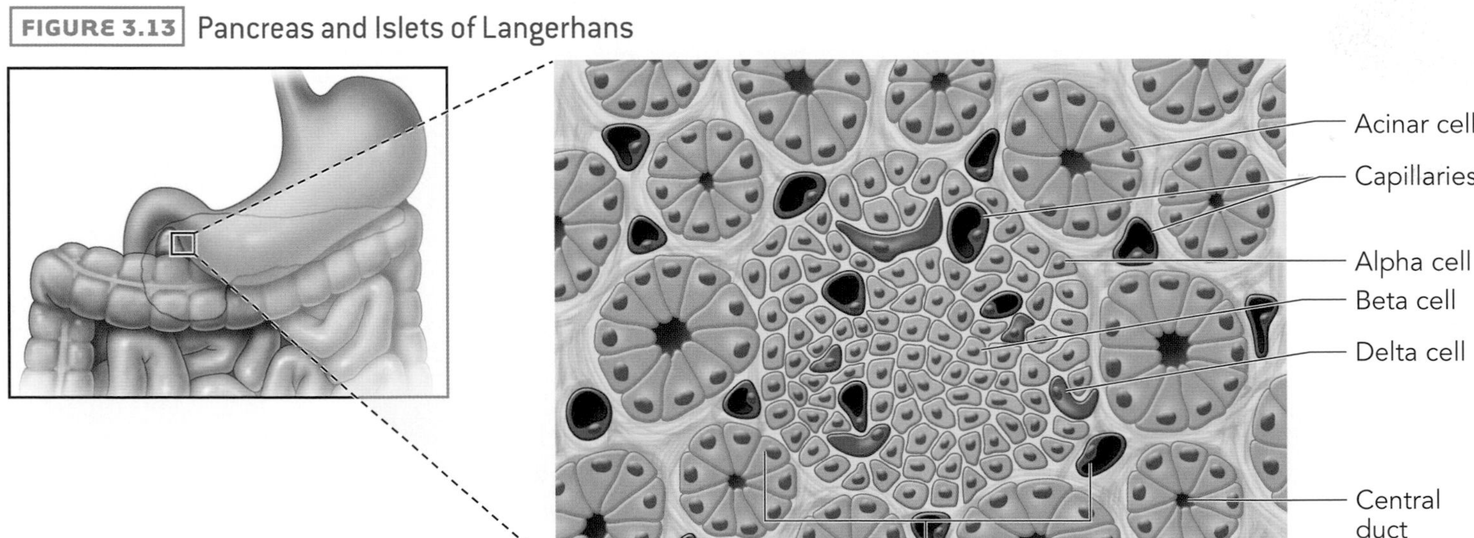

Insulin is associated high levels of blood glucose. Insulin is released by the β *cells* of the pancreas when blood levels of carbohydrates or proteins are high. It affects carbohydrate, fat, and protein metabolism. In the presence of insulin, carbohydrates are stored as glycogen in the liver and muscles, fat is stored in adipose tissue, and amino acids are taken up by the cells of the body and made into proteins. The net effect of insulin is to lower blood glucose levels.

When insulin binds to a membrane receptor, a cascade of reactions is triggered inside the cell. Except for neurons in the brain and a few other cell types that are not affected by insulin, the cells of the body become highly permeable to glucose upon the binding of insulin. The binding of insulin causes the insulin receptor to activate the opening of a membrane bound glucose transporter. The permeability of the membrane to amino acids is also increased. In addition, intracellular metabolic enzymes are activated and, much more slowly, even translation and transcription rates are affected.

The effects of **glucagon** are nearly opposite to those of insulin. Glucagon is released by the α *cells* of the pancreas. Glucagon stimulates glycogenolysis (the breakdown of glycogen), and gluconeogenesis in the liver. It acts via the second messenger system of cAMP. In higher concentrations, glucagon breaks down adipose tissue, increasing the fatty acid level in the blood. The net effect of glucagon is to raise blood glucose levels.

The pancreatic hormones can be remembered by the mnemonic BAD PIGS: β cells, α cells, and δ cells of the Pancreas release Insulin, Glucagon, and Somatostatin.

FIGURE 3.14 The Antagonistic Effects of Insulin and Glucagon

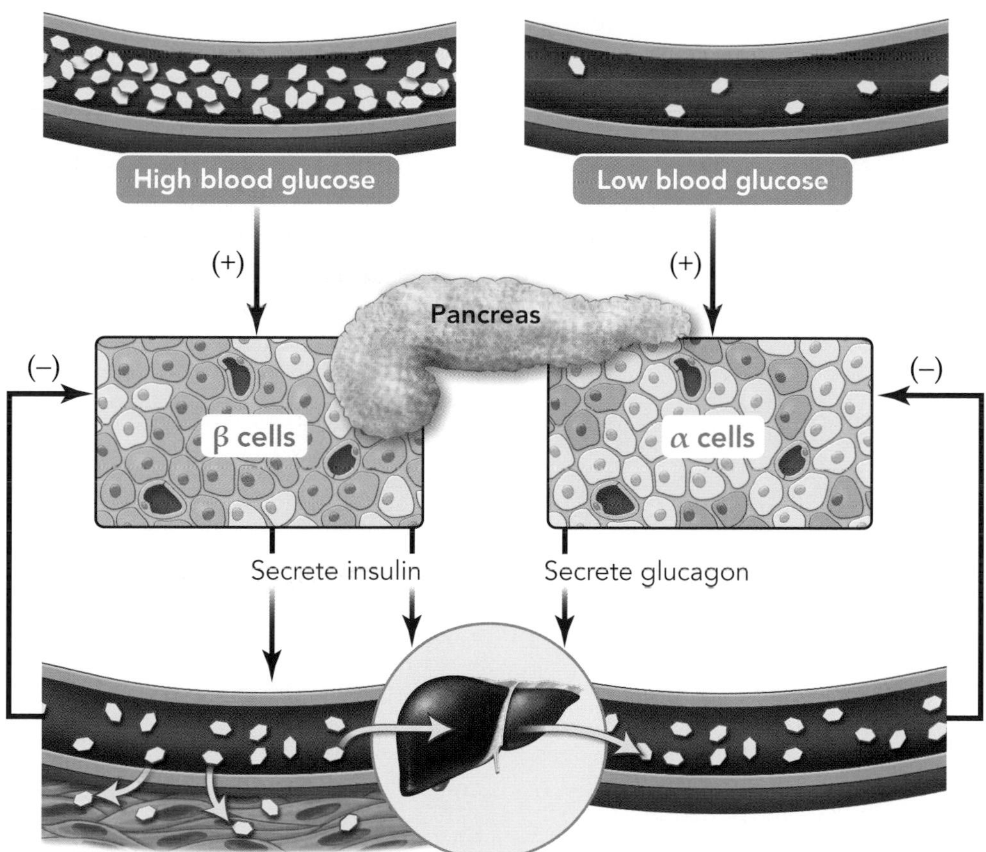

Type I diabetes, or autoimmune diabetes, occurs when the immune system attacks the β cells of the pancreas. Without β cells, the pancreas is unable to secrete insulin and the person cannot regulate blood glucose levels effectively. Confusion, seizures, and even coma and death can be the result of uncontrolled type I diabetes.

These questions are NOT related to a passage.

Question 49

Aldosterone exerts its effects on target cells by:

- O **A.** binding to a receptor at the cell surface, setting off a second-messenger cascade.
- O **B.** diffusing into adrenal cortical cells, where it influences transcription of certain DNA sequences.
- O **C.** flowing across the synapse, where it binds and initiates an action potential.
- O **D.** entering into target cells, where it increases the rate of production of sodium-potassium pump proteins.

Question 50

A patient develops an abdominal tumor resulting in the secretion of large quantities of aldosterone into the bloodstream. Which of the following will most likely occur?

- O **A.** Levels of renin secreted by the kidney will increase.
- O **B.** Levels of oxytocin secreted by the pituitary will increase.
- O **C.** Levels of aldosterone secreted by the adrenal cortex will decrease.
- O **D.** Levels of aldosterone secreted by the tumor will decrease.

Question 51

Which of the following is true for all endocrine hormones?

- O **A.** They act through a second messenger system.
- O **B.** They bind to a protein receptor.
- O **C.** They dissolve in the blood.
- O **D.** They are derived from a protein precursor.

Question 52

All of the following act as second messengers for hormones EXCEPT:

- O **A.** cyclic AMP.
- O **B.** calmodulin.
- O **C.** acetylcholine.
- O **D.** cyclic GMP.

Question 53

Which of the following is true of all steroids?

- O **A.** The target cells of any steroid include every cell in the body.
- O **B.** Steroids bind to receptor proteins on the membrane of their target cells.
- O **C.** Steroids are synthesized on the rough endoplasmic reticulum.
- O **D.** Steroids are lipid soluble.

Question 54

The pancreas is a unique organ because it has both exocrine and endocrine function. The exocrine portion of the pancreas releases:

- O **A.** digestive enzymes directly into the blood.
- O **B.** digestive enzymes through a duct.
- O **C.** hormones straight into the blood.
- O **D.** hormones through a duct.

Question 55

Most steroid hormones regulate enzymatic activity at the level of:

- O **A.** replication.
- O **B.** transcription.
- O **C.** translation.
- O **D.** the reaction.

Question 56

Which of the following side effects might be experienced by a patient who is administered a dose of thyroxine?

- O **A.** An increase in endogenous TSH production
- O **B.** A decrease in endogenous TSH production
- O **C.** An increase in endogenous thyroxine production
- O **D.** A decrease in endogenous parathyroid hormone production

3.6 Homeostasis and Hormones

The hormones that have been discussed in this lecture can be grouped into four categories according to their functions in maintaining homeostasis in the body: tropic hormones, blood chemistry hormones, hormones that determine metabolic rate, and hormones that affect reproduction and development.

Tropic hormones are released from one gland and cause downstream release of other hormones from their target endocrine glands. The **blood chemistry hormones** control the concentrations of sodium, calcium, and glucose in the bloodstream. For the purposes of the MCAT®, **osmolarity** in the bloodstream is primarily determined by sodium concentrations. While insulin and glucagon are the main hormones involved in **blood glucose regulation**, it is also important to note that human growth hormone, cortisol, and epinephrine can all increase blood glucose transiently. The **stress hormones** control how the body reacts to stress, including the short-term and long-term response, while the **determinants of metabolic rate** control the body's use of energy on a day to day basis. Finally, the **reproduction and development hormones** control the body's ability to reproduce effectively and maintain secondary sexual characteristics.

Table 3.2 presents the hormones as they function in maintaining homeostasis. Spend some time learning how deviations away from homeostasis trigger hormone release and how these hormones act to restore balance.

Apply Key 2: Hormones respond to imbalances that have moved the body away from homeostasis. The hormone insulin responds to higher than normal levels of blood glucose to lower them back to normal, while glucagon responds to lower than normal levels of blood glucose to raise them.

MCAT® THINK

Try these questions to see if you understand the difference between ADH and aldosterone.

1. A person eats a bag of potato chips for lunch and now has a high concentration of sodium in the bloodstream. What will be the primary hormonal response?
2. A person enters a water-drinking contest on the local radio station. She drinks two gallons of water over the course of two hours, and now has a low concentration of sodium in the bloodstream. What will be the primary hormonal response?
3. A person has studied hard for four hours without anything to drink. His blood volume is now low. What will be the primary hormonal response?
4. A person has been in the desert for two days and ran out of water 10 hours ago. He now feels like he will pass out. What will be the primary hormonal response?

Answer on page 106.

TABLE 3.2 > **Hormones Maintain Homeostasis**

Function		Hormone	Trigger	Action	Source	Target	Type
Tropic hormones (target other endocrine organs)		TSH	Low thyroid hormone levels	Stimulates thyroid hormone release	Anterior pituitary	Thyroid	Peptide
		ACTH	Low cortisol levels	Stimulates adrenal cortex		Adrenal cortex	
		LH	Low estrogen or testosterone levels, GnRH release during the menstrual cycle, rising estrogen during luteal surge	Ovulation and estrogen or testosterone release		Gonads	
		FSH	Low inhibin, GnRH release during the menstrual cycle	Promotes growth of follicles or sperm production			
		hCG	Pregnancy	Maintenance of the corpus luteum	Placenta		
Blood chemistry	**Sodium (Osmolarity)**	Aldosterone	Low sodium levels, low blood osmolarity	Increases sodium reabsorption, increased blood pressure	Adrenal cortex	Nephron distal tubule	Steroid
		ADH	Low blood pressure	Increases water reabsorption, increased blood pressure	Posterior pituitary	Nephron collecting duct	Peptide
	Calcium	Parathyroid hormone	Low blood calcium	Increases blood calcium	Parathyroid	Bones, kidneys, GI tract	Peptide
		Calcitonin	High blood calcium	Decreases blood calcium	Thyroid		
	Glucose	Insulin	High blood glucose	Decreases blood glucose	Pancreas	Systemic	Peptide
		Glucagon	Low blood glucose	Increases blood glucose	Pancreas	Liver, adipocytes	
		hGH	Periods of growth and development	Increases blood glucose, long bone growth	Anterior pituitary	Systemic	
		Epinephrine	Stress	Increases blood glucose, transcription of target genes	Adrenal medulla	Systemic	Tyrosine derivative
		Cortisol	Stress		Adrenal cortex	Systemic	
Determinants of metabolic rate		T_3	Low levels of thyroid hormone	Increases basal metabolic rate	Thyroid	Systemic	Tyrosine derivative
		T_4					
Reproduction and development		Estrogen	LH	Female secondary sex characteristics, menstrual cycle regulation	Ovaries	Systemic	Steroid
		Testosterone		Male secondary sex characteristics	Testes	Systemic	
		Progester-one	hCG	Maintains the uterus for pregnancy	Ovaries, placenta	Endometrium	
		Oxytocin	Birth, breastfeeding	Stimulates uterine, mammary gland contractions	Posterior pituitary	Breasts, uterus	Peptide

These questions are NOT related to a passage.

Question 57

Sympathetic stimulation results in responses most similar to the effects of which of the following hormones?

- O **A.** Insulin
- O **B.** Acetylcholine
- O **C.** Epinephrine
- O **D.** Aldosterone

Question 58

When compared with the actions of the nervous system, those of the endocrine system are:

- O **A.** quicker in responding to changes, and longer-lasting.
- O **B.** quicker in responding to changes, and shorter-lasting.
- O **C.** slower in responding to changes, and longer-lasting.
- O **D.** slower in responding to changes, and shorter-lasting.

Question 59

Insulin shock occurs when a patient with diabetes self-administers too much insulin. Typical symptoms are extreme nervousness, trembling, sweating, and ultimately loss of consciousness. The physiological effects of insulin shock most likely include:

- O **A.** a pronounced increase in gluconeogenesis by the liver.
- O **B.** a rise in blood fatty acid levels, leading to atherosclerosis.
- O **C.** a dramatic rise in blood pressure.
- O **D.** dangerously low blood glucose levels.

Question 60

Vasopressin, a hormone involved in water balance, is produced in the:

- O **A.** hypothalamus.
- O **B.** posterior pituitary.
- O **C.** anterior pituitary.
- O **D.** kidney.

Question 61

Osteoporosis is an absolute decrease in bone tissue mass, especially trabecular bone. All of the following might be contributory factors to the disease EXCEPT:

- O **A.** increased sensitivity to endogenous parathyroid hormone.
- O **B.** defective intestinal calcium absorption.
- O **C.** menopause.
- O **D.** abnormally high blood levels of calcitonin.

Question 62

All of the following hormones are produced by the anterior pituitary EXCEPT:

- O **A.** thyroxine.
- O **B.** growth hormone.
- O **C.** prolactin.
- O **D.** luteinizing hormone.

Question 63

Which of the following hormonal and physiological effects of stress would NOT be expected in a marathoner in the last mile of a marathon?

- O **A.** Increased glucagon secretion
- O **B.** Increased heart rate
- O **C.** Decreased ACTH secretion
- O **D.** Decreased blood flow to the small intestine

Question 64

Parathyroid hormone is an important hormone in the control of blood calcium ion levels. Parathyroid hormone directly impacts:

I. bone density.
II. renal calcium reabsorption.
III. blood calcium concentration.

- O **A.** I only
- O **B.** I and II only
- O **C.** I and III only
- O **D.** I, II and III

3.7 Reproduction and Development

Be aware of basic male anatomy but don't stress too much about it.

Except for FSH, LH, HCG, and *inhibin*, which are peptides, the reproductive hormones discussed in this section are steroids released from the testes, ovaries, and placenta.

The Male Reproductive System

Remember Key 3: Both male and female reproductive hormones are derived from cholesterol. They are similar in structure and process, but they act along different timelines. Male hormones tend to be secreted continually, while female hormones cycle depending upon the woman's age, pregnancy status, and point in the menstrual cycle.

FSH stimulates Sertoli cells and LH stimulates Leydig cells.

The MCAT® requires knowledge of the basic anatomy of the male and female reproductive systems (Figures 3.15 and 3.18). The male **gonads**, meaning organs involved in the production of gametes, are called the **testes**. Production of sperm (Figure 3.16 and 3.17) occurs in the **seminiferous tubules**, a set of long, twisted tubes in the testes that are lined by Sertoli cells and spermatogonia. Spermatogonia located in the seminiferous tubules arise from epithelial tissue to become spermatocytes, spermatids, and then spermatozoa. *Sertoli cells* stimulated by **follicle-stimulating hormone (FSH)** surround and nurture the spermatocyte and spermatids. *Leydig cells*, located in the interstitium between the tubules, release **testosterone** when stimulated by **luteinizing hormone (LH)**. Testosterone is the primary **androgen** (male sex hormone), and stimulates the germ cells to differentiate into sperm. Testosterone is also responsible for the development of secondary sex characteristics such as pubic hair, enlargement of the larynx, and growth of the penis and seminal vesicles. While testosterone helps to initiate the growth spurt at puberty, it also stimulates closure of the epiphyses of the long bones, ending growth in height. Sertoli cells secrete *inhibin*, a peptide hormone (actually a glycoprotein) that acts on the pituitary gland to inhibit FSH secretion.

A **spermatogonium** is a sperm stem cell before it undergoes meiosis. Once the cell enters meiosis, it is referred to as a *spermatocyte*. At the end of meiosis, it is known as a *spermatid*. The spermatid loses its cytoplasm and forms the *head*, *mid-piece*, and *tail*, as shown in Figure 3.16, to become a spermatozoon (sperm). The

FIGURE 3.15 Male Reproductive Anatomy

Helpful mnemonic for the Male Reproductive Anatomy: SEVEN UP

Seminal Vesicle
Epididymis
Vas deferens
Ejaculatory Duct
N (nothing)
Urethra
Penis

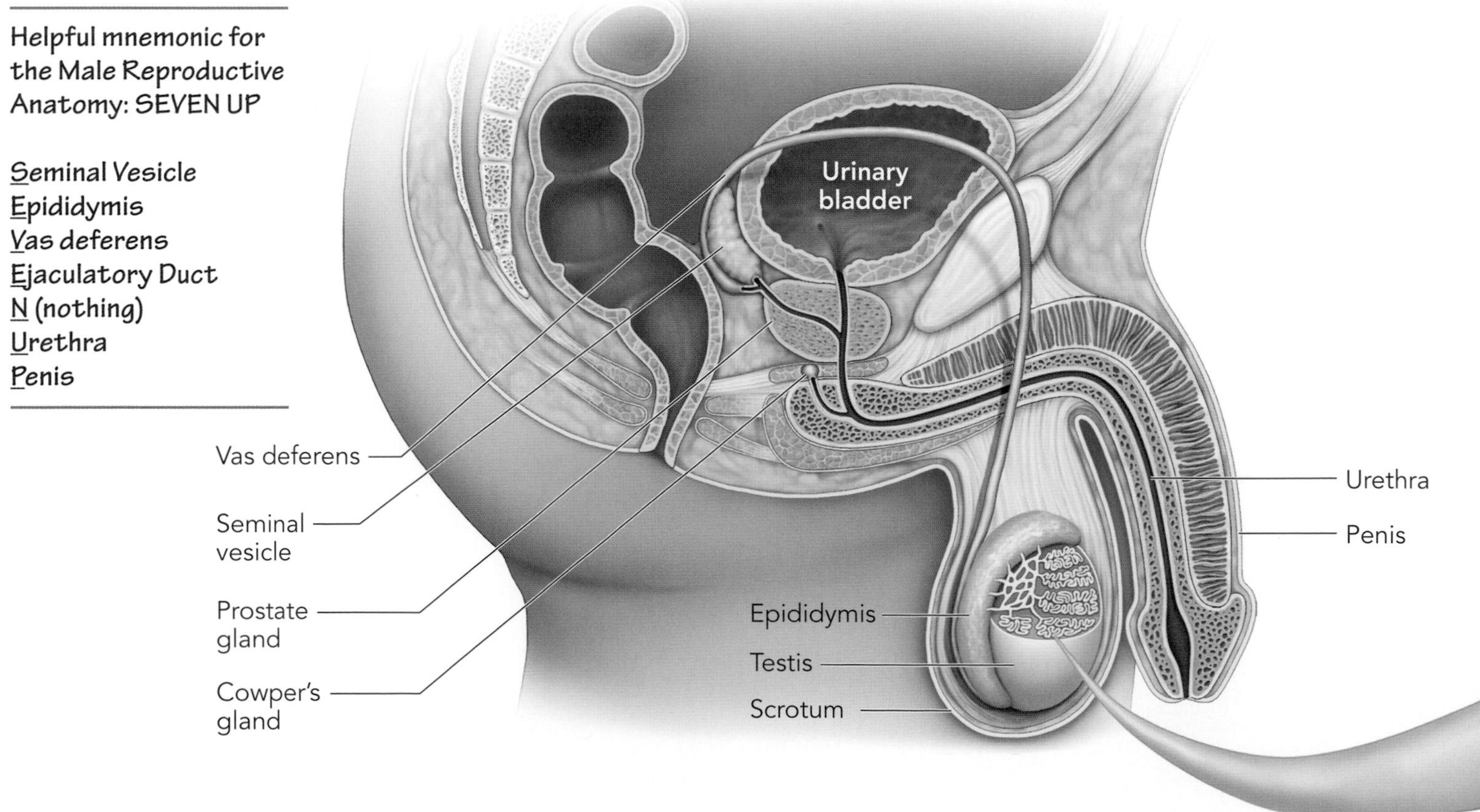

FIGURE 3.16 Spermatozoon

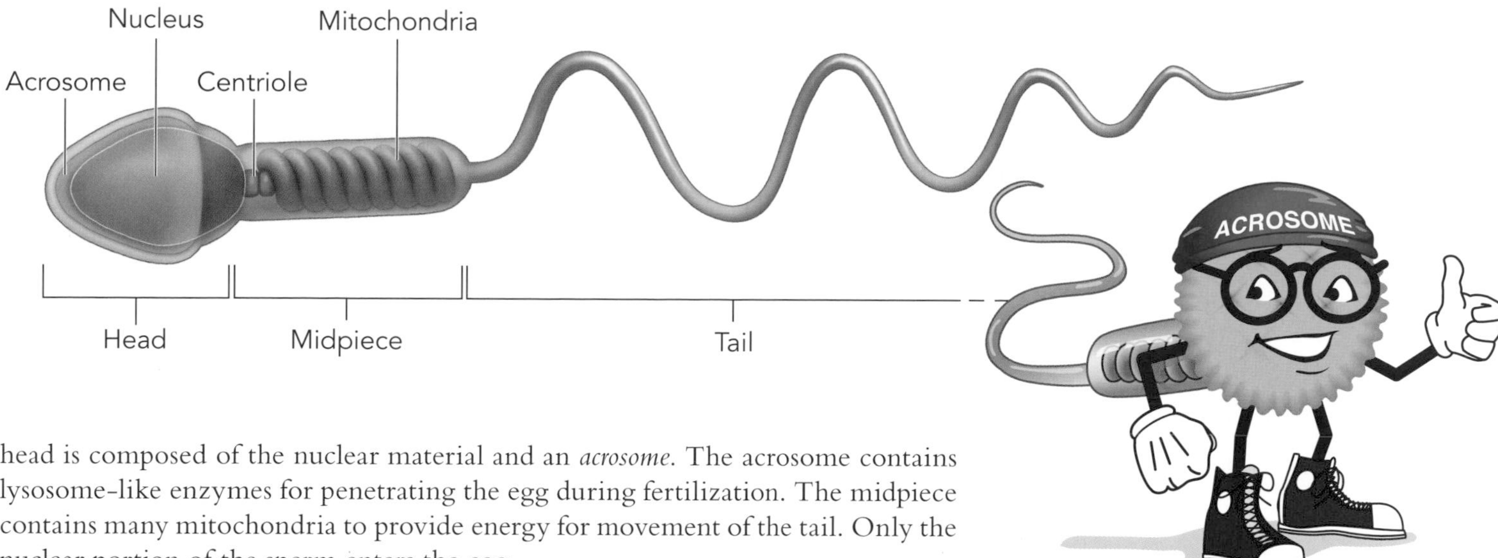

I am ready to go on my trip! I have only the things I need. I have my tail to swim. I have my mitochondria to give me lots of energy. And I have my acrosome to let me get inside an egg. Anything else would just slow me down!

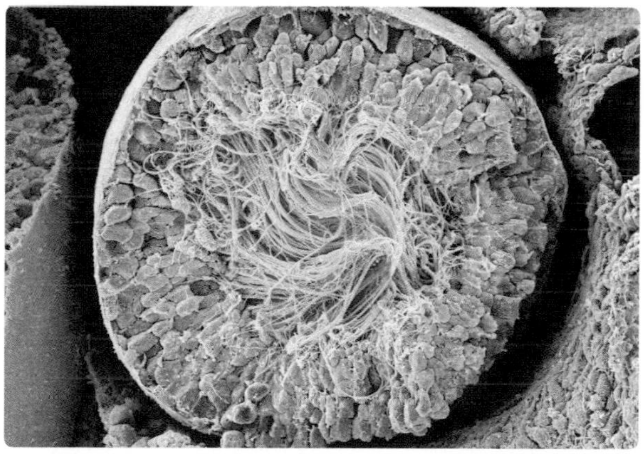

head is composed of the nuclear material and an *acrosome*. The acrosome contains lysosome-like enzymes for penetrating the egg during fertilization. The midpiece contains many mitochondria to provide energy for movement of the tail. Only the nuclear portion of the sperm enters the egg.

After a spermatozoon exits the seminiferous tubules, it is carried to the **epididymis** to mature. Upon ejaculation, spermatozoa are propelled through the **vas deferens** into the **urethra** and out of the penis. **Semen** is the complete mixture of spermatozoa and fluid that leaves the penis upon ejaculation. Semen is composed of fluid from the **seminal vesicles**, the **prostate**, and the **bulbourethral glands** (also called **Cowper's glands**). Spermatozoa become activated for fertilization in a process called *capacitation*, which takes place in the vagina.

The seminiferous tubules are the site of spermatogenesis (sperm production). Each sperm cell consists of a head (green) and a tail (blue). The heads of the sperm are buried in Sertoli cells (yellow and orange), which nourish the developing sperm.

FIGURE 3.17 Spermatogenesis

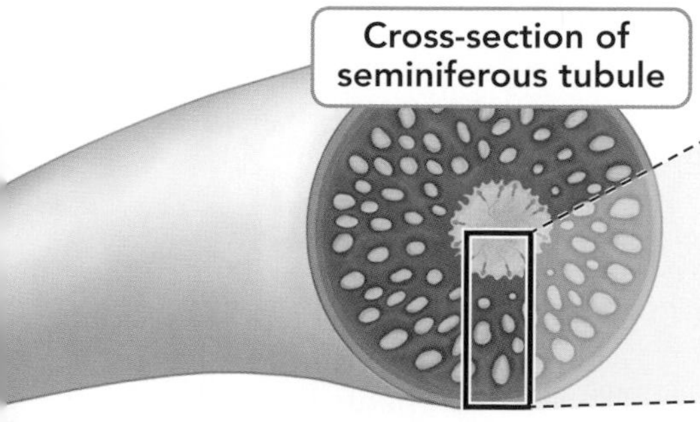

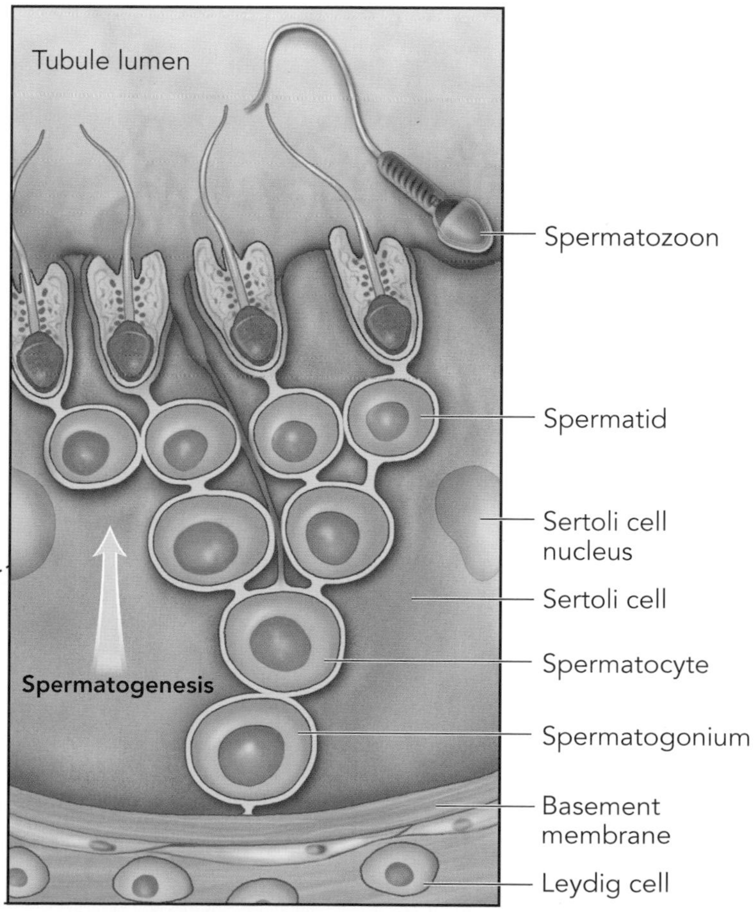

The Female Reproductive System

The male reproductive cycle is relatively linear. Sperm are produced at a constant rate every day, and male hormones have no true monthly cycle. By contrast, the female reproductive cycle is indeed a cycle. In fact, there are two intricately related cycles: one in the ovaries and one in the uterus.

The foundations of the ovarian cycle begin at birth with the start of oogenesis. All of the eggs of the female are arrested as primary oocytes, in prophase I of meiosis, at birth. At puberty, the ovarian cycle begins. FSH stimulates the growth of *granulosa cells* around the primary oocyte (Figure 3.18). The granulosa cells secrete a viscous substance around the egg called the **zona pellucida**. The structure at this stage is called a *primary follicle*. Next, theca cells differentiate from the interstitial tissue and grow around the follicle to form a *secondary follicle*. Upon stimulation by LH, theca cells secrete androgen, which is converted to **estradiol** (a type of **estrogen**) by the granulosa cells in the presence of FSH and secreted into the blood.

Eventually, the follicle grows and bulges from the ovary. Typically, estradiol inhibits LH secretion by the anterior pituitary. However, just before ovulation, the estradiol level rises rapidly and causes a dramatic increase in LH secretion. This increase is called the **luteal surge**. The luteal surge results from a positive feedback loop, where rising estrogen levels increase LH levels, which increase estrogen. The luteal surge causes **ovulation**, the bursting of the follicle and release of the egg (now a secondary oocyte, having just completed meiosis I) into the body cavity. The egg is swept into the **Fallopian (uterine) tube** by the *fimbriae*. The remaining portion of the follicle is left behind to become the **corpus luteum**. The corpus luteum secretes estradiol and **progesterone** throughout pregnancy, or, in the case of no pregnancy, for about 2 weeks until the corpus luteum degrades into the **corpus albicans**.

FIGURE 3.18 Female Reproductive Anatomy and Ovulation

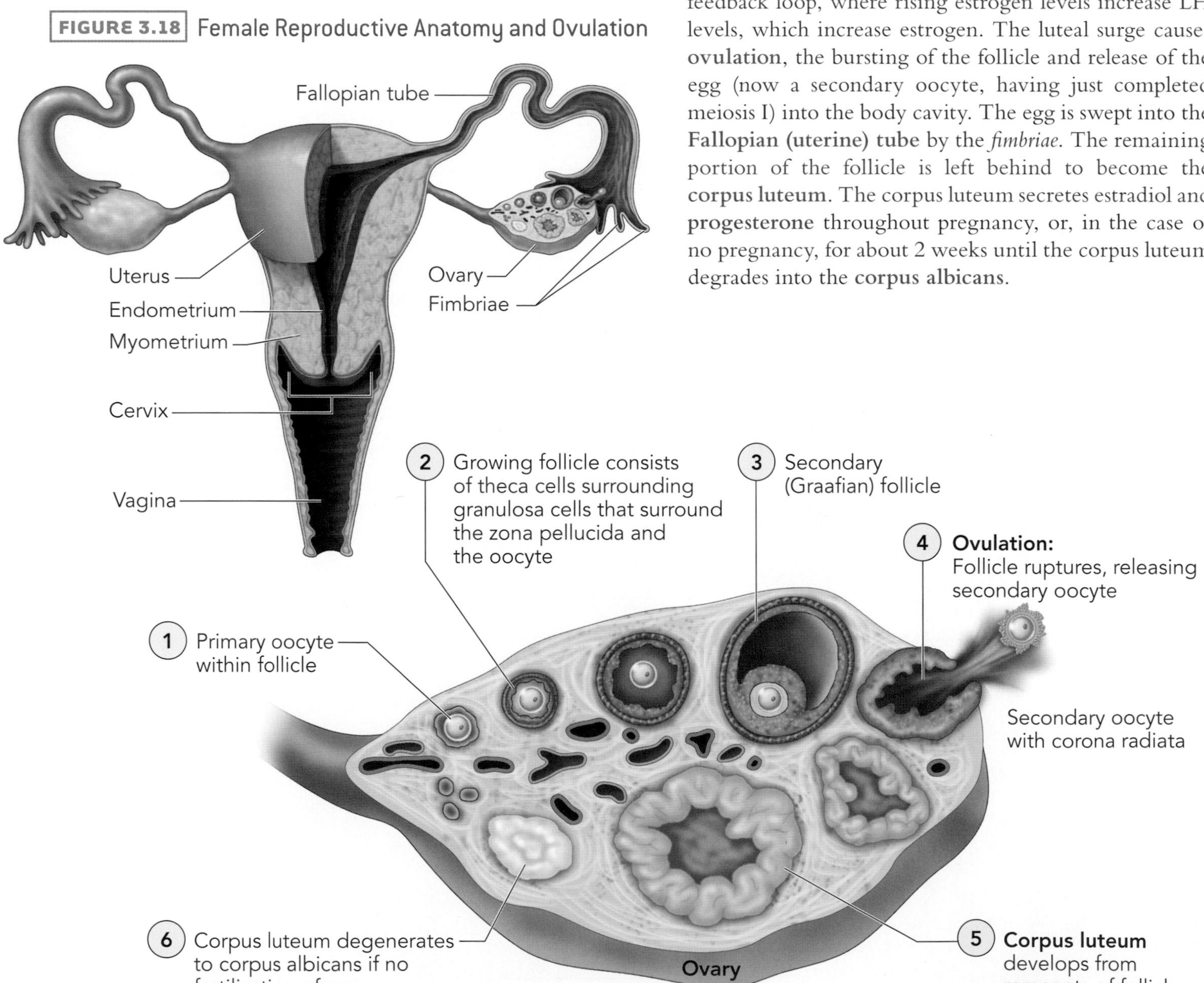

FIGURE 3.19 The Menstrual Cycle

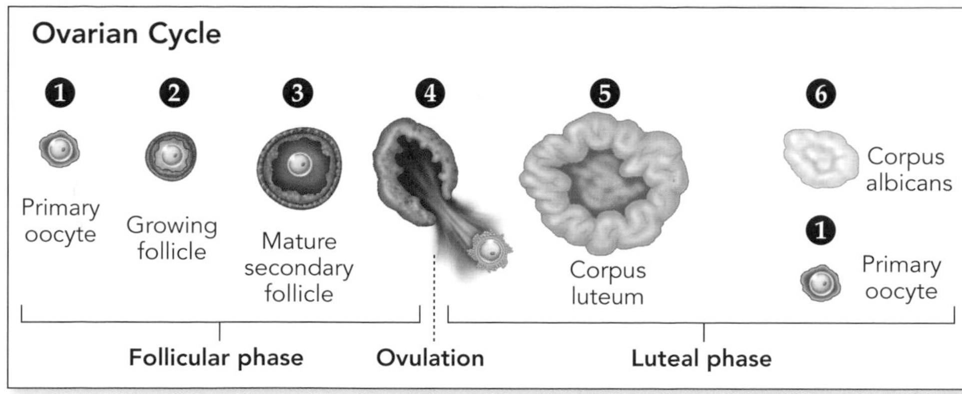

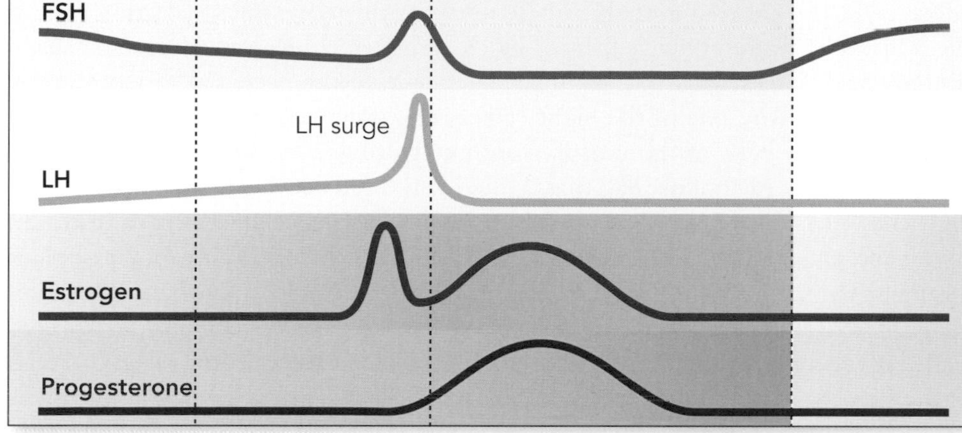

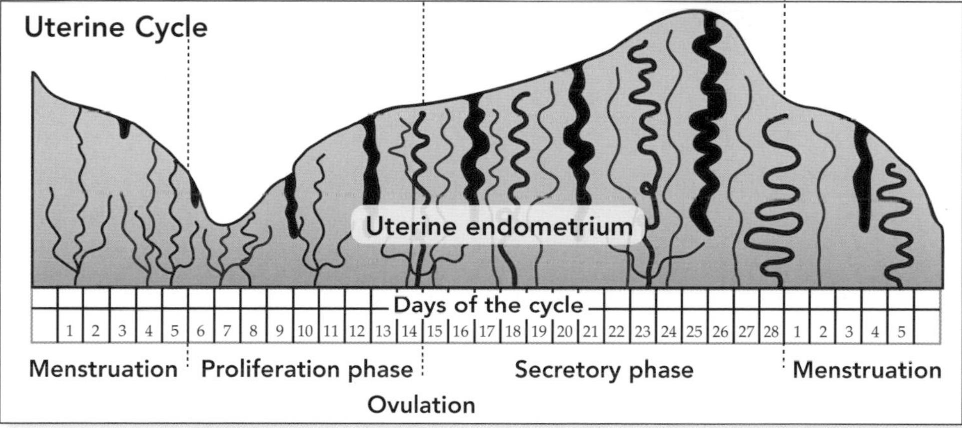

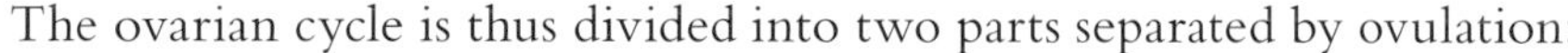

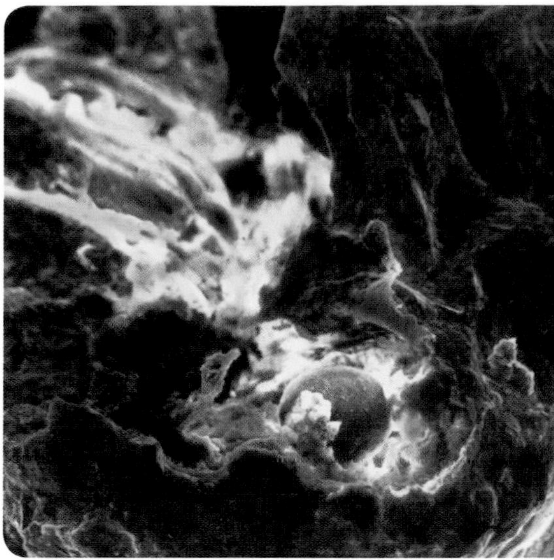

Shown above is a mature ovum (Graafian follicle) at ovulation. The ovum (red) is surrounded by remnants of corona cells and liquid from the ruptured ovarian follicle.

Notice that both menstruation and the proliferation phase of the uterus occur during the follicular phase of the ovarian cycle.

The ovarian cycle is thus divided into two parts separated by ovulation:

1. The *follicular phase*, which begins with the development of the follicle and ends at ovulation; and
2. The *luteal phase*, which begins with ovulation and ends with the degeneration of the corpus luteum into the corpus albicans.

The hormones produced during the ovarian cycle set the course of the cycle that takes place in the uterus. The uterine cycle begins on day 0, which marks the first day of *menstruation*. The first day of the uterine cycle corresponds is also the first day of the follicular phase in the ovary. As the follicle matures and begins to release more estradiol, menstruation stops, and the uterine wall enters the *proliferation phase*. The proliferative phase is the building phase, and lasts until ovulation. After ovulation, the corpus luteum begins to secrete progesterone, which acts as a maintenance hormone for the uterus. The uterine wall is held intact in the event of pregnancy. As the corpus luteum degrades into the corpus albicans, it is no longer able to secrete progesterone to maintain the uterine wall. The uterine wall sloughs off and produces menstruation, starting a new cycle.

Pregnancy and Embryology

Sperm cell fertilizing an egg cell. The sperm cell (brown) is trying to penetrate the surface of the egg cell (blue). Once a sperm has fertilized the egg, rapid chemical changes make the outer layer (zona pellucida) of the egg cell thicken, preventing other sperm cells from entering. The head of the successful sperm cell releases genetic material that mixes with the genetic material in the egg cell.

Once in the Fallopian tube, the egg is swept toward the uterus by cilia. Fertilization normally takes place in the Fallopian tubes. The enzymes of the acrosome in the sperm are released upon contact with the egg, and digest a path for the sperm through the granulosa cells and the zona pellucida. The cell membranes of the sperm head and the oocyte fuse upon contact, and the sperm nucleus enters the cytoplasm of the oocyte. The entry of the sperm causes the *cortical reaction*, which prevents other sperm from fertilizing the same egg. Now the oocyte goes through the second meiotic division to become an **ovum** and releases a second polar body. Fertilization occurs when the nuclei of the ovum and sperm fuse to form the **zygote**.

Cleavage begins while the zygote is still in the Fallopian tube. The zygote goes through many cycles of mitosis; when the zygote is comprised of sixteen or more cells, it is called a **morula**. The embryo at this stage does not grow during cleavage. The first eight cells formed by cleavage are equivalent in size and shape and are said to be *totipotent*, meaning that they have the potential to express any of their genes. Any one of the eight cells at this stage could produce a complete individual. The cells of the morula continue to divide for four days, forming a **blastocyst**, a mostly hollow ball that is filled with fluid and has a small cell mass on one side. The blastocyst lodges in the uterus in a process called **implantation** on about the seventh day after fertilization. The outer cells of the blastocyst implant in the uterine wall and fuse with the uterine tissue to form the **placenta**. The small mass of cells on the inside of the blastocyst will become the embryo. The inner cell mass is made up of **stem cells**, which are *pluripotent*, meaning that each has the ability to develop into most of the types of cells in the human body depending on its position in the mass. Upon implantation, the female is said to be pregnant.

FIGURE 3.20 Early Cleavages in Animal Development

After implantation, the placenta begins secreting the peptide hormone **human chorionic gonadotropin (HCG)**. HCG prevents the degeneration of the corpus luteum and maintains its secretion of estrogen and progesterone. HCG in the blood and urine of the mother is the first outward sign of pregnancy. The placenta reaches full development by the end of the first trimester and begins secreting its own estrogen and progesterone while lowering its secretion of HCG.

As the embryo develops past the eight cell stage, the cells become different from each other due to **cell-cell communication**. In the blastocyst, the inner cells become the embryo and the outer cells become the placenta. This process where a cell becomes committed to a specialized developmental path is called **determination**. Cells undergo determination such that they will give rise to a particular tissue early on. The specialization that occurs at the end of development, forming a specialized tissue cell, is called **differentiation**. The fate of a cell is typically determined early on, but that same cell usually does not differentiate into a specialized tissue cell until much later, at the end of the developmental process.

The process of differentiation involves a complex series of intercellular interactions. It also provides an example of **gene regulation in development**. Differentiated cells do not lose the genetic material associated with features of other types of cells. Instead, parts of the genome that are not required for the specialized function of the differentiated cell are not expressed, due to epigenetic regulatory factors such as signals from other cells (Figure 3.21).

Not all cells in an adult organism are terminally differentiated. Skin, liver, and blood cells all have *multipotent stem cells* that can regenerate these systems as needed, since these types of tissues commonly experience injury (skin, liver) or loss (blood). Recent research has focused on the ability to generate pluripotent stem cells from multipotent stem cells, so that organs without **regenerative capacity** might one day be able to regenerate following injury or removal.

The formation of the **gastrula** occurs in the third week after fertilization in a process called **gastrulation**. The **first cell movements** occur as cells begin to slowly move about the embryo. In mammals, a *primitive streak* is formed, which is analogous to the blastopore in aquatic vertebrates. Cells destined to become mesoderm migrate into the primitive streak. During gastrulation, the three primary germ layers are formed:

1. the **ectoderm**,
2. the **mesoderm**, and
3. the **endoderm**.

FIGURE 3.21 Influence of Neighboring Cells

Cell-cell communication influences the differentiation of neighboring cell types.

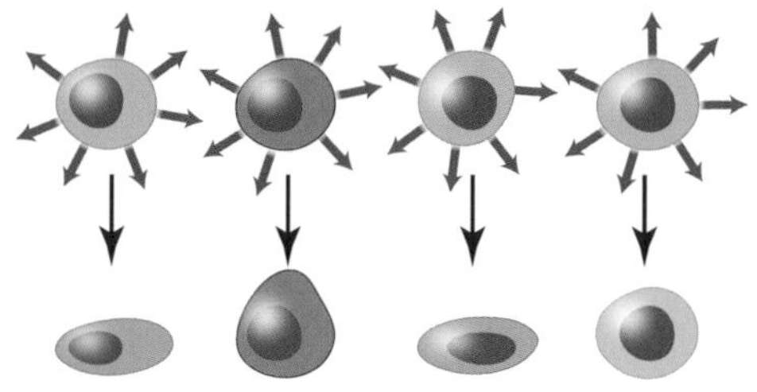

Totipotent stem cells can become any cell, placenta or embryo. Pluripotent stem cells can become any one of the initial three germ layers, all of which are a part of the embryo. Multipotent stems cells can replace cells of a particular lineage, such as skin, liver, and blood.

FIGURE 3.22 Gastrulation

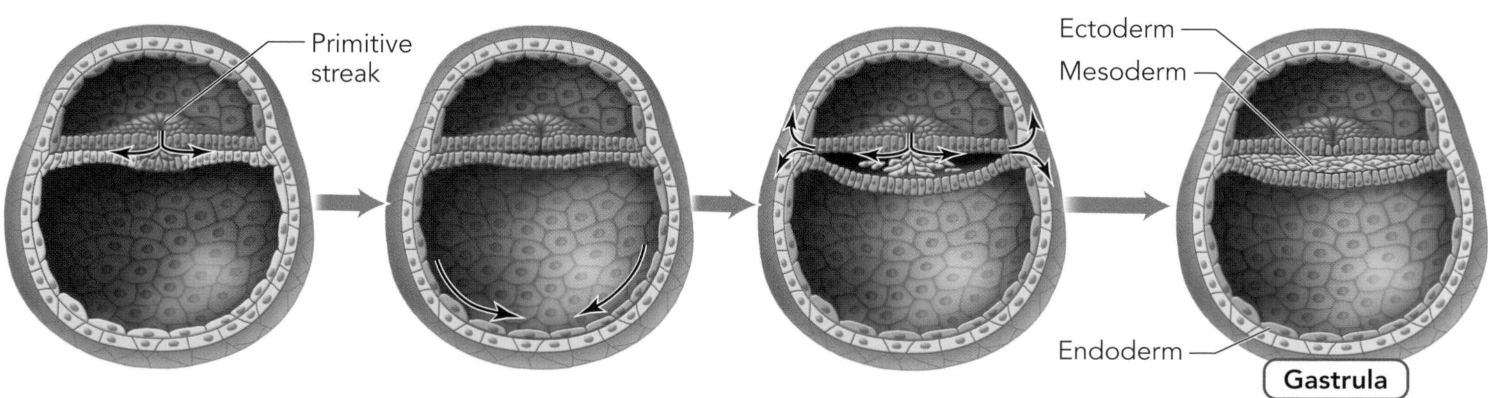

TABLE 3.3 > Fates of the Primary Germ Layers

Ectoderm	Epidermis of skin, nervous system, sense organs
Mesoderm	Skeleton, muscles, blood vessels, heart, blood, gonads, kidneys, dermis of skin
Endoderm	Lining of digestive and respiratory tracts, liver, pancreas, thymus, thyroid

Although there are no absolute rules for memorizing which tissues arise from which germ layer, certain guidelines can be followed for the MCAT®. The ectodermal cells develop into the outer coverings of the body, such as the outer layers of skin, nails, and tooth enamel, and into the cells of the nervous system and sense organs. The endodermal cells develop into the lining of the digestive tract, and into much of the liver and pancreas. The mesoderm develops into the structures that lie between the inner and outer coverings of the body: the muscle, bone, and the rest.

In the third week, the gastrula develops into a **neurula** in a process called **neurulation**. Through **induction**, the **notochord** (made from mesoderm) causes overlying ectoderm to thicken and form the *neural plate*. The notochord eventually degenerates, while a *neural tube* forms from the neural plate to become the spinal cord, brain, and most of the nervous system. The cells of the ectoderm that are close to the neural tube are known as the **neural crest**. The cells of the neural crest function mostly as accessory cells to the nervous system, such as Schwann cells. They form the second order neurons of the autonomic nervous system. They also give rise to parts of the endocrine organs discussed in this lecture, including the adrenal medulla and the C cells of the thyroid. In order to travel to various locations in the body, neural crest cells undergo **cell migration**. Throughout embryological development, this regulated process of cell movement ensures that cells end up in the correct locations of the structures to which they contribute.

For the MCAT®, know that induction occurs when one cell type affects the direction of differentiation of another cell type. It is also important to know that the neural tube becomes the nervous system and is derived from ectoderm.

Part of normal cell development is **programmed cell death** or **apoptosis**. Apoptosis is essential for development of the nervous system, operation of the immune system, and destruction of tissue between fingers and toes to create hands and feet in humans. Damaged cells may undergo apoptosis as well. Failure to do so may result in cancer. Apoptosis is a complicated process in humans, but it is essentially regulated by protein activity as opposed to regulation at the level of transcription or translation. The proteins involved in apoptosis are present but inactive in a normal healthy cell. In mammals, mitochondria play an important role in apoptosis.

Senescence is the process by which cells stop proliferating in response to environmental stressors and are ultimately cleared away by immune cells. It is another developmental mechanism that allows the shaping of tissues. Senescence is necessary for normal embryological development, but it is also implicated in the aging process later in life.

Physiological Development

So far developmental mechanisms have been discussed on the cellular level, but the MCAT® also requires an understanding of human physiological development at the level of the whole organism prenatally, in early youth, and during puberty. **Prenatal development**, which is development that occurs before birth, begins with the development of the embryo. Following the earliest stages of development described above—cleavage, blastula formation, gastrulation, and neurulation—the process of organ formation begins. By the ninth week of pregnancy, major organs are developing in their final locations, and the developing organism is now considered to be a fetus. Development continues throughout the second trimester, but the fetus is not yet capable of independent survival. The third trimester is

Throughout the process of prenatal development, the fast growth of multiple structures means that harmful substances (teratogens) can have particularly negative consequences. Teratogens and some other aspects of development will be discussed in the *Psychology & Sociology* manual. Much of the information in this section could appear on the Psychosocial section of the MCAT®.

characterized by continued growth, particularly in the brain. The brain is not yet fully developed at the time of **parturition** (birth); a completely developed brain would be too large to pass through the human pelvis.

Prenatal development can be roughly categorized by trimester. The 1st semester is mostly development, the 2nd trimester is growth and development, and the 3rd trimester is mostly growth.

A clever mnemonic to remember the order of development for an embryo is Many Babies Get Naps (M = morula, B = blastula, G = gastrula, N= neurula)

Development in childhood and adolescence, like prenatal development, follows a predictable sequence. Consider **motor development**, which takes the infant from the earliest stage—the immobility of a newborn—to independent walking. Although children progress at different rates, the basic milestones of progress are generally the same. The direction of development is from head to toe (a *cephalocaudal pattern*) and from the midline of the body to the periphery (a *proximodistal pattern*). In other words, progress starts with the head, moves down to the trunk, and then moves down and out as limb movement is mastered. More specifically, lifting of the head is followed by lifting of the chest. Next the child is able to sit with support, and then without support. Then the ability to stand with support is developed, followed by crawling. The last three steps are walking with support, standing independently, and finally walking independently.

To remember how motor development progresses, think "head to toe and in to out."

Another important period of rapid and dramatic physical development is puberty. **Puberty** refers to the biological changes that ultimately lead to sexual maturity. **Adolescent development** includes the psychosocial processes that accompany puberty. Adolescence, then, is a socially designated period of rapid psychological and social change that is intertwined with pubertal development. Like prenatal development, adolescent development involves rapid physical growth. At the same time, sexual maturation occurs in the context of primary sex characteristics (i.e. development of the reproductive organs) and the secondary sex characteristics.

In addition to influencing physiological development, reproductive hormones influence behavior during pubertal development. This is just one of the many ways that the endocrine system affects behavior.

It is less important to know each step than it is to understand that motor development is predictable. You can't skip straight from sitting up to walking!

TABLE 3.4 > Major Hormones of the Endocrine System

Gland	Hormone	Solubility	Effect
Anterior pituitary	HGH	Polar	Growth of nearly all cells
	ACTH	Polar	Stimulates adrenal cortex
	FSH	Polar	Growth of follicles in female; Sperm production in male
	LH	Polar	Causes ovulation; stimulates estrogen and testosterone secretion
	TSH	Polar	Stimulates release of T_3 and T_4 in the thyroid
	Prolactin	Polar	Promotes milk production
Posterior pituitary	Oxytocin	Polar	Milk ejection and uterine contraction
	ADH	Polar	Water absorption by the kidney; increases blood pressure
Adrenal cortex	Aldosterone	Nonpolar	Reduces Na^+ excretion; increases K^+ excretion; raises blood pressure
	Cortisol	Nonpolar	Increases blood levels of carbohydrates, proteins, and fats
Adrenal medulla	Epinephrine	Polar	Stimulates sympathetic actions
	Norepinephrine	Polar	Stimulates sympathetic actions
Thyroid	T_3, T_4	Nonpolar	Increase basal metabolic rate
	Calcitonin	Polar	Lowers blood calcium
Parathyroid	PTH	Polar	Raises blood calcium
Pancreas	Insulin	Polar	Promotes glucose entry into cells, decreasing blood glucose levels
	Glucagon	Polar	Increases gluconeogenesis, increasing blood glucose levels
Ovaries	Estrogen	Nonpolar	Growth of female sex organs; causes LH surge
	Progesterone	Nonpolar	Prepares and maintains uterus for pregnancy
Testes	Testosterone	Nonpolar	Secondary sex characteristics; closing of epiphyseal plate
Placenta	HCG	Polar	Stimulates corpus luteum to grow and release estrogen and progesterone
	Estrogen	Nonpolar	Enlargement of mother's sex organs; stimulates prolactin secretion
	Progesterone	Nonpolar	Maintains uterus for pregnancy

These questions are NOT related to a passage.

Question 65

A drug that causes increased secretion of testosterone from the interstitial cells of a physically mature male would most likely:

- O **A.** cause the testes to descend prematurely.
- O **B.** delay the onset of puberty.
- O **C.** cause enhanced secondary sex characteristics.
- O **D.** decrease core body temperature.

Question 66

During the female menstrual cycle, increasing levels of estrogen cause:

- O **A.** a positive feedback response, stimulating LH secretion by the anterior pituitary.
- O **B.** a positive feedback response, stimulating FSH secretion by the anterior pituitary.
- O **C.** a negative feedback response, stimulating a sloughing off of the uterine lining.
- O **D.** a negative feedback response, stimulating decreased progesterone secretion by the anterior pituitary.

Question 67

Which of the following statements are true of development?

I. Programmed cell death is a necessity for normal development.
II. The three germ layers form during implantation.
III. HCG prevents degeneration of the corpus luteum.

- O **A.** I only
- O **B.** III only
- O **C.** I and III only
- O **D.** II and III only

Question 68

Decreasing progesterone levels during the luteal phase of the menstrual cycle are associated with:

- O **A.** thickening of the endometrial lining in preparation for implantation of the zygote.
- O **B.** increased secretion of LH, leading to the luteal surge and ovulation.
- O **C.** degeneration of the corpus luteum in the ovary.
- O **D.** increased secretion of estrogen in the follicle, leading to the flow phase of the menstrual cycle.

Question 69

The inner linings of the Fallopian tubes are covered with a layer of cilia. The purpose of this layer is to:

- O **A.** remove particulate matter that becomes trapped in the mucus layer covering the Fallopian tubes.
- O **B.** maintain a layer of warm air close to the inner lining, protecting the ovum from temperature changes occurring in the external environment.
- O **C.** kill incoming sperm, thereby preventing fertilization.
- O **D.** facilitate movement of the ovum towards the uterus.

Question 70

Which of the following endocrine glands produces testosterone?

- O **A.** The anterior pituitary
- O **B.** The pancreas
- O **C.** The adrenal cortex
- O **D.** The adrenal medulla

Question 71

Which of the following does NOT accurately describe cleavage in human embryos?

- O **A.** The solid ball of cells produced during cleavage is called a morula.
- O **B.** The size of the embryo remains constant throughout the cell divisions of cleavage.
- O **C.** Cell division occurs in one portion of the egg in meroblastic cleavage.
- O **D.** Daughter cells are genetically identical to parent cells.

Question 72

The heart, bone and skeletal muscle most likely arise from which of the following primary germ layers?

- O **A.** The ectoderm
- O **B.** The endoderm
- O **C.** The gastrula
- O **D.** The mesoderm

STOP

TERMS YOU NEED TO KNOW

Adolescent development
Adrenal cortex
Adrenal glands
Adrenaline
Adrenocorticotropic hormone (ACTH)
Aging
Aldosterone
Androgen
Anterior pituitary
Antidiuretic hormone (ADH or vasopressin)
Apoptosis
Basal metabolic rate
Blastocyst
Blood chemistry hormones
Blood glucose regulation
Bulbourethral glands (Cowper's glands)
Calcitonin
Cell-cell communication
Cell migration
Cleavage
Corpus albicans
Corpus luteum
Cortisol
Determinants of metabolic rate
Determination
Differentiation
Ectoderm
Effector
Endocrine glands
Endoderm
Epididymis
Epinephrine
Estradiol
Estrogen
Exocrine glands
Fallopian (uterine) tube
First cell movement
Follicle-stimulating hormone (FSH)
Gastrula
Gastrulation
Gene regulation in development
Glucagon
Glucocorticoids
Gluconeogenesis
Gonads
Human chorionic gonadotropin (HCG)
Human growth hormone (HGH)
Hypothalamus
Implantation
Induction
Insulin
Luteal surge
Luteinizing hormone (LH)
Mesoderm
Mineralocorticoids
Morula
Motor development
Neural crest
Neurula
Neurulation
Noradrenaline
Norepinephrine
Notochord
Osmolarity
Ovulation
Ovum
Oxytocin
Parathyroid hormone (PTH)
Parturition
Peptide hormones
Placenta
Posterior pituitary
Prenatal development
Progesterone
Programmed cell death
Prolactin
Prostate
Puberty
Receptors
Regenerative capacity
Reproduction and development hormones
Semen
Seminal vesicles
Seminiferous tubules
Senescence
Sperm
Spermatogonium
Stem cells
Steroid hormones
Stress hormones
Target cell
Testes
Testosterone
Thyroid hormones: T_3, T_4
Thyroid-stimulating hormone (TSH)
Tropic hormones
Tyrosine derivatives
Urethra
Vas deferens
Zona pellucida
Zygote

MCAT® THINK ANSWERS

Pg. 93:

1. ADH would be high, to absorb free water and decrease plasma osmolarity.
2. ADH would be low, to promote the excretion of water in the urine.
3. Aldosterone would initially be high, to promote the increase of blood volume without affecting blood osmolarity.
4. This is a tricky question. Blood volume is low, so levels of aldosterone will increase. However, this question is different from question three because it involves more extended and severe dehydration. Since the body is desperate to maintain its intravascular volume, ADH will be high. In emergency situations, the body will sacrifice blood sodium levels for volume. An MCAT® question will always make it clear that blood volume is very low (here, words like "desert" and "feels like he will pass out"), and it will not require you to choose between high levels of aldosterone and ADH.

DON'T FORGET YOUR KEYS

1. See a hormone, think of its gland, function, polarity, and target(s).

2. Hormones respond to imbalance in the body to restore homeostasis. Normally functioning hormones do not create imbalance.

3. Reproductive hormones and cells in males and females are similar in both structure and process but act along different timelines.

The Circulatory, Respiratory, and Immune Systems

THE 3 KEYS

1. The goal of respiration and circulation is to meet metabolic needs.
2. Differential pressures and concentrations drive fluid flow and reactions.
3. The innate immune response is non-specific, fast, and short-acting. The acquired immune response is specific, slow, and long-lasting.

4.1 Introduction

This lecture covers multiple systems with interrelated functions that all concern the transport and delivery of important substances throughout the body. The unique features that allow each system to carry out its function will be considered. The respiratory and cardiovascular systems work together to deliver oxygen to the tissues, allowing cells to carry out metabolism and gain energy to perform their necessary functions. Both of these systems make use of branching structures that generate a huge surface area for the exchange of materials. Differential pressures drive the movement of substances in both of these systems, as will be described throughout the lecture.

The blood is the link between the respiratory and cardiovascular systems. Oxygen diffuses into the bloodstream, binds to a carrier molecule, and is then carried throughout the body for delivery to the tissues according to their metabolic needs. The cardiovascular system routinely deposits excess fluid into the tissues, which is transported away by the lymphatic system. The immune system, which protects the organism from potentially harmful invading substances, involves all of the systems described above.

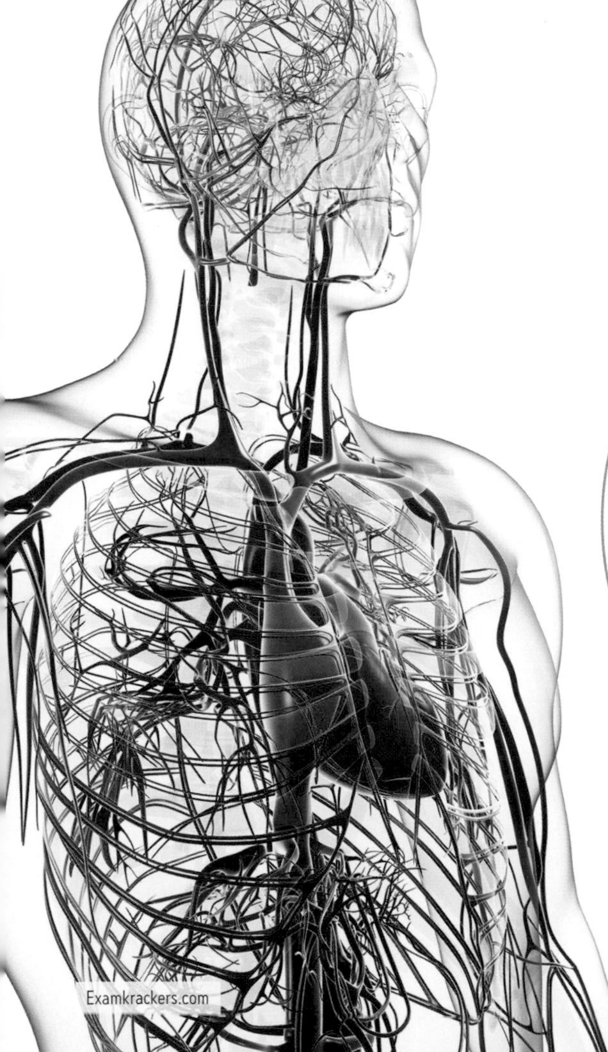

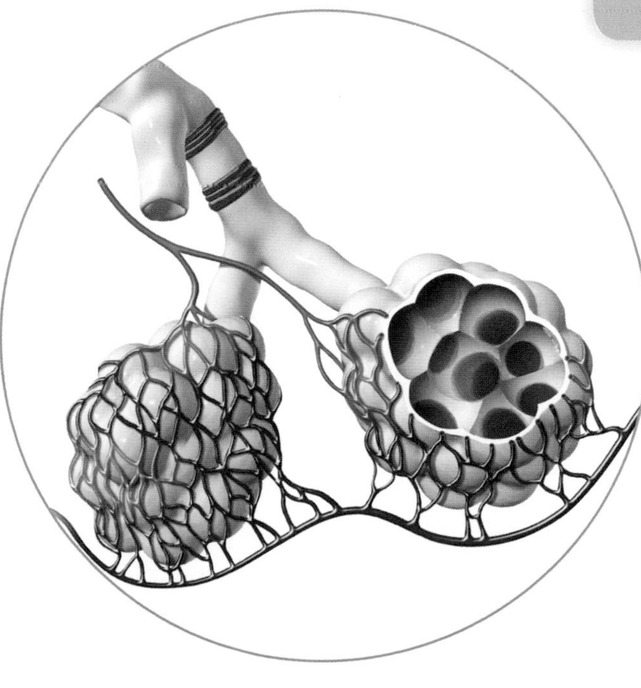

4.2 The Respiratory System and Gas Exchange

The respiratory system is used by the body to ensure that the cells have a constant supply of oxygen and are able to get rid of the waste product carbon dioxide. Both oxygen and carbon dioxide are involved in aerobic respiration: oxygen is necessary for the last step in chemiosmotic synthesis of ATP, while carbon dioxide is produced by the Krebs cycle. Cells need a constant supply of oxygen, and they also need a way to prevent the buildup of carbon dioxide and other waste products. The respiratory system is key to both of these cellular requirements.

You need to know that the primary job of the respiratory system is to deliver oxygen to the blood and expel carbon dioxide. It is also important to know the basic anatomy of the respiratory system given here. Another part of the respiratory tract's function is to prepare the air by warming, moistening, and cleaning it. Ciliated cells in the respiratory tract play a role in filtering the air by trapping foreign particles. Since microtubules are found in cilia, and ciliated cells are found in the respiratory tract (and the Fallopian tubes and ependymal cells of the spinal cord), a defect in microtubule production might result in a problem with breathing (or fertility or circulation of cerebrospinal fluid).

Because the respiratory system is a site of exchange between the external environment and the body, it plays a role in protection against disease by trapping potentially harmful incoming **particulate matter** and ushering it back out of the body.

The respiratory system is also important for **thermoregulation**, the regulation of body temperature such that it stays within a narrow range. **Panting** increases respiration rate, bringing more water into the upper part of the respiratory tree that can then evaporate. **Evaporation** is an exothermic process, so it has a cooling effect on the body. (See the Solutions and Electrochemistry Lecture in the *Chemistry* manual to learn about evaporation in general, and the Muscle, Bone and Skin Lecture in this manual to learn more about how evaporation cools the body). In addition, **nasal and tracheal capillary beds** participate in a method of thermoregulation used by the cardiovascular system, described later in this lecture.

Anatomy of the Respiratory System

The respiratory system provides a path for gas exchange between the external environment and the blood (Figure 4.1). The movement of oxygen and carbon dioxide through the respiratory system is governed by changes in the pressure differential between the chest cavity and external environment, as will be described. Air enters through the nose and then moves through the pharynx, larynx, trachea, bronchi, bronchioles, and into the alveoli, where oxygen is exchanged for carbon dioxide from the blood.

The *nasal cavity* is the space inside the nose. Various structures and substances within the nasal cavity act to filter, moisten, and warm incoming air. **Nasal hairs** at the front of the cavity trap large dust particles. **Mucus** secreted by goblet cells traps smaller dust particles that were able to bypass the coarse nasal hair and also moistens the air. Capillaries within the nasal cavity warm the air. **Cilia** move mucus and dust back toward the pharynx to be removed by spitting or swallowing. In this way, the nasal cavity participates in the immune function of the respiratory system, preventing some potentially harmful substances from entering the body.

The *pharynx* (or throat) functions as a passageway for food and air.

The *larynx* contains the vocal cords. It sits behind the *epiglottis*, which is the cartilaginous structure that rises to block the opening of the trachea during swallowing, preventing the entry of food into the airway. Any nongaseous material that enters the larynx triggers a coughing reflex and is forced back out.

The **trachea** (or windpipe) lies in front of the esophagus. It is composed of ringed cartilage covered by ciliated mucous cells. Like in the nasal cavity, mucus and cilia in the trachea collect particulate matter and usher it back out, in this case toward the pharynx. Before entering the lungs, the trachea splits into the right and left **bronchi**. Each bronchus branches many more times to become tiny **bronchioles**. Bronchioles terminate in grape-like clusters called *alveolar sacs*, composed of tiny **alveoli**. Oxygen diffuses from each alveolus into an adjacent capillary, where it is picked up by red blood cells. The red blood cells release carbon dioxide, which diffuses into the alveolus and is expelled upon exhalation.

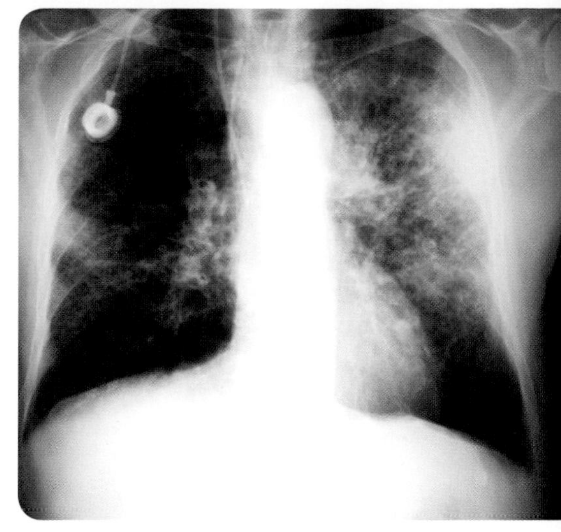

X-ray of a lung cancer.

Why does the airway undergo so much branching into smaller structures? The answer is that branching creates a huge amount of surface area for gas exchange. The cardiovascular system branches into capillaries for similar reasons.

FIGURE 4.1 Respiratory System

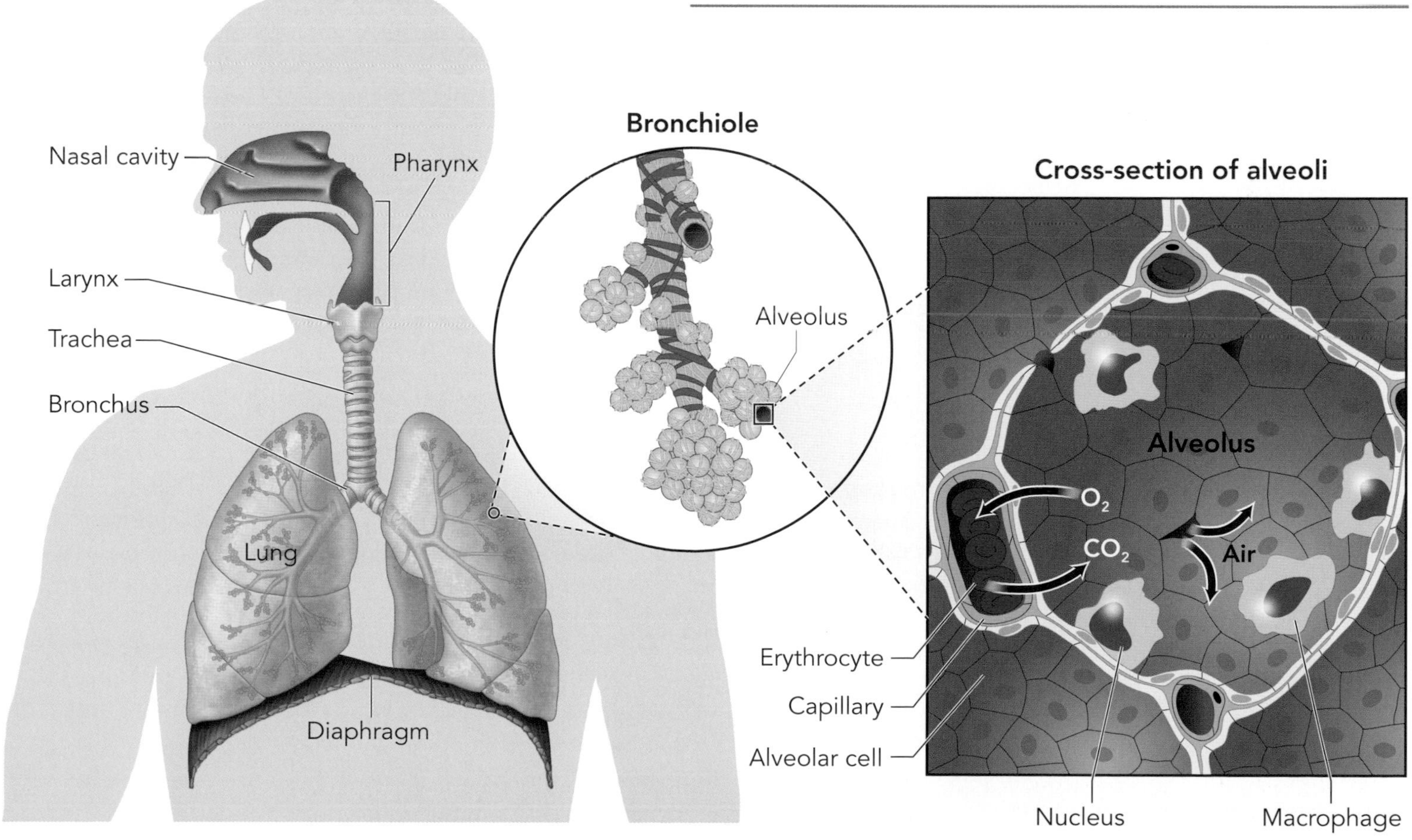

Mechanics of Respiration

Inspiration and expiration of air is governed by changing **differential pressures** in the chest cavity and airway. Air flows from the external environment into the body and back out again according to changes in the pressure of the airway relative to atmospheric pressure. When the airway and alveoli are at a negative gauge pressure, air flows inwards; when the airway and alveolar pressure become greater than atmospheric pressure, air flows back out to the environment.

To understand the mechanisms of inspiration and expiration, it is first necessary to examine the pressures in the respiratory system at rest. The lungs have a natural tendency to collapse inward due to their elasticity, while the **rib cage** tends to expand outwards, in the opposite direction. The combination of these effects causes the pressure of the chest cavity at rest to be negative compared to atmospheric pressure. However, alveolar pressure is equal to atmospheric pressure, so air does not flow inwards.

Inspiration occurs when the **medulla oblongata** of the midbrain signals the **diaphragm** to contract. The diaphragm is a thin sheet of skeletal muscle that is innervated by the phrenic nerve. The diaphragm is dome-shaped in a relaxed state, but it flattens upon contraction, expanding the chest cavity. *Intercostal muscles* (rib muscles) also help to expand the chest cavity. As a result, pressure in the air cavity becomes even more negative than it is at rest, and pressures in the airway and alveoli also become negative. Atmospheric pressure forces air into the lower-pressure area of the lungs.

When the medulla stops signaling the diaphragm to contract, it relaxes, causing the chest cavity to shrink. The **elasticity** (also called **resiliency**) of the lungs, along with the increased pressure in the chest cavity, forces air out of the body. Note that expiration as described here is a passive process. When forced expiration is needed, such as during exercise, the shrinking of the chest cavity is aided by the abdominal muscles and a set of intercostal muscles distinct from those involved in inspiration.

During inspiration, the ability of the lungs to expand in response to changing pressure is counteracted by **surface tension** in the alveoli. Each tiny alveolus has a high ratio of surface area to volume, providing a large surface for the diffusion of a small amount of gas. This ratio facilitates the diffusion of gas into the surrounding capillaries, but it also leads to the problem of surface tension. The strong intermolecular forces of the thin layer of water that cover the inner surface of the alveolus create high surface tension, which in turn generates pressure that tends to collapse the alveolus and opposes expansion of the lungs. However, the alveoli contain a type of cell that produces *surfactant*, a material composed of amphipathic phospholipids. Surfactant coats the alveolar surface and breaks up the intermolecular forces between water molecules, reducing surface tension. During inspiration, surfactant decreases surface tension and assists the expansion of the lungs.

FIGURE 4.2 Respiration and the Diaphragm

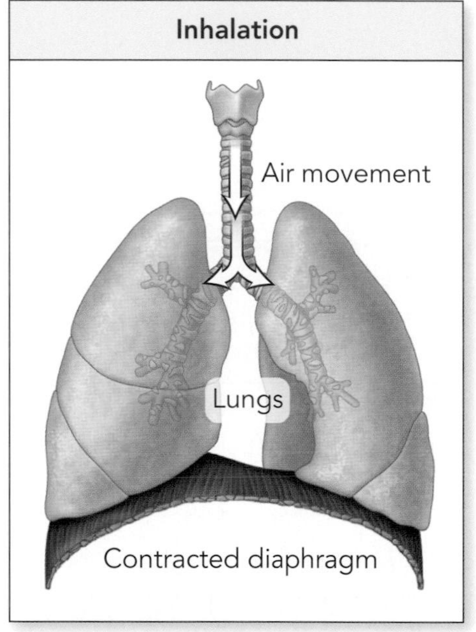

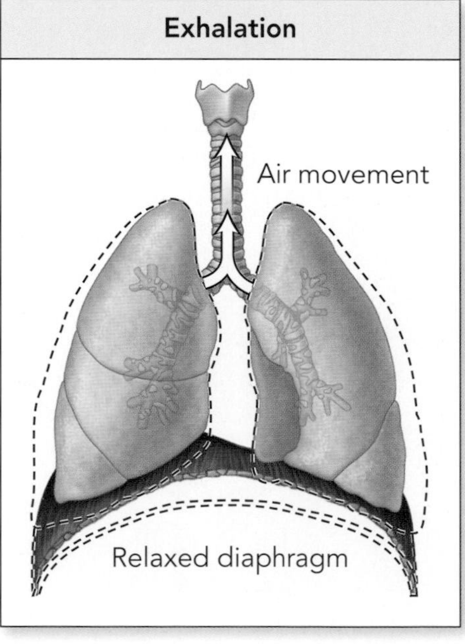

You know from the universal gas law ($PV = nRT$) that a change in volume causes an inverse change in pressure when other factors are held constant, so the changes in volume of the chest cavity during respiration must cause corresponding pressure changes. Just remember that air, like any other fluid, will generally flow to an area of lower pressure. During inspiration, the chest cavity volume expands, decreasing the air pressure to a lower pressure than that of the atmosphere, with the result that air flows into the lungs. During expiration, the chest cavity volume decreases, increasing the pressure in the chest cavity until it becomes greater than atmospheric pressure, with the result that air flows back out of the body.

FIGURE 4.3 Alveolar and Intrapleural Pressure

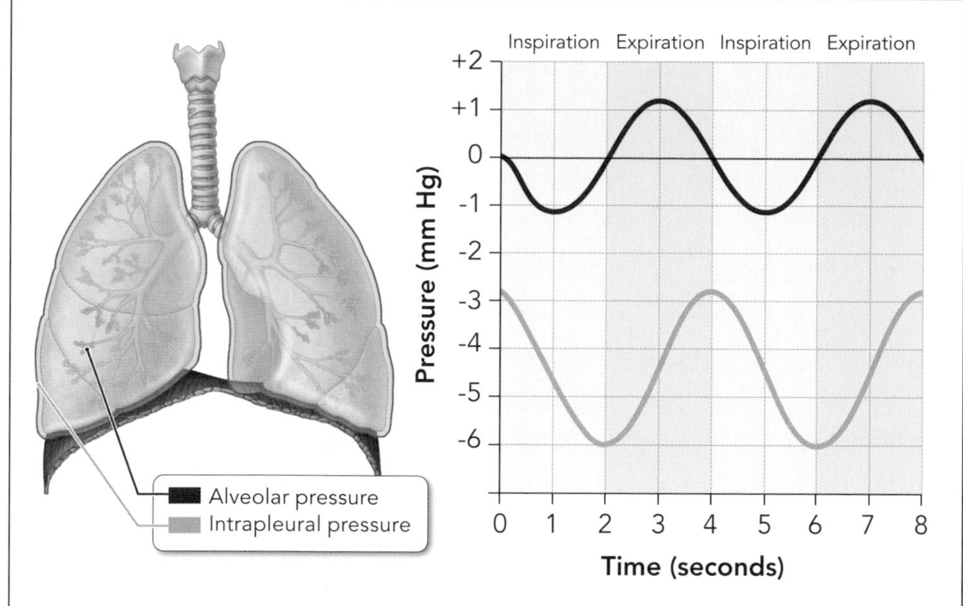

Apply Key 2: The differences in pressures drive the flow of fluids like air and blood. When you breathe in, the pressure inside the lungs is lower than the pressure of the outside air, allowing air to flow into the body. When you exhale, the pressure inside the lungs is greater than the pressure outside, expelling the air from the body.

The Chemistry of Gas Exchange

Alveolar gas exchange is governed by principles of gas and solution chemistry. It occurs passively, through **diffusion**, the tendency of a gas to travel towards an area of lower concentration or pressure. This section will describe how differing partial pressures of oxygen and carbon dioxide ensure their movement in the correct direction. In addition, given the fact that these gases must be transported in the liquid substance of the blood, gas solubility and transport will be considered.

Typically, the air we breathe in is 79% nitrogen and 21% oxygen, with negligible amounts of other trace gases. Exhaled air is 79% nitrogen, 16% oxygen, and 5% carbon dioxide, along with trace gases. Inside the lungs, the partial pressure of oxygen (pO_2) is approximately 110 mm Hg, while the partial pressure of carbon dioxide (pCO_2) is approximately 40 mm Hg. The deoxygenated blood contained in the pulmonary capillaries has a much lower pO_2, at 40 mm Hg, and a higher pCO_2 of 46 mm Hg. The **differential partial pressures** of oxygen and carbon dioxide cause the diffusion of oxygen into the capillaries and the diffusion of carbon dioxide into the alveoli. The high partial pressure of oxygen in the alveoli compared to the low partial pressure in the deoxygenated blood in adjacent capillaries promotes the diffusion of oxygen into the capillaries. Meanwhile, because the blood has transported carbon dioxide away from the cells as waste, it has a high partial pressure of carbon dioxide compared to that of the alveoli; as a result, CO_2 diffuses into the alveoli for expiration. Remember, small nonpolar molecules like O_2 and CO_2 can diffuse passively through a cellular membrane.

The rate at which gases diffuse across a membrane is governed by *Fick's Law*. Fick's Law states that the rate of diffusion is directly proportional to the surface area and differential partial pressure across the membrane, and is inversely proportional to the thickness of the membrane across which diffusion occurs. The thin membranes of the alveoli, together with the branching of the respiratory system, create a large surface area that encourages diffusion.

To understand how gases diffuse in the alveoli, just remember that gases will diffuse to an area of lower pressure. You can even think of this in terms of Le Châtelier's Principle, where dissolved gas in solution is on one side of the reaction and the gas vapor is on the other. When pressure increases on one side, more of the gas shifts in the reverse direction.

Once oxygen has diffused across the alveolar and blood vessel membranes, the amount of the gas that can be dissolved into the blood is described by Henry's Law. Henry's Law states that the amount of a gas that can be dissolved in solution is directly proportional to the partial pressure of gas in equilibrium with the liquid: $C = P \times \text{solubility}$ (where C is the concentration of the dissolved gas, and P is the partial pressure). As the partial pressure of a gas increases, the concentration of the gas dissolved in solution also increases. As oxygen diffuses into the capillaries from the alveoli, the pO_2 of the blood rises until it reaches the high partial pressure of O_2 found in the alveoli. This process facilitates the dissolution of oxygen in the blood, as described by Henry's Law. Still, the solubility of oxygen in the blood is not nearly high enough for sufficient oxygen to be dissolved for transport to the body's cells. The majority of oxygen in the blood is transported by hemoglobin.

98% of oxygen in the blood binds rapidly and reversibly with the protein **hemoglobin** found inside the erythrocytes, forming **oxyhemoglobin**. Hemoglobin is composed of four polypeptide subunits, each with a single heme cofactor. The heme cofactor is an organic molecule with an atom of iron at its center. Each of the four iron atoms in hemoglobin can bind with one O_2 molecule. When one of the iron atoms in hemoglobin binds an O_2 molecule, oxygenation of the other heme groups is accelerated. Similarly, release of an O_2 molecule by any of the heme groups accelerates release by the others. This phenomenon is called **cooperativity**.

Biological Molecules and Enzymes
BIOLOGY 1

The oxygen dissociation curve of hemoglobin (Figure 4.4) shows the percent of hemoglobin that has bound oxygen at various partial pressures of oxygen. Although hemoglobin is not an enzyme, the dissociation curve can be viewed as a modified Michaelis-Menten enzyme curve. Recall that the rate of an enzyme-catalyzed reaction increases with increasing concentration of the substrate and then levels off, due to a lack of enzymes that are available to bind the substrate. In a similar fashion, the percent saturation of hemoglobin increases as the partial pressure of oxygen increases and then levels off when all of the subunits have bound oxygen. (In the arteries of a typical person breathing room air, the oxygen saturation is 97%; the flat portion of the curve in this region shows that small fluctuations in oxygen pressure have little effect.) The dissociation curve demonstrates the **oxygen affinity** of hemoglobin by showing the percentage bound at a given partial pressure, just as the slope of a Michaelis-Menten curve indicates the enzymatic affinity for the substrate. However, the shape of the oxygen dissociation curve is impacted by the cooperativity effect, which causes the affinity of hemoglobin to increase as the subunits bind O_2. As a result, the O_2 saturation of hemoglobin increases sigmoidally (in an S-shape) rather than linearly.

Myoglobin is a single chain protein that stores oxygen in muscle cells. It does not experience the Bohr effect like hemoglobin does and does not exhibit cooperativity.

In addition to the cooperativity affect, environmental conditions affect the affinity of hemoglobin for oxygen, just as the presence of competitors and other types of environmental influences can change the affinity of an enzyme for its substrate. The oxygen saturation of hemoglobin depends on the carbon dioxide pressure, pH, and temperature of the blood. A **rightward shift of the oxygen dissociation curve** occurs in response to an increase in carbon dioxide pressure, hydrogen ion concentration, or temperature. In each case, the rightward shift reflects hemoglobin's lowered affinity for oxygen and can also be viewed as a downward shift. An increased demand for oxygen leads to a shift that allows for a lower affinity of hemoglobin for oxygen, which leads to the release of oxygen for increased delivery to the tissues. When the shift is due to increasing hydrogen ion concentration (i.e. decreased pH), it is called the *Bohr shift*. As will be described, increasing CO_2 concentration causes lowered pH in the bloodstream, so the Bohr shift encompasses both increasing CO_2 and increasing hydrogen ion concentration. CO_2 and hydrogen ions affect the oxygen dissociation curve through allosteric effects: they bind to deoxygenated hemoglobin and cause a change in shape that then discourages the binding of oxygen.

It is easy to remember that increases in CO_2, hydrogen ion concentration, and temperature all cause the oxygen dissociation curve to shift to the right if you realize that they all reflect the body's increased need for oxygen. When metabolic demands are high, such as during exercise, more oxygen is needed for aerobic respiration. At the same time, CO_2 builds up as a byproduct of metabolism (causing hydrogen ion concentration to increase). Finally, heat is released during the process of metabolism, so increased temperature also signals the need for increased oxygen delivery.

2,3-DPG, also known as 2,3-BPG, is a chemical found in red blood cells that also shifts the curve to the right/downwards by binding to deoxygenated hemoglobin and decreasing its affinity for oxygen. 2,3-DPG increases in response to low-oxygen environments, such as high altitudes, to ensure that the tissues still receive sufficient oxygen. Note that by lowering the affinity of hemoglobin for oxygen when oxygen is scarce, 2,3-DPG contributes to the flattened slope of the oxygen dissociation curve at low pO_2.

FIGURE 4.4 Oxygen Dissociation Curve

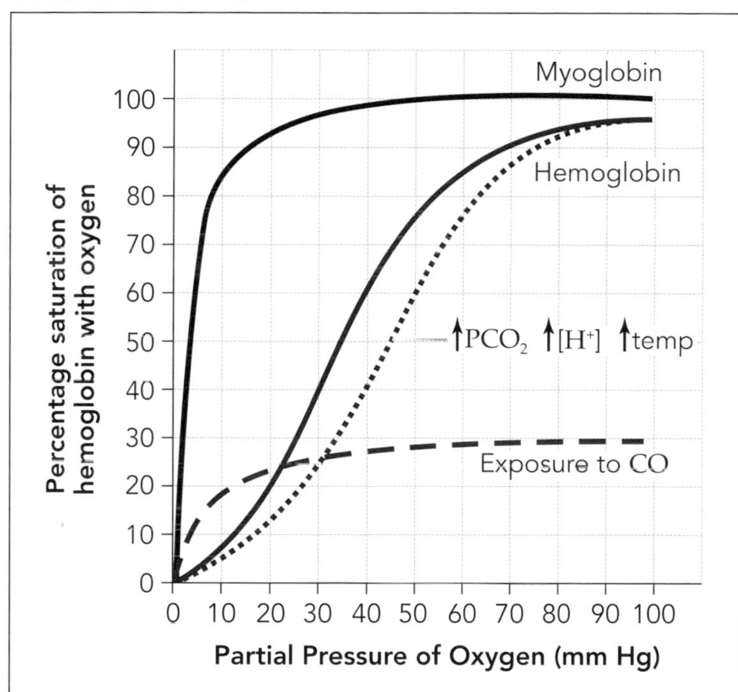

MCAT® THINK

In contrast to carbon dioxide, which binds to a different site on hemoglobin from oxygen, carbon monoxide acts as a competitive inhibitor and prevents the binding of oxygen. The affinity of hemoglobin for carbon monoxide is more than 200 times greater than its affinity for oxygen. However, the presence of CO shifts the oxygen-dissociation curve to the left, reflecting the remaining sites' heightened affinity for oxygen while the maximum saturation percentage of hemoglobin is reduced relative to the amount of CO in the blood. The overall ability of oxygen to be carried bound to hemoglobin in the blood is decreased, while the oxygen that is able to bind does not unload at the tissues as it should, with the net result that oxygen delivery to the tissues is severely limited. In cases of carbon monoxide poisoning, pure oxygen can be administered to displace the CO from hemoglobin.

Carbon dioxide has its own dissociation curve, which relates blood content of carbon dioxide to carbon dioxide pressure. The greater the partial pressure of carbon dioxide, the greater the concentration of carbon dioxide dissolved in the blood, increasing the amount capable of binding to hemoglobin. However, when hemoglobin becomes saturated with oxygen, it has a lowered affinity for carbon dioxide. In other words, just as the binding of carbon dioxide to hemoglobin reduces its affinity for oxygen, the oxygenation of hemoglobin lowers its affinity for carbon dioxide. This is called the *Haldane effect*. The Haldane effect facilitates the transfer of carbon dioxide from the blood to the lungs, and from the tissues to the blood. Recall that oxygen pressure, and therefore the percent of hemoglobin saturated with oxygen, is higher in the lungs. As a result, carbon dioxide is released from hemoglobin in the lungs, facilitating its exhalation from the body. On the other hand, hemoglobin will have a lower oxygen saturation percentage at the lower oxygen pressure found in the tissues, facilitating the binding of carbon dioxide for transport to the lungs. The off-loading of oxygen allows reduced hemoglobin (that is, hemoglobin without oxygen) to act as a blood buffer by accepting excess protons produced by the conversion of carbon dioxide to bicarbonate ion.

Use your understanding of enzyme inhibition to remember how the oxygen dissociation curve shifts. Just as inhibitors that lower enzyme affinity for the substrate cause a downward shift, environmental conditions that decrease the affinity of hemoglobin for oxygen cause a rightward/downward shift in the oxygen dissociation curve.

Oxygen pressure is typically 40 mm Hg in body tissues, lower than in the lungs because oxygen has been used up by aerobic respiration. Carbon dioxide, however, is built up as a waste product and is at a higher partial pressure in the tissues than in the blood. As the blood moves through the systemic capillaries, oxygen diffuses into the tissues, and carbon dioxide diffuses into the blood. Carbon dioxide is carried by the blood in three forms:

1. dissolved in solution,
2. as bicarbonate ion, and
3. in carbamino compounds (combined with hemoglobin and other proteins).

Ten times as much carbon dioxide is carried as bicarbonate as in either of the other forms. The formation of bicarbonate ion is catalyzed by the enzyme carbonic anhydrase in the reversible reaction:

$$CO_2 + H_2O \rightleftharpoons HCO_3^- + H^+$$

For the MCAT®, memorize this equation and understand how it proceeds in the forward and reverse directions! Myoglobin is a single chain protein that stores oxygen in muscle cells. Notice that is does not shift similarly to the hemoglobin curve because it cannot exhibit cooperativity.

As with any reversible reaction, the predominant direction can be predicted according to Le Châtelier's Principle, where an increased ratio of reactant to product favors the forward direction and vice versa. The reaction above is best understood by considering the amount of CO_2 present. In the tissues, where there is a high concentration of CO_2 created as a metabolic waste product (and therefore a high partial pressure of CO_2), the forward reaction dominates, forming bicarbonate ion. The low partial pressure of CO_2 in the lungs then shifts the reaction in the reverse direction, allowing the conversion of the bicarbonate ion to CO_2 for expiration. The carbonic anhydrase reaction is also important for buffering blood pH. When the blood becomes too acidic, the reverse reaction is favored, raising the pH of the blood by decreasing the concentration of hydrogen ions. When the blood becomes too basic, the forward reaction is favored, lowering the pH of the blood by increasing the concentration of hydrogen ions.

Carbonic anhydrase is present inside the red blood cells but not in the plasma. When carbon dioxide is absorbed from the tissue by erythrocytes, bicarbonate ion builds up in the red blood cells. As a result, bicarbonate diffuses out of the erythrocytes and into the plasma. To prevent a buildup of negative charge in the plasma, chloride ions move into the red blood cell in exchange for bicarbonate ions in a phenomenon called the *chloride shift* (Figure 4.5). The entry of chloride ions offsets the buildup of hydrogen ions in the erythrocyte, contributing to electrical neutrality and helping to prevent the buildup of acidity. The chloride shift occurs in the lungs as well, but in the opposite direction. Chloride ions flow into the plasma in exchange for bicarbonate ions, which diffuse out of the plasma and back into the red blood cells. Bicarbonate ion is then converted back into CO_2 that can be expired.

The overall process of carbon dioxide transport can be summarized as follows: once CO_2 has diffused into red blood cells from the tissues, carbonic anhydrase catalyzes its conversion to bicarbonate. Bicarbonate then diffuses down its concentration gradient into the plasma. Once the blood reaches the lungs, bicarbonate diffuses back into the erythrocytes and is converted back into CO_2. In both the lungs and the tissues, chloride ions diffuse in the opposite direction of bicarbonate ions to prevent a buildup of negative charge.

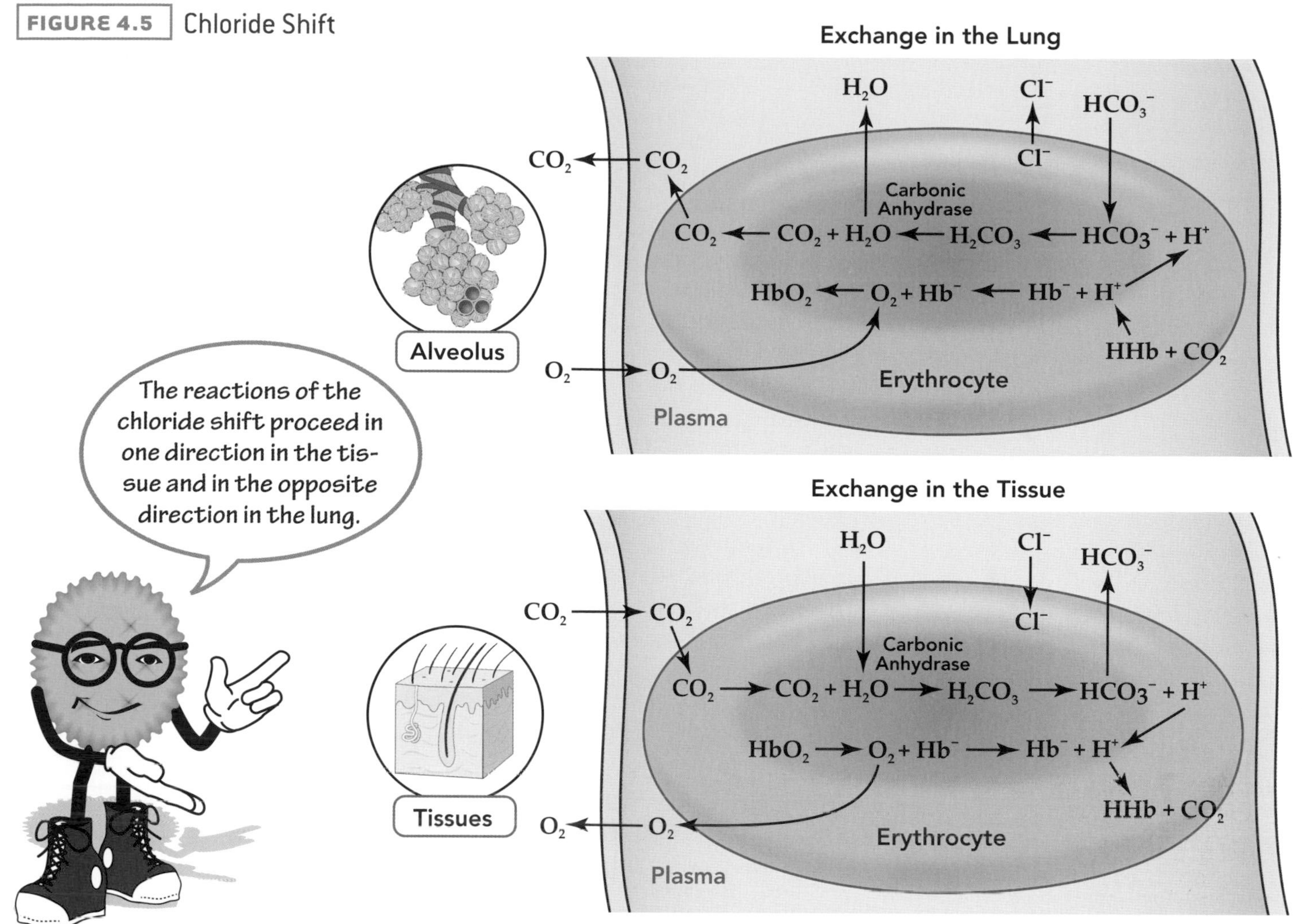

FIGURE 4.5 Chloride Shift

You don't need to memorize too many of the details of gas exchange; however, it is crucial to understand the mechanisms. Know the effects that temperature, pH, and carbon dioxide pressure have on hemoglobin. Be sure that you can read a dissociation curve and predict shifts.

Now that you have a greater understanding of how hemoglobin interacts with oxygen and carbon dioxide, let's revisit the shifts of the oxygen dissociation curve discussed earlier in this section. Recognize that the Bohr shift is associated with both pH and CO_2, since the hydrolysis of carbon dioxide leads to an increased concentration of hydrogen ions. Also note that the decreased affinity of hemoglobin for oxygen at the high carbon dioxide level found in the tissues assists in the release of oxygen that then diffuses into the cells. In a similar effect, 2,3-DPG increases when the body is at a high altitude, shifting the curve to the right and facilitating increased oxygen delivery to the tissues when oxygen is low.

Respiratory centers in the medulla regulate breathing rate and can respond to changes in the chemical composition of the blood, particularly changes in the partial pressure of CO_2. An inspiratory center stimulates inspiration by signaling the diaphragm to contract. When necessary due to exercise or other causes of active breathing, an expiratory center signals muscles that assist in forced exhalation. Control of breathing rate by the medulla is affected by the input of **central chemoreceptors** located in the medulla and **peripheral chemoreceptors** in the carotid arteries and aorta. (Chemoreceptors are a type of sensory receptor that bind chemicals and signal higher centers in the nervous system to produce a response.) Central and peripheral chemoreceptors monitor carbon dioxide concentration in the blood and increase the breathing rate when the level is too high. Nervous system regulation of respiration rate has high **CO_2 sensitivity**; a small increase in the partial pressure of CO_2 quickly triggers an increased respiration rate to allow more carbon dioxide to be expired. Since increased CO_2 and decreased pH go hand in hand, CO_2 levels must be tightly controlled to avoid a dangerous decrease in the pH of the blood. By contrast, the partial pressure of oxygen dissolved in the bloodstream must drop significantly in order to change respiration rate.

Understand the connection between rate of breathing and carbon dioxide, pH, and oxygen levels in the blood. When you see carbon dioxide in the biological environment, always think of increased acidity due to the hydrogen ions that are produced by the hydration of CO_2.

Oxygen concentration is monitored mainly by peripheral chemoreceptors, while CO_2 concentration is monitored by central chemoreceptors. Both peripheral and central chemoreceptors respond to pH changes. The respiratory system exerts **pH control** by adjusting breathing rate, and thus the partial pressure of CO_2 in the body, in response to potentially dangerous disturbances to the pH of the bloodstream. In the case of acidosis (too much acid in the blood), the body compensates by increasing the breathing rate, thereby expelling carbon dioxide and raising the pH of the blood.

What about all that nitrogen? What effect does nitrogen have on the body?

Remember your chemistry. Nitrogen is extremely stable due to its strong triple bond. It diffuses into the blood, but doesn't react with the chemicals in the blood. However, people that go diving must be careful. As the pressure increases with depth, more nitrogen diffuses into the blood. When divers come back up, the pressure decreases and the gas volume increases. If they don't allow enough time for the nitrogen to diffuse out of the blood and into the lungs, the nitrogen will form bubbles. Among other problems, these bubbles may occlude (block) vessels, causing decompression sickness, also known as 'the bends.'

These questions are NOT related to a passage.

Question 73

Alkalosis is increased blood pH resulting in a leftward shift of the oxyhemoglobin dissociation curve. Which of the following might cause alkalosis?

- O **A.** Hypoventilation
- O **B.** Hyperventilation
- O **C.** Breathing into a paper bag
- O **D.** Adrenal steroid insufficiency

Question 74

Which of the following would most likely occur in the presence of a carbonic anhydrase inhibitor?

- O **A.** The blood pH would increase.
- O **B.** The carbamino hemoglobin concentration inside erythrocytes would decrease.
- O **C.** The rate of gas exchange in the lungs would decrease.
- O **D.** The oxyhemoglobin concentration inside erythrocytes would increase.

Question 75

Which of the following will most likely occur during heavy exercise?

- O **A.** Blood pH will decrease in the active tissues.
- O **B.** Less oxygen will be delivered to the tissues due to increased cardiac contractions that result in increased blood velocity.
- O **C.** Capillaries surrounding contracting skeletal muscles will constrict to allow increased freedom of movement.
- O **D.** The respiratory system will deliver less nitrogen to the blood.

Question 76

An athlete can engage in blood doping by having blood drawn several weeks before an event, removing the blood cells, and having them reinjected into her body a few days before an athletic activity. Blood doping is most likely an advantage to athletes because:

- O **A.** the increased concentration of immune cells in the blood after reinjection can decrease the chances of becoming ill just before the competition.
- O **B.** the increased red blood cell count in the blood after reinjection can facilitate greater gas exchange with the tissues.
- O **C.** the increased blood volume after reinjection can ensure that the athlete maintains adequate hydration during the event.
- O **D.** the decreased red blood cell count of the blood in the weeks before the competition can facilitate training by decreasing the viscosity of the blood.

Question 77

Carbon dioxide partial pressure:

- O **A.** increases in the blood as it travels from the systemic venules to the inferior vena cava.
- O **B.** increases in the blood as it travels from the pulmonary arteries to the pulmonary veins.
- O **C.** is greater in the blood in the systemic capillary beds than in the alveoli of the lungs.
- O **D.** is greater in the blood in the systemic capillary beds than in the systemic tissues.

Question 78

At high altitude, water vapor pressure in the lungs remains the same and carbon dioxide pressure falls slightly. Oxygen pressure falls. The body of a person remaining at high altitudes for days, weeks, and even years will acclimatize. All of the following changes assist the body in coping with low oxygen EXCEPT:

- O **A.** increased red blood cells.
- O **B.** decreased vascularity of the tissues.
- O **C.** increased pulmonary ventilation.
- O **D.** increased diffusion capacity of the lungs.

Question 79

In an asthma attack, a patient suffers from difficulty breathing due to constricted air passages. The major causative agent is a mixture of leukotrienes called slow-reacting substance of anaphylaxis. During an asthma attack, slow reacting substance of anaphylaxis most likely causes:

- O **A.** smooth muscle spasms of the bronchioles.
- O **B.** cartilaginous constriction of the trachea.
- O **C.** edema in the alveoli.
- O **D.** skeletal muscle spasms in the thorax.

Question 80

Sustained heavy exercise results in all of the following changes to blood chemistry except:

- O **A.** lowered pH.
- O **B.** raised CO_2 tension.
- O **C.** increased temperature.
- O **D.** decreased carboxyhemoglobin.

4.3 The Blood

FIGURE 4.6 Blood Composition

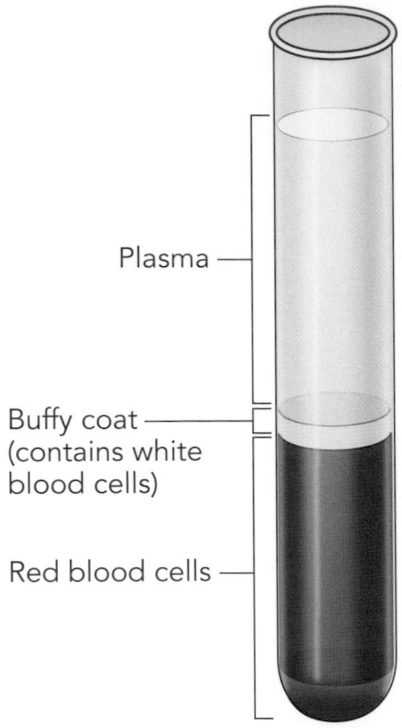

Plasma

Buffy coat (contains white blood cells)

Red blood cells

Blood has many important functions in the body. It plays a crucial role in the respiratory system by transporting oxygen to the tissues and carrying away carbon dioxide. It also regulates the extracellular environment of the body by transporting nutrients, waste products, and hormones. Additionally, it is important for the thermoregulation of the body. Dilation of peripheral blood vessels allows the body to radiate heat and cool off, while constriction of peripheral blood vessels prevents heat loss in cold temperatures. The blood is **connective tissue**, so like any connective tissue, it contains cells and a matrix. It travels throughout the body via the cardiovascular system, as will be described in the following section.

Blood contains three main components, which separate when a blood sample is placed in a centrifuge (Figure 4.6): 1. the plasma, 2. red blood cells, and 3. the buffy coat (white blood cells).

Think of the blood as the body's highway for the transport of various substances. It is a solution that exists for the purpose of carrying cells and molecules to their destinations.

Plasma contains the matrix of the blood, which includes water, ions, urea, ammonia, proteins, and other organic and inorganic compounds. The body regulates the overall volume of blood by altering the amount of water in the plasma. This regulation of blood volume contributes to the control of blood pressure. Important proteins contained in the plasma are albumin, immunoglobulins, and clotting factors. **Albumins** transport fatty acids and steroids. They also help regulate the osmotic pressure of the blood, facilitating transfer of substances across capillary walls. Immunoglobulins (also called antibodies) are a major component of the immune system and are discussed later in this lecture. Plasma from which the clotting protein fibrinogen has been removed is called **serum**. Albumin, fibrinogen, and most other plasma proteins are formed in the liver. Gamma globulins that constitute antibodies are made in the lymph tissue. An important function of plasma proteins is to act as a source of amino acids for tissue protein replacement.

Know that the job of erythrocytes is to deliver oxygen to the tissues and to remove carbon dioxide. Also, remember that red blood cells do not have nuclei.

Erythrocytes (red blood cells) are essentially bags of hemoglobin. They have no organelles, not even nuclei, which means they do not undergo mitosis. They are disk-shaped vesicles whose main function is to transport O_2 and CO_2. Squeezing through capillaries wears out their plasma membranes in about 120 days. Most worn out red blood cells are removed from circulation and destroyed as they squeeze through channels in the spleen or, to a lesser extent, in the liver. The percentage by volume of red blood cells is called the **hematocrit**. The hematocrit is typically 35-50% and is greater in men than women.

Unlike erythrocytes, **leukocytes** (white blood cells) do contain organelles, but do not contain hemoglobin. Their function is to protect the body from foreign invaders.

FIGURE 4.7 Granulocytes

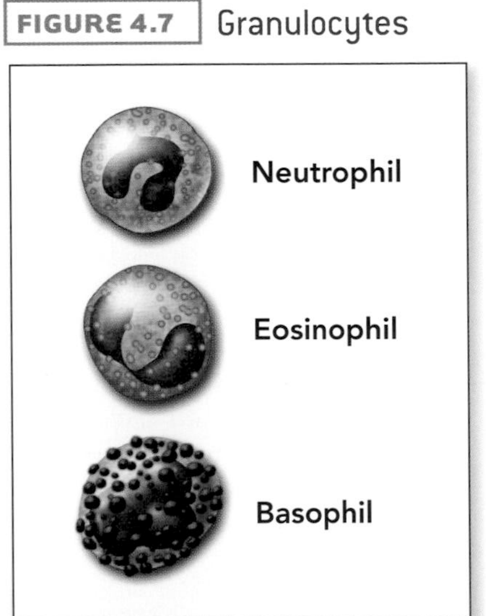

All blood cells differentiate from the same type of precursor, a **stem cell** residing in the bone marrow. Erythrocytes lose their nuclei while still in the marrow. After entering the bloodstream as reticulocytes, they lose the rest of their organelles within 1 or 2 days. Leukocyte formation is more complex due to the existence of many different types. The *granular leukocytes* (*granulocytes*) are neutrophils, eosinophils, and basophils. With respect to dyeing techniques, neutrophils are neutral to acidic and basic dyes, eosinophils stain in acidic dyes, and basophils stain in basic dyes. Generally, granulocytes remain in the blood for only 4 to 8 hours before they are deposited in the tissues, where they live for 4 to 5 days. Neutrophils are the most abundant granular leukocyte. They serve as first responders to the scene of infection to kill foreign pathogens and recruit other immune cells. *Agranular leukocytes* (*agranulocytes*) include monocytes, lymphocytes, and megakaryocytes. Once deposited in the tissues, monocytes become macrophages and may live for months to years. Lymphocytes may also live for years.

Platelets are small portions of membrane-bound cytoplasm torn from megakaryocytes. Like red blood cells, platelets do not have nuclei. However, they do contain actin and myosin, mitochondria, and residual pieces of the Golgi body and endoplasmic reticulum. Platelets are capable of making prostaglandins and some important enzymes. Healthy individuals have many platelets in their blood. Platelets have a half-life of 8–12 days in the blood.

Platelets have an important function in **coagulation**, which functions to minimize blood loss and facilitate healing when blood vessels are damaged. When platelets come into contact with injured endothelium, they become sticky and begin to swell, releasing various chemicals. They also activate other platelets. The platelets stick to the endothelium and to each other, forming a loose *platelet plug*. The accumulated platelets assist in triggering an enzymatic cascade that ultimately results in the formation of a **blood clot**. The most important step in coagulation is the polymerization of the plasma protein **fibrinogen** to form fibrin threads that attach to the platelets and form a tight plug. Coagulation begins to appear in seconds in small injuries and 1 to 2 minutes in larger injuries.

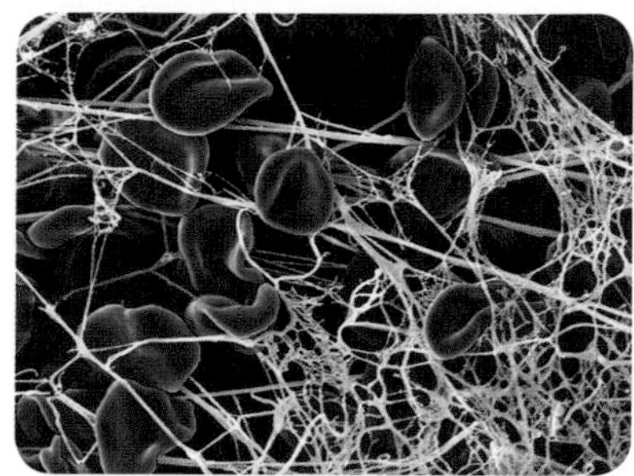

When blood clots, erythrocytes are trapped in a fibrin mesh (yellow). The production of fibrin is triggered by platelets, which are activated when a blood vessel is damaged. The fibrin binds the various blood cells together, forming a solid blood clot.

The leukocyte composition in the blood is shown below:

Neutrophils.........62%
Lymphocytes......30%
Monocytes..........5.3%
Eosinophils..........2.3%
Basophils.............0.4%

Notice that granulocytes live for a very short time, whereas agranulocytes live for a very long time. This is because granulocytes function nonspecifically against all infective agents, whereas most agranulocytes work against specific agents of infection. Agranulocytes need to hang around in case the same infective agent returns; granulocytes multiply quickly against any infection, and then die once the infection is gone. Make sure you understand this distinction, but don't worry too much about the other details.

4.4 The Cardiovascular System and Blood Pressure

The cardiovascular system consists of the heart, blood, and blood vessels. It is a transport system that pumps blood throughout the body, carrying out the functions of extracellular environment regulation and the delivery of nutrients, oxygen, hormones, and immune cells to the tissues. The cardiovascular system also transports metabolic waste such as carbon dioxide away from the cells. As described in the previous section, the cardiovascular system works closely with the respiratory system to meet cells' metabolic needs by delivering oxygen and preventing a buildup of metabolic byproducts.

Understand Key 1: The respiratory and circulatory systems provide the way for cells to meet their metabolic needs. The respiratory system brings in oxygen and gets rid of carbon dioxide. The circulatory system transports nutrients like glucose and fatty acids to cells and takes away waste products to be eliminated in the bile and urine.

A significant feature of the cardiovascular system is its ability to react to internal and external conditions by directing the flow of blood in response to **nervous and endocrine control**. Epinephrine and norepinephrine, released by neurons of the sympathetic nervous system and by the adrenal medulla of the endocrine system, can direct blood by causing the constriction of blood vessels. The division of the cardiovascular system into ever smaller branches allows strict regulation of the flow of blood according to the metabolic needs of the tissues. The physical structures of the blood vessels of each section of the cardiovascular system are particularly suited to specific functions, as will be described.

Use the names of the atria and ventricles to help you remember their functions. An atrium is a big room that could be used as a waiting area; the atria of the heart are like waiting rooms where blood collects before moving on to the ventricles. The VENTricles are pumps that VENT blood to the rest of the body.

The cardiovascular system is said to be a **closed circulatory system** because it contains no openings for the blood to leave the vessels. For the MCAT®, it is important to be able to trace the circulatory path of the blood through two distinct sections of the cardiovascular system that are separated by the heart. The first half of the circulation, called the **systemic circulation**, directs oxygenated blood to the tissues and then returns deoxygenated blood to the heart. Beginning with the **left ventricle**, blood is pumped through aortic valve into the **aorta**. From the aorta branch many smaller **arteries**, which themselves branch into still smaller **arterioles**, which branch into still smaller **capillaries**. Blood from the capillaries is collected into **venules**, which collect into larger **veins** that collect again into the **superior** and **inferior venae cavae**. The venae cavae empty into the **right atrium** of the heart.

The second half of the circulation is called the **pulmonary system** because it transports blood to the lungs for oxygenation. From the right atrium, blood is squeezed through the tricuspid valve into the **right ventricle**. The right ventricle pumps blood through the pulmonary valve to the **pulmonary arteries** and arterioles and then to the capillaries of the lungs. The pulmonary arteries are the only arteries in the adult body that carry deoxygenated blood. From the capillaries of the lungs, blood collects in venules, then in veins, and finally in the **pulmonary veins** leading to the heart. (True capillaries branch off arterioles, and do not represent the only route between an arteriole and venule.) The pulmonary veins empty into the **left atrium**, which leads to the left ventricle through the mitral valve.

FIGURE 4.8 Circulation of Blood

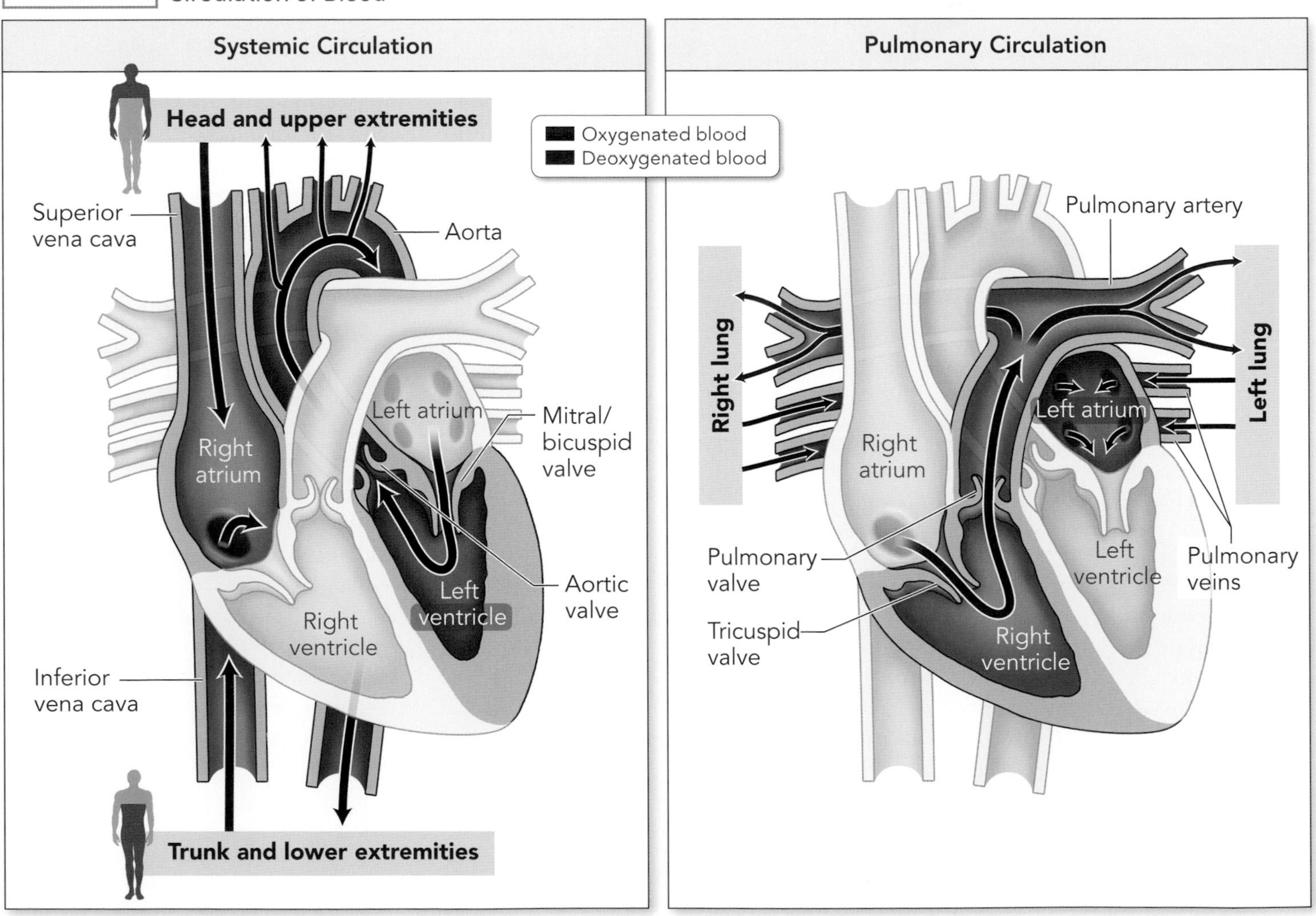

It is crucial to know that the heart has four chambers and to know the bolded names of the different parts of the heart and types of blood vessels, but don't be too concerned with memorization. Instead, concentrate on function. You can think of the ventricles as two neighboring pumps with differing functions and corresponding physical features. Each pump serves one of the sections of the circulatory system. The left ventricle has to propel the blood through the systemic circulation, so it contracts with more force; its muscular wall is thicker and stronger than that of the right ventricle. The RIGHT ventricle pumps blood to the pulmonary circulation to send waste RIGHT out of the body! The pulmonary circulation also functions to oxygenate blood that has transferred oxygen to the tissues, as described earlier in this lecture.

The Cardiac Impulse

The heart is a large muscle with regular contractions that are stimulated by electrical activity, i.e. action potentials. Specialized groups of cells in the heart initiate the cardiac impulse. Nervous system regulation of these cells can slow down or speed up the heart rate according to the body's needs. Conducting cells gather into fibers that ensure the coordinated spread of the action potential throughout the heart. Unlike skeletal muscle, cardiac muscle is not attached to bone. Instead, its fibers form a net that contracts upon itself, squeezing blood into the arteries. Systole occurs when the ventricles contract; diastole occurs during relaxation of the entire heart and then contraction of the atria.

The blood is propelled by hydrostatic pressure created by the contractions of the heart. The rate of these contractions is controlled by the autonomic nervous system, but the autonomic nervous system does not initiate the contractions. Instead the heart contracts according to the pace set by the **sinoatrial (SA) node**, a group of specialized cardiac muscle cells located in the right atrium. The SA node is *autorhythmic* (meaning it contracts by itself at regular intervals), spreading its contractions to the surrounding cardiac muscles via **electrical synapses** formed by **gap junctions**. The pace of the SA node is faster than that of normal heartbeats, but the **parasympathetic vagus nerve** innervates the SA node, slowing the contractions to produce the typical resting heart

FIGURE 4.9 Transmission of the Cardiac Impulse

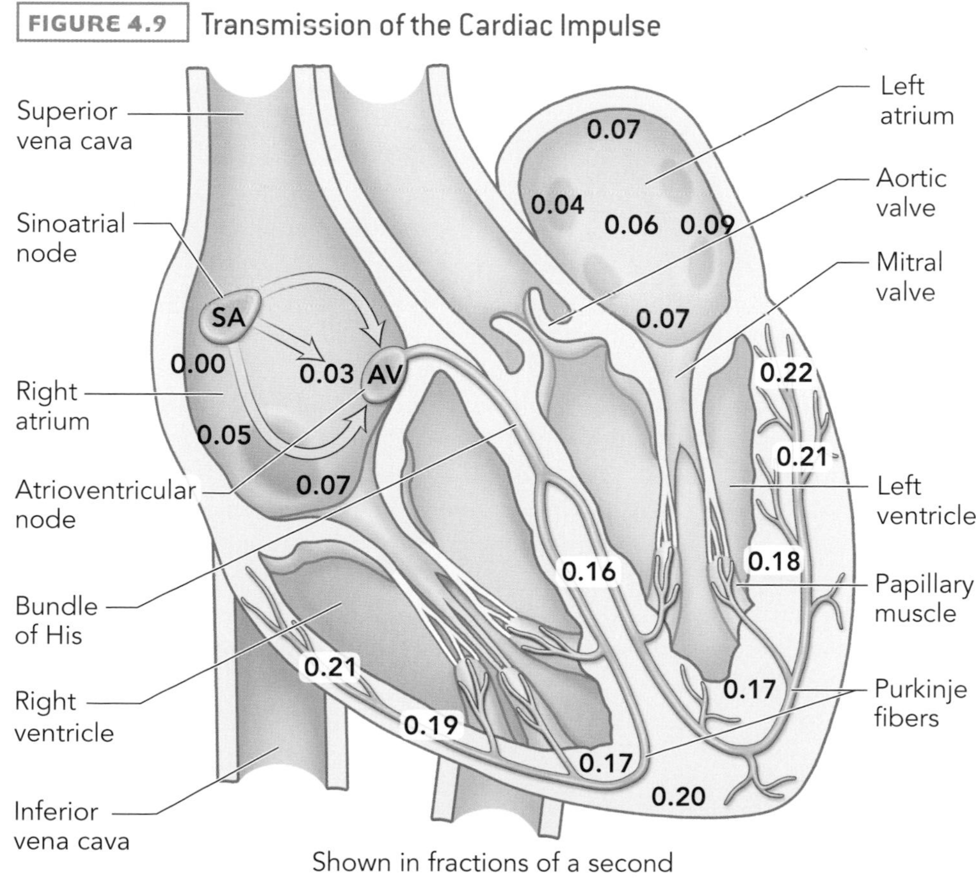

rate. Increased or decreased parasympathetic input can trigger quick changes in heart rate in response to the organism's internal and external environment.

The action potential generated by the SA node spreads around both atria, causing them to contract. While the atria are contracting, the action potential spreads to the **atrioventricular (AV) node** located in the interatrial septum (the wall of cardiac muscle between the atria). The AV node is slower to depolarize than the SA node, creating a delay which allows the atria to finish their contraction and squeeze their contents into the ventricles before the ventricles begin to contract. From the AV node, the action potential moves down conductive fibers called the **bundle of His**, located in the wall separating the ventricles. The action potential then branches out through the ventricular walls via conductive fibers called **Purkinje fibers**. From the Purkinje fibers, the action potential spreads through gap junctions from one cardiac muscle cell to the next. The Purkinje fibers in the ventricles allow for a more unified, and stronger, contraction.

The Blood Vessels: Specialization of Function

The three major types of blood vessels in the cardiovascular system are arteries, capillaries, and veins. This section describes the unique function and corresponding physical features of each type. Arteries act as a pressure store, capillaries are the site of the exchange of materials with the tissues, and veins act as a volume store for blood travelling back towards the heart.

The structure of arteries allows them to maintain a high pressure, facilitating the travel of blood to the lower-pressure venous system. Arteries have thick elastic walls that stretch as they fill with blood during systole. When the ventricles finish their contraction, the stretched arteries recoil, keeping the blood moving smoothly. Arteries are wrapped in smooth muscle that is typically innervated by the sympathetic nervous system. Epinephrine is a powerful vasoconstrictor, meaning that it causes arteries to narrow. Larger arteries have less smooth muscle per volume than medium-sized arteries, so they are less affected by sympathetic innervation. Medium-sized arteries, on the other hand, can constrict enough under sympathetic stimulation to reroute blood.

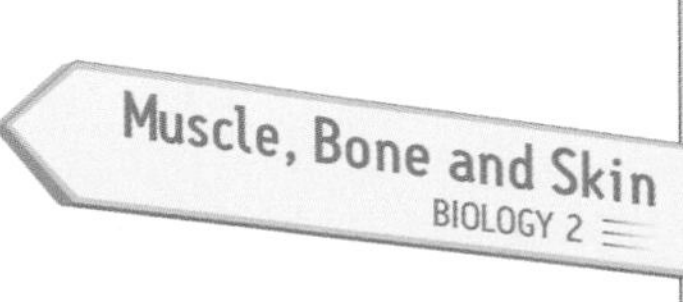

Arterioles are very small. Like arteries, they are wrapped by smooth muscle. Constriction and dilation of arterioles in response to stimulation from the nervous and endocrine systems can be used to regulate blood pressure and reroute blood. The arterioles participate in thermoregulation by controlling the flow of warm blood to capillary beds in the skin. When body temperature is too high, they dilate to allow blood flow to increase so that heat can dissipate. Blood can instead be directed away from the surface through constriction of the arterioles to prevent heat loss when the body's temperature is too low.

Capillaries are microscopic blood vessels. Much like alveoli, they have a huge total surface area created by the branching of larger structures. Capillary walls are composed of **endothelial cells** and are only one cell thick. The diameter of a capillary is roughly equal to that of a single red blood cell. The thin walls of capillaries are well-suited for the transport of materials. Nutrient and gas exchange with any tissue other than vascular tissue takes place only across capillary walls, not across arterioles or venules. As described in the context of thermoregulation, **heat exchange** across the walls of capillaries can also occur. When substances travel across the capillary wall, they enter the *interstitium*, which is the fluid surrounding cells in the cellular networks known as *tissues*. Differential pressures between the interstitium and blood in the adjacent capillary guide the transfer of materials, as described later in this lecture.

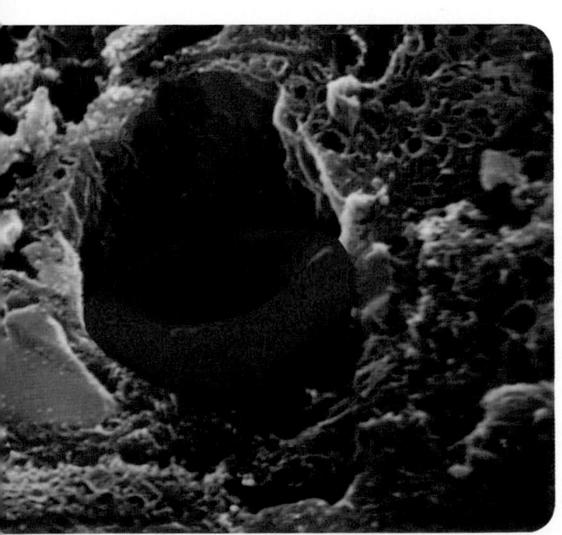

A red blood cell enters a capillary.

FIGURE 4.10 Types of Blood Vessels

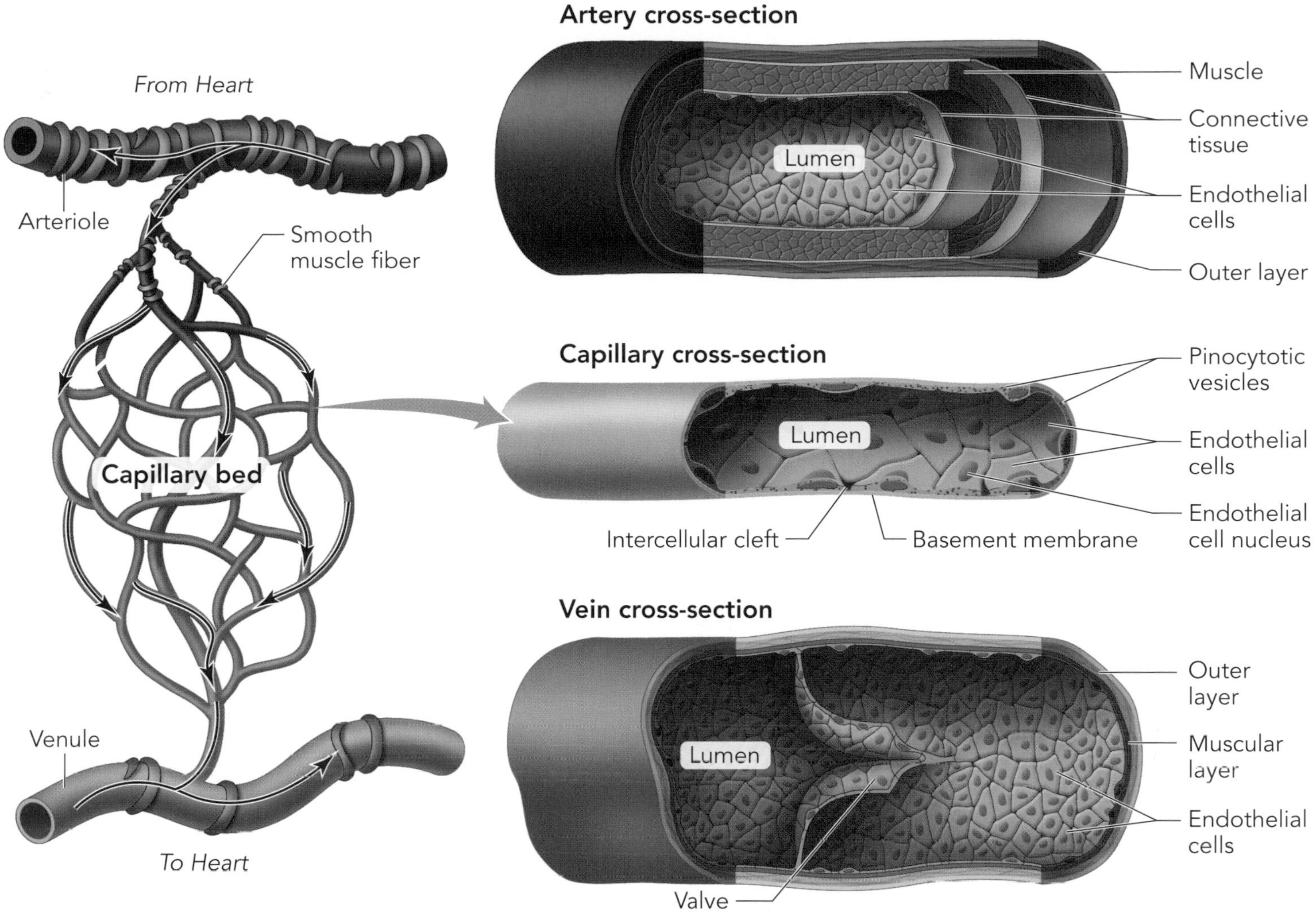

From Heart
Arteriole
Smooth muscle fiber
Capillary bed
Venule
To Heart

Artery cross-section

Lumen
Muscle
Connective tissue
Endothelial cells
Outer layer

Capillary cross-section

Lumen
Pinocytotic vesicles
Endothelial cells
Endothelial cell nucleus
Intercellular cleft
Basement membrane

Vein cross-section

Lumen
Outer layer
Muscular layer
Endothelial cells
Valve

For the MCAT®, you need to know the methods by which substances cross the walls of capillaries. The thin walls of capillaries are particularly important for gas exchange within the respiratory system, since the rate of diffusion across a membrane is inversely proportional to the thickness of the membrane.

There are four methods by which materials cross capillary walls:

1. pinocytosis,
2. diffusion or transport through capillary cell membranes,
3. movement through pores in the cells called fenestrations, and
4. movement through the spaces between the cells.

Substances use different methods to cross capillary walls according to their chemical and physical characteristics. Small lipid-soluble molecules, such as oxygen and carbon dioxide, can easily diffuse through the membranes of the endothelial cells. Water-soluble substances are forced to pass through gaps between the cells rather than travelling through the cell membranes. Proteins are even more restricted in their passage in or out of capillaries, since they are large and charged. When they are able to cross capillary walls, they do so via pinocytosis or fenestrations.

Venules and veins are similar in structure to arterioles and arteries, with a few key differences relating to their role of providing volume storage. The lumen of veins is larger than the lumen of comparable arteries, and the walls are much thin-

FIGURE 4.11 Blood Volume

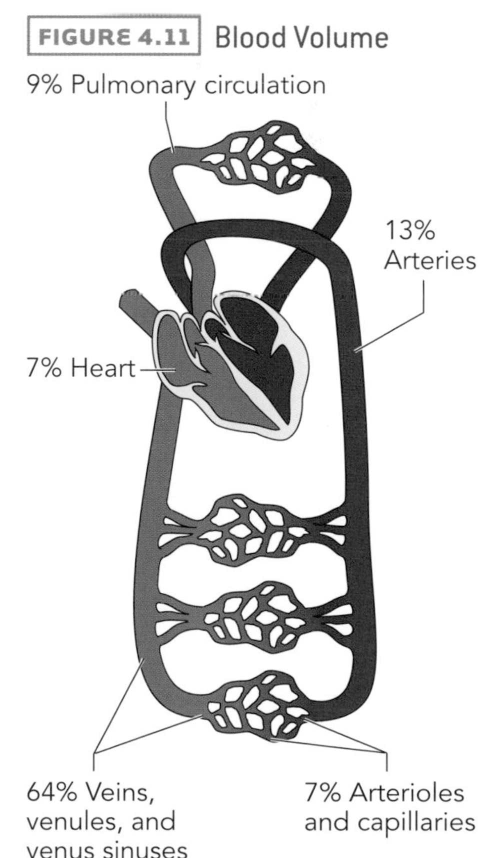

ner and have less elasticity. Veins can stretch to hold a far greater volume of blood. Veins, venules, and venus sinuses in the systemic circulation hold about 64% of the blood in a body at rest, acting as a reservoir. By contrast, arteries, arterioles, and capillaries in the systemic circulation contain about 20% of the blood. Because of the lower pressure in the venous system, veins and venules contain one-way valves to prevent the backflow of blood.

An artery carries blood away from the heart; a vein carries blood toward the heart. Don't make the mistake of defining arteries as blood vessels that carry oxygenated blood. The pulmonary arteries contain blood that has already traveled through the entire systemic circulation: this is the most deoxygenated blood in the body.

Blood Pressure and Flow

The two pressures, systolic and diastolic, correspond to the two numbers that represent your blood pressure at the doctor's office. The systolic pressure is always the greater number and goes on top, while the lower number, the diastolic pressure, goes on the bottom.

The physical characteristics of the different types of blood vessels described in the previous section impact flow and pressure within the cardiovascular system. The relationship between pressure difference and fluid flow (described in the Fluids Lecture of the *Physics* manual) is analogous to the relationship between voltage and current flow. Pressure encourages flow throughout the system, while resistance in the blood vessels affects pressure and flow.

At the level of the heart, pressure is generated to pump blood to the body. Recall that the pumping of the heart can be considered in two phases: systole and diastole. The pressure of blood in the arteries can be measured during these two periods, providing measures of the highest and lowest pressures throughout the cardiac cycle. Systole refers to the period in which the ventricles are contracting to eject blood. **Systolic pressure** is the highest pressure, and is measured in the arteries during systole. **Diastolic pressure** is the pressure during the relaxation of the ventricles and filling of the atria and is the lowest pressure in the cardiac cycle.

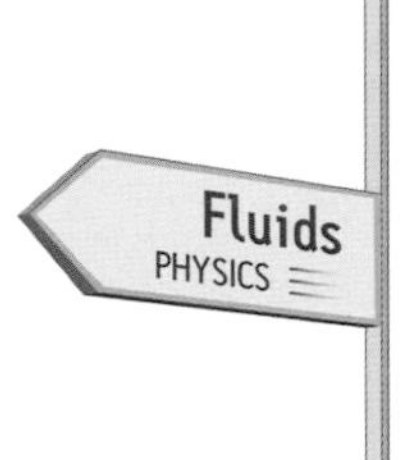

The relationship between pressure and flow in the cardiovascular system can be roughly described by the following equation: $Q = \frac{\Delta P}{R}$. This is analogous to the equation for current in an electric circuit, $I = \frac{V}{R}$. Pressure difference and voltage follow the same basic principles. Just as current flows from higher to lower voltage, fluids flow from higher to lower pressure. This equation demonstrates the directly proportionate relationship between pressure and resistance, which has implications for the regulation of blood pressure.

As in other systems of the body, the smooth functioning of the cardiovascular system requires homeostatic mechanisms. Blood pressure in the arteries must be kept at a constant high level, about 100 mm Hg, to provide a sufficient driving pressure for blood to travel through the systemic circulation and back to the heart. The maintenance of high pressure in the arteries is also important because blood pressure drops as it travels through the cardiovascular system (as described below). If blood pressure were too low at the first set of blood vessels, the arteries, the lower pressure at the level of the capillaries would be insufficient for the exchange of nutrients and waste.

To understand regulation of blood pressure, remember that the cardiovascular system as a whole is being considered. The inverse relationship between pressure and resistance assumes that flow remains constant. This is true in the cardiovascular system because it is a closed circulatory system. Even though flow through various blood vessels can be altered according to the metabolic needs of the surrounding tissues, the total flow through the system is relatively constant.

The two main methods used by the body to regulate blood pressure are the baroreceptor reflex (involving quick nervous system control) and the renin-angiotensin-aldosterone system (involving slower hormonal control). Both systems involve the detection of changes in blood pressure by mechanoreceptors, and they both alter blood pressure by changing the output of blood from the heart. In addition, both systems can affect blood pressure by changing the **total peripheral resistance**, meaning the overall resistance of the entire systemic circulatory system. Changes in total peripheral resistance are primarily achieved through constriction or dilation of smooth muscle surrounding arterioles, which are the blood vessels that contribute the most to peripheral resistance. When the cross-sectional area of a blood vessel decreases due to constriction of smooth muscle surrounding the arteriole, the resistance increases. Increasing the resistance of arterioles throughout the peripheral circulation increases the total peripheral resistance.

Even though capillaries are smaller, they have less resistance than arterioles. The overall resistance of the capillaries can be thought of as a set of resistors arranged in parallel. When arterioles branch into capillaries, the blood is able to flow through many capillaries that are arranged in parallel. This decreases the overall resistance of the capillary bed.

The *baroreceptor reflex* regulates blood pressure by altering both cardiac output and blood vessel resistance to flow. Baroreceptors (a type of mechanoreceptor) located within arteries detect changes in blood pressure. They respond by signaling centers in the brainstem to alter sympathetic and parasympathetic nervous system (SNS and PNS) output to the heart and blood vessels. Increased PNS activation counteracts an increase in blood pressure by slowing the heart's rate of contraction, whereas increased SNS output counteracts a decrease in blood pressure by causing heart rate to increase. SNS signals also cause blood vessels to constrict, increasing their resistance. Both of these effects of SNS activation increase blood pressure.

The *renin-angiotensin-aldosterone system* regulates blood pressure through the **regulation of plasma volume**. This system is activated when mechanoreceptors in arteries leading to the kidneys detect a decrease in blood pressure. A cascade of enzymatic effects triggered by the secretion of renin (itself an enzyme) leads to increased intake and retention of water, which in turn increases plasma volume. The increase in the volume of blood flowing through the cardiovascular system causes an increase in blood pressure. Two hormones, aldosterone and antidiuretic hormone, are involved in the mechanism of this effect.

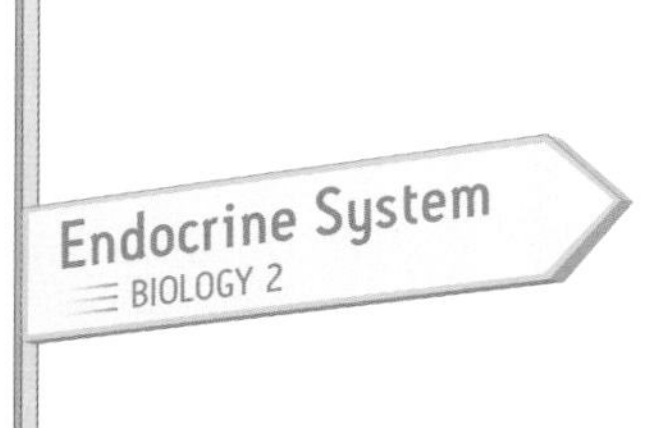

As described above, the flow of blood through the cardiovascular system is greatly affected by changes in blood pressure. The characteristics of different blood vessels also cause variations in blood flow in different areas of circulation. Cross-sectional area is a particularly significant factor. The total cross-sectional area of the veins is about four times that of the arteries, but the total cross-sectional area of the capillaries is far greater than the cross-sectional area of either the arteries or veins. Since the blood volume flow rate is approximately constant, the blood velocity is inversely proportional to the cross-sectional area. For this reason, the movement of blood is slowest in the capillaries. This allows more time for exchange of oxygen and waste products with the tissues.

Remember that while each individual capillary is tiny, the total cross-sectional area of all of the capillaries is much greater than that of any of the other types of blood vessels.

The flow rate of blood can be more precisely described using *Poiseuille's Law*, as discussed in the Fluids Lecture of the *Physics* manual:

$$Q = \Delta P \frac{\pi r^4}{8\eta L}$$

This equation can be used to calculate the flow of a real fluid. Poiseuille's Law again demonstrates that flow is directly proportional to pressure difference. The equation also shows how significantly radius affects flow rate, since radius is raised to the fourth power. Note that this equation, like the continuity equation, predicts a low blood flow rate in the capillaries. (In this case it is the influence of the small radii of individual capillaries, rather than their large total cross-sectional area, that is considered.) Poiseuille's Law also demonstrates how significantly nervous input to smooth muscle around vessels can affect blood flow by inducing even tiny changes in radius.

To understand the movement of blood through the cardiovascular system, it is necessary to understand how pressure changes throughout the different types of blood vessels. Pressure is highest in the aorta, right next to the ventricular pump, and continues decreasing to reach the lowest level in the veins. The pumping force of the heart is the major contributor to pressure in the blood vessels. To compensate for their lower pressure, veins have a valve system that prevents back flow of blood.

It is easy to remember that pressure is highest in the aorta, lower in the capillaries, and lowest in the veins if you remember the principle that fluid travels from high to low pressure. Since blood travels from the heart to the aorta, through the capillaries, and back to the heart via the veins, pressure must be dropping at each subsequent set of blood vessels to encourage flow. You can also remember that the arteries are a pressure store, while the veins are a volume store. This is partly because the lower pressure allows blood to pool in the veins.

Contraction of skeletal muscle helps blood move through veins; however, the major propulsive force moving blood through the veins is the pumping of the heart.

FIGURE 4.12 Cross-sectional Area vs. Velocity

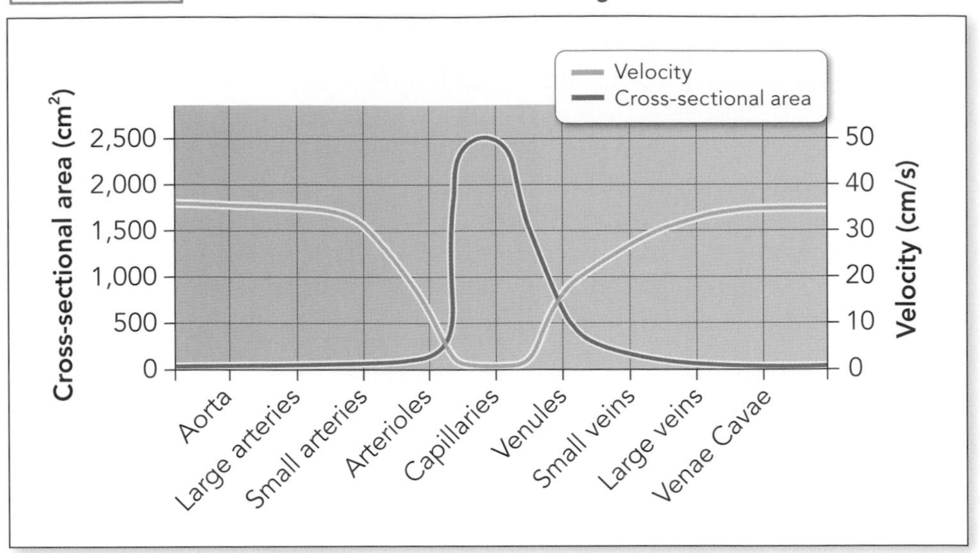

FIGURE 4.13 Blood Pressure

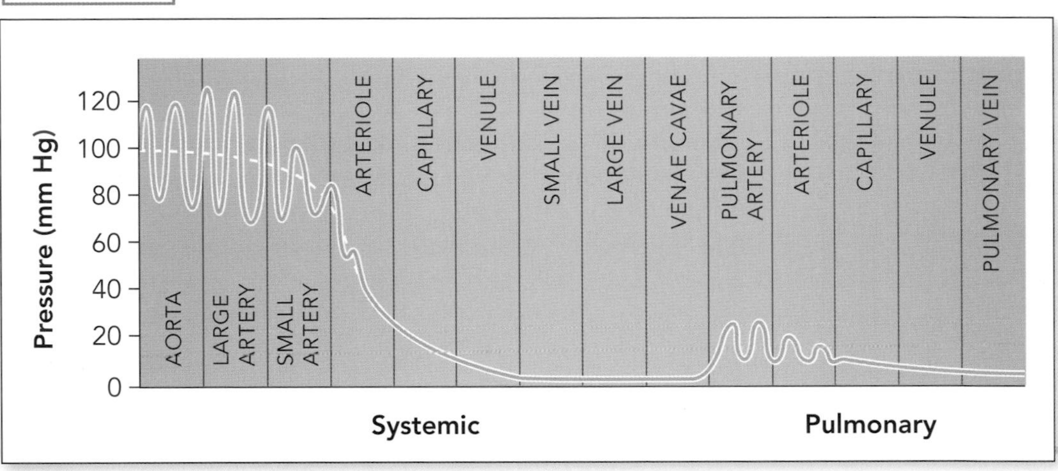

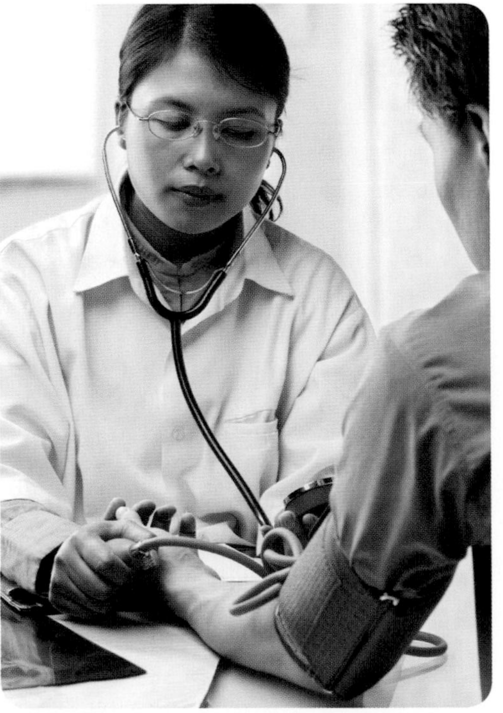

Information on blood pressure, volume, velocity, and cross-sectional area of vessels is likely to appear in a physics passage. Don't memorize anything, but take a moment here to consider the relationships between variables.

Pressure: Blood pressure is high near the heart and decreases as it gets farther away from the heart.

Velocity: A single artery is much bigger than a single capillary, but there are many more capillaries than arteries. The total cross-sectional area of all those capillaries put together is much greater than the cross-sectional area of a single aorta or a few arteries. Blood flow follows the Continuity Equation, $Q = Av$, reasonably well, so velocity is greatest in the arteries, where cross-sectional area is smallest, and velocity is lowest in the capillaries, where cross-sectional area is greatest.

According to Bernoulli's equation, the sum of potential energy, kinetic energy, and pressure of a flowing fluid is equal to a constant. In other words, a decrease in the velocity of blood should correspond to an increase in pressure. This would imply that blood pressure should be higher in the capillaries than in the aorta, since blood travels much more slowly in the capillaries. This is not the case because Bernoulli's equation only applies to ideal fluid flow, where energy is conserved. Blood is not an ideal fluid; energy (and thus pressure) is lost due to friction against the walls of the blood vessels.

In addition to the important role that pressure plays in the flow of blood through the vessels and back to the heart, a counterbalance of pressures governs the exchange of materials at the level of the capillaries. Even though pressure is very low in the capillaries, there is still sufficient hydrostatic pressure against the walls to facilitate the transfer of materials from capillaries to the tissues, as described below.

Capillaries are found close to all cells of the body. *Precapillary sphincters* regulate the flow of blood into capillary beds according to the metabolic needs of the tissues. There are two types of pressure across the wall of the capillary: osmotic pressure and hydrostatic pressure. **Osmotic pressure**, as discussed in the Cell Lecture, is the "pulling" pressure of solutes in solution that leads to the diffusion of solvent across a membrane. In the case of the capillaries, osmotic pressure encourages the flow of fluid from the tissues into the capillaries. Because the fluid is moving so slowly in the capillaries, it exerts a high hydrostatic pressure on the walls (in the opposite direction of the osmotic pressure), encouraging the flow of fluid into the interstitium. When blood first enters a capillary (Figure 4.14), hydrostatic pressure is greater than osmotic pressure, so net fluid flow is out of the capillary and into the interstitium. Osmotic pressure remains relatively constant throughout the capillary because it is set by proteins in the blood that are too large to move out of the capillaries. However, hydrostatic pressure drops from the arteriole end to the venule end as fluid flows out of the capillary. Osmotic pressure overcomes hydrostatic pressure near the venule end of a capillary, and net fluid flow is out of the interstitium and into the capillary. The net result of fluid exchange by the capillaries is a 10% loss of fluid from the capillaries to the interstitium. The next section of this lecture will describe how the lymphatic system acts to prevent excess fluid buildup in the tissues.

The discussion of how pressure contributes to the exchange of fluids across capillary walls should sound familiar: it's the same principle as the differential partial pressures that drive gas exchange in the respiratory system! Just as O_2 and CO_2 each diffuse to areas of lower partial pressure, in the circulatory system fluid travels to the area of lowest pressure. Movement of fluid into the interstitium is called filtration, while movement of fluid back into blood vessels is called reabsorption.

FIGURE 4.14 Fluid Exchange in the Capillaries

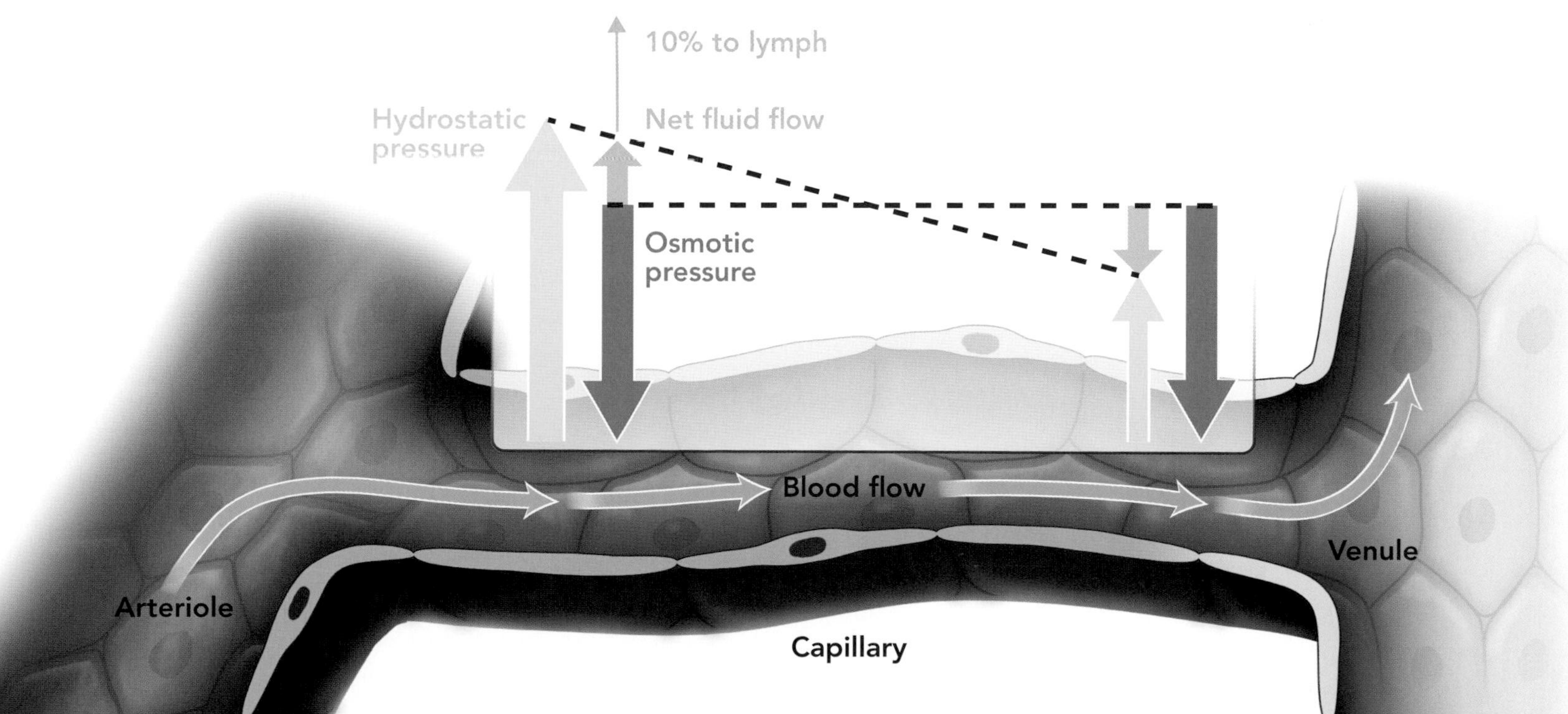

These questions are NOT related to a passage.

Question 81

The atrioventricular node:

- O **A.** is a parasympathetic ganglion located in the right atrium of the heart.
- O **B.** conducts an action potential from the vagus nerve to the heart.
- O **C.** sets the rhythm of cardiac contractions.
- O **D.** delays the contraction of the ventricles of the heart.

Question 82

Cardiac output, which is the product of the heart rate and the stroke volume (the amount of blood pumped per contraction by either the left or the right ventricle), would most likely be:

- O **A.** greater if measured using the stroke volume of the left ventricle.
- O **B.** greater if measured using the stroke volume of the right ventricle.
- O **C.** the same regardless of which stroke volume is measured.
- O **D.** dependent on the viscosity of the blood.

Question 83

Which of the following is responsible for the spread of the cardiac action potential from one cardiac muscle cell to the next?

- O **A.** Gap junctions
- O **B.** Desmosomes
- O **C.** Tight junctions
- O **D.** Acetylcholine

Question 84

In the congenital heart defect known as patent ductus arteriosus, the ductus arteriosus, which connects the aorta and the pulmonary artery during fetal development, fails to close at birth. This will likely lead to all of the following EXCEPT:

- O **A.** equal, or increased, oxygen concentration in the blood that reaches the systemic tissues.
- O **B.** increased oxygen concentration in the blood that reaches the lungs.
- O **C.** increased work load imposed on the left ventricle.
- O **D.** increased work load imposed on the right ventricle.

Question 85

A birth defect is detected in a patient, and tests reveal that there is a connection of the aorta to the pulmonary artery near the heart. Which of the following is true in this patient?

- O **A.** There is increased oxygen uptake in the lungs due to an increased oxygen gradient.
- O **B.** There is reduced oxygen uptake in the lungs due to a reduced oxygen gradient.
- O **C.** There is reduced oxygen uptake in the lungs due to an increased oxygen gradient.
- O **D.** There will be a reduced total volume of blood.

Question 86

Hypovolemic shock represents a set of symptoms that occur when a patient's blood volume falls abruptly. Hypovolemic shock is most likely to occur during:

- O **A.** arterial bleeding.
- O **B.** venous bleeding.
- O **C.** low oxygen intake.
- O **D.** excess sodium consumption.

Question 87

The capillary network comprises the greatest cross-sectional area of blood vessels in the body with high resistance to blood flow. In a healthy individual, the highest blood pressure would most likely be found in:

- O **A.** the aorta.
- O **B.** the venae cavae.
- O **C.** the systemic capillaries.
- O **D.** the pulmonary capillaries.

Question 88

Which of the following conditions would cause the hemoglobin oxygen dissociation curve to move to the left?

- O **A.** Elevated pCO_2
- O **B.** Elevated pH
- O **C.** Elevated 2,3-bisphosphoglycerate
- O **D.** Elevated temperature

Questions 81-88 of 144

STOP

4.5 The Lymphatic System as Drainage

One of the primary functions of the lymphatic system is to collect excess interstitial fluid that results from fluid exchange in the capillary beds, as described in the previous section, and return it to the blood. The lymphatic system also removes proteins and other particles that are too large to be taken up by the capillaries, including **glycerides**. The pathway back to the blood takes the excess fluid through lymph nodes, which are well prepared to elicit an immune response if necessary. In this way, the lymph system recycles interstitial fluid while also monitoring it for infection. In addition, the lymph system reroutes low-solubility fat digestates around the small capillaries of the intestine and into the large veins of the neck. (As discussed in the Digestive and Excretory Systems Lecture, fats, unlike other biological molecules, enter the lymphatic system rather than traveling into the capillaries.) Most tissues are drained by lymphatic channels.

Lymph circulates one way through the lymphatic vessels, eventually dumping into the thoracic duct and the vena cava.

The lymphatic system is an **open system**. In other words, fluid enters at one end and leaves at the other. This is in contrast to the closed circulatory system. Lymph capillaries are like tiny fingers protruding into the tissues. To enter the lymph system, interstitial fluid flows between overlapping endothelial cells (Figure 4.15c). Large particles literally push their way between the cells into the lymph. The cells overlap in such a fashion that, once inside, large particles cannot push their way out.

Typically, the interstitial fluid has a slightly negative gauge pressure. As the interstitial pressure rises toward zero, lymph flow increases. Factors that affect interstitial pressure include: blood pressure, plasma osmotic pressure, interstitial osmotic pressure (e.g. from proteins and infection response), and permeability of capillaries. Like veins, lymph vessels contain intermittent valves, which allow fluid to flow in only one direction.

Fluid is propelled through these valves in two ways. First, smooth muscle in the walls of larger lymph vessels contracts when stretched. Second, the lymph vessels may be squeezed by adjacent skeletal muscles, body movements, arterial pulsations, and compression from objects outside the body. Lymph flow in an active individual is considerably greater than in an individual at rest.

The lymph system empties into large veins at the *thoracic duct* and the *right lymphatic duct*. Lymph from the right arm and head enters the blood through the right lymphatic duct. The rest of the body is drained by the thoracic duct. In this way, fluid is returned back into the cardiovascular system.

Throughout the lymphatic system there are many lymph nodes, containing large quantities of lymphocytes. Lymph nodes filter and trap particles; they are also where lymphocytes are stimulated to respond to pathogens, as will be discussed in the following section.

FIGURE 4.15 The Lymphatic System

The lymphatic system is an open system; fluid enters at one end and leaves at the other. Fluid returns to the blood at the right lymphatic duct and the thoracic duct.

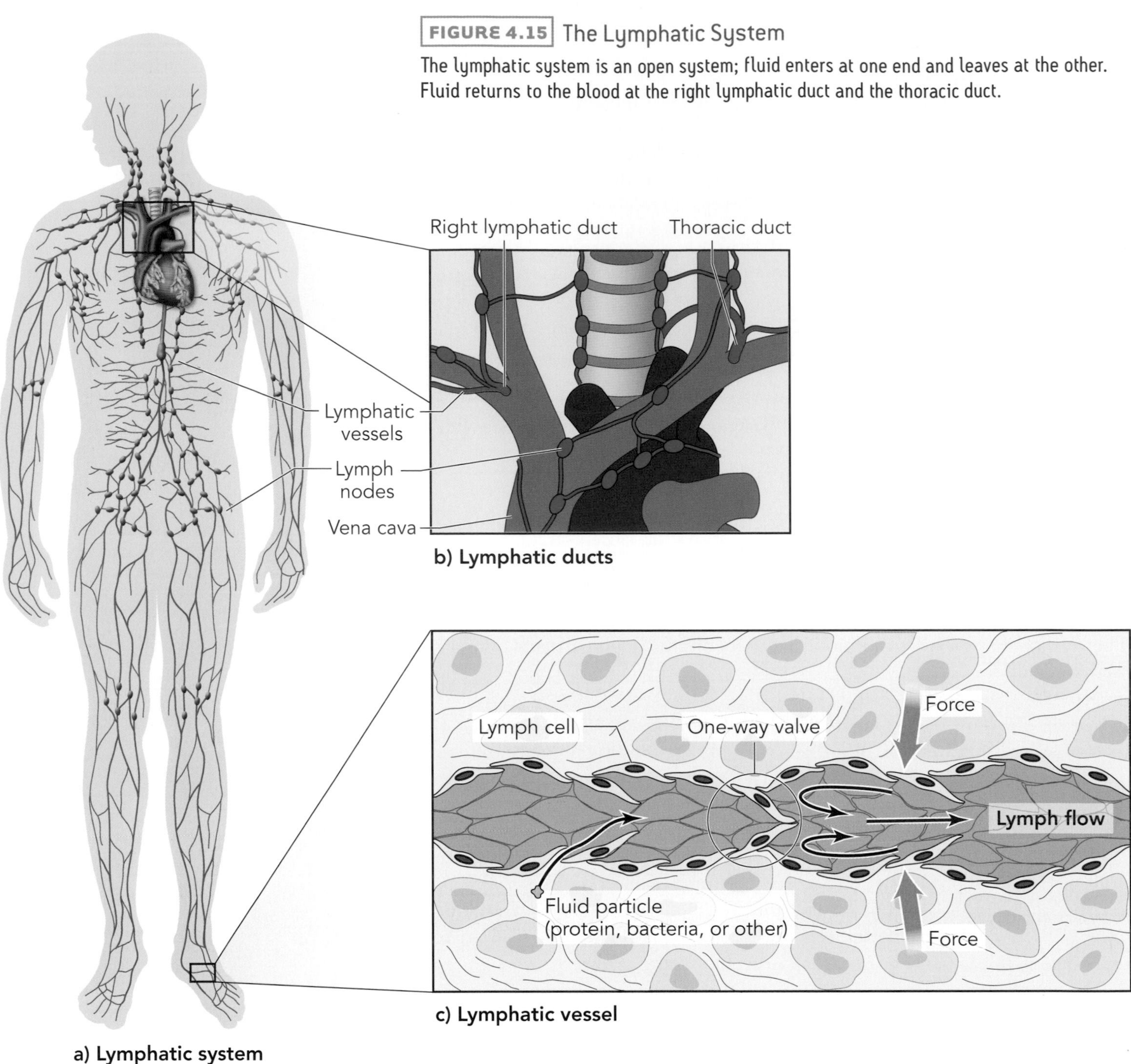

4.6 The Immune System: Components and Communication

Immune cells either eat or secrete.

The purpose of the immune system is to protect the human body from toxins and from invasion by other organisms, including bacteria, viruses, fungi, and parasites. The immune system involves many cell types and is present in nearly all tissues of the body. Each of the systems described previously plays a role in the immune system. Generally speaking, immune cells use one of two strategies to accomplish the primary goals of the immune system. They either phagocytose ("eat") dangerous material or secrete effector molecules that have downstream functions.

Some parts of the immune system are front line defenders, acting swiftly but non-specifically. Others act more slowly, but with strong targeted responses. The immune system also learns over time (has "memory") such that following exposure to a particular pathogen, subsequent exposures will lead to a quicker and stronger response. The immune system is also tightly regulated. Excessive regulation causes susceptibility to infection, while insufficient regulation leads to autoimmune destruction of the organism's own cells.

Innate immunity includes our foot soldiers; acquired immunity includes our tactical strikes! When you think about the immune system, organize the different components into the categories of fast or slow and general or specific.

FIGURE 4.16 Innate and Acquired Immunity

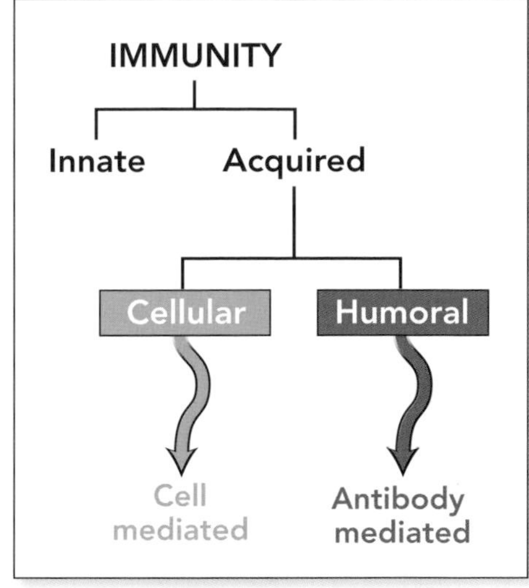

The immune system is made up of two major interconnected branches: innate immunity and acquired immunity. **Innate immunity** is quick and **non-specific**, providing a generalized protection from most intruding organisms and toxins. **Acquired (adaptive) immunity** develops more slowly and only after the body has experienced the initial attack. This branch provides protection against **specific** organisms or toxins. Acquired immunity can be broken down into B-cell mediated and T-cell mediated immunity, as will be discussed.

Many different cells, tissues, and organs are involved in the immune system. In addition to the lymphatic system, the spleen, thymus, and bone marrow all play important roles (Figure 4.17).

FIGURE 4.17 Organs of the Immune System

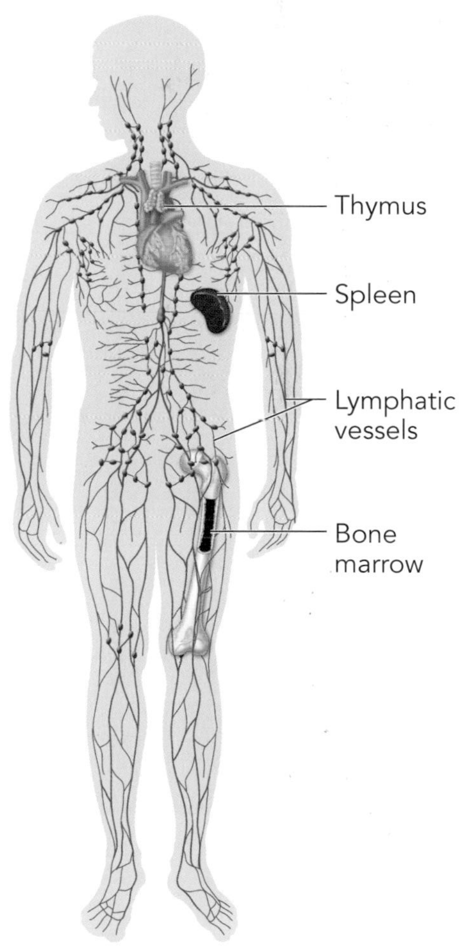

Innate Immunity

Innate immunity has two key features: it occurs quickly at the start of an infection, and it is non-specific (it does not adapt to a particular pathogen). Innate immunity includes a variety of cellular and non-cellular defenses, such as:

1. the skin as a barrier to organisms and toxins,
2. stomach acid and digestive enzymes to destroy ingested organisms and toxins,
3. phagocytotic cells such as neutrophils and macrophages, and
4. chemicals in the blood.

Injury to tissue results in **inflammation**. Inflammation functions to "wall-off" affected tissue and local lymph vessels from the rest of the body, impeding the spread of the infection. It includes dilation of blood vessels, increased permeability of capillaries, swelling of tissue cells, and migration of granulocytes and macrophages to the inflamed area (Figure 4.18). *Histamine*, *prostaglandins*, and *lymphokines* are just some of the causative agents of inflammation that are released by the tissues.

MCAT® THINK

How can innate immune cells recognize that there is a foreign invader? One way is by sensing Pathogen-Associated Molecular Patterns, or PAMPs. In this mechanism, innate immune cells look for molecules that are expressed by pathogens but not by humans. For example, many gram-negative bacteria express a carbohydrate called lipopolysaccharide, or LPS. The human body does not produce LPS, and certain innate immune cells recognize LPS as foreign. In fact, purified LPS alone, in the absence of bacteria, could send a person into septic shock due to the innate immune response.

FIGURE 4.18 The Inflammatory Response

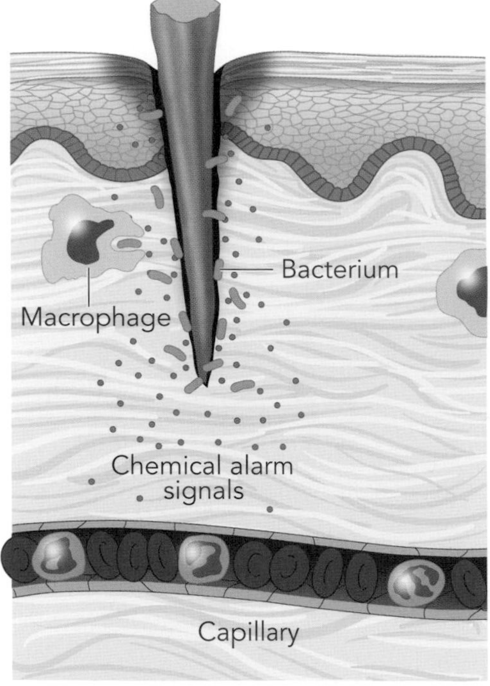

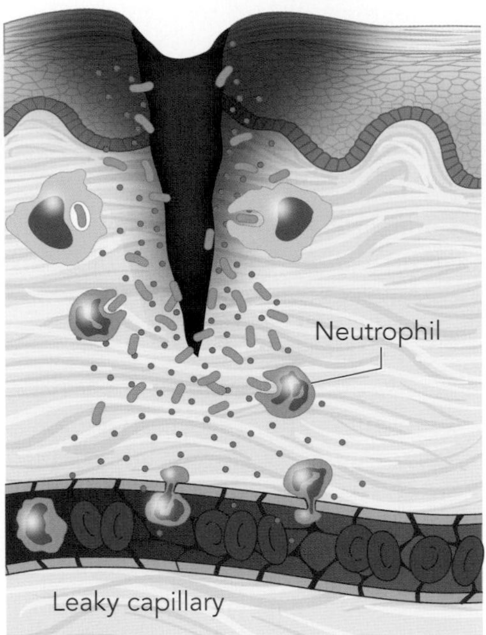

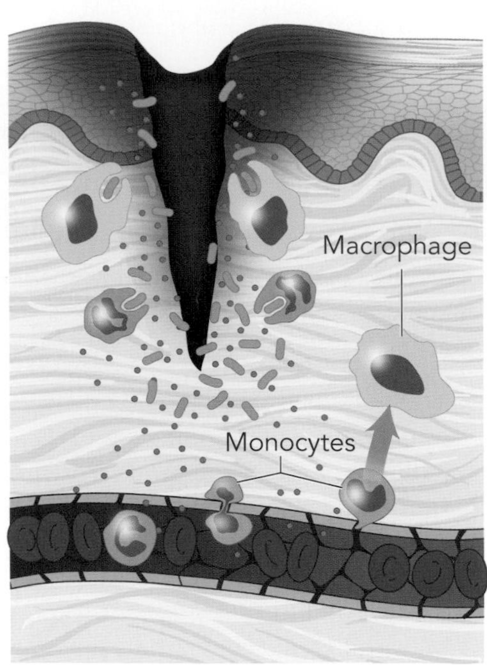

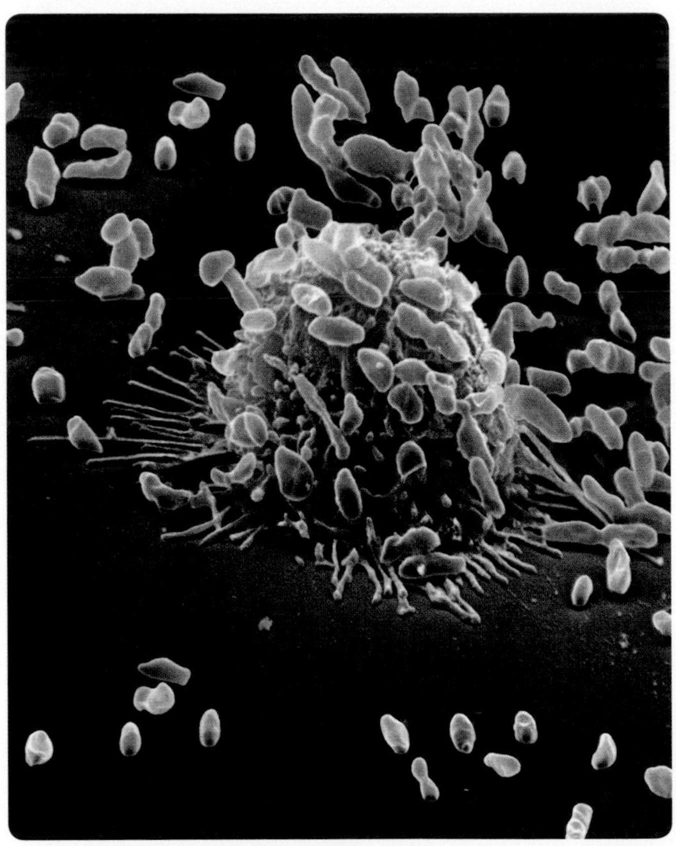

Macrophages ingest bacteria as part of the innate immune response to infection.

Multiple types of cells are involved in the innate immune system. Many of the earliest responders are **phagocytes**, meaning that they ingest ("eat") dangerous substances and then destroy them.

Infectious agents that pass through the skin or other barrier defenses and enter the body are first attacked by local **macrophages**. These phagocytotic giants can engulf as many as 100 bacteria. *Neutrophils* are next on the scene. Most neutrophils are stored in the bone marrow until needed, but some can be found circulating in the blood or in the tissues. Neutrophils move toward infected or injured areas, drawn by chemicals released from damaged tissue or by the infectious agents themselves. To enter the tissues, neutrophils slip between endothelial cells of the capillary walls, using an amoeboid like process called *diapedesis*. A single neutrophil can phagocytize from 5 to 20 bacteria.

Monocytes circulate in the blood until they, too, move into the tissues by diapedesis. Once inside the tissues, monocytes mature to become macrophages.

Neutrophils and macrophages die after engulfing bacteria and dead tissue. These dead leukocytes, along with tissue fluid and dead tissue, make up the substance known as *pus*. These first responders are expendable because they are non-specific and as such can be easily replaced.

Eosinophils and *basophils* are two additional innate immune cells related to neutrophils. Eosinophils work mainly against parasitic infections, while basophils release many of the chemicals of the inflammation reaction.

FIGURE 4.19 Cell Lineage

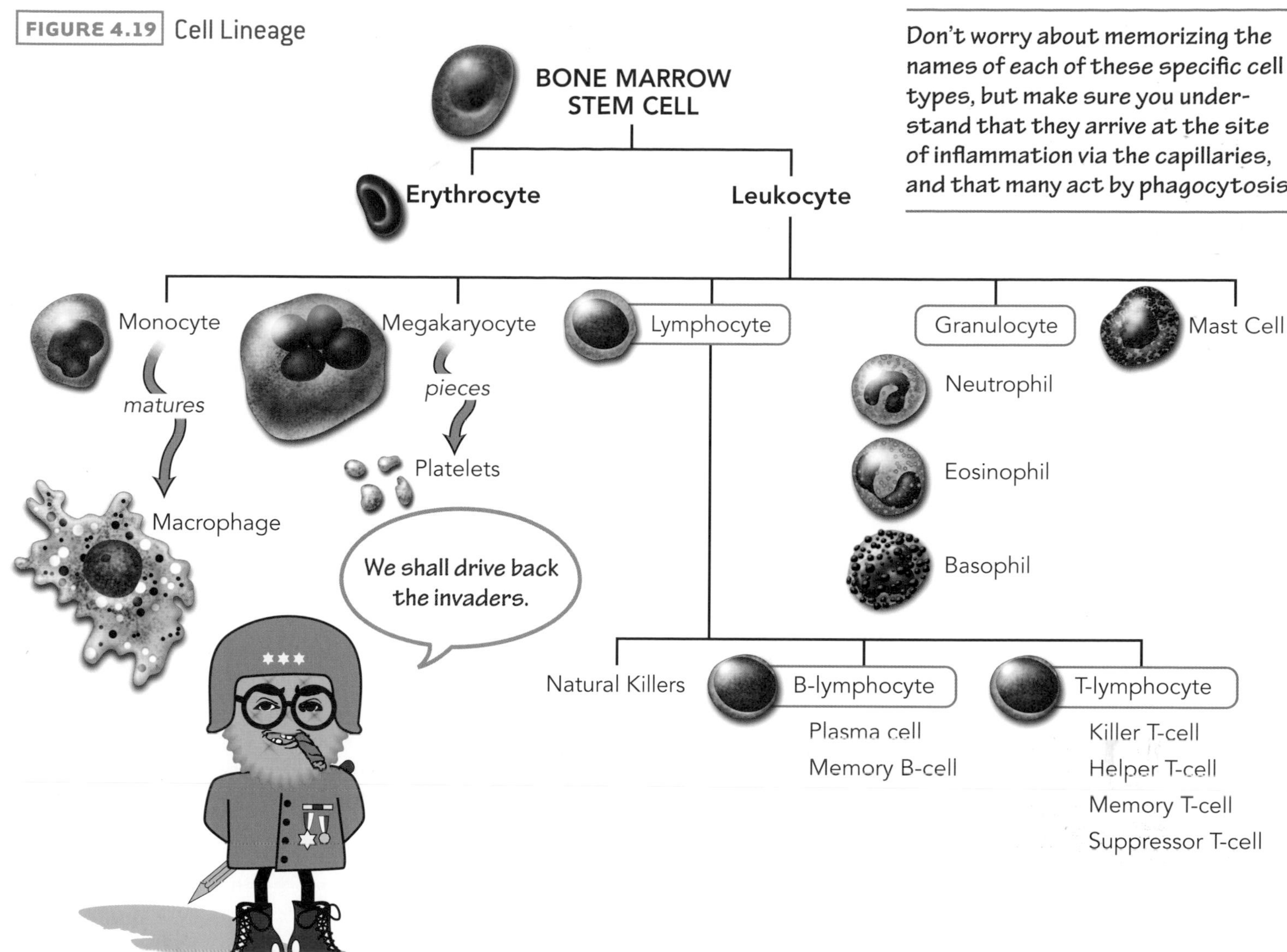

Don't worry about memorizing the names of each of these specific cell types, but make sure you understand that they arrive at the site of inflammation via the capillaries, and that many act by phagocytosis.

Acquired Immunity

Acquired or adaptive immunity is geared towards a specific pathogen. A key feature of this branch of the immune system is immunological memory, which allows the immune system to respond more quickly and strongly to each successive encounter with the same pathogen. Vaccination also harnesses the adaptive immune system by stimulating it with harmless components of dead bacteria and viruses to form protective antibodies. There are two types of acquired immunity:

1. **B-cell immunity**, also called **humoral** or **antibody-mediated immunity**
2. **T-cell immunity**, also called **cell-mediated immunity**

Remember Key 3: The innate immune system provides a non-specific but rapid response against pathogens, mediated by the work of neutrophils, macrophages, basophils, and eosinophils. The adaptive immune system is made up of B-cells, T-cells, and antigen-presenting cells. It mounts a sustained and specific response to a pathogen. It can even store the memory of the pathogen for a more rapid response in the event of reinfection.

B-cell Immunity

Humoral or antibody-mediated immunity is effective against bacteria, fungi, parasitic protozoans, viruses, and blood toxins. However, it cannot act against invading substances that have already made their way into cells.

Antibody-mediated immunity is promoted by **B-lymphocytes**, which differentiate and mature in the adult bone marrow and the fetal liver. Each B-lymphocyte makes a single type of **antibody** or *immunoglobulin*, which can recognize and bind to a particular potentially harmful foreign particle (an **antigen**). Initially, a B-lymphocyte displays this antibody on its membrane, and the antibody is called a *B-cell receptor (BCR)*. Later, many antibodies are produced as secreted proteins. The process by which an antibody (or BCR) recognizes a foreign particle is called antigen-antibody recognition. The portion of the antibody that binds to an antigen is highly specific for that antigen and is called an *antigenic determinant*.

The potency of the secondary response explains how vaccines, which allow exposure to a non-active version of a particular infectious agent, protect the body from future infection. It also explains why it's said that some diseases, like chicken pox, "can't be caught twice."

The immune response that results from the first exposure to an antigen is known as the **primary response**, which requires 20 days to reach its full potential. During the primary response, the BCR recognizes the appropriate antigen under the right conditions. When this happens, the B-lymphocyte, assisted by a helper T-cell, differentiates into plasma cells and memory B-cells. **Plasma cells**, which can survive for decades or even a lifetime, begin synthesizing free antibodies and releasing them into the blood. **Memory B-cells** proliferate and remain in the body. Like plasma cells, they are long-lived. In the case of re-infection, each of these cells can be called upon to synthesize antibodies, resulting in a faster acting and more potent effect called the **secondary response**. The secondary response requires approximately 5 days to reach its full potential.

Circulating antibodies in the blood, produced by plasma cells, have a variety of functions that promote the destruction of harmful invading substances:

1. First, the antibodies may mark the antigen for phagocytosis by macrophages and *natural killer cells*.
2. Once bound, the antibodies may begin a cascade of reactions involving blood proteins (called *complement*) that cause the antigen bearing cell to be perforated.
3. The antibodies may cause the antigenic substances to *agglutinate* (stick together) or even precipitate, or, in the case of a toxin, the antibodies may block its chemically active portion.
4. Finally, free antibodies may attach their bases to *mast cells*. When an antibody whose base is bound to a mast cell also binds to an antigen, the mast cell releases histamine and other chemicals. This mechanism plays a role in allergies and in anaphylactic shock.

The variable region is composed of 110-130 amino acids from both the heavy and the light chains. Since this region is variable, it can be used to bind different antigens and thus determines antigen specificity.

FIGURE 4.20 Antibody Structure

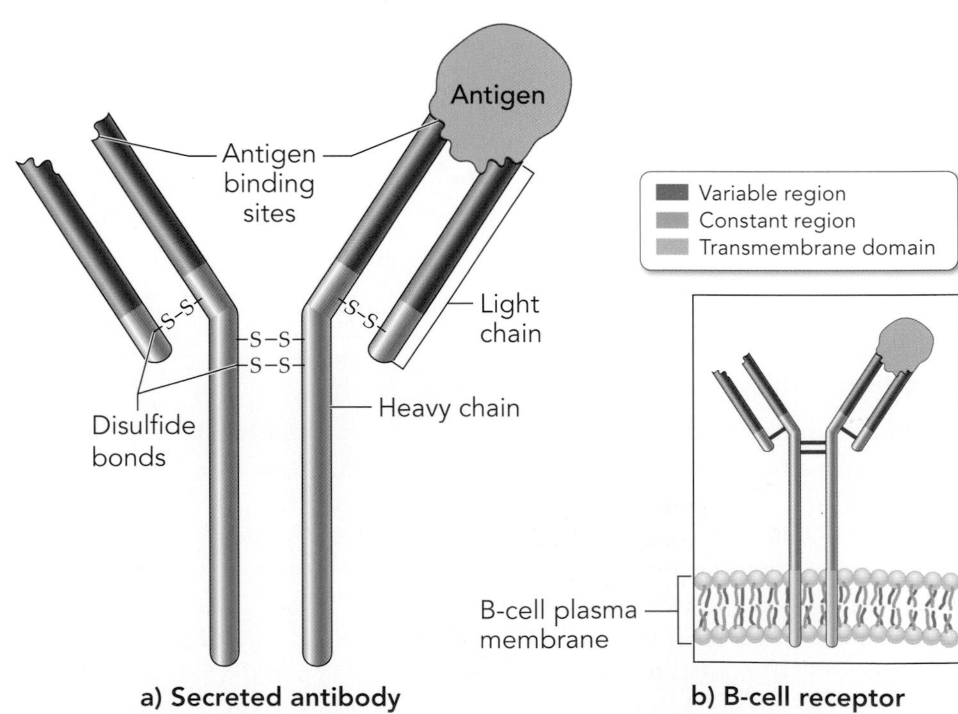

T-cell Immunity

Cell-mediated immunity is effective against cells that have already been infected because it is not restricted to free-floating substances, unlike antibody-mediated immunity. It involves **T-lymphocytes**, which are made in the bone marrow and mature in the thymus. Similar to a B-lymphocyte, a T-lymphocyte has an antibody-like protein at its surface that recognizes antigens, termed the *T-cell receptor (TCR)*. However, T-lymphocytes never make free antibodies.

During a primary immune response, when a TCR recognizes the appropriate antigen under the right conditions, T-lymphocytes differentiate into *helper T-cells*, *memory T-cells*, *suppressor T-cells*, and *killer T-cells* (also called *cytotoxic T-cells*). As discussed previously, helper T-cells assist in activating B-lymphocytes. They also activate other types of T-lymphocytes, including killer and suppressor T-cells. Long-lived memory T-cells have a similar function to memory B-cells in the secondary immune response. Suppressor T-cells play a negative feedback and regulatory role in the immune system. Killer T-cells bind to the antigen-carrying cell and release *perforin*, a protein that punctures the antigen-carrying cell. Killer T-cells can attack many cells because they do not phagocytize their victims; unlike macrophages and neutrophils, they are not themselves destroyed when they kill invading pathogens. Killer T-cells are responsible for fighting some forms of cancer and for attacking transplanted tissue.

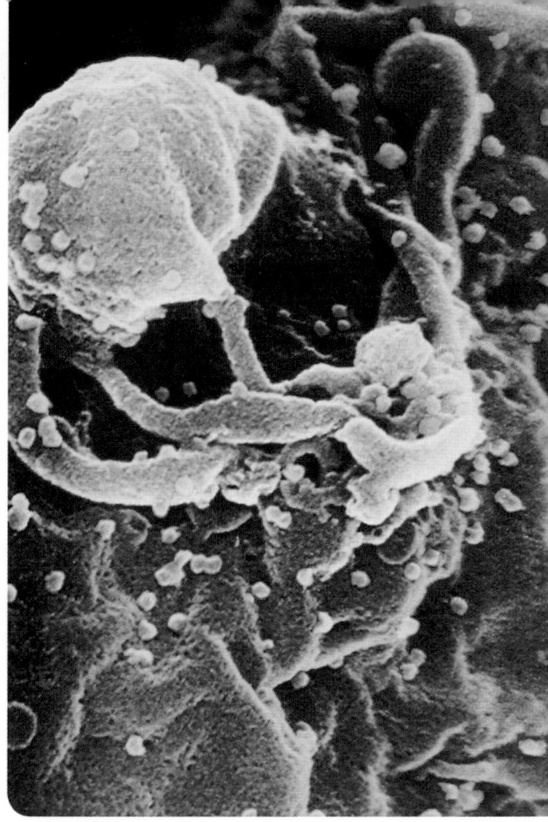

This scanning electron micrograph shows HIV particles budding from an infected T-lymphocyte.

Helper T-cells are the cells attacked by HIV. This is why the virus is so devastating to the body's ability to produce an effective immune response.

Antigen Recognition by B-cells and T-cells

The complex process by which B-cells and T-cells recognize antigens and are activated (stimulated to multiply and differentiate) has great importance for the functioning of the immune response. Errors in this process can cause immunodeficiencies or autoimmune disease.

All human cells, except for those that do not contain nuclei, express **major histocompatibility complex (MHC)** molecules on their cell surface. The function of these membrane bound proteins is to display antigens for recognition. There are two major types of MHC molecules. **MHC class I** molecules display antigens derived from intracellular pathogens such as viruses and some bacteria. Since all cells can be infected by these pathogens, all nucleated cells have MHC class I molecules. The process by which intracellular antigens are processed and displayed on the cell surface is termed the *endogenous pathway*. **MHC class II** molecules display antigens derived from extracellular pathogens. Since these pathogens must be phagocytosed, MHC class II molecules are displayed by phagocytic cells. Cells that phagocytose extracellular bacteria or other pathogens in order to display their antigens are termed *professional antigen presenting cells (APCs)*. Professional APCs include macrophages, dendritic cells, and some B-cells. The process by which these antigens are processed and displayed by MHC class II molecules on the cell surface is termed the *exogenous pathway*. Figure 4.21 compares the exogenous and endogenous pathways.

B-cell and T-cell activation

B- and T-cells need two signals to be activated, termed *Signal 1* and *Signal 2*. The requirement for two separate signals protects against an autoimmune response. Signal 1 is provided when the BCR or TCR recognizes its appropriate antigen. Signal 2 is often called a "danger signal" because B-cells and T-cells receive it when there is an active infection. The methods by which B- and T-cells receive these signals are slightly different.

FIGURE 4.21 Endogenous vs. Exogenous Pathways

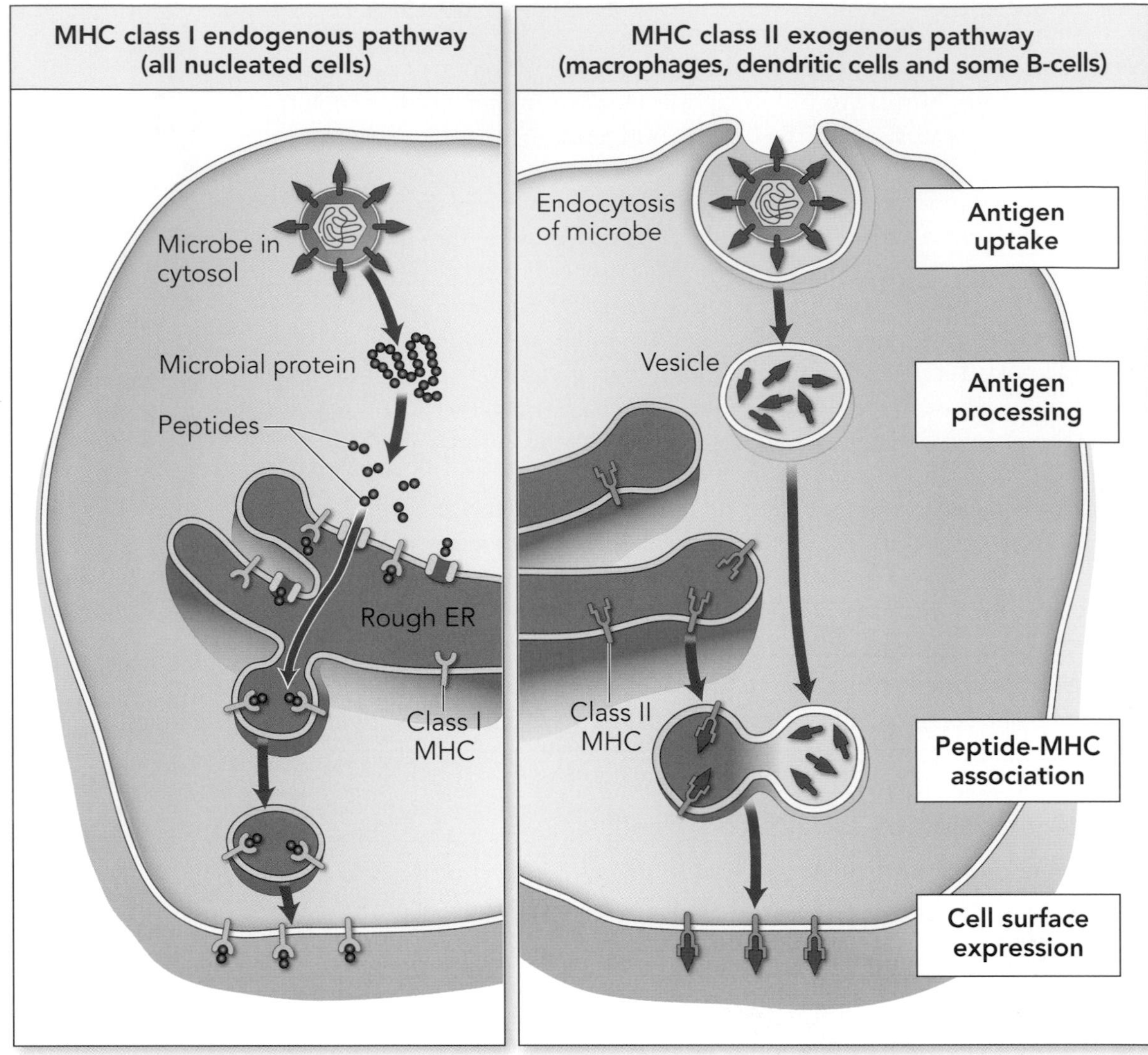

The goal of both the endogenous and exogenous pathways is to process an antigen from a pathogen or molecule and display that antigen on an MHC molecule on the cell surface. Generally speaking, both pathways have the same four steps:

1. Antigen uptake: In the case of the endogenous pathway, the pathogen or molecule is already inside the cell. In the case of the exogenous pathway, the antigen is extracellular and must be phagocytosed.

2. Antigen processing: The pathogen or molecule must be processed into smaller peptides (antigens). This occurs in the cytosol in the endogenous pathway, and in vesicles in the exogenous pathway.

3. Peptide-MHC association: The antigens then associate with the MHC molecules, which have undergone folding in the ER. In the endogenous pathway, the antigens are transported into the ER. In the exogenous pathway, a vesicle containing MHC Class II fuses with a vesicle containing antigens.

4. Cell surface expression: The end result of both pathways is that the antigen-MHC complex is expressed on the cell surface and can interact with the appropriate immune cells.

T-cells receive Signal 1 from professional APCs. The APCs have engulfed a pathogen, processed antigens via the exogenous pathway, and are now displaying antigenic determinants on the MHC class II molecules on their surfaces. The T-cell recognizes this combination of MHC and antigen, which provides Signal 1 via TCR stimulation and downstream signaling. Meanwhile, Signal 2 is provided via different cell surface receptors in a variety of ways. This signal is often provided by components of the innate immune system in response to infectious insults, either by circulating cytokines or up-regulation of membrane-bound surface proteins. If a T-cell receives both Signal 1 and Signal 2, it begins to proliferate and differentiate.

B-cells receive Signal 1 when a free floating antigen cross-links multiple BCRs on the same cell. Signal 2 for B cells is normally provided by helper T-cells. Again, both signals are required for proliferation and differentiation.

Infection often occurs in connective tissues, so B- and T-cell activation often takes place in the nearest lymph node. However, there are numerous sites with lymphatic tissues in the body, including the spleen, the gut, and other locations. Lymphocyte activation and proliferation can occur in any of these sites.

The reason your lymph nodes may swell during infection is that B- and/or T-lymphocytes are rapidly proliferating in response to the pathogen.

FIGURE 4.22 Antigen Presentation and Cell Communication

T-cells cannot be stimulated unless two different signals are present. This prevents T-cell activation (and a potentially harmful autoimmune response) in the absence of active infection.

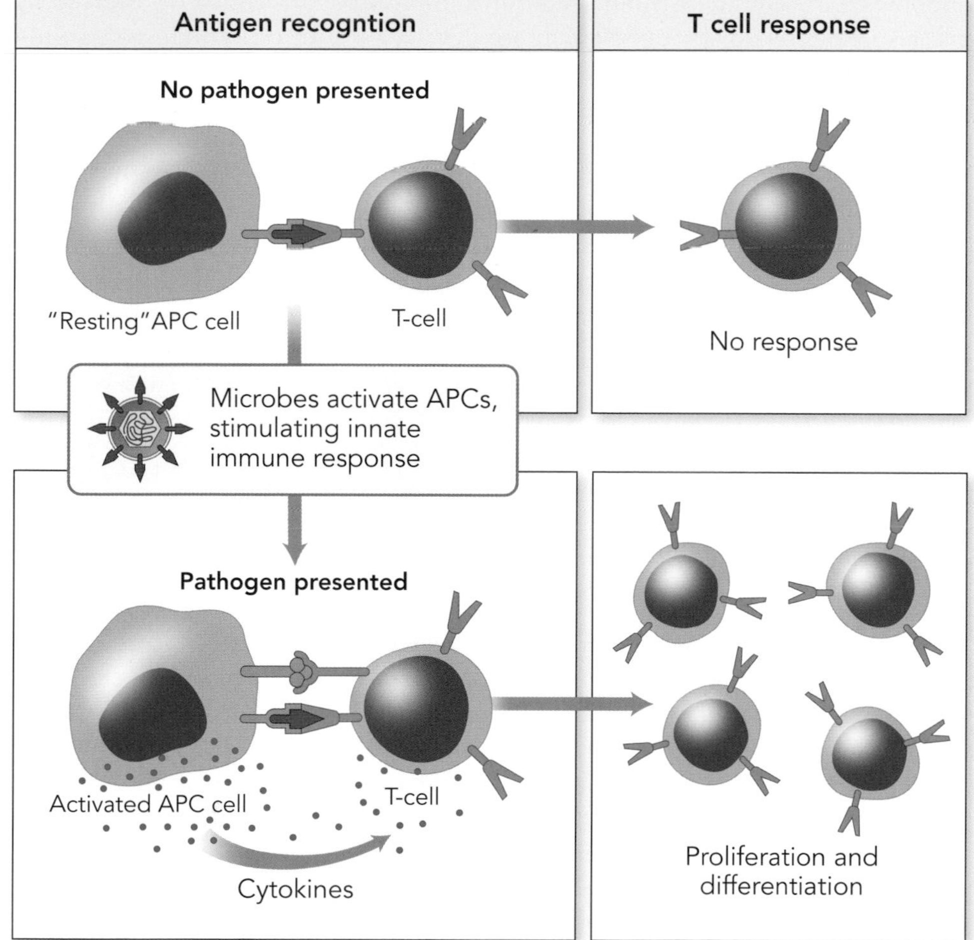

Regulation of the Immune System

Numerous processes regulate the immune system, and errors in these processes can cause pathologic conditions. Too much regulation renders the immune system deficient and unable to properly protect against foreign invaders. Too little regulation can result in **autoimmune diseases**, where a hyperactive immune system attacks the body's own tissues. Autoimmune diseases sometimes occur due to failures of clonal selection. To understand how the immune system is regulated, it is necessary to discuss the concept of clonal selection, which includes both positive and negative selection.

> **MCAT® THINK**
>
> Some autoimmune diseases are among the most common reasons for seeking medical intervention. For instance, Type I diabetes results from autoimmune destruction of insulin-producing pancreatic beta cells. Although failure of negative selection can result in autoimmune disease, there are many other immunoregulatory mechanisms in place. Failure of these mechanisms can also result in autoimmunity.

Each lymphocyte expresses BCRs or TCRs of a single specificity. However, due to processes that generate genetic diversity during development, the total population of B- and T-cells can express millions of different BCRs and TCRs. Immature B- and T-cells undergo the process of **clonal selection**, in which only certain types are permitted to mature and proliferate.

First of all, cells must show that they are capable of recognizing antigens in the context of host MHC molecules. MHC molecules hold the antigens that come from bacterial, viral, and fungal pathogens. If a B- or T-cell cannot recognize the MHC molecule, it will be unable to determine whether the antigen is foreign or not. B- and T-cells that cannot recognize MHC Class I or II undergo apoptosis. Those that cannot undergo apoptosis. This process is called **positive selection**.

Secondly, the cells must show that they are not inappropriately activated by host cells in the absence of infection. In addition to displaying antigens derived from pathogens in the context of an invasion, MHC Class I molecules normally display antigens from endogenous normal proteins (termed *self-antigens*). An immune response to these proteins would result in autoimmune disease, so B- and T-cells that respond too strongly to MHC molecules with self-antigens undergo apoptosis. This is called **negative selection**.

Collectively, positive and negative selection allow for the survival of only those B- and T-cells that are able to distinguish between self and non-self antigens. These opposing processes eliminate both lymphocytes that are ineffective in the production of an immune response and those that present the danger of overreacting and attacking the organism's own cells. T-cells undergo clonal selection in the thymus, while B-cells undergo selection in the bone marrow. Cells that survive the selection process are released to lodge in lymphoid tissue or to circulate between the blood and the lymph fluid. Once these surviving cells recognize the appropriate foreign antigen, they undergo differentiation and proliferation, completing the process of clonal selection.

In clonal selection, the appropriate B- or T-cell is positively and negatively selected, and then it is cloned in response to a pathogen.

Let's imagine a bacterial infection. First we have inflammation. Macrophages, then neutrophils, engulf the bacteria. Interstitial fluid is flushed into the lymphatic system, where lymphocytes wait in the lymph nodes. Macrophages process and present the bacterial antigens on MHC molecules to T-lymphocytes. With the assistance of helper T-cells, B-cells differentiate into memory cells and plasma cells. The memory cells are preparation in the event that the same bacteria attacks again (the secondary response). The plasma cells produce antibodies, which are released into the blood to attack the bacteria. It is important to know that a single antibody is specific for a single antigen, and that a single B-lymphocyte produces only one antibody type.

FIGURE 4.23 Humoral vs. Cell-Mediated Immunity

This figure provides a summary of adaptive immunity. Note the many ways in which cells from both branches of the adaptive immune system interact with each other. When a cell is stimulated, either by a molecule or by another cell, it undergoes proliferation and/or differentiation. Also note the difference between the primary antigen exposure (top) and the secondary antigen exposure (bottom). Remember that the secondary exposure will generate a much quicker adaptive immune response.

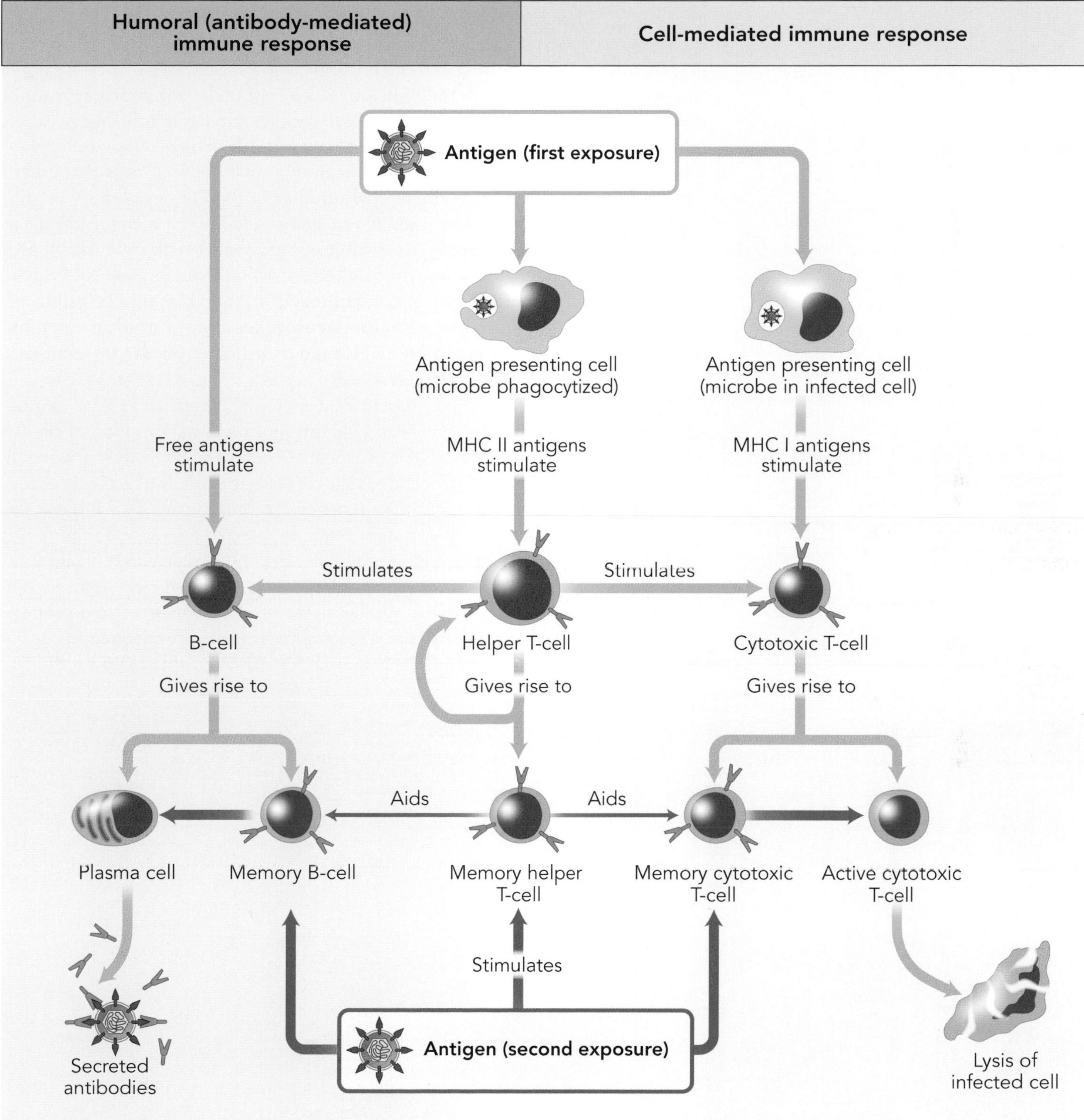

Blood Types

Blood can be categorized into different types according to the presence of certain types of antigens on the surfaces of red blood cells. Because heredity determines an individual's blood type, a full understanding of blood types requires knowledge of both genetics and immunology. Blood types are identified by A and B surface antigens. For instance, in type A blood, the red blood cell membrane has A antigens and does not have B antigens. Of course, if the erythrocytes have A antigens, the immune system does not make A antibodies. Type O blood has neither A nor B antigens, so the immune system makes both A and B antibodies. A blood donor can donate blood only to an individual that does not make antibodies against the donor blood. Table 4.1 shows a '+' sign for donor/recipient combinations where blood agglutinates (is rejected), and a '–' sign for combinations where no agglutination occurs. Notice that an individual with type O blood is a "universal donor" who can generally donate to anyone (all minuses in the donor column), because type O blood does not contain any blood group antigens that could trigger an immune response in a recipient. Meanwhile, an individual with type AB blood is a "universal recipient" who can generally receive blood from anyone (all minuses in the recipient row) because no antibodies for A or B antigens will be produced.

The genes that produce the A and B antigens are co-dominant, so an individual having type A or B blood can be heterozygous or homozygous. An individual with type O blood has two recessive alleles.

Rh factors are surface proteins on red blood cells, first identified in Rhesus monkeys. Individuals with genotypes that code for nonfunctional products of the Rh gene are said to be *Rh-negative*. All others are *Rh-positive*. Transfusion reactions involving the Rh factor, if they occur at all, are usually mild. Rh factor is more of a concern during the pregnancy of an Rh-negative mother with an Rh-positive fetus. In the first pregnancy, the mother is not exposed to fetal blood until giving birth, and problems are rare. Upon exposure, the mother develops an immune response against the Rh-positive blood. In a second pregnancy, the fetus, if Rh-positive, may be attacked by the antibodies of the mother, which are small enough to pass through the placental barrier. This problem can be life threatening.

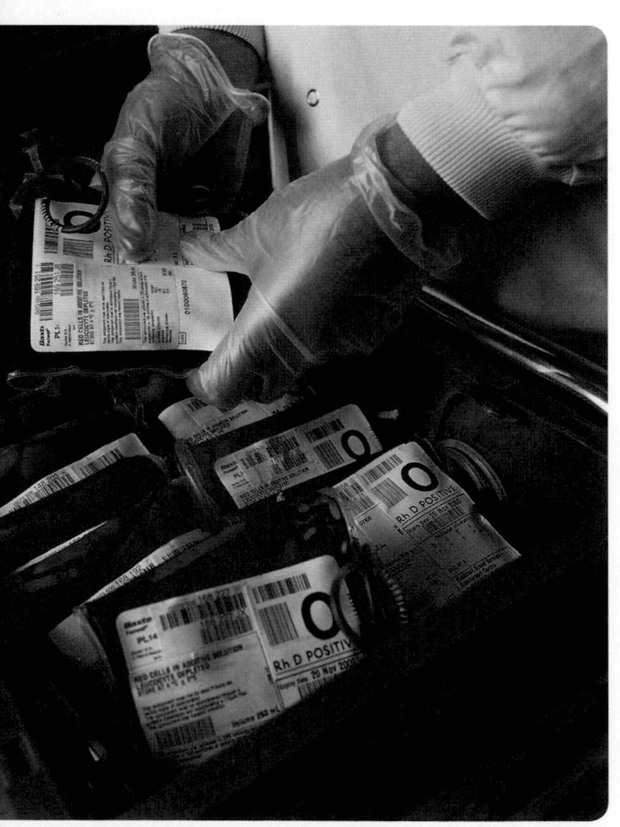

TABLE 4.1 > **Blood Types**

		Donor			
		A	B	AB	O
Recipient	A	–	+	+	–
	B	+	–	+	–
	AB	–	–	–	–
	O	+	+	+	–

Blood type	Genotype
A	I^AI^A or I^Ai
B	I^BI^B or I^Bi
AB	I^AI^B
O	ii

These questions are NOT related to a passage.

Question 89

In Waldenstrom's macroglobulinemia, a select clone of B cells becomes dysfunctional and cancerous, and begins to take over space in bone marrow that was previously dedicated to stem cells. Which of the following most likely occurs in Waldenstrom's macroglobulinemia?

I. Anemia
II. Inability to fight infection
III. Reduced clotting

- O **A.** I only
- O **B.** III only
- O **C.** I and II only
- O **D.** I, II, and III

Question 90

Humoral immunity involves the action of:

- O **A.** cytotoxic T-lymphocytes.
- O **B.** stomach acid.
- O **C.** pancreatic enzymes.
- O **D.** immunoglobulins.

Question 91

Antibodies function by:

- O **A.** phagocytizing invading antigens.
- O **B.** adhering to circulating plasma cells and marking them for destruction by phagocytizing cells.
- O **C.** preventing the production of stem cells in the bone marrow.
- O **D.** attaching to antigens via their variable portions.

Question 92

Lymphatic vessels absorb fluid from the interstitial spaces and carry it to the:

- O **A.** kidneys, where it is excreted.
- O **B.** large intestine, where it is absorbed and returned to the bloodstream.
- O **C.** lungs, where the fluid is vaporized and exhaled.
- O **D.** lymphatic ducts, which return it to the circulation.

Question 93

Negative selection takes place in the corticomedullary junction of the thymus where developing T cells interact with medullary Thymic Epithelial Cells (mTECs). The immature T cells would be expected to display:

- O **A.** self-antigen on Class I MHCs.
- O **B.** self-antigen on Class II MHCs.
- O **C.** non-self antigen on Class II MHCs.
- O **D.** self-antigen on Class I and Class II MHCs.

Question 94

An organism exposed to a pathogen for the first time will exhibit an innate immune response involving:

- O **A.** B-lymphocytes.
- O **B.** T-lymphocytes.
- O **C.** granulocytes.
- O **D.** an organism exposed to a pathogen for the first time must acquire immunity before it can respond.

Question 95

The molecule shown below is most likely to:

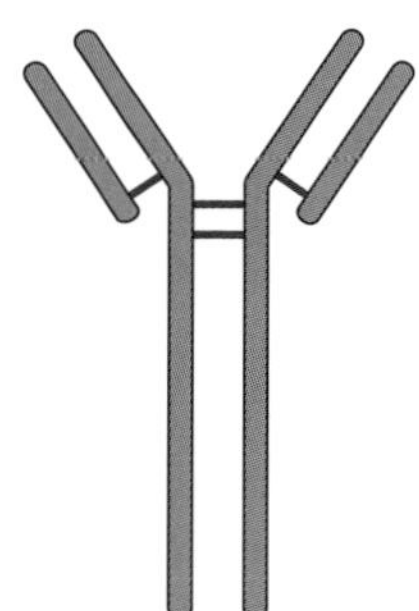

I. increase the presentation of viral proteins on MHC Class II molecules.
II. cause fluid movement from the blood vessels to the extracellular space.
III. increase clearance of viral particles in the spleen.

- O **A.** I only
- O **B.** I and II only
- O **C.** I and III only
- O **D.** I, II, and III

Question 96

In type I diabetes, self-reactive T-cells erroneously attack and destroy pancreatic beta cells that display self-antigens. This autoimmune process can be most precisely described as a failure of:

- O **A.** antigen processing.
- O **B.** B-cell activation.
- O **C.** positive selection.
- O **D.** negative selection.

TERMS YOU NEED TO KNOW

Acquired (adaptive) immunity
Albumin
Alveolar gas exchange
Alveoli
Antibody
Antigen
Antigen-antibody recognition
Aorta
Arteries
Arterioles
Atrioventricular (AV) node
Autoimmune diseases
B-cell immunity (humoral/antibody-mediated immunity)
B-lymphocytes
Blood clot
Bronchi
Bronchioles
Bundle of His
Capillaries
Carbonic anhydrase
Central chemoreceptors
Cilia
Clonal selection
Closed circulatory system
CO_2 sensitivity
Coagulation
Connective tissue
Cooperativity
Diaphragm
Diastolic pressure
Differential partial pressures
Differential pressures
Diffusion
Elasticity (resiliency)
Electrical synapses
Endothelial cells
Erythrocytes
Evaporation
Fibrinogen
Gap junctions
Glycerides
Heat exchange
Hematocrit
Hemoglobin
Henry's Law
Inflammation
Innate immunity
Left atrium
Left ventricle
Leukocytes
Macrophage
Major histocompatibility (MHC)
Medulla oblongata
Memory B-cells
MHC Class I
MHC Class II
Mucus
Nasal and tracheal capillary beds
Nasal hairs
Negative selection
Nervous and endocrine control
Non-specific
Open system
Osmotic pressure
Oxygen affinity
Oxyhemoglobin
Panting
Particulate matter
Peripheral chemoreceptors
Phagocytes
pH control
Plasma
Plasma cells
Platelets
Positive selection
Primary response
Pulmonary arteries
Pulmonary system
Pulmonary veins
Purkinje fibers
Regulation of plasma volume
Rib cage
Right atrium
Right ventricle
Rightward shift of oxygen dissociation curve
Secondary response
Serum
Sinoatrial (SA) node
Specific
Stem cell
Superior and inferior venae cavae
Surface tension
Systemic circulation
Systolic pressure
T-cell immunity (cell-mediated immunity)
T-lymphocytes
Thermoregulation
Total peripheral resistance
Trachea
Vagus nerve
Veins
Venules

DON'T FORGET YOUR KEYS

1. The goal of respiration and circulation is to meet metabolic needs.

2. Differential pressures and concentrations drive fluid flow and reactions.

3. The innate immune response is non-specific, fast, and short-acting. The acquired immune response is specific, slow, and long-lasting.

The Digestive and Excretory Systems

THE 3 KEYS

1. The three processes of the digestive system are digestion, absorption, and excretion. Surface area and length promote absorption.

2. From the kidney, substances travel in one of only two directions: "into the toilet" (filtration, secretion) and "back to the body" (reabsorption).

3. Diffusion down concentration gradients and active pumps in the ascending loop of Henle set up the concentration gradient in the nephron and kidney.

5.1 Introduction

This lecture discusses the structure and function of the human digestive and excretory systems. Humans use the macromolecules that make up food to power the body through the metabolic processes of glycolysis and cellular respiration. The body's ability to extract useful nutrition from food is central in sustaining life.

Digestion, the process by which the body extracts nutrition, allows macromolecules to enter the body by passing through the membranes lining the digestive tract. Digestion consists of three main processes: macromolecule breakdown, absorption, and excretion. Once food is broken down into its smaller constituent parts, nutrients can be absorbed into the bloodstream and made available for metabolic processes. Following digestion and absorption, the excretory system serves to process toxins and eliminate the waste products of metabolism and other bodily functions from the body. Although excretion involves the liver, which processes toxins, the excretory system primarily involves the kidney and the smallest operating unit of the kidney, the nephron. The manipulation of permeability and concentration gradients at the level of the nephron ensures that waste products are properly eliminated. All of these processes are mediated by enzyme activity.

When studying digestion for the MCAT®, keep the following organizing principles in mind: location, pH, and surface area. Within the digestive tract, each location has a specific function. The environments of different locations have different pH values, which help facilitate their various functions. Also, remember that digestion is primarily a system of exchange between the external and internal environments of the body. As in other systems of exchange, such as the respiratory and circulatory systems, surface area is critical. Surface area is increased by structural arrangements in different regions of the digestive tract. For a review of the basic properties of the macromolecules that are digested and absorbed, review Lecture 1 of the *Biology: Molecules* book.

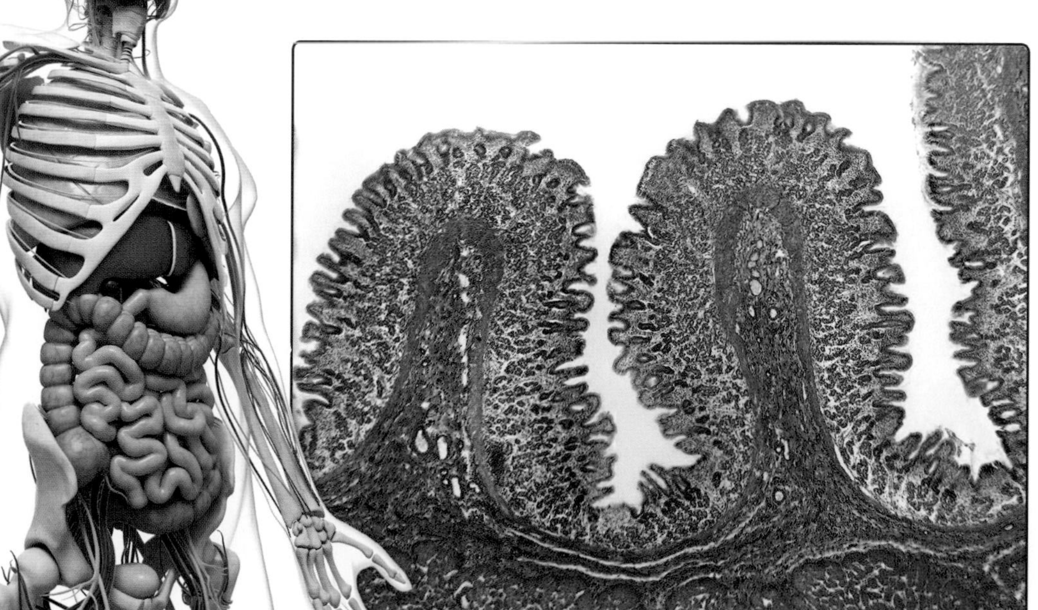

The small intestines have multiple small projections, called villi, that increase the surface area through which absorption of nutrients can occur.

5.2 Anatomy of the Digestive System

Be familiar with the basic anatomy of the digestive system and consider how location is related to function.

It is important to know the basic anatomy of the digestive tract (Figure 5.1). In the order in which food passes, the main components of the digestive tract are the **mouth**, **esophagus**, **stomach**, **small intestine**, **large intestine**, and **anus**. The entire digestive tract is considered to be outside the body because it is open to the environment at the mouth and the anus. Nutrients passing through the digestive tract do not enter into the body until they pass through the membrane of one of the digestive organs and are absorbed into the bloodstream.

FIGURE 5.1 Anatomy of the Digestive System

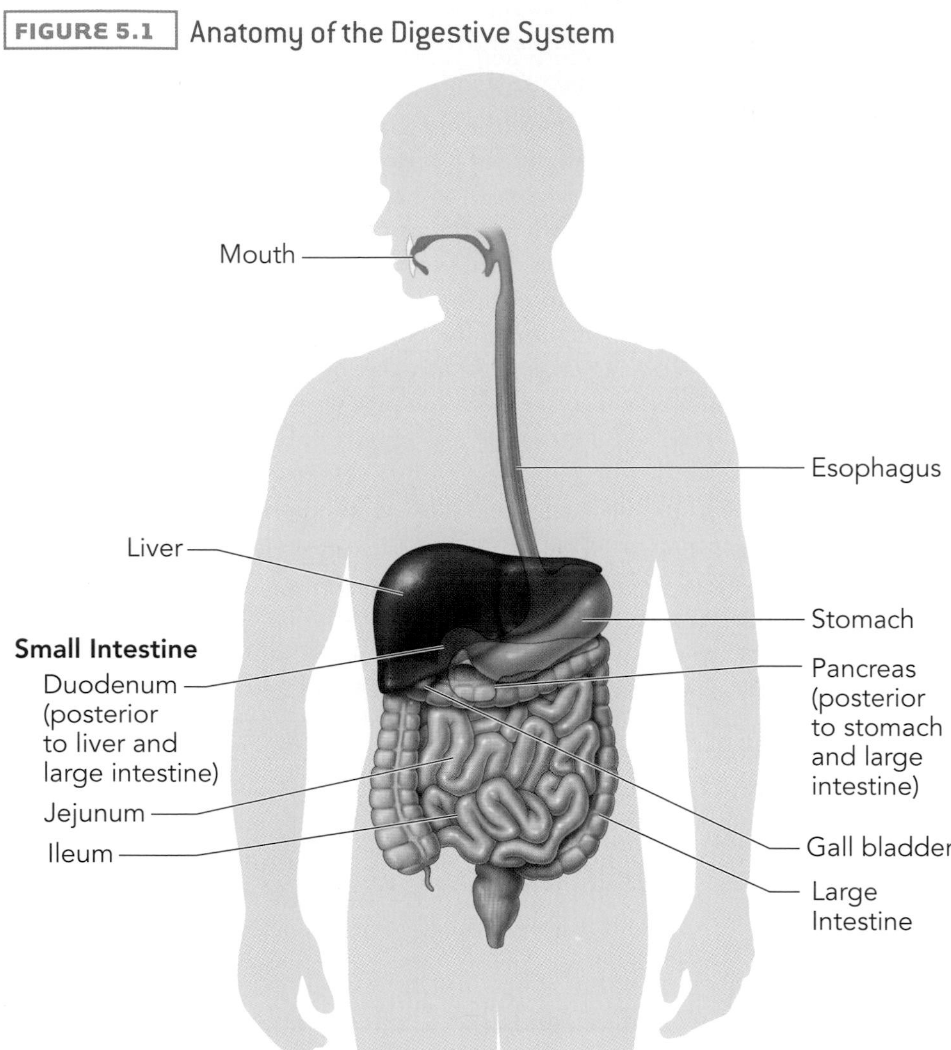

The Mouth and Esophagus

The first step in digestion is **ingestion**, the process of taking in food through the mouth. Two kinds of digestion begin in the mouth: physical and chemical. The physical breakdown of food occurs through the process of chewing, while the chemical breakdown of food is initiated primarily by the enzyme **α-amylase**, contained in saliva. Starch is the major carbohydrate in the human diet and is composed of many polymers of glucose and other sugar molecules. α-amylase begins to break down the long straight chains of starch into polysaccharides by cleaving the α-1,4 glycosidic bonds. Chewing increases the surface area of food, enabling more enzymes to act on the food at any one time. Chewed food forms a clump in the mouth called a *bolus*. The bolus is pushed into the esophagus by swallowing and then moves down the esophagus. (Technically, swallowing includes the movement of the bolus from the esophagus into the stomach, and is composed of a voluntary and involuntary stage.) **Peristalsis**, a contraction of the smooth muscle in the digestive tract, creates a wave motion that pushes along the partially digested food. Peristalsis is a common feature of the digestive tract, also occurring in the small and large intestines. Saliva acts as **lubrication** for the food, helping it to move down the esophagus. No digestion occurs in the esophagus. After food travels through the esophagus, it passes through the lower esophageal sphincter, a collection of smooth muscle that helps seal off the stomach from the esophagus, into the stomach for digestion.

Remember that digestion begins with carbohydrates in the mouth via α-amylase, and that there is no digestion in the esophagus. Also understand peristalsis, the wavelike motion of smooth muscle that moves food through the digestive tract. It is similar to squeezing a tube of toothpaste at the bottom and sliding your fingers toward the top to expel the toothpaste.

The Stomach

In the stomach, some of the major processes of digestion begin to take place so that the macromolecules in food can be broken down sufficiently for use by the body. The stomach is a rounded flexible pouch that serves to both mix and store food. The outer surface of the stomach is smooth, while the inner surface has many folds that allow the stomach to expand as it fills. The major anatomical regions of the stomach include the **fundus**, **body**, and **pylorus**. The fundus collects excess gas produced by digestion and can expand to store food ingested during a large meal prior to digestion. The body is the primary site of digestion, while the pylorus acts to prevent the passage of undigested food into the small intestines. The bolus moves into the stomach through the *lower esophageal sphincter* (or *cardiac sphincter*). A **sphincter** is a ring of muscle that is normally contracted so that there is no opening at its center. Both physical and chemical digestion continue in the stomach, and the digestion of proteins begins. The enzyme pepsin catalyzes the chemical breakdown of proteins in the stomach. The stomach also breaks food into smaller pieces using physical means. It is composed of several layers of smooth muscle that powerfully churn the contents of the stomach and reduce food to a semi-fluid mass called *chyme*.

FIGURE 5.2 Gross Anatomy of the Stomach

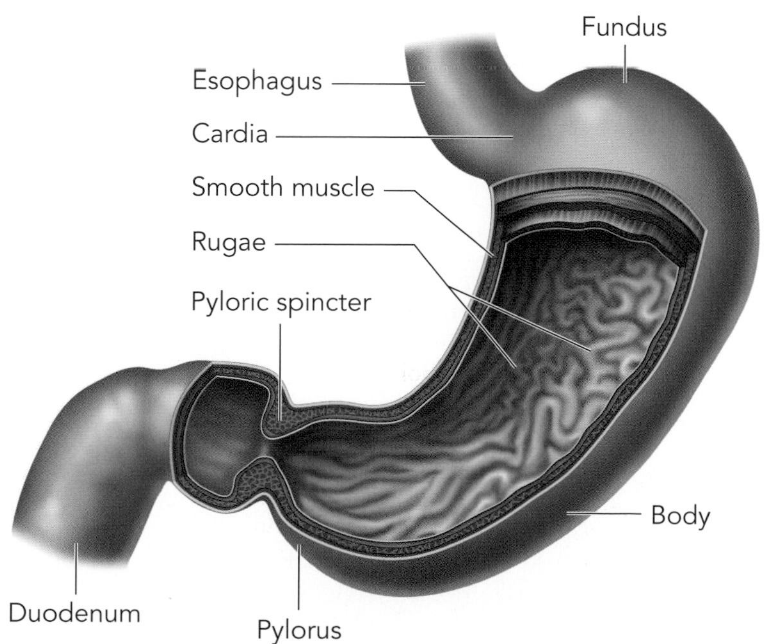

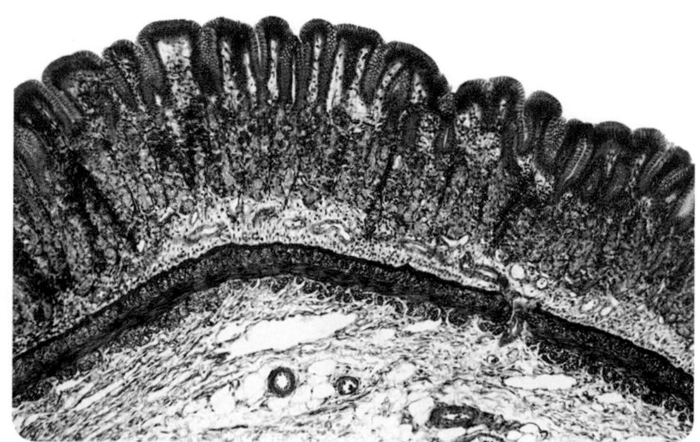

The lining of the stomach is called the mucosa.

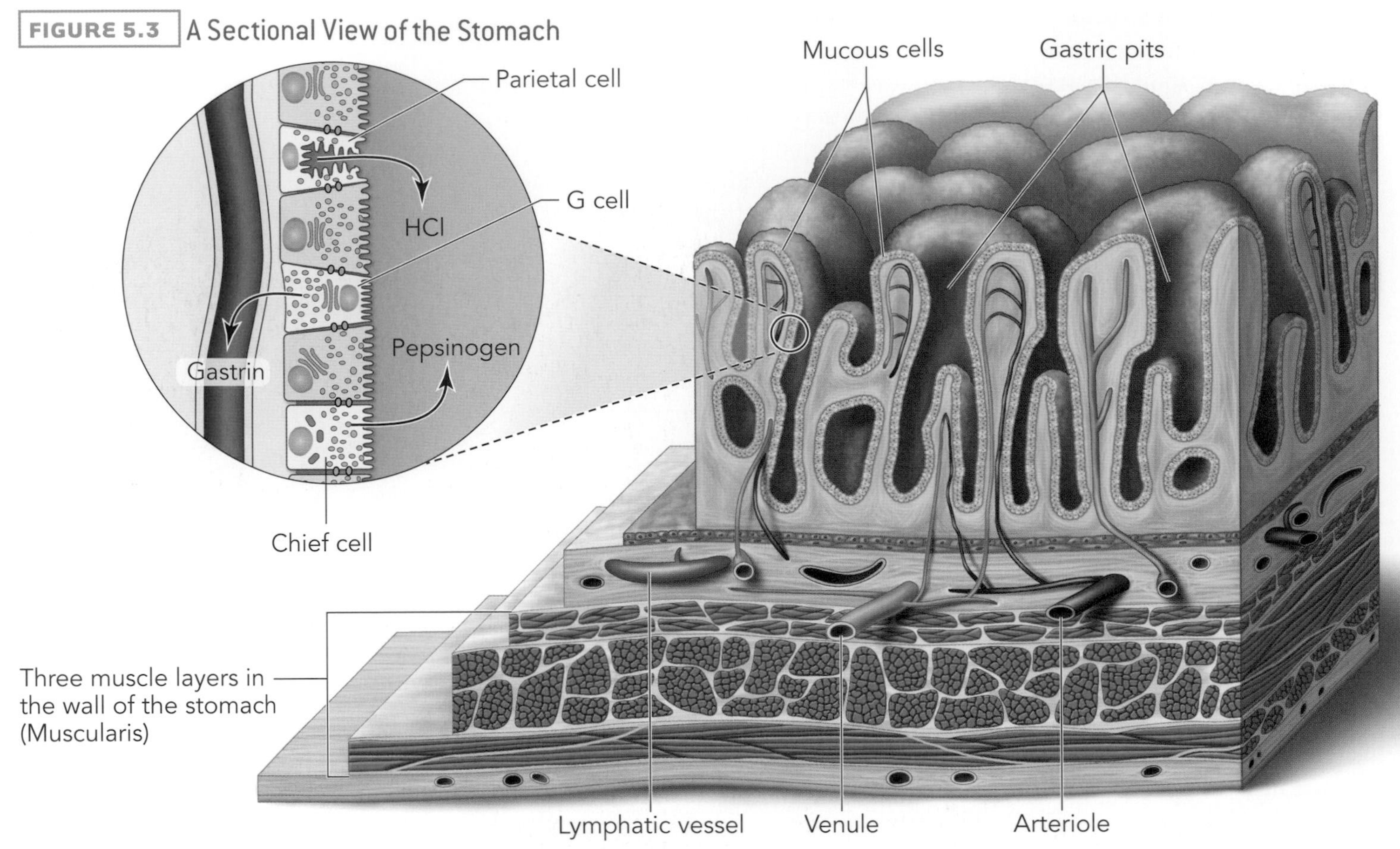

FIGURE 5.3 A Sectional View of the Stomach

The inside of the stomach is highly acidic. The acidic environment is maintained by **gastric juice**, a combination of acid, enzymes, and hormones released by cells in the lining of the stomach. A full stomach has a pH of 2. The low pH of the stomach contributes to protein digestion by denaturing proteins. It also helps to kill ingested bacteria.

The stomach contains **exocrine glands** in recesses called *gastric pits*, as shown in Figure 5.3. Remember that exocrine glands secrete molecules onto epithelial surfaces instead of the blood stream. In the stomach, the exocrine glands use ducts to deliver their secretions to specific locations in the external environment – in this case, to the lumen of the digestive tract. Other examples of exocrine glands are those that produce sweat, saliva, milk, and earwax. The exocrine system is understood in contrast to the endocrine system, which is a system of glands that do not have ducts and instead secrete their products directly into the internal environment of the bloodstream to circulate throughout the body. The exocrine cells in the lining of the stomach serve a variety of purposes, including maintenance of the acidity of the environment and aiding in digestion.

There are four major cell types in the stomach (Figure 5.4):

1. **mucous cells**,
2. **chief (peptic) cells**,
3. **parietal (oxyntic) cells**, and
4. **G cells.**

There are multiple types of mucous cells, but all of them perform the same basic function of secreting mucus. The mucous cells line the stomach wall and the necks of the exocrine glands. Mucus, composed mainly of a sticky glycoprotein and electrolytes, lubricates the stomach wall so that food can slide along its surface without causing damage. Mucus also protects the epithelial lining of the stomach wall from the acidic environment inside the stomach. Some mucous cells secrete a small amount of pepsinogen.

Chief cells are found deep in the exocrine glands. They secrete **pepsinogen**, the zymogen precursor to **pepsin**. Zymogens (also called proenzymes) are inactive precursors to enzymes that can be activated to become functional enzymes. The low pH in the stomach activates pepsinogen to become pepsin. Once activated, pepsin begins protein digestion.

Parietal cells are also found in the exocrine glands of the stomach. Parietal cells secrete **hydrochloric acid (HCl)** into the lumen of the stomach through active transport, which requires a large input of energy. Carbon dioxide is involved in the process, making carbonic acid inside the cell. The hydrogen from the carbonic acid is expelled to the lumen side of the cell, while the bicarbonate ion is expelled to the interstitial fluid side. The net result is that the pH of the stomach is lowered and the pH of the blood is raised. Parietal cells also secrete *intrinsic factor*, which, further into the digestive process, helps the ileum absorb the vitamin B12.

G cells secrete **gastrin**, a large peptide hormone that stimulates parietal cells to secrete HCl. G cells communicate with both the outside (the stomach lumen) and the inside (the bloodstream) of the body. The G cells in the stomach wall are activated through a variety of methods, including the presence of polypeptides in the stomach, stomach distension, and input from the parasympathetic nervous system through the vagus nerve. When an activated G cell releases gastrin into the bloodstream, the signal causes increased HCl production in stomach parietal cells. The major hormones that affect the secretion of the stomach juices are acetylcholine, gastrin, and *histamine*. Acetylcholine increases the secretion of all cell types. Gastrin and histamine mainly increase HCl secretion.

Remember that the main purpose of the stomach is to break down macromolecules for absorption in the small intestine. Once food is adequately broken down into chyme, it leaves the stomach through the pyloric sphincter, which leads into the small intestine.

The function of pepsin is to break down proteins. The body, which is made out of proteins, has to be careful about where pepsin is secreted to avoid digesting its own cells! The pepsin precursor, pepsinogen, is inactive, so it doesn't harm the cells of the stomach as it is released. It only becomes active in response to the acidic environment of the stomach.

MCAT® THINK

The cell types in the stomach wall are a good example of how form follows function. Parietal cells use a lot of energy to create HCl for digestion, so these cells contain large numbers of mitochondria. Cells specializing in secretion, such as mucous cells, chief cells, and G cells, contain larger amounts of rough endoplasmic reticulum and Golgi bodies.

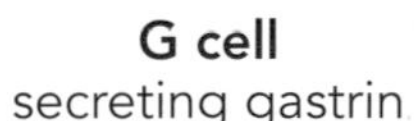
FIGURE 5.4 Gastric Gland Cell Types

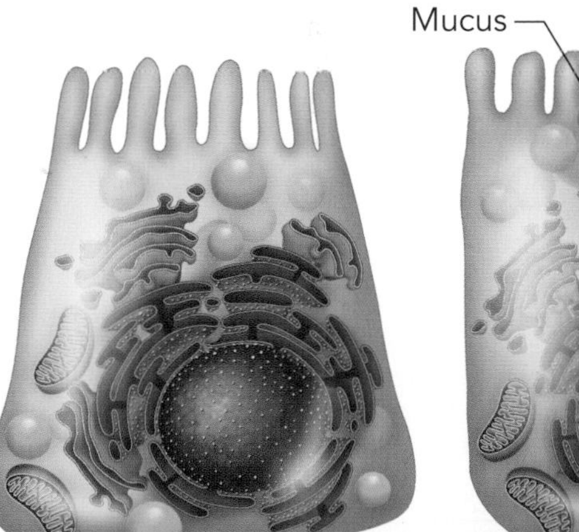

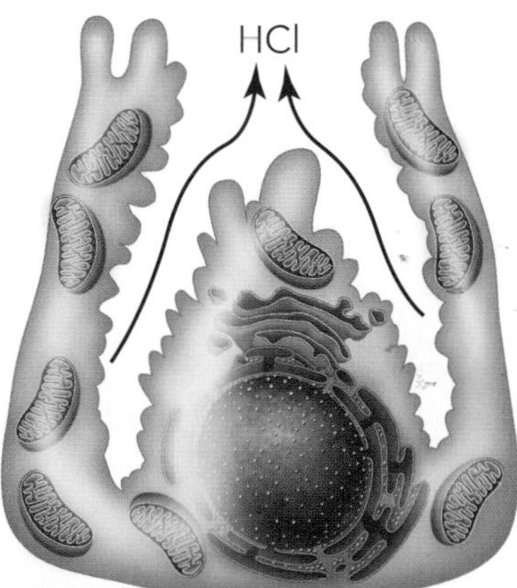

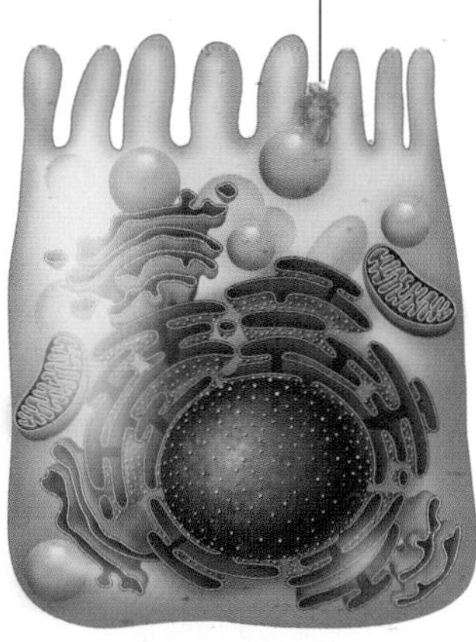

For the MCAT®, be familiar with the gastric cell types and their products.

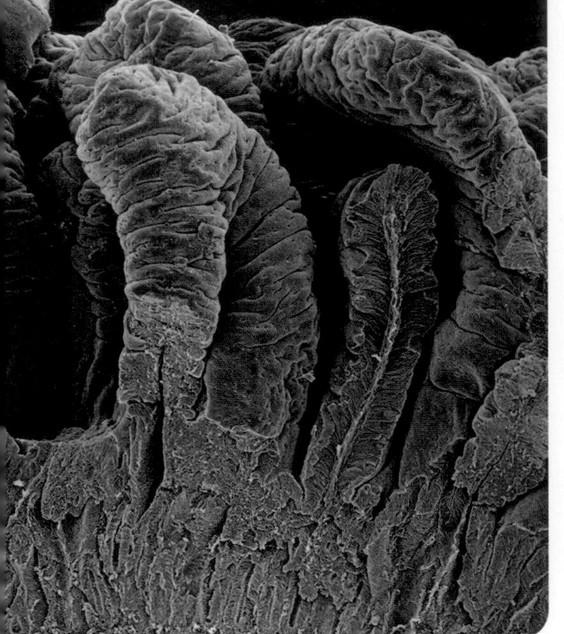

False-color scanning electron micrograph of a section through the wall of the human duodenum, showing the villi, which project 0.5 to 1 mm out into the intestinal lumen. Villi greatly increase the effective absorptive and secretory surface of the mucosa (mucus membrane) lining the small intestine. Each villus contains a central core of connective tissue (yellowish orange). This tissue contains large blood vessels, capillaries, some smooth muscle cells, and a lymph vessel known as a lacteal.

The Small Intestine

About 90% of digestion and absorption occurs in the small intestine. In a living human, the small intestine is about 3 meters in length. As food is digested, it passes through the three parts of the small intestine: the **duodenum**, the **jejunum**, and then the **ileum**. Most digestion occurs in the duodenum, and most absorption occurs in the jejunum and ileum. The wall of the small intestine is similar to the wall of the stomach except that the outermost layer contains finger-like projections called **villi** (Figure 5.5). The villi increase the surface area of the intestinal wall, allowing for greater digestion and absorption. Within each villus are a capillary network and a lymph vessel, called a **lacteal**. Lacteals are responsible for absorbing fats. Nutrients absorbed through the wall of the small intestine pass into the capillary network or the lacteal, depending on the type of macromolecule.

On the apical (lumen side) surface of the cells of each villus (cells called *enterocytes*) are much smaller finger-like projections called **microvilli**. The microvilli increase the surface area of the intestinal wall still further. Ample surface area is essential to facilitate the absorption of nutrients through the wall of the digestive tract and into the body. Under a light microscope, the microvilli appear as a fuzzy covering, called the brush border. The **brush border** contains membrane bound digestive enzymes, such as carbohydrate digesting enzymes (*dextrinase*, *maltase*, *sucrase*, and *lactase*), protein-digesting enzymes called peptidases, and nucleotide-digesting enzymes called *nucleosidases*. Some of the epithelial cells are **goblet cells** that secrete mucus to lubricate the intestine and help protect the brush border from mechanical and chemical damage. Similar to the mucous cells of the stomach, goblet cells secrete a sticky mix of glycoproteins and electrolytes. Dead cells regularly slough off into the lumen of the intestine and are replaced by new cells.

Digestion occurs in the small intestine on both a small and large scale. The breakdown of macromolecules into smaller pieces takes place within the lumen of the small intestine. These pieces are then broken down ever further within the

FIGURE 5.5 Small Intestine and Villus

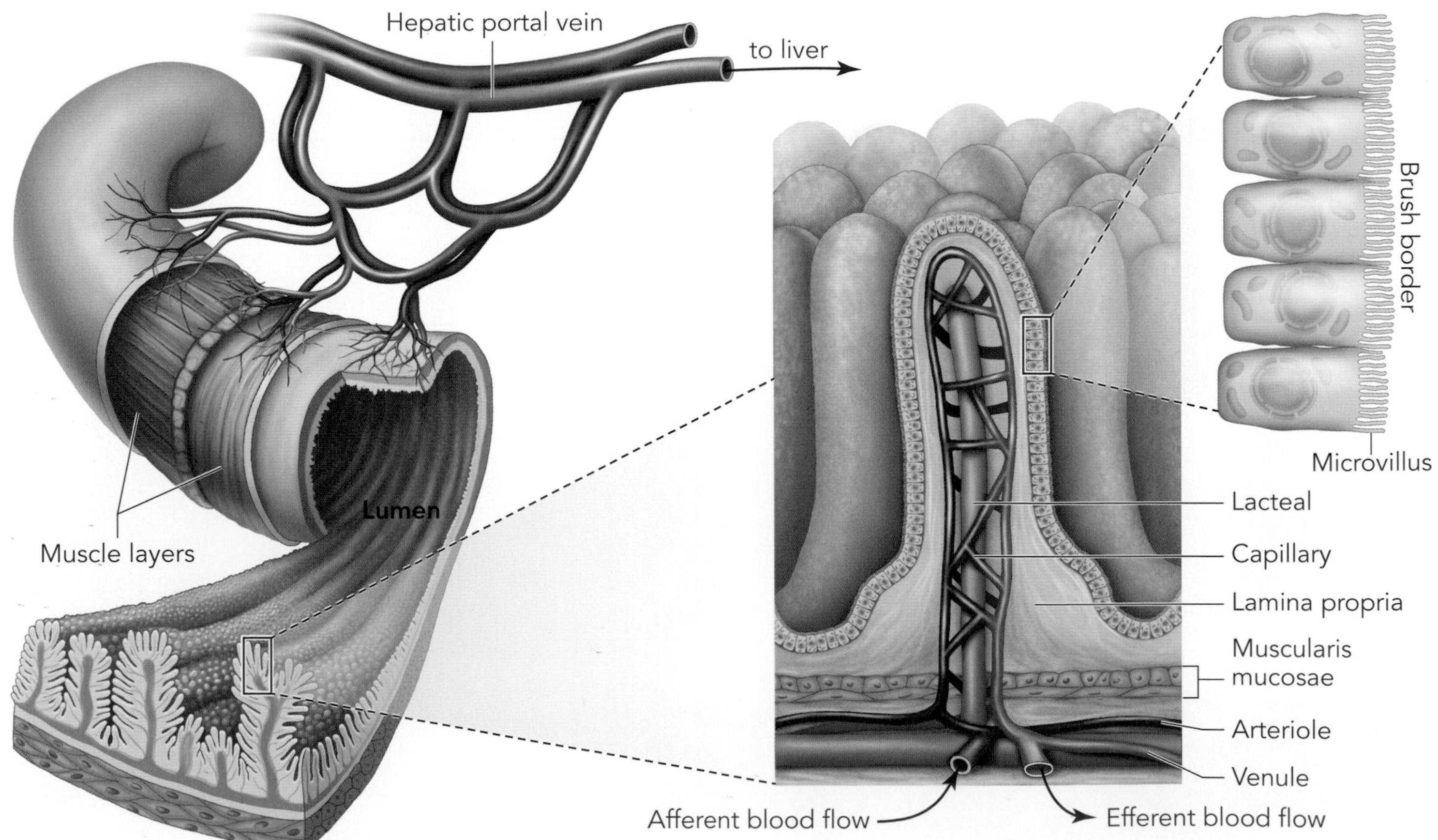

brush border. Micro-digestion within the brush border is an example of how the size of a structure matches the task that it carries out. The microvilli of the brush border are very small and are located directly adjacent to the channels through which nutrients will be transported. Microvilli are ideally sized to capture the pieces of macromolecules that have been broken down in the lumen. Enzymes in the microvilli further break down these pieces into their smallest constituent parts (e.g. amino acids, dipeptides, or tripeptides, in the case of proteins). The constituent parts are then transported through the membrane and into the body.

Located deep between the villi are the intestinal exocrine glands, called *crypts of Lieberkuhn*. These glands secrete an intestinal juice with a pH of 7.6 and *lysozyme*. Lysozyme contributes to regulation of bacteria within the intestine by weakening the bacterial cell wall.

As discussed throughout this section, the cells lining the small intestine, called enterocytes, are highly specialized for absorption. As a tradeoff, these cells are not as well suited to the secretion of digestive enzymes. Digestion within the small intestine is aided by digestive enzymes from the pancreas that continue to break down macromolecules and facilitate their absorption.

The Pancreas

The **pancreas** aids the digestive process in a number of ways. The fluid inside the duodenum has a pH of 6 because the hydrochloric acid from the stomach is neutralized by **bicarbonate ion** secreted by the pancreas. The pancreas is an endocrine gland that secretes insulin and glucagon, regulators of carbohydrate and fat metabolism. The pancreas also acts as an exocrine gland, creating enzymes that aid the digestive processes in the small intestine. The *acinar cells* of the pancreas release digestive enzymes into the *main pancreatic duct*, which carries the enzymes to the duodenum. The major enzymes released by the pancreas are trypsin, chymotrypsin, pancreatic amylase, lipase, ribonuclease, and deoxyribonuclease. All pancreatic enzymes are released as zymogens. Trypsin is activated by the enzyme *enterokinase* located in the brush border. Activated trypsin then activates the other enzymes.

Trypsin and **chymotrypsin** degrade proteins into small polypeptides. Another pancreatic enzyme, *carboxypolypeptidase*, cleaves amino acids from the sides of these peptides. Most proteins reach the brush border as small polypeptides. Here they are reduced to amino acids, dipeptides, and tripeptides before they are absorbed into the enterocytes. Enzymes within the enterocytes reduce the dipeptides and tripeptides to amino acids.

Like salivary amylase, **pancreatic amylase** hydrolyzes polysaccharides to disaccharides and trisaccharides; however, pancreatic amylase is much more powerful. Pancreatic amylase degrades nearly all of the carbohydrates from the chyme into oligosaccharides. The brush border enzymes finish degrading these polymers to their respective monosaccharides before they are absorbed.

Just as zymogens are released in the stomach, digestive enzymes from the pancreas are released as zymogens so that they do not digest the pancreas on the way to their destination – the food in the small intestine.

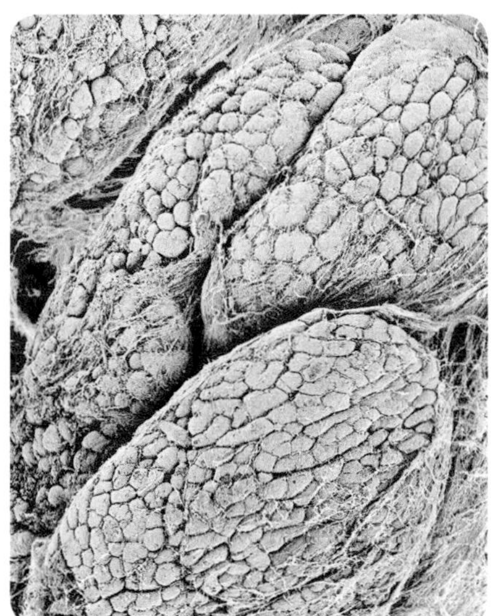

Several lobes of the pancreas are seen here, separated by fissures. The smaller sections seen on each lobe are clusters of acinar cells. The enzymes secreted by acinar cells drain into a highly branched system of ducts of increasing size, terminating in the main pancreatic duct, which feeds into the duodenum. The other function of the pancreas is the endocrine secretion of hormones, particularly insulin. Fragments of connective tissue and blood vessels are also seen.

FIGURE 5.6 The Pancreas

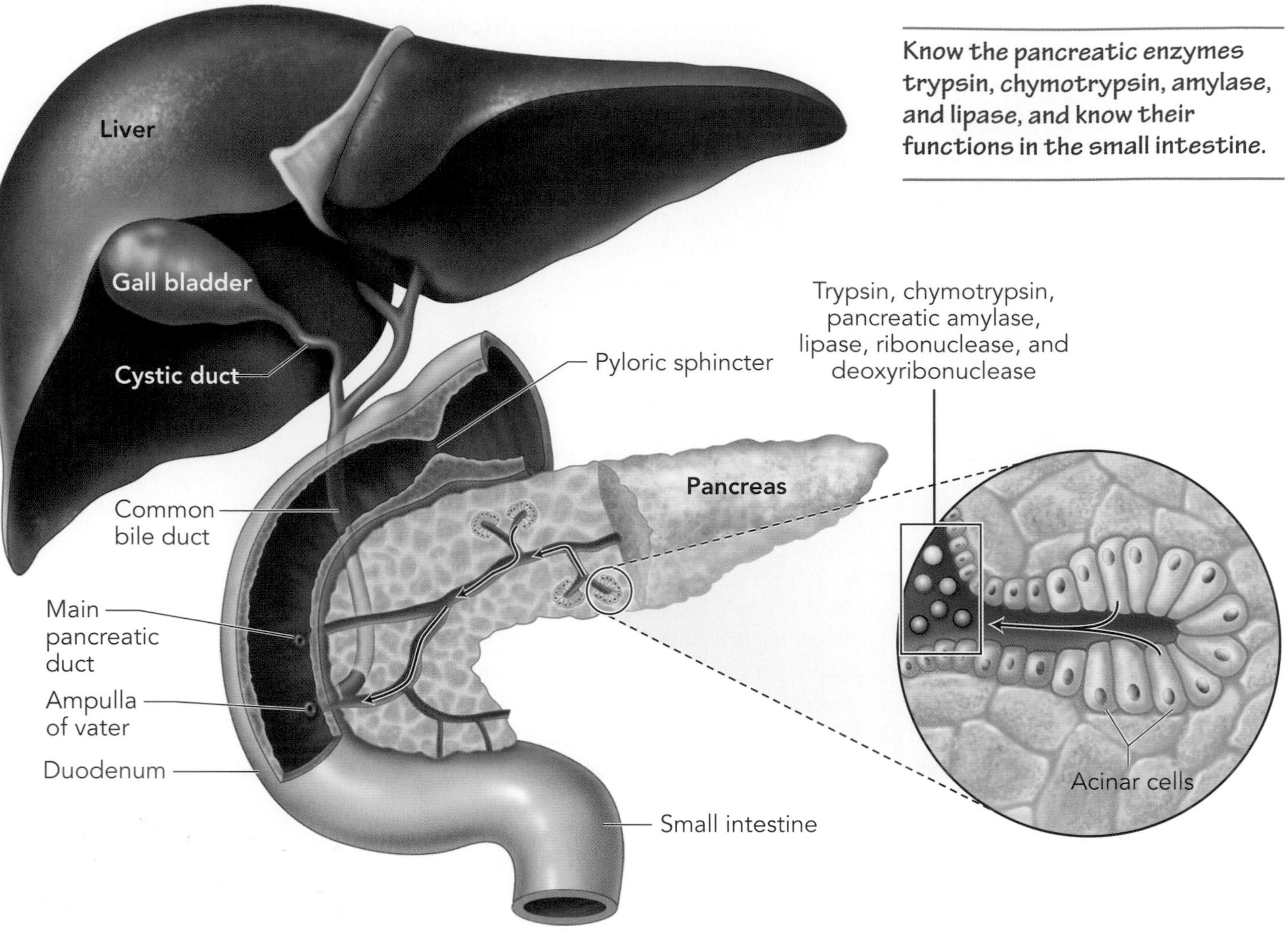

Know the pancreatic enzymes trypsin, chymotrypsin, amylase, and lipase, and know their functions in the small intestine.

Bile is necessary to increase the surface area of fat, but it does not digest the fat. In other words, bile physically separates fat molecules, but does not break them down chemically.

Lipase degrades fat, specifically **triglycerides**. However, since the intestinal fluid is an aqueous solution, the fat clumps together, reducing the surface area upon which lipase can act. This problem is solved by the addition of **bile**. Bile is produced in the liver and stored in the **gall bladder**. The gall bladder releases bile through the *cystic duct*, which empties into the *common bile duct* shared with the liver. The common bile duct empties into the *pancreatic duct* before connecting to the duodenum. In the small intestine, bile emulsifies the fat, breaking it up into small particles without changing it chemically. Emulsification increases the surface area of the fat so that lipase can degrade it into mainly fatty acids and monoglycerides. These products are shuttled to the brush border in bile micelles and then absorbed by the enterocytes. Much of the bile is reabsorbed by the small intestine and transported back to the liver.

Once the majority of digestion and absorption have occurred in the small intestine, the chyme exits the ileum through the cecum into the ascending colon of the large intestine.

The Large Intestine

The large intestine has five parts:

1. ascending colon,
2. transverse colon,
3. descending colon,
4. sigmoid colon, and
5. rectum.

The large intestine is involved in excretion, absorption, and some digestion. The major functions of the large intestine are water absorption and electrolyte absorption. When these functions fail, diarrhea or constipation results. The large intestine also contains a variety of bacterial flora. The bacteria are in a symbiotic relationship with the human that they inhabit. They live off of partially digested food within the digestive tract and, in turn, produce vitamin K, B12, thiamin, and riboflavin that can be used by the body.

Healthy feces are composed of 75% water. The remaining solid mass is 30% dead bacteria, 10-20% fat (mainly from bacteria and sloughed enterocytes), 10-20% inorganic matter, 2-3% protein, and 30% roughage (i.e. cellulose) and undigested matter. The rectum acts as a storage receptacle for feces until the waste is eliminated through the anus.

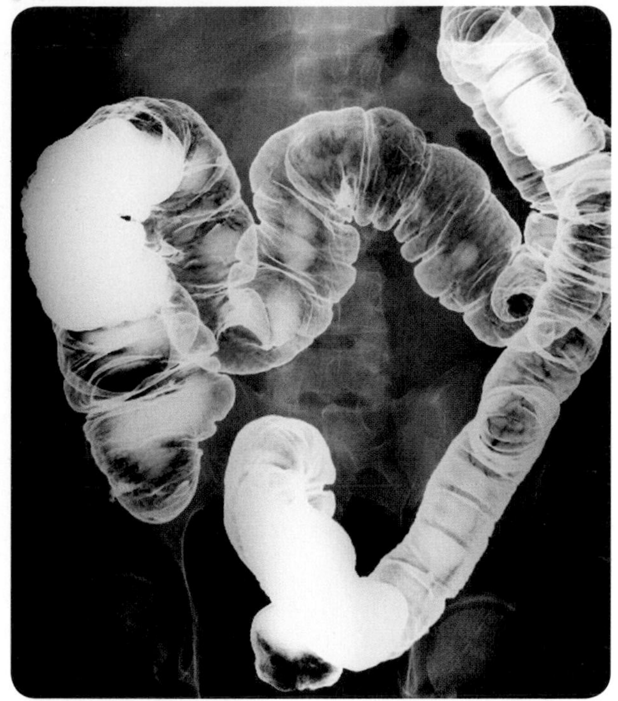

The folded structure of the large intestine is shown above.

FIGURE 5.7 Large Intestine

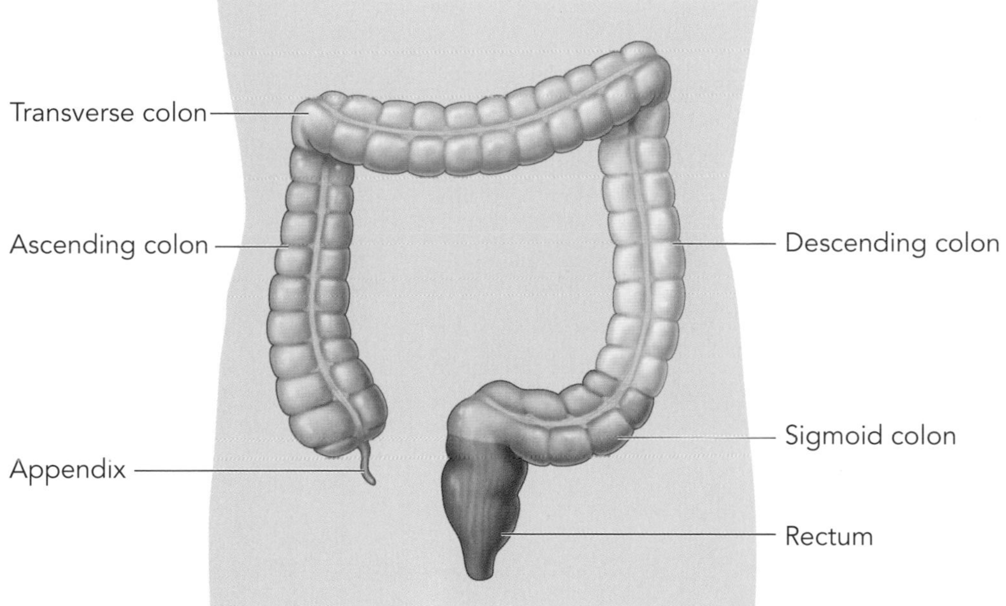

Whenever you get an MCAT® question involving the large intestine, think of water absorption. Profuse water loss in the form of diarrhea often results when there is a problem with the large intestine. Be aware that there is a symbiotic relationship between humans and bacteria in the large intestine. Bacteria get our leftovers, and we get certain vitamins from them. We have more bacteria in our gut than we have cells. Bacteria can sometimes cause disease, but in the large intestine they are necessary for our survival. See the Cell Lecture for more about symbiotic relationships and bacteria.

5.3 Gastrointestinal Hormones

The hormones involved in digestion are part of the **endocrine** (not exocrine) system and are a means of communication between organs of the digestive system. **Hormones** circulate within the bloodstream, so this kind of communication takes place within, rather than outside, the body.

Although there is inter-communication between the organs of digestion and the brain, the hormonal regulation in digestion tends to function according to an orderly chain of command. The brain stimulates the stomach to begin the digestive process, the stomach signals the small intestine, and the small intestine releases hormones that act on the pancreas. All of these processes promote normal digestion and also provide feedback to the brain on how digestion is progressing in the body.

Much of the nervous input to the digestive system takes place through the enteric nervous system. The **enteric nervous system** consists of a large network of neurons surrounding the digestive organs, helping to regulate processes such as smooth muscle contraction (peristalsis), fluid exchange, blood flow to the digestive organs, and hormone release. Nervous input from the brain also helps start the production of hormones. Parasympathetic input via the vagus nerve helps prepares the stomach and small intestine for digestion. As described in the context of the stomach, parasympathetic nervous stimulation causes the stomach to produce gastrin in anticipation of digestion. (For detailed information on the parasympathetic system, see the Nervous System Lecture.) Gastrin is one of the major hormones of digestion because of its action in causing increased HCl production. The stomach can also use gastrin to send signals to itself. The fuller the stomach becomes, the more acid it needs for digestion. When G cells sense that the stomach is full, they release more gastrin to increase acid production.

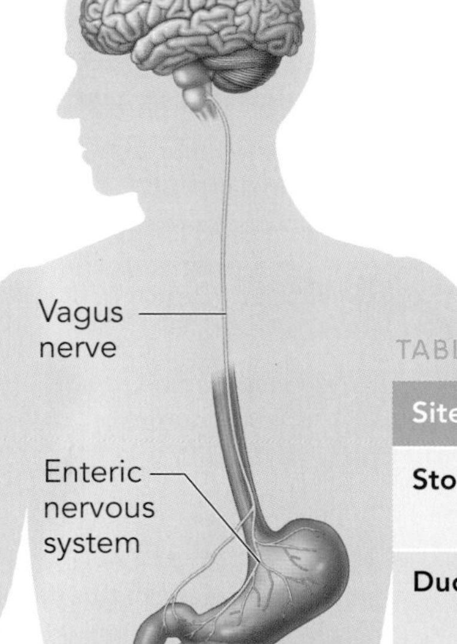

TABLE 5.1 > **Vagus Nerve and Digestive Hormones**

Site	Hormone	Stimulus	Target	Effects
Stomach	Gastrin	ACh release from vagus nerve	Stomach	Stimulates production of HCl
Duodenum	Secretin	Arrival of HCl in chyme	Pancreas	Stimulates secretion of sodium bicarbonate and enzymes
	Gastric inhibitory polypeptide	Arrival of fat and protein digestates in chyme	Pancreas	Stimulates enzyme secretion
			Stomach	Decreases motor activity
	Cholecystokinin (CCK)	Arrival of fat digestates in chyme	Pancreas	Stimulates enzyme secretion
			Stomach	Decreases motor activity

Signals from the stomach to the small intestine usually take the form of the release of chyme. The presence of chyme in the duodenum is both the message and the messenger telling the duodenum to start digestion.

Other important digestive hormones include *secretin*, *cholecystokinin*, and *gastric inhibitory polypeptide*, local peptide hormones secreted by the small intestine after a meal that act on the pancreas. Each of these hormones increases blood insulin levels, especially in the presence of glucose.

Secretin acts as a messenger from the small intestine to the pancreas. It lets the pancreas know that chyme has reached the small intestine and that pancreatic enzymes are needed for digestion. Hydrochloric acid in the duodenum causes secretin release. When it reaches the pancreas, secretin stimulates the release of sodium bicarbonate, which helps control pH, along with the pancreatic enzymes that aid in the digestive processes of the duodenum.

The contents of the duodenum may also influence what type of hormones are released. *Gastric inhibitory polypeptide* is released in response to fat and protein digestates in the duodenum, and to a lesser extent, in response to carbohydrates. It has a mild effect in decreasing the motor activity of the stomach. Similarly, food in the upper duodenum, especially fat digestates, causes the release of *cholecystokinin (CCK)*. Cholecystokinin causes gallbladder contraction and pancreatic enzyme secretion. It also decreases the motor activity of the stomach. These changes occur because fats are more difficult to digest than carbohydrates, due to the fact that they are hydrophobic and the digestive tract is an aqueous environment. The decreased motility of the stomach resulting from the action of both of these hormones causes the stomach to release chyme into the duodenum at a slowed pace, giving the pancreatic enzymes in the duodenum more time to emulsify fats.

You don't have to remember all of these gastrointestinal hormones, although they may appear in a passage. Instead, study the major ideas of digestion. The body eats in order to gain energy from food. The digestive system breaks down the food so it can be absorbed into the body. If the food moved too fast through the digestive tract, it could come out undigested. The stomach stores food and releases small amounts at a time to be digested and absorbed by the intestine. This way, the body can take in (eat) a large amount of food at a single time and take a long time to digest it. Some of the gastrointestinal hormones described in this section help regulate this process.

TABLE 5.2 > **Major Digestive Enzymes in Humans**

Source/Enzyme	Action
Salivary Glands	
Salivary amylase	Starch → Maltose
Stomach	
Pepsin	Proteins → Peptides; autocatalysis
Pancreas	
Pancreatic amylase	Starch → Maltose
Lipase	Fats → Fatty acid and glycerol
Nuclease	Nucleic acids → Nucleotides
Trypsin	Proteins → Peptides; Zymogen activation
Chymotrypsin	Proteins → Peptides
Carboxypeptidase	Peptides → Shorter peptides and amino acids
Small intestine	
Aminopeptidase	Peptides → Shorter peptides and amino acids
Dipeptidase	Dipeptides → Amino acids
Enterokinase	Trypsinogen → Trypsin
Nuclease	Nucleic acids → Nucleotides
Maltase	Maltose → Glucose
Lactase	Lactose → Galactose and glucose
Sucrase	Sucrose → Fructose and glucose

Digestion is the breakdown of food. The next section will discuss absorption, the process by which byproducts of digestion enter the body.

These questions are NOT related to a passage.

Question 97

As chyme is passed from the stomach to the small intestine, the catalytic activity of pepsin:

- O **A.** increases because pepsin works synergistically with trypsin.
- O **B.** increases because pepsin is activated from its zymogen form.
- O **C.** decreases in response to the change in pH.
- O **D.** decreases because pepsin is digested by pancreatic amylase in the small intestine.

Question 98

Which of the following is the best explanation for why pancreatic enzymes are secreted in zymogen form?

- O **A.** A delay in digestion is required in order for bile to increase the surface area of chyme.
- O **B.** Enzymes are most active in zymogen form.
- O **C.** Zymogens will not digest bile in the pancreatic duct.
- O **D.** Pancreatic cells are not as easily replaced as intestinal epithelium.

Question 99

Immediately following a meal, the pH of the blood leaving the capillaries of the stomach would be expected to:

- O **A.** increase due to an increase in hydrogen ion concentration.
- O **B.** increase due to a decrease in hydrogen ion concentration.
- O **C.** decrease due to an increase in hydrogen ion concentration.
- O **D.** decrease due to a decrease in hydrogen ion concentration.

Question 100

Which of the following reaction types is common to the digestion of all macronutrients?

- O **A.** Hydrolysis
- O **B.** Reduction
- O **C.** Glycolysis
- O **D.** Phosphorylation

Question 101

Researchers are studying a rare medical condition that causes food to move through the digestive system at three times the normal rate. Which of the following hypotheses would be inconsistent with what is known about digestion?

- O **A.** These individuals will suffer from multiple vitamin deficiencies.
- O **B.** Individuals with this condition will be obese.
- O **C.** Certain digestive hormones may be down-regulated.
- O **D.** Communication between the stomach and duodenum may be impaired.

Question 102

Where in the gastrointestinal tract does disaccharide digestion primarily take place?

- O **A.** In the HCl-filled pylorus of the stomach
- O **B.** In the villi-covered lumen of the colon
- O **C.** At the brush border of the jejunum and ileum
- O **D.** Via salivary amylase in the mouth

Question 103

Supplemental digestive enzymes are given to people who struggle to break down polymeric macromolecules. If a patient is prescribed a lipid digestive enzyme, in what location of the digestive system are they lacking a fully functional enzyme?

- O **A.** Duodenum
- O **B.** Stomach
- O **C.** Large intestine
- O **D.** Ileum

Question 104

In humans, most chemical digestion of food occurs in the:

- O **A.** mouth.
- O **B.** stomach.
- O **C.** duodenum.
- O **D.** ileum.

5.4 Molecular Digestion and Absorption

The function of the entire digestive tract is to convert ingested food into basic nutrients that can be absorbed by the small intestine. Once absorbed into the enterocytes, nutrients are processed and carried to individual cells throughout the body for use. The following section describes the process of absorption and the fates of the major nutrients: carbohydrates, proteins, and fats. The polarity of nutrients is an important factor when considering how nutrients pass through digestive membranes and how they travel once inside the body. This section is mainly background knowledge to provide the bigger picture of the digestive system; very little of this information will be tested directly on the MCAT®.

Secondary active transport drives a majority of macromolecule absorption. The Na^+/K^+ ATPase is important for maintaining sodium and potassium gradients that allow molecules like glucose to be transported into enterocytes. Remember that secondary active transport occurs when the movement of one ion, such as Na^+, down its concentration gradient is coupled with the movement of a second ion, such as glucose, up its concentration gradient.

Recall Key 1: Digestion, absorption, and excretion are the three main functions of the digestive system. Macromolecules are broken down into individual molecules and absorbed into the body through diffusion and facilitated transport. Waste products are excreted from the body into the bile or urine for elimination. Increased surface area and length promote increased digestion, absorption, and excretion.

The Cell
BIOLOGY 2

It is not necessary to memorize all of the details of the mechanisms for digestion of carbohydrates, proteins, and fats. Instead, understand the general principles. Recall that what substances are able to pass through plasma membranes is highly dependent on size and polarity. The inside of a plasma membrane is non-polar. In order to pass through a membrane, polar molecules must be broken down into smaller pieces. Non-polar molecules can pass through the membrane even if they are large. In the aqueous, polar environment of the cytoplasm or bloodstream, the opposite is true. Larger polar molecules travel freely, while non-polar molecules require assistance to travel through these polar environments. Use these principles to guide your understanding of digestion and absorption of macromolecules.

Carbohydrates

Carbohydrates provide a quick, easily accessible energy source for humans. As discussed in detail in the Metabolism Lecture, the body can metabolize carbohydrates through glycolysis and the citric acid cycle to produce energy needed to power cellular processes. The major carbohydrates in a human diet are sucrose, lactose, and starch. Cellulose, the polysaccharide making up the cell wall of plants, cannot be digested by humans and is considered *roughage*. Sucrose and lactose are disaccharides made from glucose and fructose, and from glucose and galactose, respectively. Starch is a straight chain of glucose molecules. Typically, 80% of the end product of carbohydrate digestion is glucose, and 95% of the carbohydrates in the blood are glucose.

A schematic of carbohydrate absorption is shown in Figure 5.8. By the time sugars reach the surface across which absorption occurs, they have been broken down into their smallest component parts (usually monosaccharides like glucose) so that they can pass more easily through the cell membrane. Glucose is absorbed via a secondary active transport mechanism down the concentration gradient of sodium. Sodium is actively pumped out of the enterocyte on the basolateral side. The resulting low sodium concentration inside the enterocyte drags sodium from the intestinal lumen into the cell through a transport protein, but only after glucose has also attached itself to the protein. In this way, glucose is dragged into the enterocyte by sodium. As the concentration of glucose inside the cell builds, glucose moves out of the cell on the basolateral side via facilitated transport. When

You can ignore most of the details here and concentrate on the big picture of carbohydrate digestion, absorption, and metabolism. Relate this information to glycolysis and the Krebs cycle in the Metabolism Lecture of *Biology 1: Molecules* for a complete picture. Notice that most of the glucose is stored for later use. When glycogen stores are full, glucose is converted to fat for long-term energy storage. The conversion of glucose to fat takes place in the liver and adipocytes and is stored in the adipocytes. Keep in mind the role of the liver in processing carbohydrates.

FIGURE 5.8 Carbohydrate Digestion and Absorption

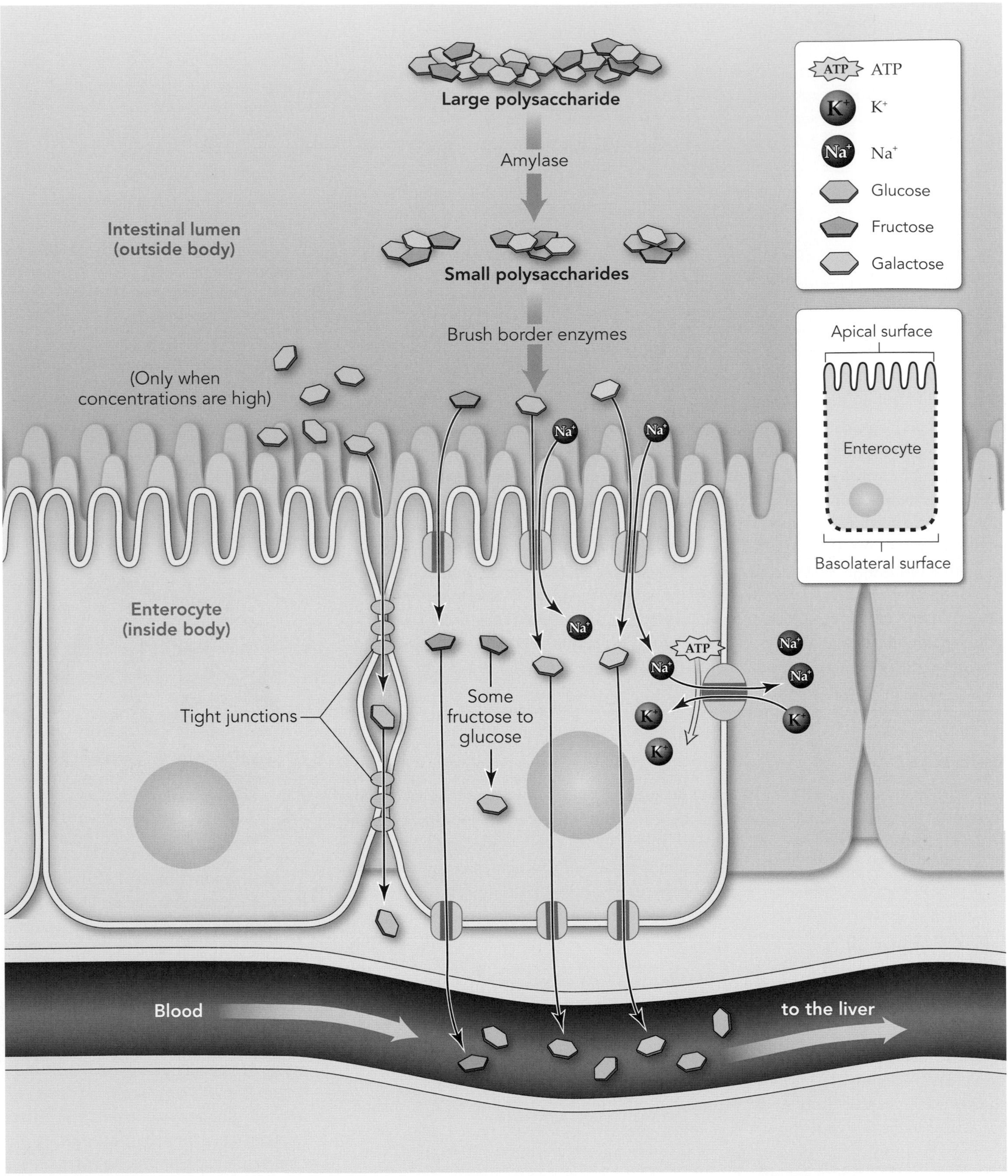

For the MCAT®, make sure you know how larger molecules are broken down into their smaller macromolecules and how individual monomers, like glucose, are absorbed. It is also important to known how secondary active transport facilitates absorption and whether absorbed molecules are exported into the blood or into the lymphatics. Think through the flow of a large molecule all the way to the systemic circulation for Figures 5.8, 5.9, and 5.10.

there are high concentrations of glucose in the intestinal lumen, glucose may also be absorbed by a second mechanism. At high concentrations, glucose builds up in the space around the cells and raises the osmotic pressure there. The aqueous solution of the lumen is dragged into the space through the tight junctions, pulling glucose along with it.

Galactose follows a similar absorption path to glucose. Fructose is absorbed via facilitated diffusion, and much of it is converted to glucose inside the enterocyte. All carbohydrates are absorbed into the bloodstream and carried by the portal vein to the liver. One of the functions of the liver is to maintain a fairly constant blood glucose level by releasing and storing glucose. The liver absorbs the carbohydrates and converts nearly all the galactose and fructose into glucose. If blood glucose levels are sufficient, the liver converts excess glucose into glycogen for storage. The formation of glycogen is called **glycogenesis**. When the blood glucose level decreases, the reverse process, **glycogenolysis**, takes place in the liver and glucose is returned to the blood.

In all cells except enterocytes and the cells of the renal tubule, glucose is transported from high concentration to low concentration through facilitated diffusion. Nearly all cells are capable of producing and storing some glycogen, but only muscle cells and especially liver cells store large amounts. When the cells have reached their saturation point with glycogen, carbohydrates are converted to fatty acids and then triglycerides in a process requiring a small amount of energy.

Proteins

Like carbohydrates, proteins tend to be hydrophilic and cannot easily pass through the cell membrane. They are also broken down into small pieces during absorption. Protein digestion results in amino acids, dipeptides, and tripeptides. Absorption of many of these products occurs via a co-transport mechanism down the concentration gradient of sodium, similar to the mechanism used by glucose. A few amino acids are transported by facilitated diffusion. Because the chemistry of amino acids varies greatly, each transport mechanism is specific to a few amino acids or polypeptides.

Nearly all polypeptides that are absorbed into an enterocyte are hydrolyzed to their amino acid constituents by enzymes. From the enterocytes, amino acids are absorbed directly into the blood and then are taken up by all cells of the body, especially the liver. Transport into the cells may be facilitated or active, but is never simple diffusion, since amino acids are too large and polar to diffuse through the membrane. The cells immediately create proteins from the amino acids so that the intracellular amino acid concentration remains low, preserving the concentration gradient. Most proteins are easily broken down and returned to the blood when needed. When cells reach their upper limit for protein storage, amino acids can be burned for energy or converted into fat for storage. Approximately 4 Calories of energy can be gained per gram of protein. By contrast, carbohydrates produce about 4.5 Calories per gram, and fats produce about 9 Calories per gram. Ammonia, a nitrogen containing compound, is a byproduct of gluconeogenesis from proteins. Nearly all ammonia is converted to **urea** by the liver and then excreted in the urine.

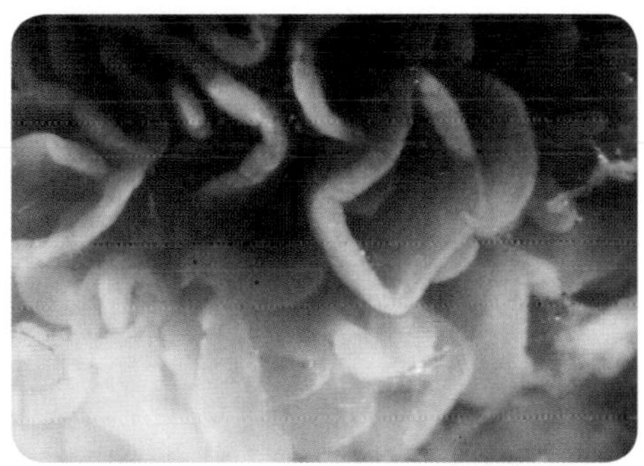

This photo shows the normal surface pattern of the jejunal mucosa.

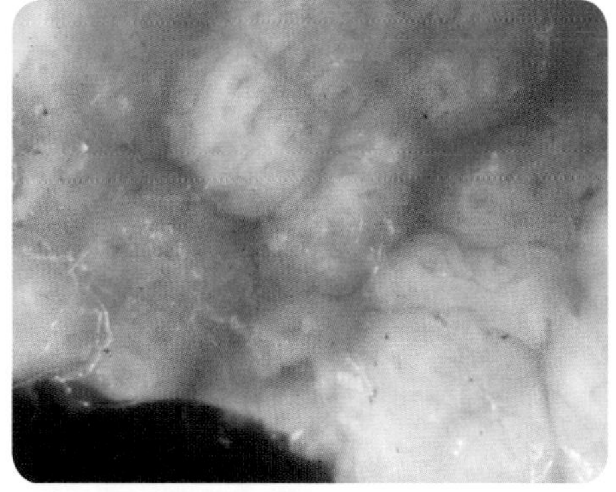

This photo shows the jejunal mucosa of a person with celiac disease. Celiac disease is an uncommon condition caused by hypersensitivity to a component of gluten, a protein in wheat flour.

Again, concentrate on the big picture. Remember that virtually all dietary protein is completely broken down into its amino acids before being absorbed into the blood. In fact, any protein that is not broken down completely may cause allergic reactions. Also, when you think of proteins, think nitrogen.

FIGURE 5.9 Protein Digestion and Absorption

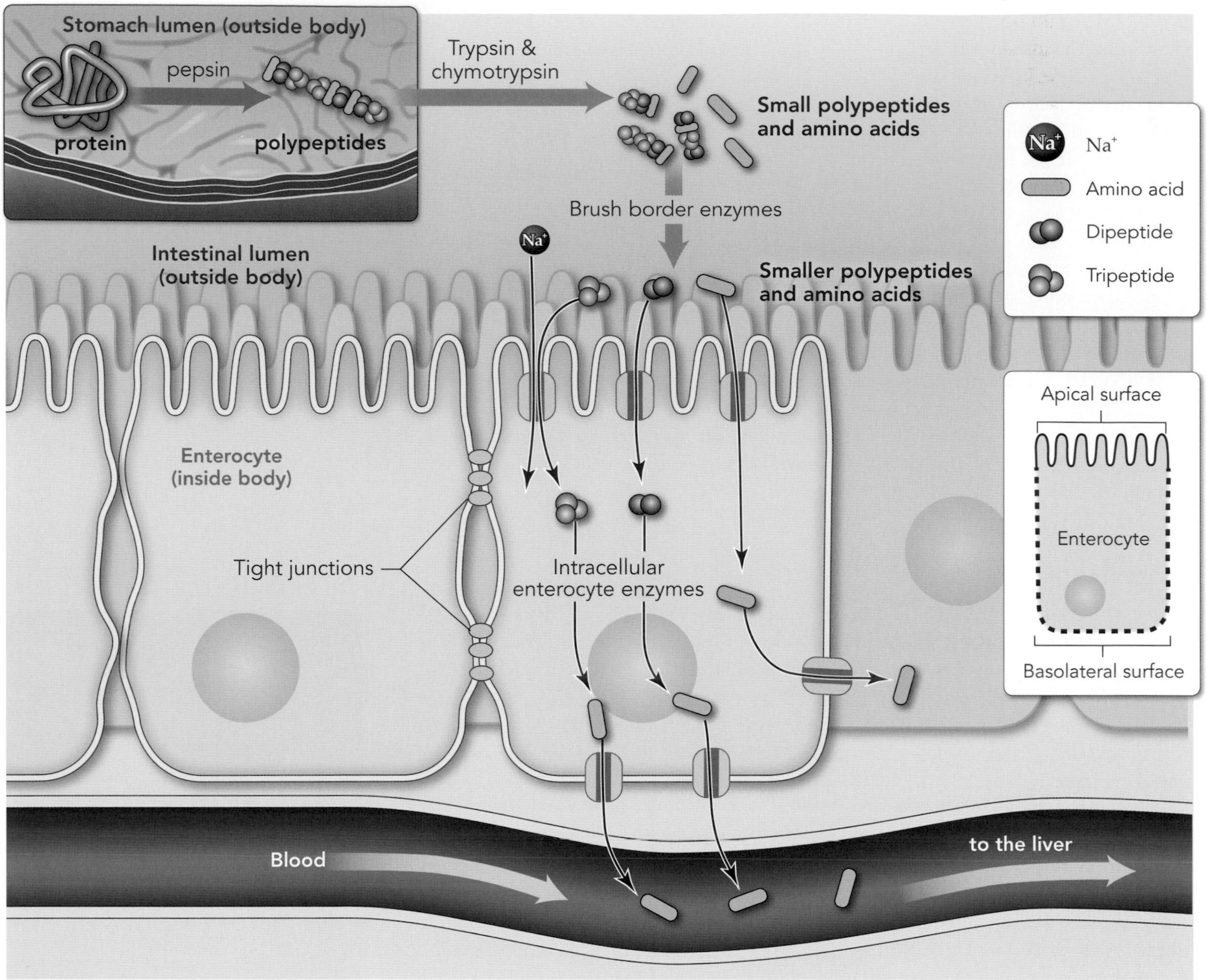

Fats

Because they are hydrophobic molecules, the absorption of fats encounters many challenges that are opposite to those of proteins and carbohydrates. Fats are not easily transported in the aqueous environments of the digestive lumen and the intracellular space, but they easily pass through hydrophobic cell membranes. Most dietary fat consists of triglycerides, which are broken down to monoglycerides and fatty acids in the digestive process. These components are then shuttled to the brush border by bile micelles. At the brush border, fats easily diffuse through the enterocyte membrane (Figure 5.10). After delivering their fats, the micelles return to the chyme to pick up more fat digestates. Micelles also carry small amounts of hydrolyzed phospholipids and cholesterol, which, like monoglycerides and fatty acids, diffuse through the enterocyte membrane.

Fat absorption is also required for the proper uptake of some vitamins, called *fat-soluble vitamins*. These vitamins are hydrophobic and are co-absorbed with fat. The fat soluble vitamins are A, D, E, and K.

Once inside the enterocytes, fats must be altered such that they can travel through the aqueous environment of the cell. Monoglycerides and fatty acids are converted back into triglycerides at the smooth endoplasmic reticulum. The

FIGURE 5.10 Fat Digestion and Absorption

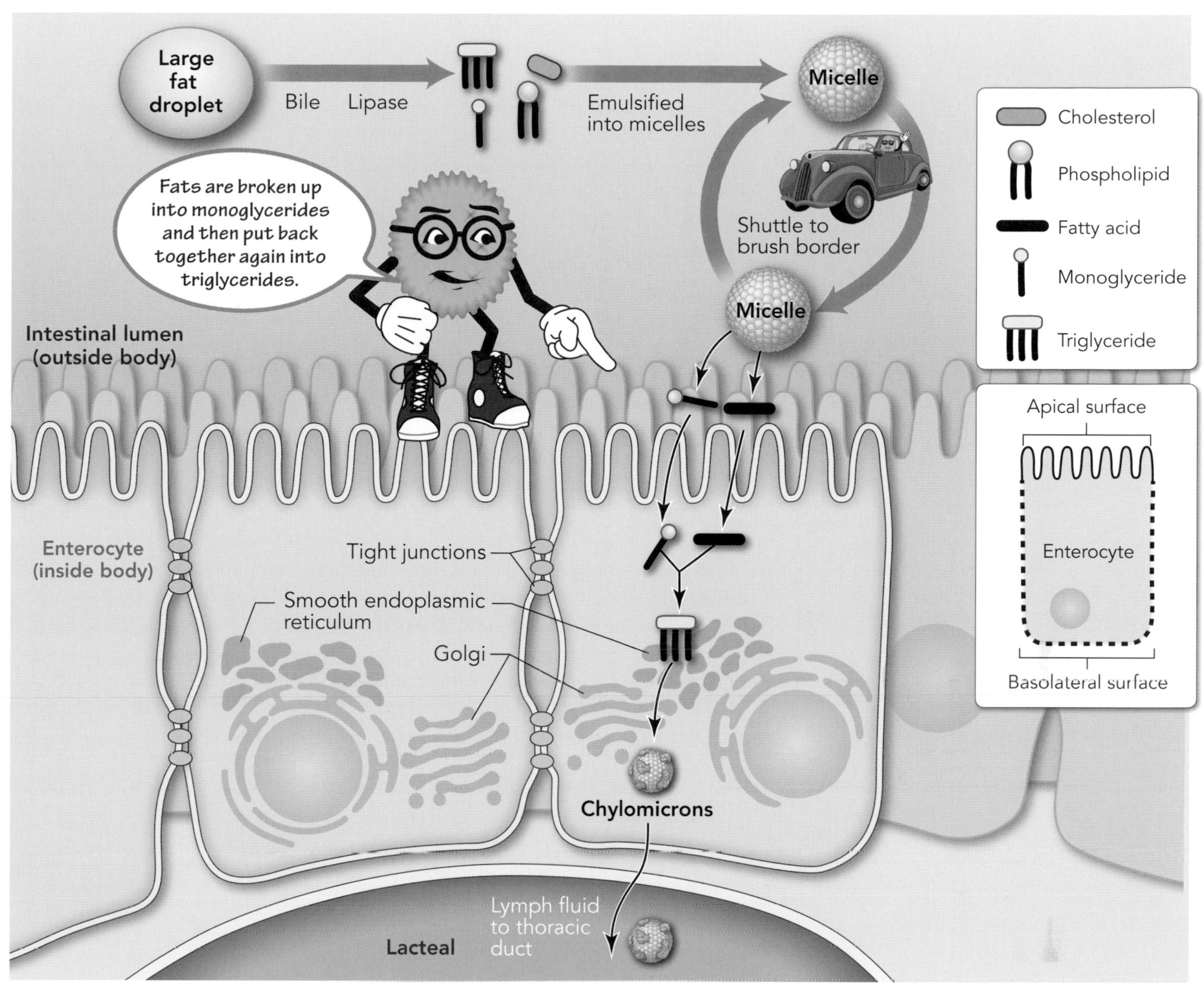

newly synthesized triglycerides aggregate within the smooth endoplasmic reticular lumen along with some cholesterol and phospholipids. These amphipathic molecules orient themselves with their charged ends pointing outward toward the aqueous solution of the lumen. Apoproteins attach to the outside of these globules. (See the Biological Molecules and Enzymes Lecture in *Biology 1: Molecules* for more on apoproteins.) The globules move to the Golgi apparatus and are released from the cell via exocytosis. Most of these globules, now called *chylomicrons*, move into the lacteals of the lymph system. Note the distinguishing feature of fat metabolism: while carbohydrates and proteins move from the enterocyte into the bloodstream, fats are absorbed into the lymph system. Travel through the lymphatics is most likely required due to the large diameter of the chylomicron that would have difficulty traversing the small capillaries and venules of the small intestine. Most ingested fat that is absorbed by this process moves through the lymph system and enters the large veins of the neck at the *thoracic duct*. Small amounts of more water soluble short chain fatty acids are absorbed directly into the blood of the villi.

The concentration of chylomicrons in the blood peaks about 1-2 hours after a meal, but falls rapidly as the fat digestates are absorbed into the cells of the body. The most significant absorption of fat occurs in the liver and adipose tissue. Chylomicrons stick to the side of capillary walls, where *lipoprotein lipase* hydrolyzes the

Keep in mind that fat is insoluble in water, so it typically requires a carrier (i.e. a lipoprotein or albumin). For the MCAT®, associate fat with efficient long-term energy storage: lots of calories (energy) per mass.

triglycerides. The products immediately diffuse into fat and liver cells, where the triglycerides are reconstituted at the smooth endoplasmic reticulum. The first stop for most of the digested fat is the liver.

From adipose tissue, most fatty acids are transported in the form of *free fatty acids*, which combines with the protein **albumin** in the blood. A single albumin molecule typically carries 3 fatty acid molecules, but is capable of carrying up to 30.

Between meals, 95% of lipids in the plasma are in the form of *lipoproteins*. Lipoproteins look like small chylomicrons, or, more precisely, chylomicrons are large lipoproteins. Besides chylomicrons, there are four types of lipoproteins:

1. *very low-density lipoproteins*,
2. *intermediate-density lipoproteins*,
3. *low-density lipoproteins*, and
4. *high-density lipoproteins*.

Biological Molecules
BIOLOGY 1

All are made from triglycerides, cholesterol, phospholipids, and protein. As the density increases, first the relative amount of triglycerides decreases, and then the relative amount of cholesterol and phospholipids decreases. Very low-density lipoproteins have many triglycerides, and high-density lipoproteins have very few triglycerides. Most lipoproteins are made in the liver. Very-low density lipoproteins transport triglycerides from the liver to adipose tissue. Intermediate and low-density lipoproteins transport cholesterol and phospholipids to the cells. The function of high-density lipoproteins is less well understood. Lower density lipoproteins seem to contribute to hardening of the arteries, while high-density lipoproteins seem to protect against such hardening. See the Biological Molecules and Enzymes Lecture for more information on lipoproteins.

FIGURE 5.11 Fat Transport Within LDL and HDL

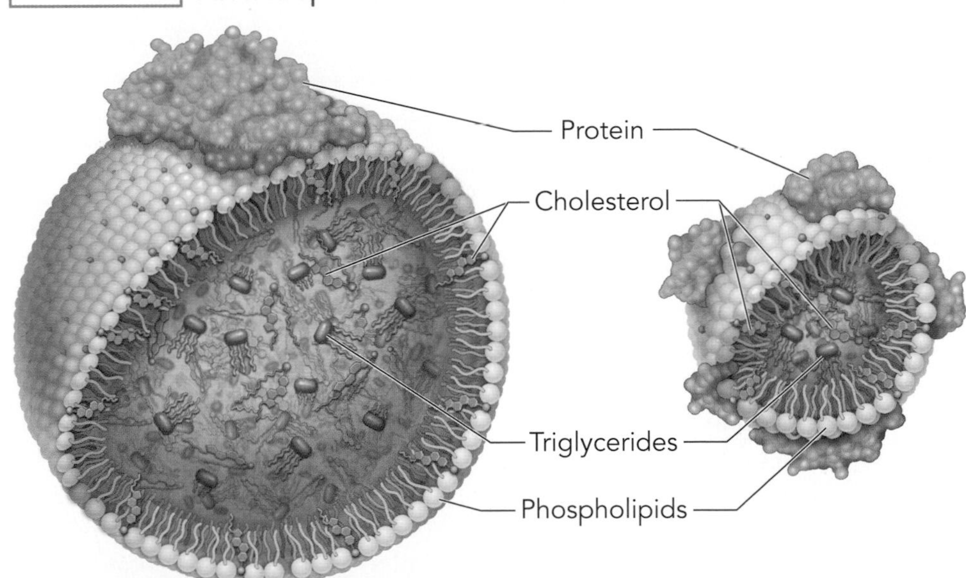

5.5 The Liver: Storage, Distribution, and Detox

The **liver** is located primarily in the upper right-hand quadrant of the abdomen adjacent to the organs of the digestive tract. Blood from the capillary beds of the intestines, stomach, spleen, and pancreas feeds into the large *hepatic portal vein*. The hepatic portal vein carries all of the blood from the digestive system to the liver so that the liver can process the blood before it is re-circulated throughout the rest of the body. Blood from the intestines must pass through the liver before it can enter the systemic circulation. By processing the blood, the liver functions as a screening mechanism for all of the nutrients, toxins, or other molecules that are absorbed from the gut into the bloodstream. The liver removes many ingested toxins from the bloodstream so that they do not enter wider circulation. A second blood supply, used to oxygenate the liver, is received through the hepatic artery. All blood received by the liver collects in the *hepatic vein*, which leads to the vena cava.

The liver serves a wide variety of functions in the body. Think of the liver as a large warehouse where substances can be recycled into new products, stored for later use, or processed for distribution.

FIGURE 5.12 Anatomy of the Liver

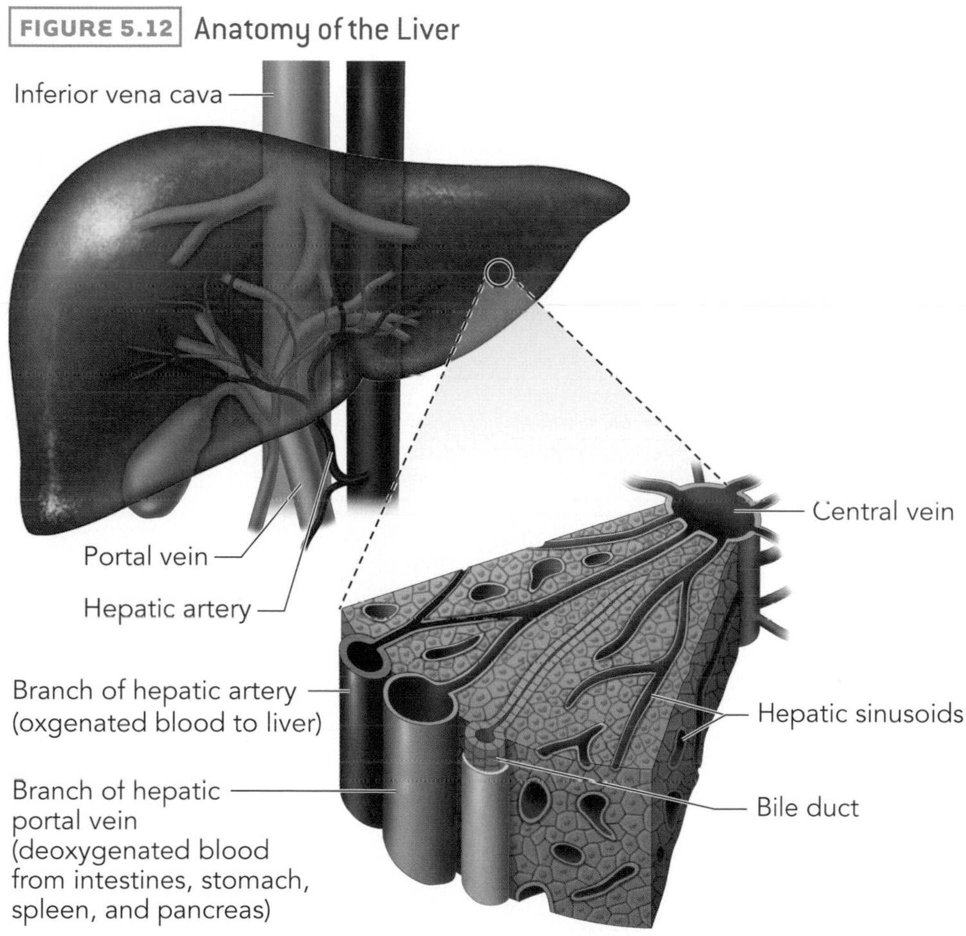

The liver has a wide variety of purposes that can be categorized as metabolic, storage, or immune functions.

Metabolic Functions of the Liver

1. **Carbohydrate metabolism**: The liver maintains normal blood glucose levels through **gluconeogenesis** (the production of glycogen and glucose from non-carbohydrate precursors), glycogenesis, and release of glucose stores according to the needs of the body.
2. **Fat metabolism**: The liver synthesizes bile from cholesterol and converts carbohydrates and proteins into fat. It oxidizes fatty acids for energy, and forms most lipoproteins. When the liver mobilizes fat for energy, it produces acids called *ketone bodies*. This often results in a condition called *ketosis* or *acidosis*. For the MCAT®, know that when the liver mobilizes fats or proteins for energy, the acidity of the blood increases.
3. **Protein metabolism**: The liver deaminates amino acids, forms urea from ammonia in the blood, synthesizes plasma proteins such as fibrinogen, prothrombin, albumin, and most globulins, and synthesizes nonessential amino acids.
4. **Detoxification**: Detoxified chemicals are secreted by the liver as part of bile or modified so that they can be excreted by the kidney.

Prothrombin and fibrinogen are important clotting factors. Albumin is the major osmoregulatory protein in the blood. Globulins are a group of proteins that include antibodies. However, antibodies are not produced in the liver. They are made by plasma cells.

Storage Functions of the Liver

1. **Blood storage**: The liver can expand to act as a blood reservoir for the body.
2. **Glycogen storage**: The liver stores large amounts of glycogen as an energy reserve that can be used to regulate blood glucose levels.
3. **Vitamin storage**: The liver stores vitamins such as vitamins A, D, and B12. The liver also stores iron.

Immune Functions of the Liver

1. **Blood filtration**: *Kupffer cells* phagocytize bacteria picked up from the intestines.
2. **Erythrocyte destruction**: *Kupffer cells* also destroy irregular erythrocytes, although most irregular erythrocytes are destroyed by the spleen.

FIGURE 5.13 Functions of the Liver

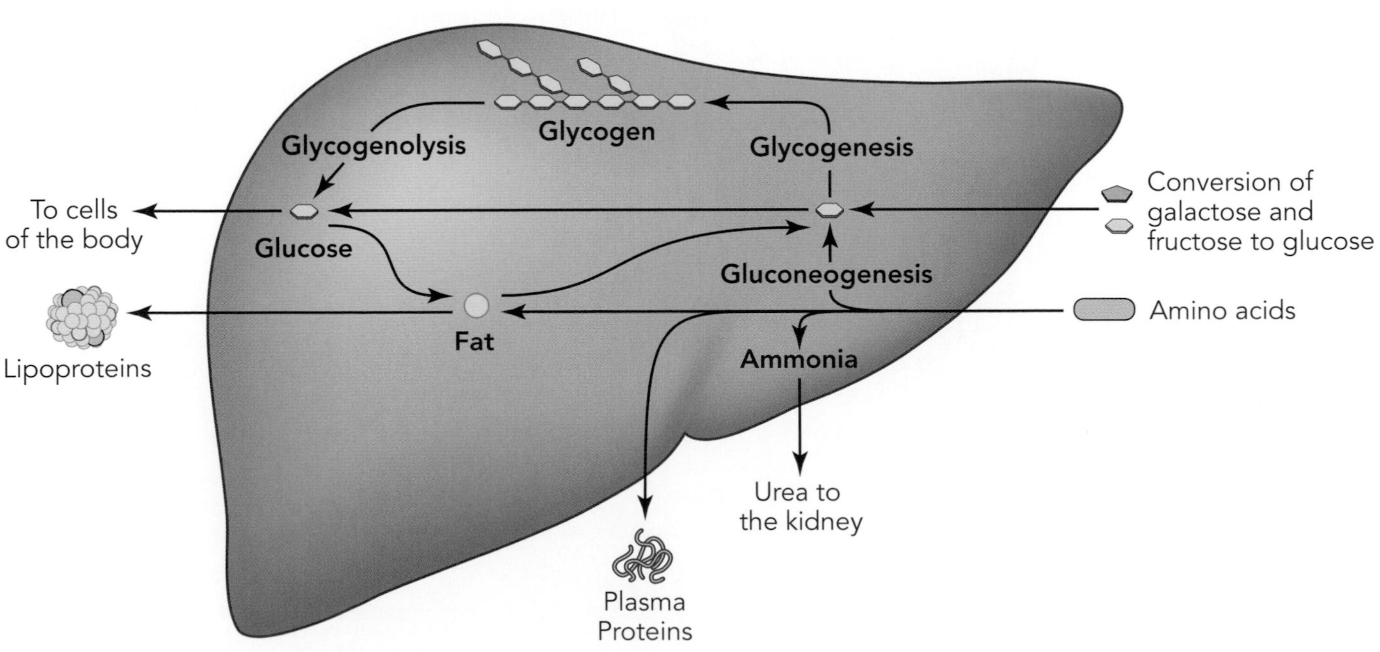

These questions are NOT related to a passage.

Question 105

When dietary triglycerides are absorbed in the small intestine, which of the following digestive processes occurs?

- O **A.** Triglycerides are absorbed by enterocytes and then transported to the bloodstream.
- O **B.** Bile acid is secreted into the small intestine to prevent the hydrolysis of triglycerides.
- O **C.** Triglycerides are hydrolyzed to monoglycerides and free fatty acids prior to absorption.
- O **D.** The hydrolyzed components of triglycerides form chylomicrons to facilitate absorption by enterocytes.

Question 106

Which of the following occurs mainly in the liver?

- O **A.** Fat storage
- O **B.** Protein degradation
- O **C.** Glycolysis
- O **D.** Gluconeogenesis

Question 107

Dietary fat consists mostly of neutral fats called triglycerides. Most digestive products of fat:

- O **A.** enter intestinal epithelial cells as chylomicrons.
- O **B.** are absorbed directly into the capillaries of the intestines.
- O **C.** are degraded to fatty acids by the smooth endoplasmic reticulum of enterocytes.
- O **D.** enter the lymph system before entering the bloodstream.

Question 108

Which of the following is LEAST likely true concerning the digestive products of dietary protein?

- O **A.** They are used to synthesize essential amino acids in the liver.
- O **B.** Some of the products are absorbed into the intestines by facilitated diffusion.
- O **C.** Energy is required for the intestinal absorption of at least some of these products.
- O **D.** Deamination of these products in the liver leads to urea in the blood.

Question 109

Cholera is an intestinal infection that can lead to severe diarrhea causing profuse secretion of water and electrolytes. A glucose-electrolyte solution may be administered orally to patients suffering from cholera. What is the most likely reason for mixing glucose with the electrolyte solution?

- O **A.** When digested, glucose increases the strength of the patient.
- O **B.** The absorption of glucose increases the uptake of electrolytes.
- O **C.** Glucose is an electrolyte.
- O **D.** Glucose stimulates secretion of the pancreatic enzyme, amylase.

Question 110

Most of the glycogen in the human body is stored in the liver and the skeletal muscles. Which of the following hormones inhibits glycogenolysis?

- O **A.** Cortisol
- O **B.** Insulin
- O **C.** Glucagon
- O **D.** Aldosterone

Question 111

Fatty acids are transported within the body bound to protein carriers. The most likely explanation for this is that:

- O **A.** blood is an aqueous solution and only hydrophobic compounds are easily dissolved.
- O **B.** blood is an aqueous solution and only hydrophilic compounds are easily dissolved.
- O **C.** blood serum contains chylomicrons which do not bind to fatty acids.
- O **D.** blood serum is lipid based and the polar region of a fatty acid will not be dissolved.

Question 112

Essential amino acids must be ingested because they cannot be synthesized by the body. In what form are these amino acids likely to enter the bloodstream?

- O **A.** Single amino acids
- O **B.** Dipeptides
- O **C.** Polypeptides
- O **D.** Proteins

Questions 105-112 of 144

STOP

I always keep track of where water and solutes are going by remembering that they will go back into my body or into the toilet. When water or a solute moves into the filtrate, it will eventually leave the body as urine going into the toilet. When water or a solute moves out of the filtrate into the medulla, the vasa recta, or the glomerulus, it's going back into my body. Waste products go into the toilet; things I want to keep stay in my body.

The kidney is all about blood composition, blood pressure, and excretion.

5.6 The Kidney: Excretion, Blood Pressure & Composition

The major functions of the **kidney** are:

1. maintaining homeostasis of body fluid volume and thereby regulating blood pressure;
2. maintaining homeostasis of plasma solute composition and helping control plasma pH; and
3. excreting waste products, such as urea, uric acid, ammonia, and phosphate.

This section will examine the structure of the kidney and the rest of the urinary system in detail, including the role of each part of the nephron and the function of the juxtaglomerular apparatus. The section also examines the mechanisms that set up and maintain a concentration gradient in the medulla, allowing the kidney to make concentrated urine. The kidney participates in the control of blood pressure in the circulatory system. It is also linked to the metabolic and detoxification functions of the liver through excretion of specific drugs and toxins as well as the products of the urea cycle.

Each one of the two kidneys is a fist-sized organ made up of an outer **cortex** and an inner **medulla**. Urine is created by the kidney and emptied into the *renal pelvis*. The renal pelvis is emptied by the *ureter*, which carries urine to the *bladder*. The bladder is drained to the external environment by the urethra. The bladder keeps the *urethra* closed through the action of the urinary sphincter muscles to avoid leaking urine. During urination, the urinary sphincter muscle relaxes and a muscle in the bladder contracts to allow the release of urine.

The functional unit of the kidney is the **nephron**. Each human kidney contains approximately one million nephrons. Blood entering a nephron first flows into a capillary bed called the **glomerulus** (Figure 5.14). Together, **Bowman's capsule** and the glomerulus make up the *renal corpuscle*. **Hydrostatic pressure** forces some plasma through **fenestrations** of the glomerular endothelium and into Bowman's capsule. The fenestrations screen out blood cells and large proteins, preventing them from entering Bowman's capsule. The fluid that enters Bowman's capsule is called *filtrate* or *primary urine*. The retained blood cells and proteins circulate through the peritubular capillary beds.

The hydrostatic pressure of the glomerulus is related to the amount of blood that is flowing through it. The diameter of the afferent and efferent arterioles can constrict or dilate to control the blood flow and thus the hydrostatic pressure. The terms "afferent" and "efferent" are used in the same way as they are in the nervous system. The afferent arteriole brings blood to the glomerulus, just like afferent nerves take information to the nervous system. The efferent arteriole takes blood away from the glomerulus, just like efferent nerves take information from the nervous system. Constriction of the afferent arteriole will decrease blood flow and hydrostatic pressure, resulting in less filtration. Dilation has the opposite effect. In contrast, contraction of the efferent arteriole increases hydrostatic pressure by preventing blood from exiting as quickly. This would increase filtration.

Filtrate moves from Bowman's capsule to the **proximal tubule**. The proximal tubule is where **secretion** and most **reabsorption** takes place. Drugs, toxins, uric acid, bile pigments, and other solutes are secreted into the filtrate by cells of the proximal tubule and will eventually leave the body as part of the urine. Hydrogen ions are secreted by the proximal tubule through an **antiport** system driven by the sodium concentration gradient. By contrast to the transport system of glucose with sodium discussed earlier in this lecture, in this antiport system the proton crosses the membrane in the opposite direction to sodium.

Reabsorption in the proximal tubule allows the kidney to retain valuable nutrients that were inadvertently filtered out and to return these substances to the rest

FIGURE 5.14 Glomerulus

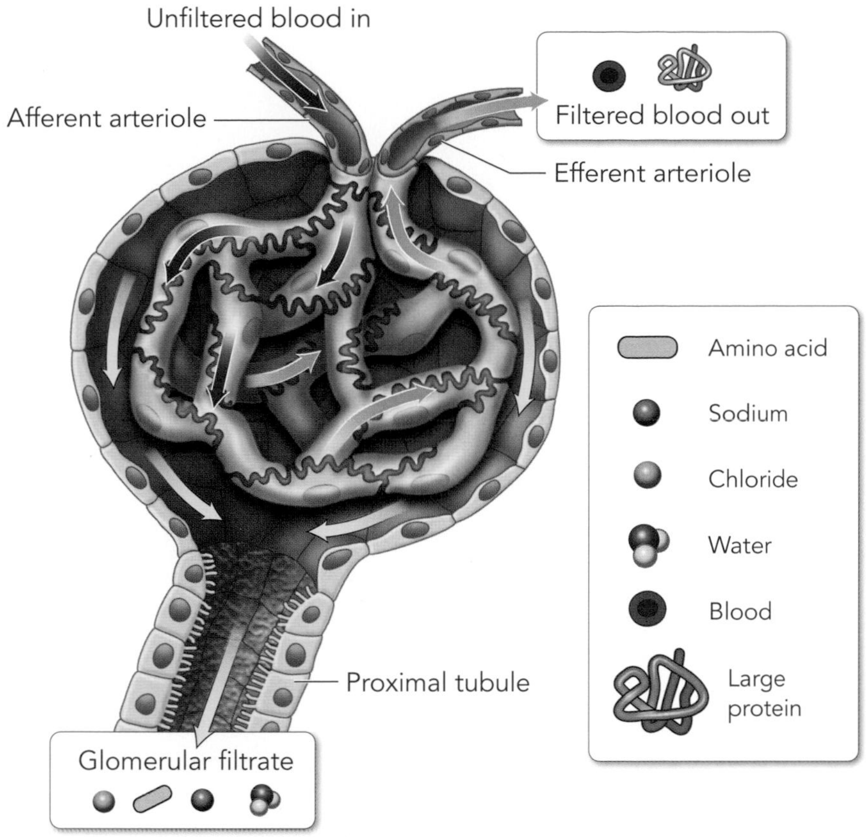

The kidneys are part of the excretory system, which eliminates waste products and also includes the intestines and the skin. Just like those organs, the kidneys have critical additional functions, including their role in maintaining homeostasis of body fluids. They also work with the lungs to maintain the plasma pH through a balance of acidic carbon dioxide and alkaline bicarbonate, and are critically important for maintaining proper blood pressure in the cardiovascular system.

of the body through the capillary circulation. Reabsorption can occur via passive or active transport. Secondary active transport proteins in the apical membranes of proximal tubule cells carry out the reabsorption of nearly all glucose, most proteins, and other solutes. These transport proteins can become saturated. Once a solute has reached its transport maximum, any more solute is washed into the urine. The concentration of a solute that saturates its transport proteins is called the *transport maximum*. Some solutes that are not actively reabsorbed are reabsorbed by simple or facilitated diffusion. Water is reabsorbed into the renal interstitium of the proximal tubules across relatively permeable tight junctions by osmosis. The net result of the proximal tubule is that the amount of filtrate in the nephron is reduced and the solute composition is altered while the overall concentration of solutes in the filtrate is unchanged.

From the proximal tubule, the filtrate flows into the **loop of Henle**, which dips into the medulla. The loop of Henle functions to increase the solute concentration, and therefore the osmotic pressure, of the medulla. At the same time, the solute concentration of the filtrate leaving the loop of Henle is decreased. This change occurs because the initial descending and final ascending segments of the loop of Henle differ in their permeability to solutes and water. The descending loop of Henle has low permeability to salt and high permeability to water, so as filtrate descends into the medulla, water passively diffuses out of the loop of Henle and into the medulla. As a result, the filtrate osmolarity in this segment of the nephron increases.

As the filtrate rises out of the medulla, solutes pass out of the ascending loop of Henle, first by passive diffusion and then through active pumps. The thick ascending loop of Henle is nearly impermeable to water, so most of the change is due to the movement of solutes from the filtrate into the medulla. This segment of the nephron increases solute concentration in the medulla while reducing it in the filtrate. A capillary bed called the *vasa recta* surrounds the loop of Henle and helps to maintain the concentration of the medulla.

Understand Key 2: Molecules that enter the kidney can only go in one of two directions: into the urine or back into the body. Water-soluble metabolites are excreted, along with excess sodium and other ions that are not needed. Water, glucose, and most ions are reabsorbed back into the blood to maintain homeostasis.

FIGURE 5.15 The Nephron and Vasa Recta

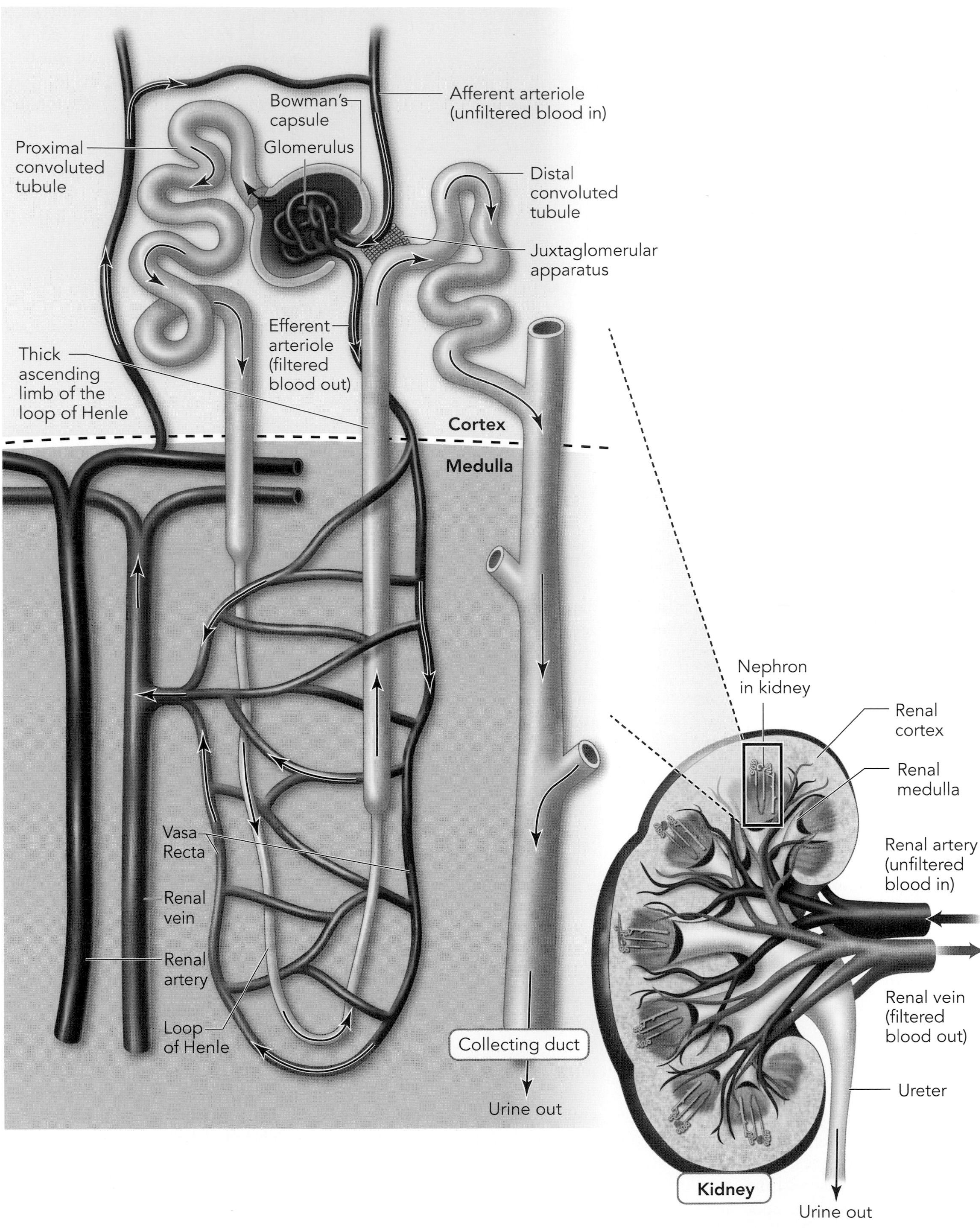

Concentration of Urine: The Osmolarity Gradient

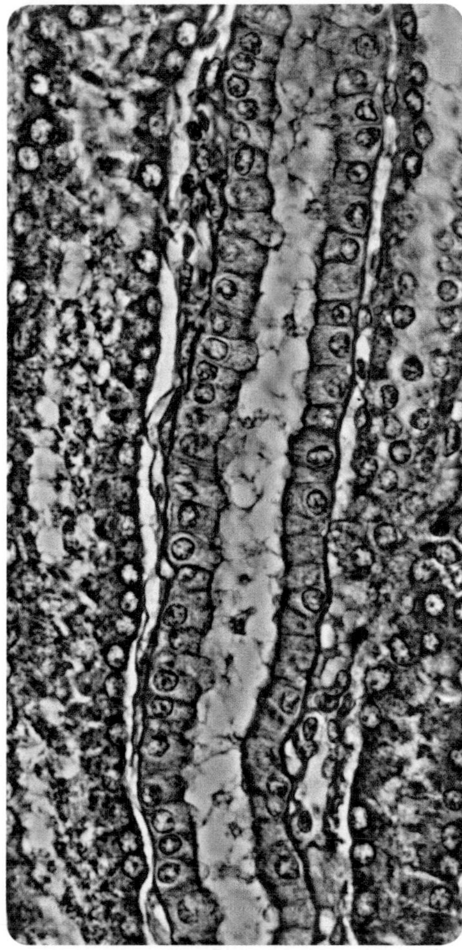
Micrograph of the kidney's tubules, which work to move water and solutes into and out of the filtrate, depending on the body's needs.

If the kidneys were not able to make concentrated urine, simply staying hydrated would require drinking gallons of water each day. This process is only possible because of the highly concentrated medulla that draws water out of the filtrate as it passes through the collecting duct. It is not necessary to understand all of the details presented here, but a general understanding of this process will be helpful for answering questions about the kidney on the MCAT®.

Processes known as the single effect, countercurrent multiplier, and countercurrent exchanger all work together to create an osmolarity gradient in the medulla. To understand how the gradient is established, it helps to imagine the nephron before it has set up the steady-state gradient within the medulla and the filtrate, as in the fetal kidneys. While reviewing this section, pay attention to differences in solute concentration in the various fluid compartments and differences in membrane permeability to water and solutes. Think about which way water or solutes would move if they were able to diffuse freely. Water will always flow to the compartment that has the highest concentration of solutes, or the highest osmolarity.

The loop of Henle has *countercurrent flow*, meaning that as filtrate moves through the descending and ascending loops it flows in opposite directions in parallel tubules that are very close together and separated only by a small amount of tissue and fluid. The vasa recta also has counter-current flow, but in the opposite direction of the loop of Henle; the blood in the vasa recta descends next to the ascending portion of the loop of Henle and ascends next to the descending portion. In addition to the main loop, the vasa recta has connecting horizontal capillaries.

The *single effect* is the process by which active transport of solute by pumps in the wall of the thick ascending loop creates a concentration gradient. The result of the single effect is that solute is more highly concentrated outside the tubule (in the medulla) than inside the tubule (in the filtrate). Water cannot diffuse across the thick wall of this segment of the ascending loop, so the gradient depends entirely on active pumping of solutes. The single effect creates a moderate osmotic gradient between the filtrate and interstitial fluid, up to a 200 mOsm concentration difference. (Recall that an osmole measures solute concentration in terms of solute particles per kilogram. Fluid with an osmolality of 1 Osm contains 1 mole of particles per kilogram, and a mOsm is 1/1000th of this concentration.) Water always tries to equalize concentrations of solutions and will diffuse down from an area with more water (less concentrated solutes) to an area with less water (more concentrated solutes). Though water cannot cross the wall of the ascending limb, it can diffuse freely through the wall of the descending loop of Henle. Since the medulla and the medullary interstitium (the fluid in the space between the cells of the medulla) have become more concentrated through the action of the pumps in the ascending limb, and since the descending limb is permeable to water, water moves out of the descending loop and into the medulla until these two areas have equal concentrations. The net result of the single effect is to dilute filtrate in the ascending limb while concentrating the medulla, and to concentrate filtrate in the descending limb of the loop of Henle.

The **counter-current multiplier mechanism** applies the single effect, which creates a static gradient, to a dynamic system where fluid is continually moving through the loop of Henle. In the body, the whole process is happening at once and is generally in homeostasis. However, to initially understand the counter-current multiplier mechanism, it is useful to imagine it as a series of discrete steps and then envision all of them happening at the same time, over and over again. These steps are: 1. pump salt from the filtrate to the medulla, 2. equilibrate water throughout the system, and 3. shift the filtrate along the tube.

In Figure 5.16 A, all of the filtrate in the loop and the entire medulla starts out at the same concentration of around 300 mOsm. In Figure 5.16 B, all of the active pumps in the ascending limb have pumped once to move solute out of the filtrate into the medullary interstitium and have set up their maximum gradient of a 200 mOsm difference across the wall (the single effect). In that instant, the concentration of filtrate in the entire ascending limb is 200 mOsm and the concentration of the entire length of the medulla is 400 mOsm. This change in concentration of the medulla affects the concentration of the filtrate in the descending limb through diffusion and equilibration, as described above. Water leaves the filtrate in the descending limb through diffusion and moves into the medulla to equilibrate the concentrations so that the concentrations of descending loop filtrate and the medulla are both 400 mOsm (Figure 5.16 C). But glomerular filtration continues to push the fluid along through the loop, so the concentrations quickly change. As shown in Figure 5.16 D, the concentrated filtrate in the descending limb gets pushed down and around (i.e., is shifted), bringing concentrated filtrate to the hairpin turn and to the bottom segment of the ascending limb. Meanwhile, the more dilute filtrate that was in the ascending limb gets pushed up, and some of it moves out of the loop.

Time out! Don't try to memorize these numbers! They are provided as an illustration. The important thing to understand about the concentration gradient in the medulla is the self-perpetuating counter-current multiplier mechanism that sets it up. Remember the purpose of the medulla's concentration gradient: facilitating reabsorption of water into the body and, correspondingly, concentrating the urine.

At this point, the cycle repeats: pump salt, equilibrate water, shift filtrate. When the pumps in the ascending limb act again (Figure 5.16 E), they are still only able to set up a maximum concentration difference of 200 mOsm across the wall. In the more dilute upper segment where the filtrate concentration was at 200 mOsm, the action of the pumps creates a filtrate concentration of 150 mOsm and a medulla concentration of 350 mOsm. Unlike in the first cycle, the concentration is not the same through the whole ascending limb. In the more concentrated lower segment, where 400 mOsm filtrate that had been in the descending limb has shifted down and been pushed up into the lower part of the ascending loop, the pumps will create the same maximum gradient of a 200 mOsm difference across the wall. In the deeper section that starts out at a higher concentration, the action of the pumps establishes a new filtrate concentration of 300 mOsm and a medulla concentration of 500 mOsm. As in the first cycle, water diffuses through the permeable walls of the descending limb, allowing the filtrate to equilibrate with the medullary interstitium (Figure 5.16 F). Filtrate in the lower segment that has already been concentrated to 400 mOsm will become further concentrated to 500 mOsm, but the newer filtrate that entered the upper segment of the descending loop will only equilibrate to 350 mOsm. This process continues, repeating the steps of pump fluid, equilibrate water, and shift filtrate. As a result, an osmolarity gradient is established. The medulla is least concentrated in the upper region near the cortex and becomes progressively more concentrated in the deeper region (Figure 5.16 G).

Like all other body tissues, the medulla requires a steady supply of oxygen and nutrients as well as removal of waste products via capillary blood flow. A process called the countercurrent exchanger, involving the vasa recta, allows the needs of the medulla to be met while avoiding disruption of the balance of solutes described above. Recall that blood entering the kidneys (via either the vasa recta or the glomerulus) has an osmolarity of approximately 300 mOsm, and that capillaries throughout the body are permeable to both solutes and water. If capillaries were randomly organized throughout the concentrated medullary interstitium, solutes would diffuse into the capillaries and be transported away from the kidney, destroying the essential concentration gradient established by the loop of Henle and the countercurrent multiplier. The hairpin loop structure of the vasa recta, combined with particularly slow blood flow through these vessels, allows blood circulation to occur without major disruption of the medulla's concentration gradient. As blood descends through the vasa recta, it moves into increasingly concentrated areas of the medulla. In these areas, water diffuses out of the vasa recta and solutes are absorbed back in so that the blood at the bottom of the vasa recta loop is as concentrated as the most concentrated area of the medulla. As this more

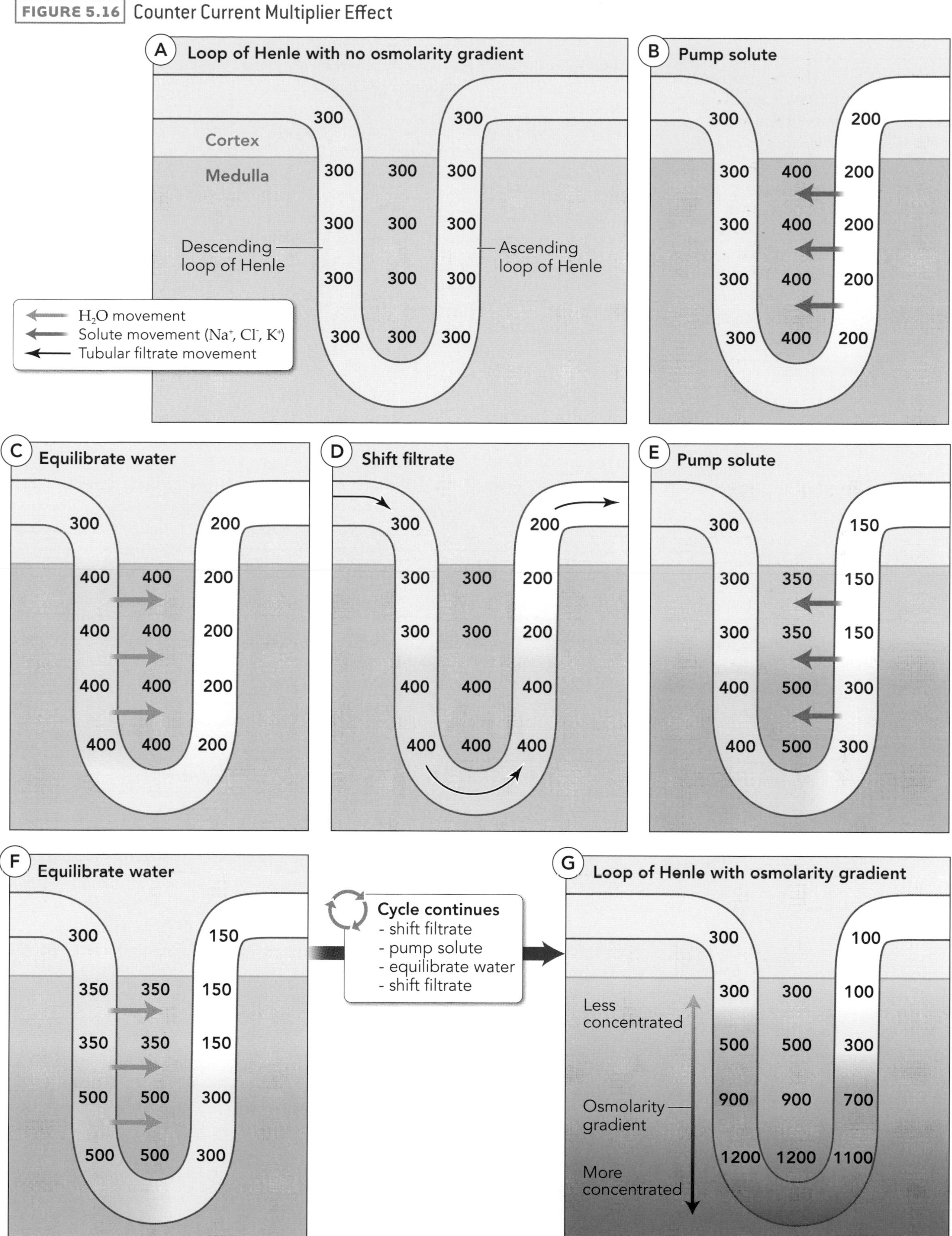

FIGURE 5.16 Counter Current Multiplier Effect

concentrated blood ascends back up through the vasa recta, it passes through less concentrated areas of the medulla, where the free flow of water and solutes means that solutes travel from the vasa recta into the medulla and water flows from the medulla into the vasa recta. As a result, the blood in the ascending segment is diluted so that is only slightly more concentrated than the surrounding medulla when it leaves the medulla to move back up into the cortex of the kidney. The vasa recta also has horizontal segments that allow blood at a certain depth of the medulla to move from the descending limb of the vasa recta to a section of the ascending limb that has the same concentration of solutes and water.

Consider Key 3: The nephron works by a series of diffusion gradients across membranes that allow only specific molecules to cross. Water is able to diffuse out of the descending loop of Henle into the tissue, but is unable to do so in the ascending loop of Henle. By contrast, the descending loop of Henle is impermeable to sodium, but sodium is actively pumped across the membrane in the ascending loop of Henle.

The concentrated inner medulla created and maintained by the processes described above provides osmotic pressure that "pulls" free water out of the filtrate, concentrating the urine before it leaves the body. The urine can be concentrated because the descending and ascending segments of the Loop of Henle differ in their permeabilites to solutes and water. The solute pumps in the ascending segment actively move solute to create an initial concentration gradient across the wall of the tubule, establishing the single effect. As new filtrate moves down into the medulla and then ascends up in the parallel tubule (countercurrent flow), it replaces the old filtrate, changing its concentration and leading to the dynamic re-equilibration and magnification of the concentration gradient known as the countercurrent multiplier. The looped vasa recta delivers oxygen and nutrients and removes waste products without disturbing the concentration gradient through the mechanism of the countercurrent exchanger.

From the Loop of Henle, filtrate moves up out of the medulla and back into the renal cortex, entering the **distal tubule**. The distal tubule reabsorbs Na^+ and Ca^{2+} while secreting K^+, H^+, and HCO_3^-. Aldosterone acts on distal tubule cells to increase the number of sodium and potassium membrane transport proteins in their membranes, causing blood pressure to increase. The net effect of the distal tubule is to lower the filtrate osmolarity. At the end of the distal tubule, in an area called the *collecting tubule* (a portion of the distal tubule, not to be confused with the collecting duct), antidiuretic hormone (ADH) causes the cells to become more permeable to water so that water flows out of the tubule back into the body, and the filtrate becomes more concentrated.

The distal tubule empties into the **collecting duct**, which carries the filtrate into the highly osmotic medulla. The collecting duct is impermeable to water by default, but like the distal tubule, it is sensitive to ADH. In the presence of ADH, the collecting duct becomes permeable to water, allowing it to enter the medulla via passive diffusion down its concentration gradient. In this stage, all of the complex steps that formed the osmotic gradient in the medulla become critically important. The osmotic pressure created by densely concentrated solutes in the medulla causes a substantial portion of the free water in the filtrate of the collecting duct to diffuse back into the body, allowing for the formation of concentrated urine.

Many collecting ducts line up side by side in the medulla to form the *renal pyramids*. The collecting ducts lead to a *renal calyx*, which empties into the renal pelvis. The renal pelvis drains urine into the ureter and down to the bladder.

Many details about the kidney are important for the MCAT®. Remember the function of each section of the nephron: 1. filtration occurs in the renal corpuscle, 2. reabsorption and secretion mostly occur in the proximal tubule, 3. the loop of Henle concentrates solute in the medulla, 4. the distal tubule empties into the collecting duct, and 5. the collecting duct concentrates the urine. Understand that the amount of filtrate is related to the hydrostatic pressure of the glomerulus. Also remember that the descending loop of Henle is permeable to water, and that the ascending loop of Henle is impermeable to water and actively transports sodium into the kidney. The loop of Henle increases the solute concentration of the medulla, while decreasing the solute concentration of the filtrate. This process makes it possible for urine to become concentrated in the collecting duct.

Don't lose sight of the big picture: the function of the kidney is homeostasis.

The Juxtaglomerular Apparatus

The **juxtaglomerular apparatus** monitors filtrate pressure in the distal tubule. Specialized cells in the juxtaglomerular apparatus secrete the enzyme renin when filtrate pressure is too low. The renin-angiotensin-aldosterone system (RAAS) regulates blood pressure and fluid balance. Renin, an enzyme secreted from the kidney, initiates a regulatory cascade that produces angiotensin I from the zymogen angiotensinogen. Angiotensin I travels to the lung through the blood and is converted into angiotensin II, which causes blood vessels to constrict, raising blood pressure. Angiotensin II also causes the adrenal cortex to secrete aldosterone. Aldosterone acts on the distal tubule, stimulating the formation of membrane proteins that absorb sodium and secrete potassium to increase blood pressure. When sodium is reabsorbed, water follows, increasing the blood volume without substantially altering blood osmolarity.

Don't worry about memorizing all of the specific steps in the RAAS. Instead focus on how blood pressure can be increased by increasing the volume of water reabsorbed.

These questions are NOT related to a passage.

Question 113

Bowman's capsule assists in clearing urea from the blood by:

- O **A.** actively transporting urea into the filtrate using ATP-driven pumps.
- O **B.** exchanging urea for glucose in an antiport mechanism.
- O **C.** allowing urea to diffuse into the filtrate under filtration pressure.
- O **D.** converting urea to amino acids.

Question 114

Tests reveal the presence of glucose in a patient's urine. This is an indication that:

- O **A.** glucose transporters in the loop of Henle are not functioning properly.
- O **B.** the patient is healthy, as glucose normally appears in the urine.
- O **C.** the proximal tubule is over-secreting glucose.
- O **D.** glucose influx into the filtrate is occurring faster than it can be reabsorbed.

Question 115

The epithelial cells of the proximal convoluted tubule contain a brush border similar to the brush border of the small intestine. The most likely function of the brush border in the proximal convoluted tubule is to:

- O **A.** increase the amount of filtrate that reaches the loop of Henle.
- O **B.** increase the surface area available for the absorption.
- O **C.** slow the rate of at which the filtrate moves through the nephron.
- O **D.** move the filtrate through the nephron with cilia-like action.

Question 116

If a patient were administered a drug that selectively bound and inactivated renin, which of the following would most likely result?

- O **A.** The patient's blood pressure would increase.
- O **B.** Platelets would be found in the urine.
- O **C.** The amount of filtrate entering Bowman's capsule would increase.
- O **D.** Sodium reabsorption by the distal tubule would decrease.

Question 117

Certain mammals in water-deprived habitats exhibit longer nephrons than mammals in habitats that are rich in fresh water. Which of the following parts of a nephron is most likely to be longer in length in mammals from the water-deprived habitat?

- O **A.** Proximal convoluted tubule
- O **B.** Distal convoluted tubule
- O **C.** Bowman's capsule
- O **D.** Collecting duct

Question 118

How are the blood levels of vasopressin and aldosterone in a dehydrated individual likely to compare with those of a healthy individual?

- O **A.** Vasopressin and aldosterone levels are likely to be lower in a dehydrated individual.
- O **B.** Vasopressin and aldosterone levels are likely to be higher in a dehydrated individual.
- O **C.** Vasopressin levels are likely to be higher in a dehydrated individual, while aldosterone levels are likely to be lower.
- O **D.** Vasopressin levels are likely to be lower in a dehydrated individual, while aldosterone levels are likely to be higher.

Question 119

An afferent arteriole in a glomerular tuft contains microscopic fenestrations which increase fluid flow. In a hypertensive patient (a patient with high blood pressure):

- O **A.** these fenestrations would constrict, resulting in decreased urinary output.
- O **B.** filtrate volume would be expected to be larger due to increased fluid pressure.
- O **C.** filtrate volume would be expected to be smaller due to increased fluid pressure.
- O **D.** urinary output would most likely be diminished due to increased solute concentration.

Question 120

In a disease state known as Conn's syndrome, an abnormally elevated level of aldosterone is observed. Which of the following best describes the blood serum electrolytes of a patient with this disease?

- O **A.** Sodium is increased and potassium is decreased.
- O **B.** Sodium is decreased and potassium is increased.
- O **C.** Sodium is increased and potassium is also increased.
- O **D.** Sodium is decreased and potassium is also decreased.

Questions 113-120 of 144

STOP

TERMS YOU NEED TO KNOW

α-amylase
Albumin
Antiport
Anus
Ascending colon
Bacterial flora
Bicarbonate ion
Bile
Blood filtration
Blood storage
Body
Bowman's capsule
Brush border
Carbohydrate metabolism
Chief cells (peptic cells)
Chymotrypsin
Collecting duct
Cortex
Counter-current multiplier mechanism
Detoxification
Descending colon
Distal tubule
Duodenum
Electrolyte absorption
Endocrine
Enteric nervous system
Erythrocyte destruction
Esophagus
Excretory system
Exocrine glands
Fat metabolism
Feces
Fenestrations
Fundus
Gall bladder
Gastric juice
Gastrin
G cells
Glomerulus
Gluconeogenesis
Glycogenesis
Glycogenolysis
Glycogen storage
Goblet cells
Hormones
Hydrochloric acid (HCl)
Hydrostatic pressure
Ileum
Ingestion
Jejunum
Juxtaglomerular apparatus
Kidney
Lacteal
Large intestine
Lipase
Liver
Loop of Henle
Lubrication
Medulla
Microvilli
Mouth
Mucous cells
Nephron
Pancreas
Pancreatic amylase
Parietal cells (oxyntic cells)
Pepsin
Pepsinogen
Peristalsis
Protein metabolism
Proximal tubule
Pyloris
Reabsorption
Rectum
Secretion
Sigmoid colon
Small intestine
Sphincter
Stomach
Transverse colon
Triglycerides
Trypsin
Urea
Villi
Vitamin storage
Water absorption

DON'T FORGET YOUR KEYS

1. The three processes of the digestive system are digestion, absorption, and excretion. Surface area and length promote absorption.

2. From the kidney, substances travel in one of only two directions: "into the toilet" (filtration, secretion) and "back to the body" (reabsorption).

3. Diffusion down concentration gradients and active pumps in the ascending loop of Henle set up the concentration gradient in the nephron and kidney.

Muscle, Bone and Skin

LECTURE 6

THE 3 KEYS

1. Somatic NS innervates voluntary muscles; autonomic NS innervates involuntary muscle.
2. The energy of an ATP molecule prepares the myosin head so that the muscle is ready and contraction is downhill in energy.
3. See Ca^{2+}, think neuromuscular function and storage in bone.

6.1 Introduction

The musculoskeletal system and the integumentary system, which includes skin and hair, consist of organs that carry out a variety of functions including protection, movement, and regulation. This lecture will present the information about muscle, bone, and skin that is required for the MCAT®. It will start by covering the three main types of muscle in the body, including how the structure of each type relates to its specific function and how different types of muscle are controlled by separate portions of the nervous system. Next the lecture will discuss the functions of bone, such as the regulation of calcium levels that makes muscle contraction possible, and the process through which new bone is made to replace old bone. The lecture will then cover the basic structure and multiple purposes of human skin. Although each performs a variety of tasks, muscle, skin, and bone are united by their roles of protecting and regulating the internal environment of the body.

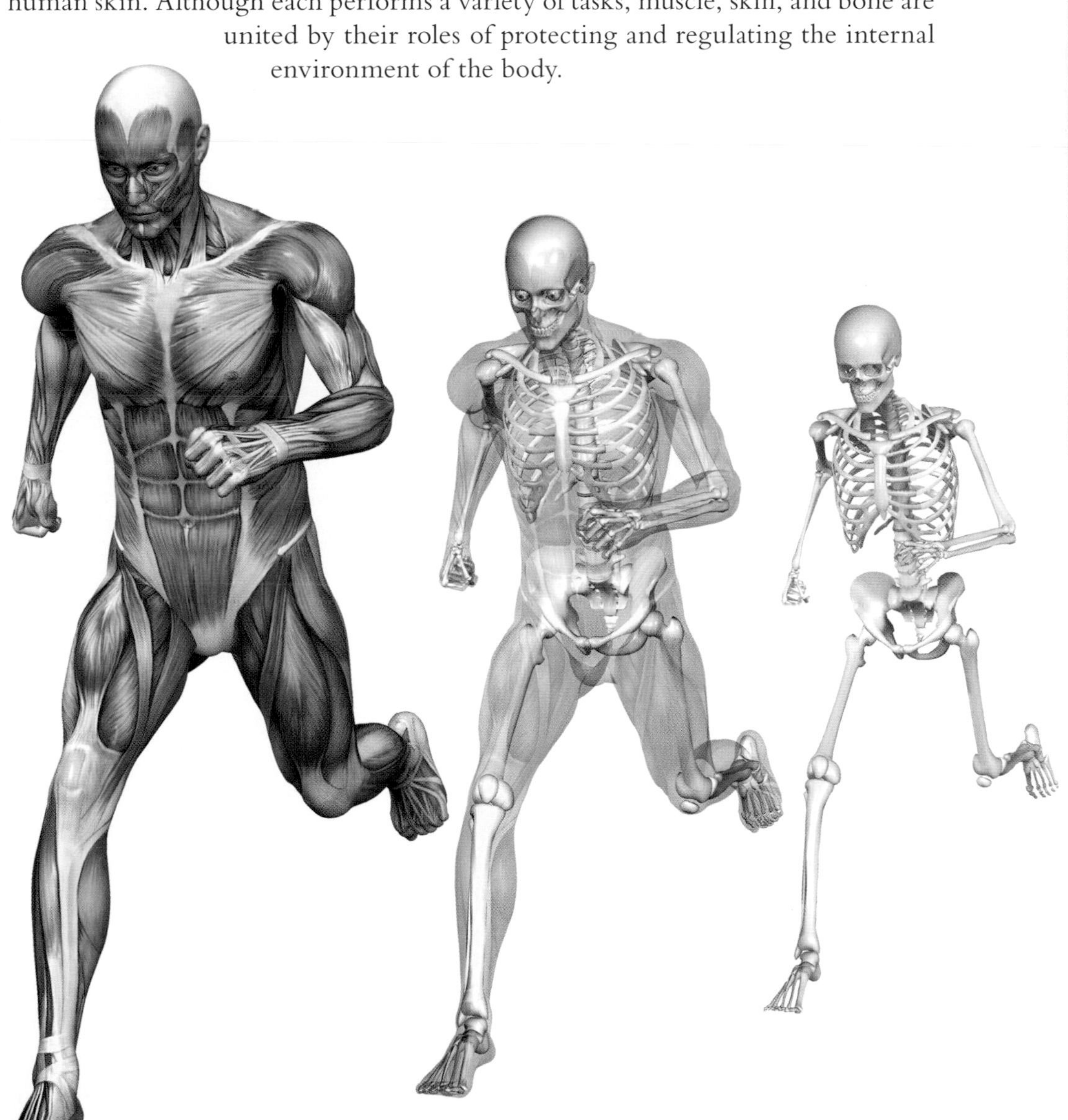

6.2 Muscle

There are three types of muscle tissue:

1. skeletal muscle,
2. cardiac muscle, and
3. smooth muscle.

The types of muscle each differ in structure, function, and nervous system innervation, but all have some properties in common. Innervation refers to how the nervous system connects to and influences the contraction pattern of a muscle. A muscle generates force by contracting its cells. The mechanisms by which muscle cells contract differ between the three types of tissue, as described later in this lecture. The major **functions of muscle** are:

1. body movement,
2. stabilization of body position,
3. movement of substances through the body, and
4. generating heat to maintain body temperature.

Know the three types of muscle and their four possible functions.

6.3 Skeletal Muscle: Physiology and Structure

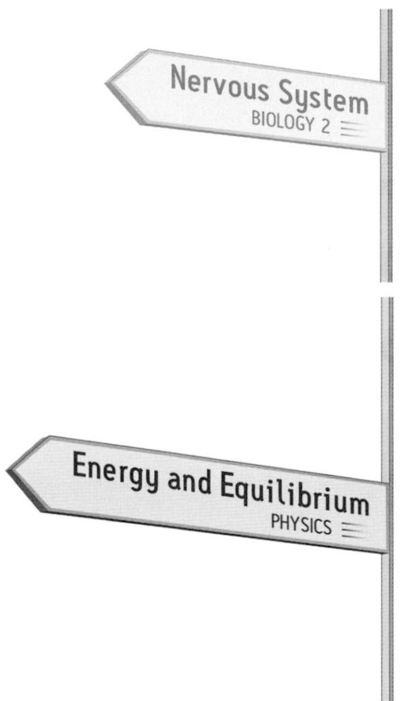

Skeletal muscle (a type of **striated muscle**) is the tissue that is commonly called "muscle" in everyday language. Skeletal muscle is **voluntary muscle**, meaning that it can be consciously controlled to produce specific desired movements. It is innervated by the somatic nervous system. Skeletal muscle moves the body and allows activities like running or lifting weights. It is also involved in thermoregulation and the movement of fluids in the cardiovascular and lymphatic systems. This section will describe the functions of skeletal muscle, as well as the mechanism used by skeletal muscle to create movement.

Skeletal muscle connects one bone to another. Rather than attaching directly to the bone, the muscle is attached by a **tendon**. Note that connective tissue connecting muscle to bone is called a tendon; connective tissue connecting bone to bone is called a **ligament**. Tendons are fibrous, allowing them to store elastic potential energy. A muscle stretches across a joint in order to create movement at that joint when the muscle contracts. The *origin* of the muscle is usually its attachment on the larger bone closer to the midpoint of the body, which remains relatively stationary. The attachment at the other end of the muscle, known as its *insertion*, is on the smaller bone farther from the midpoint, which moves relative to the larger bone as the muscle contracts. The exact placement of the origin and insertion of a muscle determines how it moves a joint and how much force, or torque, a muscle contraction can make.

Muscles work in groups to perform movements efficiently. For each movement, the muscle whose contraction is primarily responsible for the movement is called the *agonist*. A second muscle, the *antagonist*, stretches in response to the agonist's contraction and opposes the movement so that the motion is smooth and controlled. The major upper arm muscles, the biceps and the triceps, provide an example of two groups of muscles that function antagonistically to each other. In addition to antagonistic muscles, movements usually involve *synergistic* muscles. A synergistic muscle assists the agonist by stabilizing the origin bone or by positioning the insertion bone during the movement. By acting together, agonistic, antagonistic, and synergistic muscles contribute to two major functions of skeletal muscle: movement and maintenance of posture.

Understand Key 1: The somatic nervous system innervates voluntary muscles, such as skeletal muscle, that can be used to control movement. The autonomic nervous system innervates involuntary muscles, such as smooth muscle and cardiac muscle, that control processes such as heart rate and blood pressure.

FIGURE 6.1 Agonist, Antagonist, and Synergist Muscles

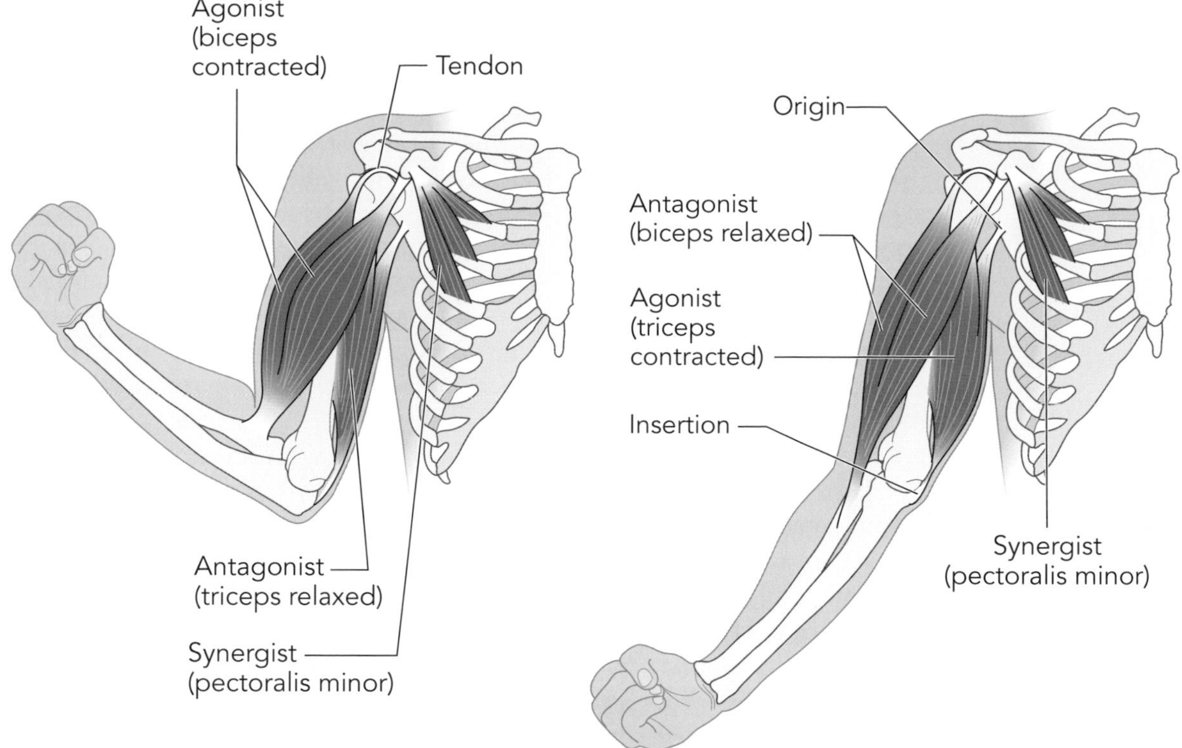

Another function of skeletal muscle is that it provides **peripheral circulatory assistance**. Contraction of skeletal muscles helps squeeze blood and lymph through their respective vessels, aiding circulation.

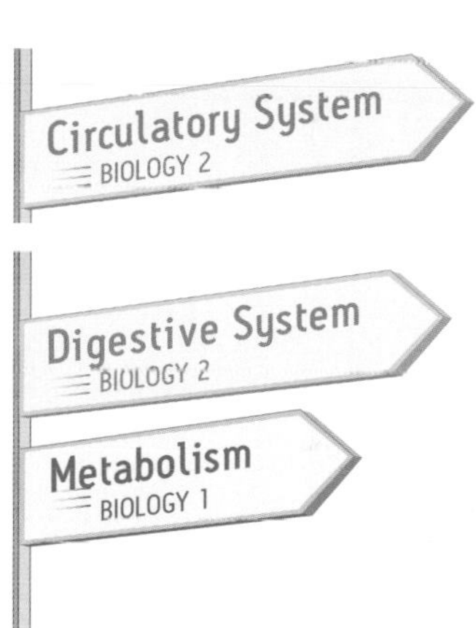

Skeletal muscle also participates in **thermoregulation**, meaning regulation of body temperature. Contraction of skeletal muscle produces large amounts of heat. The **shivering reflex** is the rapid contraction, or shaking, of skeletal muscle to warm the body. It is controlled by the hypothalamus upon stimulation by receptors in the skin and spinal cord. As described in the Digestive and Excretory Systems Lecture, skeletal muscle stores large amounts of glycogen. When necessary, the energy stored as glycogen can be metabolized and used to power muscle contractions that provide heat to the body.

A muscle uses leverage by applying a force to a bone at its insertion point and rotating the bone about the joint. This is a likely MCAT® topic because it applies the physics concept of levers and work to a biological system. Most (but not all) muscles act as class three lever systems in the body, meaning that the force applied by the contraction of the muscle (via the muscle attachment) is between the fulcrum (joint) and the load (the part of the body that moves). Class three levers, such as some muscles, use more force to perform a given amount of work than would be used if there were no lever at all. In other words, a greater force than *mg* is required to lift a mass *m*. This arrangement is advantageous because the shorter lever arm reduces the bulk of the body and increases the range and control of movement. You could see an MCAT® question or passage about the physics of levers in muscle movement.

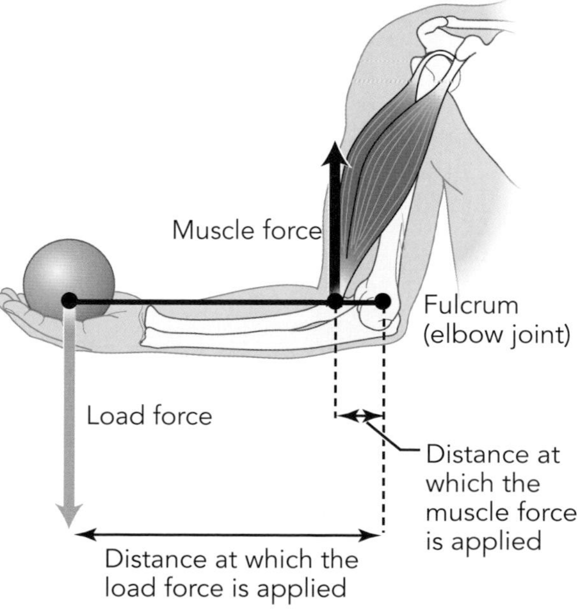

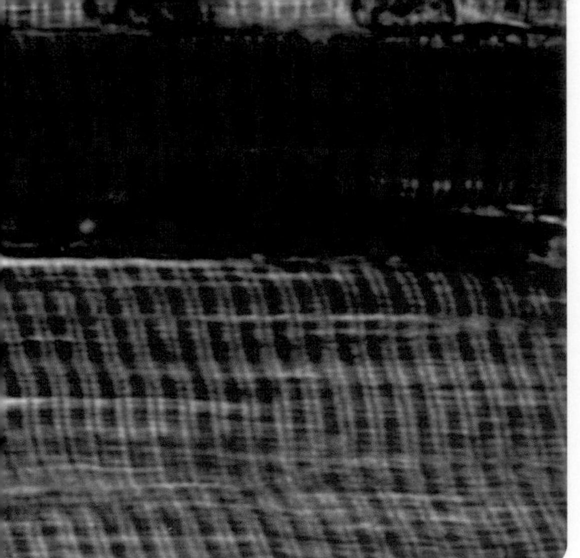

The striated (striped) banding pattern of skeletal muscle fibers can be seen in this photo. The cross striations are the arrangements of proteins (actin and myosin) that cause the fibers to contract. Multiple nuclei (fuzzy pink) can be seen along the junction of the fibers (blue). Each cylindrical fiber is a single muscle cell.

Physiology of Skeletal Muscle Contraction

Skeletal muscle cells have unique physical features that facilitate their function in contraction. The smallest functional unit of the **contractile apparatus** in skeletal muscle is the **sarcomere** (Figure 6.2). A sarcomere is composed of many strands of two kinds of protein filaments: **thick filaments** made of **myosin** and **thin filaments** made of the globular protein **actin**. Several long myosin molecules wrap around each other to form one thick filament. Globular heads, known as *myosin heads*, protrude along both ends of the thick filament. The thin filament is composed mainly of a polymer of actin. Attached to the actin are the proteins **troponin** and **tropomyosin**, which participate in the mechanism of contraction.

The myosin and actin filaments overlap and slide past each other when muscles contract and relax. The **Z line** separates one sarcomere from the next and is where actin filaments attach. Other areas of the sarcomere are defined by the presence or absence of actin and/or myosin. The area containing actin only (including the Z line) is the **I band**; the area containing myosin only is the **H zone**; and the area where myosin is present, including where it overlaps with actin, is the **A band**. The A band includes the H zone. The midline of the myosin fibers is the **M line**.

Myosin and actin filaments are laid side by side to form the cylindrical sarcomere. Sarcomeres are positioned end to end to form a *myofibril*. Each myofibril is surrounded by the specialized endoplasmic reticulum of the muscle cell, the **sarcoplasmic reticulum**. The lumen of the sarcoplasmic reticulum is filled with Ca^{2+} ions for reasons that will become clear. Lodged between the myofibrils are many mitochondria, which create the supplies of ATP required for muscular contraction. Each muscle cell contains many nuclei and is said to be *multinucleate*. A modified cellular plasma membrane called the *sarcolemma* wraps several myofibrils together to form a muscle cell (also called a myocyte or muscle fiber). Many muscle fibers are further bound into a *fasciculus*, and many fasciculi make up a single muscle. Muscles are hierarchically organized: sarcomeres, the smallest functional units, line up end to end to form myofibrils, which are wrapped together by the sarcolemma to form a muscle cell. Many muscle cells bound together form a fasciculus and many fasciculi form a muscle.

Nervous System
BIOLOGY 2

The contraction of skeletal muscle is controlled by the somatic nervous system. A **motor neuron** attaches to a muscle cell at a **motor end plate**, forming a **neuromuscular junction**. More specifically, the motor end plate is a region of highly excitable muscle, while the neuromuscular junction is the synapse between the motor neuron and the motor end plate. The action potential of the neuron releases **acetylcholine** into the synaptic cleft. The acetylcholine activates ion channels in the sarcolemma of the muscle cell, creating an action potential that propagates along the sarcolemma. The action potential moves deep into the muscle cell via small infoldings of the sarcolemma called **T-tubules**. T-tubules facilitate the uniform contraction of the muscle by allowing the action potential to spread through the muscle cell more rapidly. The action potential is transferred to the sarcoplasmic reticulum, causing voltage-gated channels on the sarcoplasmic reticulum to open. As a result, the sarcoplasmic reticulum becomes more permeable to Ca^{2+}, which it releases around the sarcomere. The presence of Ca^{2+} allows the myosin and actin fibers of the sarcomere to slide across each other, causing the shortening (contraction) of the muscle fiber.

FIGURE 6.2 Structure of Skeletal Muscle

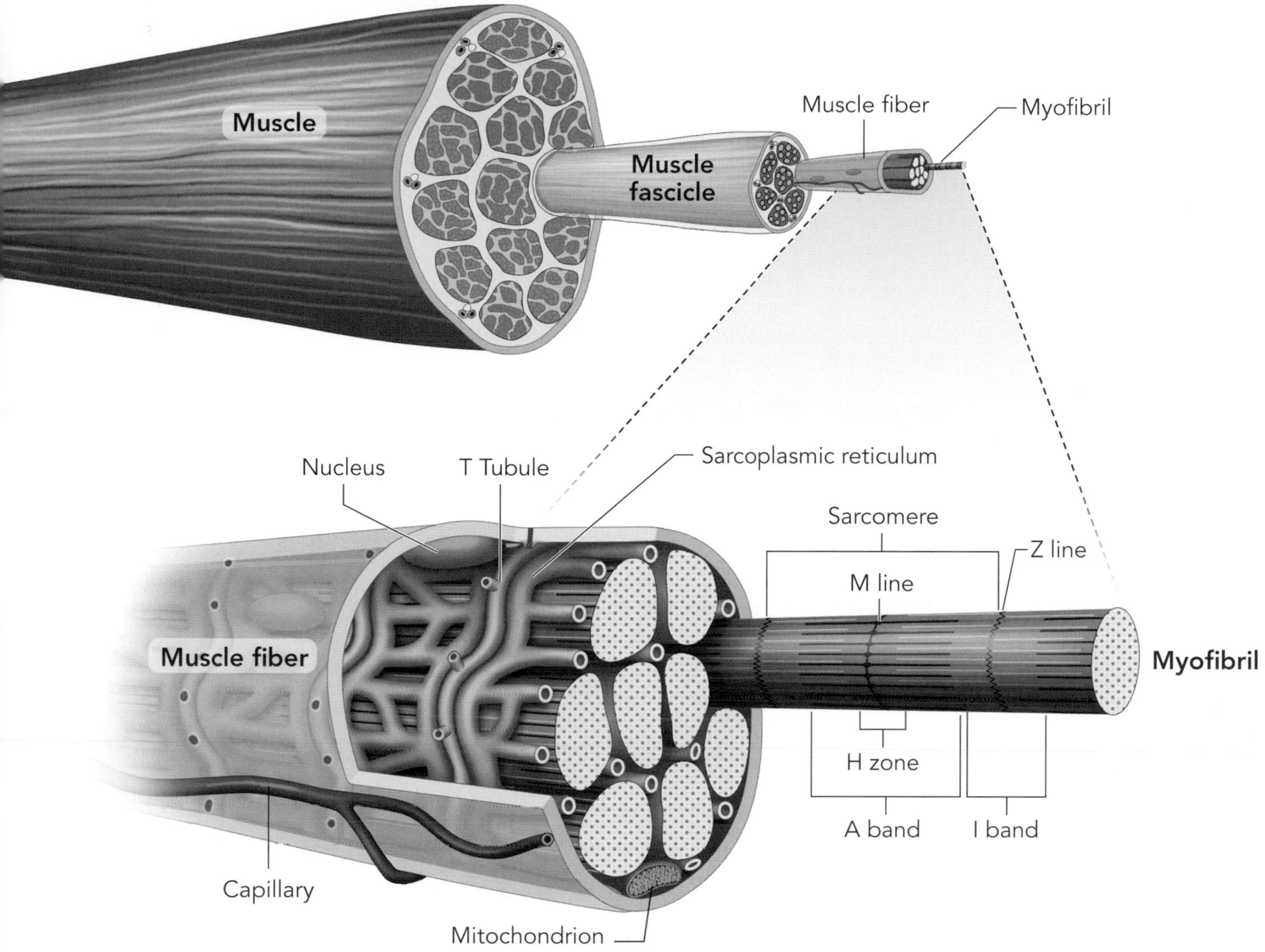

Muscle
Muscle fascicle
Muscle fiber
Myofibril
Nucleus
T Tubule
Sarcoplasmic reticulum
Sarcomere
Z line
M line
Muscle fiber
Myofibril
H zone
A band
I band
Capillary
Mitochondrion

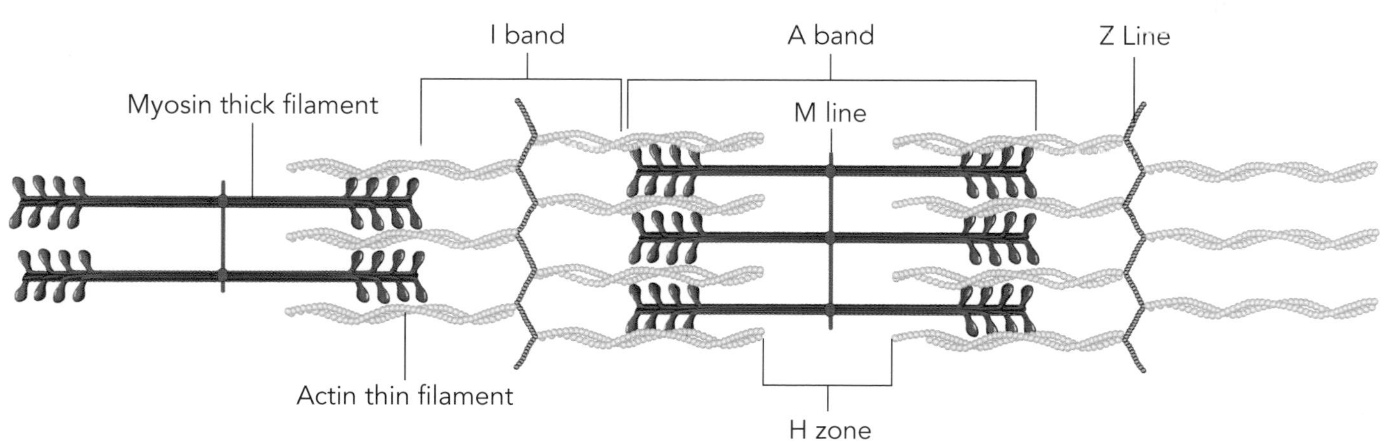

Take a minute to think about the labeled parts of the sarcomere: I band, A band, H zone, Z line, and M line. The diagram indicates that when the myosin and actin fibers slide past each other, the H zone and I band get smaller, while the A band does not change size. The A band is defined as the length of actin, which does not change during contraction. The sarcomere only shortens because the actin is pulled closer to the Z-band. The true length of actin, and therefore the length of the A band, does not change.

Recognize the importance of calcium in muscle contraction, and make sure you're familiar with the functions of the sarcoplasmic reticulum and the T-tubules. Ca^{2+} is ultimately what is responsible for allowing actin and myosin to slide across each other to generate a muscular contraction.

FIGURE 6.3 Physiology of Skeletal Muscle Contraction

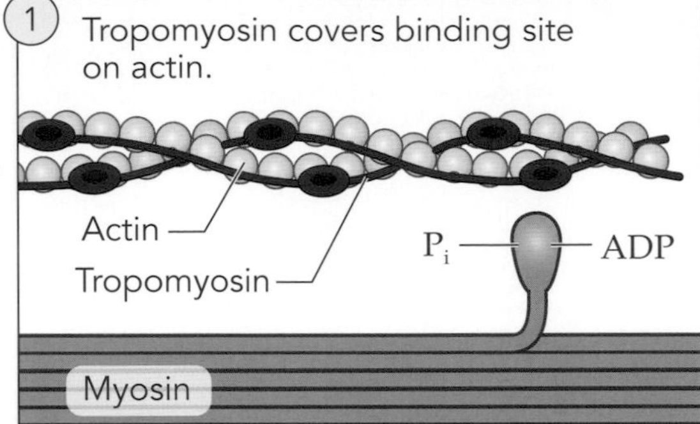

5 ATP splits to ADP and P_i and myosin cocks into high-energy position.

4 ATP binds to myosin, and actin and myosin unbind.

ATP

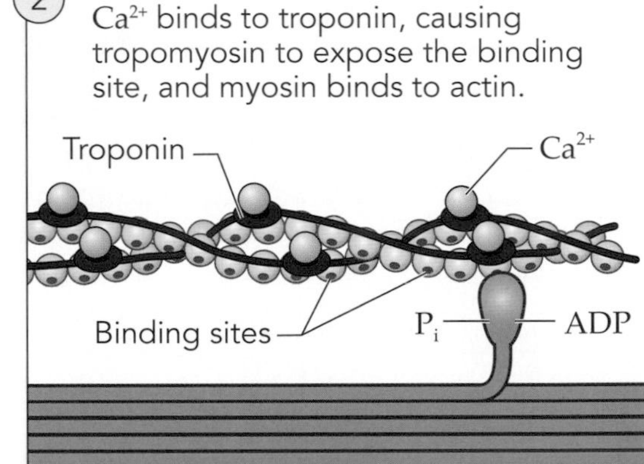

3 Myosin kicks out a P_i and ADP and bends into low-energy position, moving actin with it.

P_i

ADP

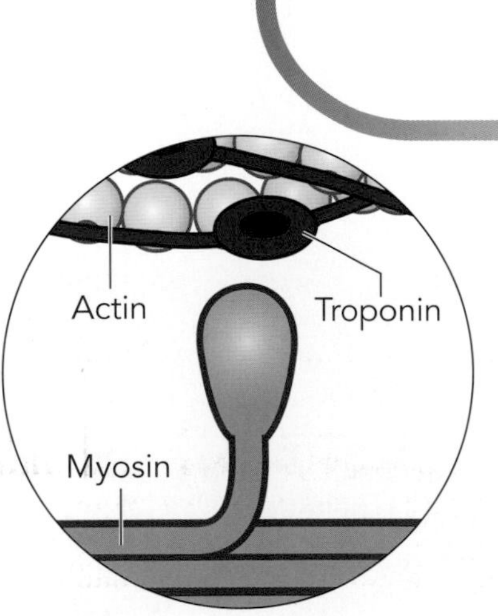

Troponin
Tropomyosin complex blocks actin-myosin interaction.

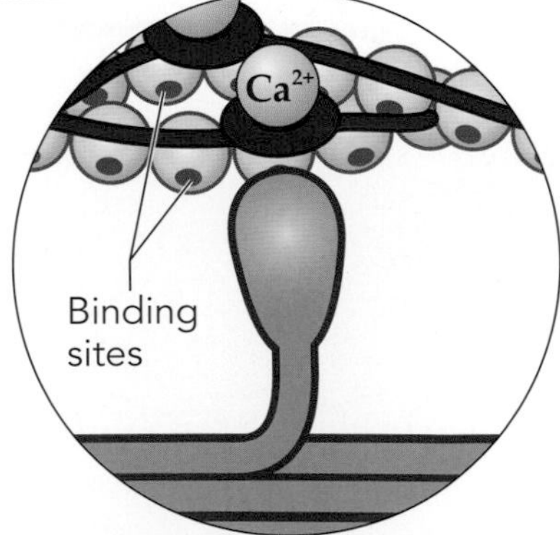

Movement of Troponin
Tropomyosin allows actin-myosin interaction.

According to the **sliding filament model**, myosin and actin work together by sliding alongside each other to create the contractile force of skeletal muscle. The process of contraction requires the expenditure of energy in the form of ATP. Each myosin head crawls along the actin in a 5 stage cycle (Figure 6.3).

1. In the resting state of muscle, the myosin heads of the thick filaments are in a high-energy "cocked" position with a phosphate and ADP group attached. Tropomyosin covers the active sites on the actin, preventing myosin heads from binding to the actin filament.
2. Somatic nervous input increases cytoplasmic Ca^{2+} levels in the muscle cells. In the presence of Ca^{2+} ions, troponin changes its configuration along the thin filament by pulling the tropomyosin back and exposing the active site. This change allows the myosin heads to bind to the actin, forming **cross-bridges** between the thick and thin filaments.
3. Next, the myosin head expels its phosphate and ADP and bends into a low energy position, dragging the actin along with it. This step is called the power stroke because it causes the shortening of the sarcomere and thus starts the muscle contraction.
4. In the fourth stage, a new ATP molecule attaches to the myosin head, causing the myosin head to be released from the active site. The active site is then covered by tropomyosin.
5. Finally, ATP splits into inorganic phosphate and ADP. The energy is used to re-cock the myosin head into the high-energy position. This cycle is repeated many times to form a contraction. At the end of each cycle, Ca^{2+} is actively pumped back into the sarcoplasmic reticulum to reset the gradient.

Consider Key 2: The energy provided by the ATP molecule powers the myosin contraction. Once ATP is bound to the myosin head, the muscle is ready for contraction. Hydrolysis of ATP into ADP + P_i is energetically downhill and powers the sliding movement of myosin and actin.

Motor Units

The muscle fibers of a single muscle do not all contract at once. If they did, the result would be an abrupt, poorly controlled movement. Instead, between 2 and 2000 fibers spread throughout the muscle are innervated by a single neuron. The neuron and the muscle fibers that it innervates are called a *motor unit*. Motor units are independent of each other. The force of a contracting muscle depends upon the number and size of active motor units, as well as the frequency of action potentials in each neuron of the motor unit. Smaller motor units are activated first, and larger motor units are recruited as needed. The result is a smooth increase in the force generated by the muscle. Muscles requiring intricate movements, like those in the finger, have small motor units. Muscles requiring greater force and less finesse, like those in the back, have larger motor units.

If I recruited my large motor units first, I would pick this mug up with so much force that all my root beer would wind up on the wall behind me. That's why I start with the small motor units and recruit larger ones as needed until I have just enough force to lift the mug to my mouth without spilling a drop.

FIGURE 6.4 Motor Unit

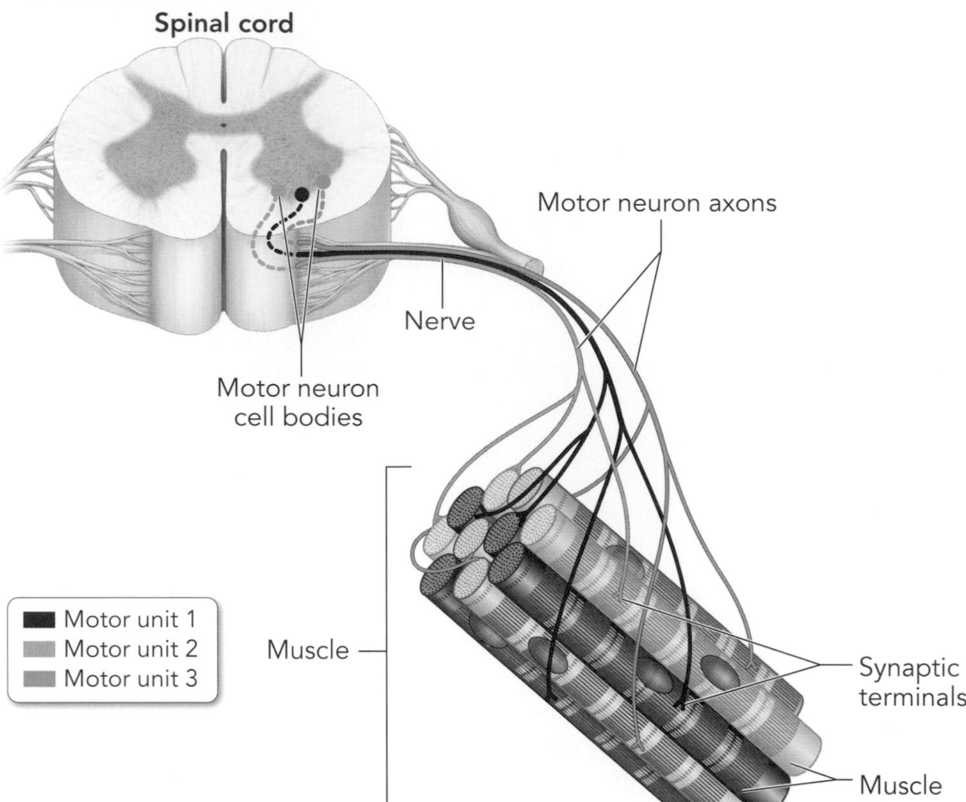

A motor unit consists of a nerve and all the muscle fibers with which it synapses.

Limitations of Skeletal Muscle and Skeletal Muscle Types

As any marathon runner or body builder knows, there are limits to how many times muscle can contract before it becomes fatigued. Fatigued muscles can no longer generate the same force of contraction. Since both nervous system input and energy from ATP are required for muscular contraction, **muscle fatigue** can result from nervous or metabolic causes. After sustained use of the same motor unit, the nerve supplying the unit can become temporarily unable to supply the signals necessary to continue frequent, high intensity contraction of the muscle. Metabolic fatigue can result from depletion of energy stores within the muscle or decreased sensitivity to the effects of calcium in the contracting muscle due to the buildup of waste products from metabolism.

Metabolic processes are constantly occurring in muscle cells to produce the ATP that is used to fuel movement. Oxygen is the final acceptor of electrons in the electron transport chain, and thus plays an important role in metabolism. During moderate exercise, there is enough oxygen in the body to meet the increased metabolic demands through continuing aerobic respiration. During strenuous exercise, however, the oxygen demands of metabolism exceed the body's supply of oxygen. Under these conditions, muscle cells can switch from the citric acid cycle and oxidative phosphorylation to anaerobic glycolysis to produce the necessary ATP. The result is an excess of lactic acid, which must later be metabolized by the body when sufficient oxygen is again available. The need for increased oxygen after exercise in order to metabolize the excess lactic acid produced is known as **oxygen debt**.

The body contains types of skeletal muscle fibers that are specialized for specific purposes and respond differently to exercise and fatigue. The two main fiber types are **type I**, which are known as **slow twitch** (or *slow oxidative*) fibers, and **type II**, which are known as **fast twitch** fibers. Fast twitch fibers are further subdivided into two categories: *fast oxidative (type II A) fibers* and *fast glycolytic (type II B) fibers*.

One way to think about muscle types is as a spectrum from entirely aerobic to entirely glycolytic. Type I fibers are highly aerobic, so they contain many mitochondria and ample myoglobin, which gives them a red color and makes them slow to fatigue. Type II fibers fatigue quickly and are heavily glycolytic.

If you're having trouble remembering the difference between type I and type II fibers, think of a rotisserie chicken. White meat (fast-twitch, type II fibers), such as the breasts, is used for short, intense bursts of activity like flapping the wings. The dark meat of the legs and thighs (slow-twitch, type I fibers) is used for sustained, low-intensity activities like wandering around the chicken coop.

Type I muscle fibers appear red because they contain large amounts of **myoglobin**, an oxygen storing protein similar to hemoglobin. Type I fibers also have large numbers of mitochondria, so they can efficiently use oxygen to generate ATP through aerobic metabolism. The fiber types differ in their **contractile velocity** (speed of contraction), maximum force production, and resistance to fatigue. Type I fibers have a relatively slow contractile velocity and produce a low amount of force. However, type I fibers have the advantage that they are slow to fatigue and can be employed for long periods of time. The large number of mitochondria and high myoglobin content of slow twitch fibers provide them with the ATP and oxygen necessary to operate for long periods without fatigue.

Myoglobin stores oxygen inside muscle cells. A molecule of myoglobin looks like one subunit of hemoglobin. It can store only one molecule of oxygen.

Type II A fibers are also red, but they have a fast contractile velocity. They are resistant to fatigue, but not as resistant as type I fibers. Type II B fibers have a low myoglobin content and thus appear white. They contract rapidly and are capable of generating great force, but they fatigue quickly. Type II B fibers contain large amounts of glycogen.

Most muscles in the body have a mixture of fiber types. The ratio of the mixture depends on the contraction requirements of the muscle and the genetics of the individual. The relative amounts of fiber types may influence a person's natural aptitude for athletic activities. Someone with a greater proportion of fast twitch fibers is likely to be naturally better at sprinting than someone with more slow twitch fibers, who would be better at running marathons. Large amounts of type I fibers are found in the postural muscles. There are large amounts of type II A fibers in the upper legs, while a greater proportion of type II B fibers is found in the upper arms.

Human muscle cells are so specialized that they have lost the ability to undergo mitosis. Only in rare cases does one muscle cell divide to form two cells.

Skeletal muscle cells are highly specialized and differentiated. Adult human skeletal muscle does not usually undergo mitosis to create new muscle cells. Instead, a number of changes occur over time to meet the need for increased strength when the muscles are exposed to forceful, repetitive contractions. These changes include increased diameter of the muscle fibers, increased numbers of sarcomeres and mitochondria, and lengthened sarcomeres. This increase in muscle cell diameter and change in muscle conformation is called *hypertrophy*.

TABLE 6.1 > Skeletal Muscle Fiber Types

	Contraction Time	Force Production	Resistance to Fatigue	Activity
Type I **Slow (oxidative) twitch**	Slow	Low	High	Aerobic
Type II A **Fast (oxidative) twitch**	Fast	High	Intermediate	Long-term anaerobic
Type II B **Fast (glycolytic) twitch**	Very fast	Very high	Low	Short-term anaerobic

These questions are NOT related to a passage.

Question 121

Upon muscle contraction the:

I. A-bands shorten.
II. I-bands shorten.
III. distances between Z-lines increase.
IV. IH-zones lengthen.

- O **A.** I only
- O **B.** II only
- O **C.** I and II only
- O **D.** I, II, III, and IV

Question 122

Irreversible sequestering of calcium in the sarcoplasmic reticulum would most likely:

- O **A.** result in permanent contraction of the muscle fibers, similar to what is seen in rigor mortis.
- O **B.** create a sharp increase in bone density as calcium is resorbed from bones to replace the sequestered calcium.
- O **C.** prevent myosin from binding to actin.
- O **D.** depolymerize actin filaments in the sarcomere.

Question 123

Which of the following statements concerning excitation-contraction coupling in skeletal muscle is false?

- O **A.** The T-tubules transfer the membrane depolarization to the sarcoplasmic reticulum.
- O **B.** The sarcoplasmic reticulum is a major Ca^{2+} storage site.
- O **C.** Ca^{2+} binds to tropomyosin to allow the myosin cross-bridges to bind to the actin filaments.
- O **D.** Excitation-contraction coupling requires voltage-gated Ca^{2+} channels.

Question 124

Muscles cause movement at joints by:

- O **A.** inciting neurons to initiate an electrical "twitch" in tendons.
- O **B.** increasing in length, thereby pushing the muscle's origin and insertion farther apart.
- O **C.** filling with blood, thereby expanding and increasing the distance between the ends of a muscle.
- O **D.** decreasing in length, thereby bringing the muscle's origin and insertion closer together.

Question 125

Studies have shown that athletes have different proportions of muscle cell types according to the type of activity they perform. Which type of muscle cell would be best suited for a marathon runner?

- O **A.** Slow oxidative fibers
- O **B.** Fast oxidative fibers
- O **C.** Slow glycolytic fibers
- O **D.** Fast glycolytic fibers

Question 126

When undergoing physical exercise, healthy adult skeletal muscle is likely to respond with an increase in all of the following EXCEPT:

- O **A.** glycolysis.
- O **B.** the citric acid cycle.
- O **C.** mitosis.
- O **D.** protein production.

Question 127

Which of the following statements regarding muscles that act as third class levers in the body is LEAST likely true?

- O **A.** They require more force to perform the allotted work than if no lever were present.
- O **B.** The force of the muscle contraction is applied between the fulcrum and the load force.
- O **C.** This arrangement increases the bulk of the body.
- O **D.** This arrangement increases the precision of movement.

Question 128

Skeletal muscle contraction may assist in all of the following EXCEPT:

- O **A.** movement of fluid through the body.
- O **B.** body temperature regulation.
- O **C.** posture.
- O **D.** peristalsis.

Questions 121–128 of 144

STOP

6.4 Features of Cardiac Muscle

The human heart is composed mainly of **cardiac muscle** (Figure 6.5). Cardiac muscle is a specialized, electrically-excitable tissue, which permits the propagation of the electrical signals that cause the heart to beat normally. Like skeletal muscle, cardiac muscle is striated, which means that it is composed of sarcomeres. However, each cardiac muscle cell contains only one nucleus. Cardiac muscle cells are separated from each other by *intercalated discs*. The intercalated discs contain gap junctions that allow an action potential to spread from one cardiac cell to the next via electrical synapses, synchronizing the contraction of groups of cardiac muscle cells. The coordination of contraction in cardiac muscle controls the efficient pumping of blood through the chambers of the heart. Because the heart must beat continuously for the entire length of a human life, the mitochondria of cardiac muscle are large and numerous to provide a constant supply of ATP. In contrast to skeletal muscle, which connects bone to bone via ligaments, cardiac muscle is not connected to bone. Instead, cardiac muscle forms a net that contracts in upon itself like a squeezing fist.

As seen through a light microscope, cardiac muscle cells form a net-like structure. The junctions between individual cells are the intercalated discs (dark lines).

Muscle is said to be involuntary when nervous control over its contractions does not reach the level of consciousness in the brain. In other words, your heart beats whether you think about it or not. For the MCAT®, it is important to know that skeletal muscle is voluntary, while cardiac and smooth muscle are not.

Cardiac muscle is **involuntary muscle**. Unlike voluntary skeletal muscle, which is controlled by the somatic nervous system, involuntary muscle such as cardiac muscle receives nervous input from the autonomic nervous system. Although cardiac muscle is self-excitatory to some extent and modulates its own rhythm through the SA and AV notes, input from the autonomic nervous system can affect contraction frequency and thus heart rate. For cardiac muscle, **sympathetic innervation** increases heart rate, while **parasympathetic innervation** via the *vagus nerve* decreases heart rate. Like skeletal muscle, cardiac muscle can grow by hypertrophy when it must work particularly hard.

Circulatory, Respiratory, and Immune Systems
BIOLOGY 2

The action potential of cardiac muscle exhibits a plateau after depolarization. The plateau is created by slow voltage-gated calcium channels, which allow calcium to enter and maintain the inside of the membrane at a positive potential difference. As a result, repolarization of cardiac muscle is slower and more frequent than that of skeletal muscle. The plateau lengthens the time of contraction, preventing a new action potential from starting in the middle of the previous contraction.

The slow repolarization of heart muscle gives the heart enough time to refill with blood and prevents tetanus, the sustained contraction that can occur in skeletal muscle and could be deadly if it occurred in the heart.

For the MCAT®, the most important part of the cardiac action potential to remember is the plateau created by the slow voltage-gated calcium channels. This plateau is not seen in neuronal action potentials.

Remember the importance of calcium in the cardiac action potential. Without it, the heart would beat far too quickly to serve as a functional pump, and death would result.

FIGURE 6.5 Cardiac Muscle

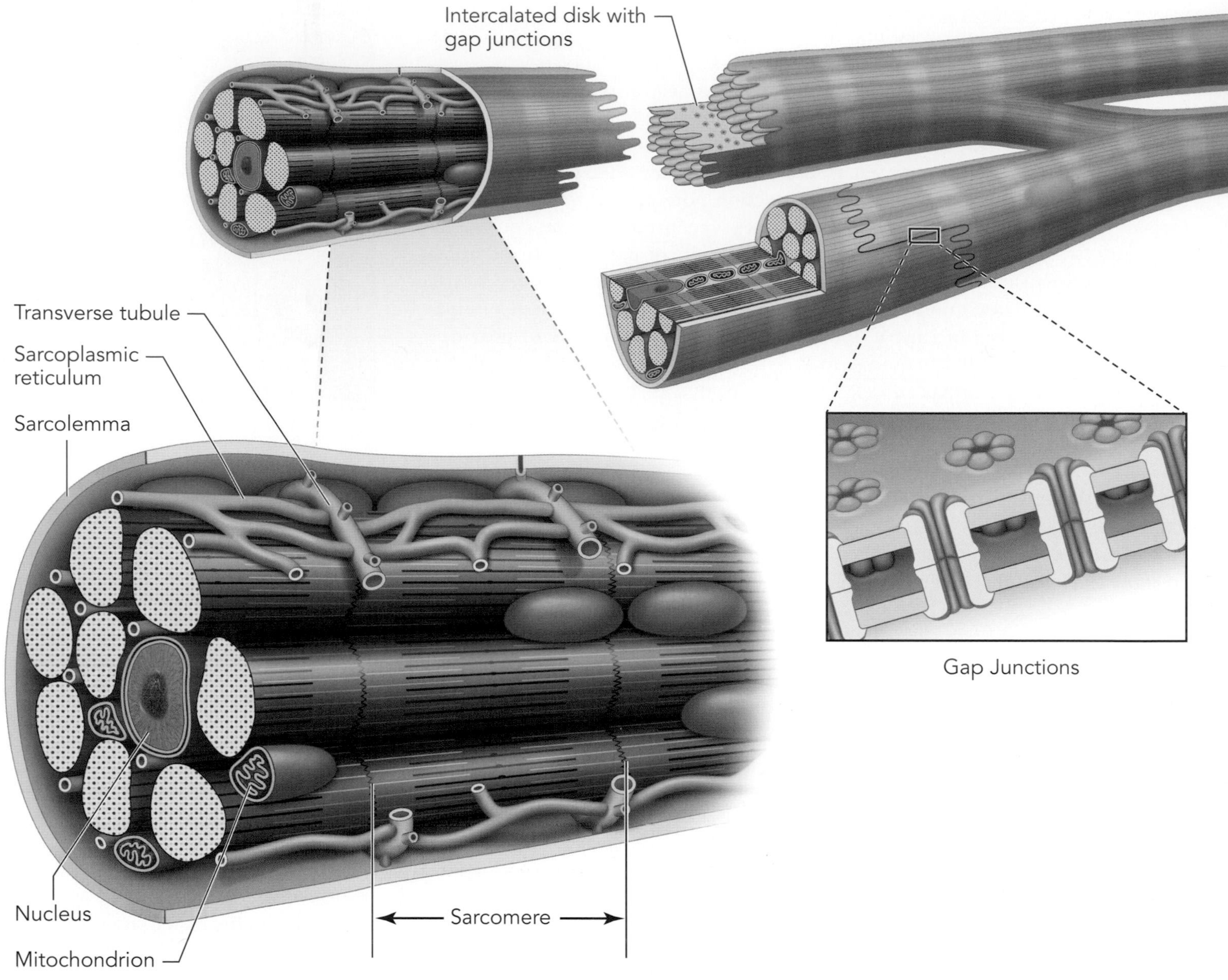

6.5 Smooth Muscle

One example of the purpose of smooth muscle is its role in digestion. Smooth muscle in the digestive organs helps push food along the digestive tract. Sympathetic innervation pauses digestion so that energy can be expended on other tasks, while parasympathetic innervation increases the rate of digestion.

Smooth muscle composes the muscular layer of internal organs and blood vessels. Like cardiac muscle, smooth muscle cells contain only one nucleus and are also involuntary, so they are also innervated by the autonomic nervous system (Figure 6.6). Smooth muscles contain thick and thin filaments, but they are not organized into sarcomeres. In addition, smooth muscle cells contain **intermediate filaments**, which are attached to *dense bodies* spread throughout the cell. Dense bodies serve a similar function to Z-lines of striated muscle. The thick and thin filaments are attached to the intermediate filaments, and, when they contract, they cause the intermediate filaments to pull the dense bodies together. Upon contraction, the smooth muscle cell shrinks length-wise.

There are two types of smooth muscle described by how they function: 1. *single-unit* and 2. *multi-unit*. Single-unit smooth muscle, also called *visceral smooth muscle*, is the most common. Single-unit smooth muscle cells are connected by gap junctions that spread the action potential from a single neuron through a large group of cells, allowing the cells to contract as a single unit. Electrical synapses via gap junctions in visceral muscle allow for faster signal transmission than would be possible with chemical synapses. The speed of transmission permits simultaneous contraction of the muscle cells in a coordinated fashion. Single-unit smooth muscle is found in small arteries and veins, the stomach, intestines, uterus, and urinary bladder.

In multi-unit smooth muscle, each fiber is attached directly to a neuron. A group of multi-unit smooth muscle fibers can contract independently of other muscle fibers in the same location. Multi-unit smooth muscle is found in the large arteries, bronchioles, pili muscles attached to hair follicles, and the iris.

In addition to responding to neural signals, smooth muscle contracts or relaxes in the presence of hormones, or in response to changes in pH, O_2 and CO_2 levels, temperature, and ion concentrations.

If you see smooth muscle on the MCAT®, remember that it is primarily found in the walls of blood vessels and internal organs, especially the digestive tract and reproductive tract. In organ walls, it functions to cause contraction of the organ, such as peristalsis in the digestive tract. In blood vessels, contraction of smooth muscle can constrict the vessel radius, helping to regulate blood flow. Smooth muscle is also present in certain other locations such as the iris of the eye and the arrector pili muscles of the skin, but if you see these examples on the MCAT®, they will likely be explained in a passage.

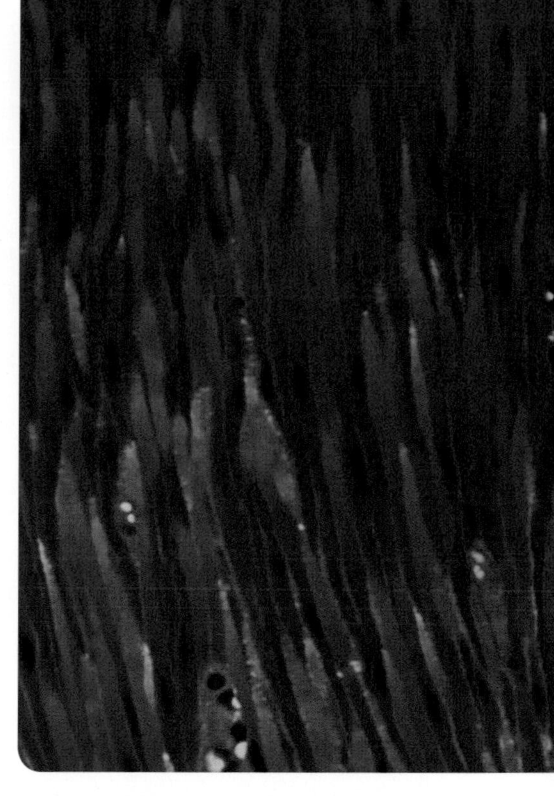

Smooth muscle is composed of spindle-shaped cells grouped in irregular bundles. Each cell contains one nucleus, seen here as a dark stained spot.

FIGURE 6.6 Smooth Muscle

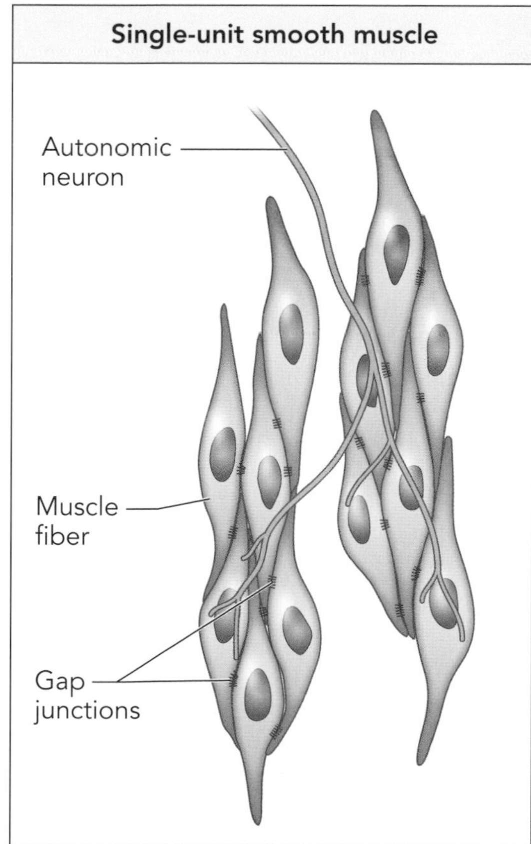

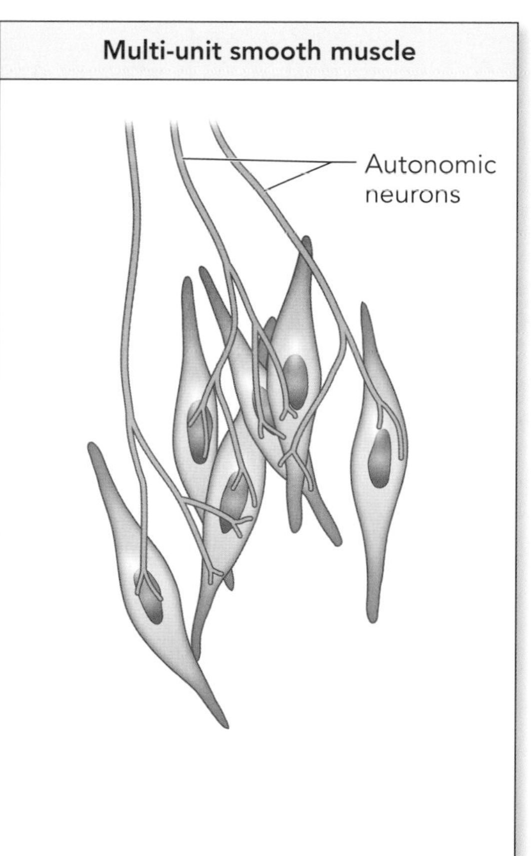

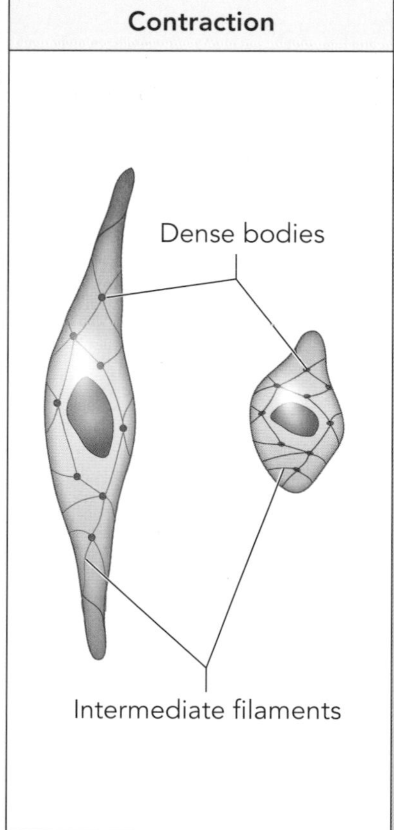

These questions are NOT related to a passage.

Question 129

The function of gap junctions in the intercalated discs of cardiac muscle is to:

- O **A.** anchor the muscle fibers together.
- O **B.** ensure that an action potential is spread to all fibers in the muscle network.
- O **C.** control blood flow by selectively opening and closing capillaries.
- O **D.** release calcium into the sarcoplasmic reticulum.

Question 130

Which of the following muscular actions is controlled by the autonomic nervous system?

- O **A.** The knee-jerk reflex
- O **B.** Conduction of cardiac muscle action potential from cell to cell
- O **C.** Peristalsis of the gastrointestinal tract
- O **D.** Contraction of the diaphragm

Question 131

During an action potential, a cardiac muscle cell remains depolarized much longer than a neuron. This characteristic of the cardiac muscle cell most likely functions to:

- O **A.** prevent the initiation of another action potential during contraction of the heart.
- O **B.** ensure that adjacent cardiac muscle cells will contract at different times.
- O **C.** keep the neuron from firing twice in rapid succession.
- O **D.** allow sodium voltage-gated channels to remain open long enough for all sodium to exit the cell.

Question 132

All of the following statements are true concerning smooth muscle EXCEPT:

- O **A.** smooth muscle contractions are longer and slower than skeletal muscle contractions.
- O **B.** a chemical change in the environment around smooth muscle may create a contraction.
- O **C.** smooth muscle does not require calcium to contract.
- O **D.** smooth muscle is usually involuntary.

Question 133

All of the following is true concerning acetylcholine EXCEPT:

- O **A.** it is used in both the peripheral and central nervous systems.
- O **B.** it binds to muscarinic receptors in both smooth and cardiac muscle.
- O **C.** it activates skeletal muscle.
- O **D.** it increases contraction in cardiac muscle.

Question 134

Cardiac muscle is excited by:

- O **A.** parasympathetic nervous excitation.
- O **B.** constriction of T-tubules.
- O **C.** increased cytosolic sodium concentration.
- O **D.** increased cytosolic calcium concentration.

Question 135

Specialized cardiac muscle cells in the SA node have the capacity for self-excitation. When the SA node is innervated by the vagus nerve, the frequency of self-excitation of these cardiac cells is likely to be:

- O **A.** slower than a normal heartbeat, because excitation by the vagus nerve decreases the heart rate.
- O **B.** slower than a normal heartbeat, because excitation by the vagus nerve increases the heart rate.
- O **C.** faster than a normal heartbeat, because excitation by the vagus nerve decreases the heart rate.
- O **D.** faster than a normal heartbeat, because excitation by the vagus nerve increases the heart rate.

Question 136

In extreme cold, just before the onset of frostbite, sudden vasodilation occurs, manifesting in flushed skin. This vasodilation most likely results from:

- O **A.** paralysis of smooth muscle in the vascular walls.
- O **B.** paralysis of skeletal muscle surrounding the vascular walls.
- O **C.** sudden tachycardia with a resultant increase in blood pressure.
- O **D.** blood shunting due to smooth muscle sphincters.

6.6 Bone: Mineral Homeostasis and Structure

Bone is living tissue and is one of the major types of connective tissue in the body. The **functions of bone** are support of soft tissue, protection of internal organs, assistance in body movement, mineral (calcium) storage, blood cell production, and energy storage in the form of adipose cells in bone marrow. This section will cover the major cell types within bone, the process by which new bone is made, and the structure and classification of bones in the human body.

Bone tissue contains four types of cells surrounded by an extensive **calcium/protein matrix** composed of inorganic minerals (notably, calcium) and proteins.

1. *Osteoprogenitor* (or *osteogenic*) cells differentiate into osteoblasts.
2. **Osteoblasts** secrete collagen and organic compounds upon which bone is formed. They are incapable of mitosis. As osteoblasts release matrix materials around themselves, they become enveloped by the matrix and differentiate into osteocytes.
3. **Osteocytes** are also incapable of mitosis. They exchange nutrients and waste materials with the blood.
4. **Osteoclasts** resorb bone matrix, releasing minerals back into the blood. They are thought to develop from the white blood cells called monocytes.

Although bones vary in size, shape and composition, some general structural features are shared by most bones. A typical long bone (Figure 6.7) has a long shaft, called the *diaphysis*, and two ends, each composed of a *metaphysis* and *epiphysis*. A sheet of cartilage between the epiphysis and the metaphysis, called the *epiphyseal plate*, is where long bones grow in length when stimulated by growth hormone (GH) during childhood and adolescence.

Each bone contains two main types of bone structure: spongy bone and compact bone. Bones typically consist of spongy bone on the inside surrounded by a shell of compact bone. **Spongy bone** (also called *trabecular* or *cancellous bone*) contains **red bone marrow**, the site of red blood cell development. **Compact bone** (also called **cortical bone**) surrounds a hollow area inside the diaphysis known as the *medullary cavity*, which holds yellow bone marrow. Yellow bone marrow contains adipose cells for fat storage.

Some organisms, such as lobsters, gain structural support and protection from having their skeleton on the outside in a shell-like exoskeleton. Humans and all other vertebrates have a skeleton on the inside, called an endoskeleton. Exo- and endoskeletons serve many of the same purposes, including structural support and protection, but it is crucial to know the difference between them.

Bone serves a wide variety of purposes in the human body. Like most other living tissue, bone must constantly replace itself. The main cell types in bone are used to break down and rebuild the structure of bone as older portions are worn out and new portions are required.

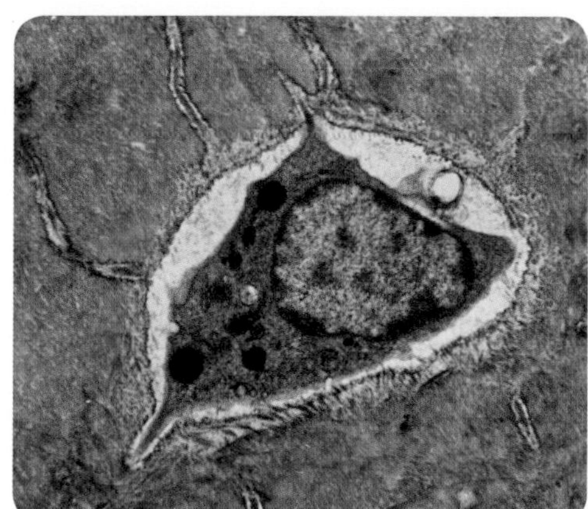

An osteocyte is an osteoblast that has become trapped within a bone cavity (lacuna). Osteocytes are responsible for bone formation, but eventually become embedded in the bone matrix and are only able to communicate with each other and the rest of the body through a system of canals and canaliculi.

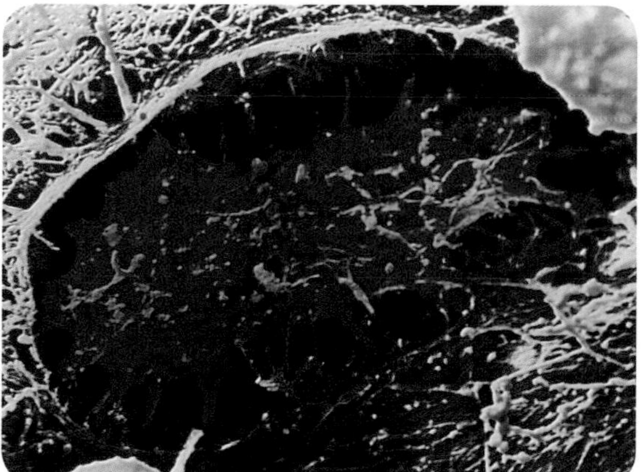

A bone-building osteoblast cell is shown here surrounded by a dense network of collagen fibers.

Know the functions of osteoblasts, osteocytes, and osteoclasts. Osteoblast and osteoclast sound similar but have opposite functions. Remember: osteoBlasts Build bone, osteoClasts Consume it.

FIGURE 6.7 Typical Long Bone Structure

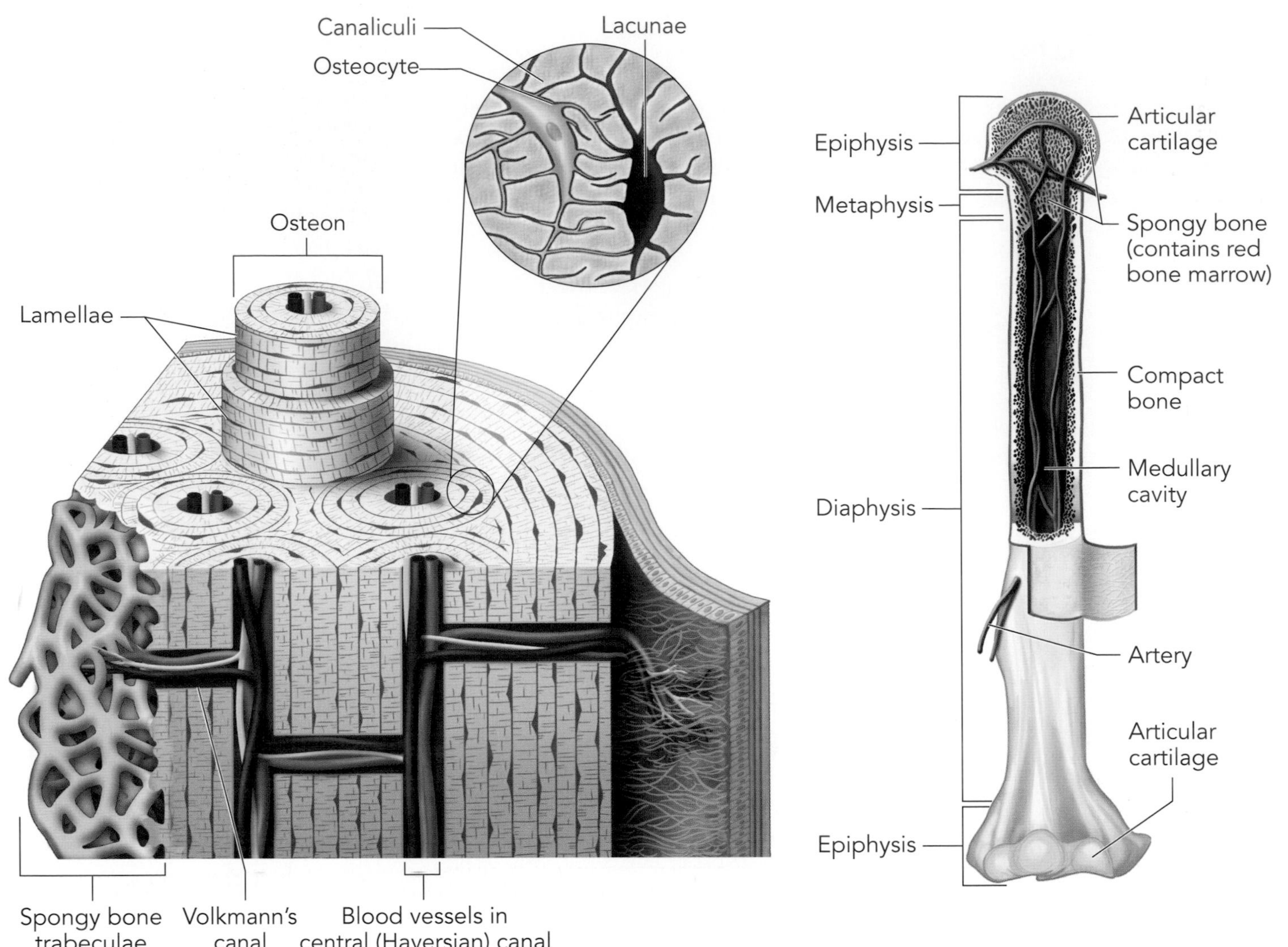

Compact bone has a highly organized microstructure. Old parcels of bone are continuously replaced with new bone through the process of *remodeling*. In remodeling, osteoclasts burrow tunnels through compact bone. The osteoclasts are followed by osteoblasts, which lay down a new bone matrix onto the tunnel walls, forming concentric rings called *lamellae*. Osteoblasts leave open spaces in the center of the lamellae known as *Haversian canals*. The entire system of lamellae and a Haversian canal is called an *osteon* or *Haversian system*. The series of spaces within bone tissue allows for communication and nutrient exchange. Osteocytes are living cells that, once they have surrounded themselves with bone matrix, are isolated from other cells. Haversian canals contain blood and lymph vessels, and are connected by crossing canals called *Volkmann's canals*. Osteocytes trapped between the lamellae exchange nutrients via spaces called *canaliculi*.

Endocrine System
BIOLOGY 2

The remodeling of bone is subject to **endocrine control**. Parathyroid hormone helps regulate Ca^{2+} levels in the bloodstream by influencing bone. High levels of parathyroid hormone signal the osteoclasts to begin breaking down bone so that calcium stored as part of the bone matrix can be released into the blood. Other hormones, such as vitamin D, help restore calcium stores in the bone by promoting calcium absorption through the digestive system. Additionally, the hormone calcitonin stimulates osteoblasts to store excess calcium as bone.

Bone's Function in Mineral Homeostasis

Osteoblasts use calcium from the blood to form new osteons. Conversely, when osteoclasts break down bone, they release that calcium into the bloodstream. One of the important functions of bone is to help regulate the mineral content of the blood, particularly calcium levels. Calcium salts are only slightly soluble, so most calcium in the blood is not in the form of free calcium ions. Instead it is bound mainly by proteins and, to a much lesser extent, by phosphates (HPO_4^{2-}) and other anions. It is the concentration of free calcium ions (Ca^{2+}) in the blood that is important physiologically. Too much Ca^{2+} causes membranes to become hypo-excitable, leading to fatigue and memory loss; too little results in cramps and convulsions. Calcium levels are maintained to generate a large concentration gradient with high calcium outside of cells and low calcium in the cytosol. Small changes in calcium distribution can result in large changes in membrane voltage. This is how cardiac action potentials are formed.

Most of the body's Ca^{2+} is stored in the bone matrix as the mineral *hydroxyapatite* [$Ca_{10}(PO_4)_6(OH)_2$], which contributes to the strength of bones. Collagen fibers give bone great tensile strength (the ability to stretch without breaking). Hydroxyapatite crystals lie alongside collagen fibers, giving bone great compressive strength (the ability to compress under the weight of a load without breaking). Some Ca^{2+} is stored in bone in the form of slightly soluble calcium salts such as $CaHPO_4$. These salts buffer plasma Ca^{2+} levels. Thus bone acts as a reservoir that can store minerals when excess Ca^{2+} and HPO_4^{2-} are present and release these minerals into the bloodstream as needed.

This computer-generated image shows multi-nucleated osteoclasts etching away trabecular bone in a process called bone resorption.

Apply Key 3: When you see Ca^{2+}, think about muscle contractions and storage in the bone. Calcium allows for the actin-myosin interactions that govern muscle contractions. This ion can be stored in the bone, strengthening it and providing a reserve supply for cellular processes that require calcium.

Bone Types and Structures

Bones can be categorized by shape. Most bones fall into one of four **specialized bone types**: 1. long, 2. short, 3. flat, or 4. irregular. **Long bones** have a shaft that is curved for strength. They are composed of compact and spongy bone. Leg, arm, finger, and toe bones are examples of long bones. **Short bones** are often roughly cuboidal in shape. Examples include the ankle and wrist bones. **Flat bones**, such as the skull, sternum, ribs, and shoulder blades, provide organ protection and large areas for muscle attachment. **Irregular bone** has an irregular shape and variable amounts of compact and spongy bone. The ossicles of the ear are irregular bones.

FIGURE 6.8 Bone Types

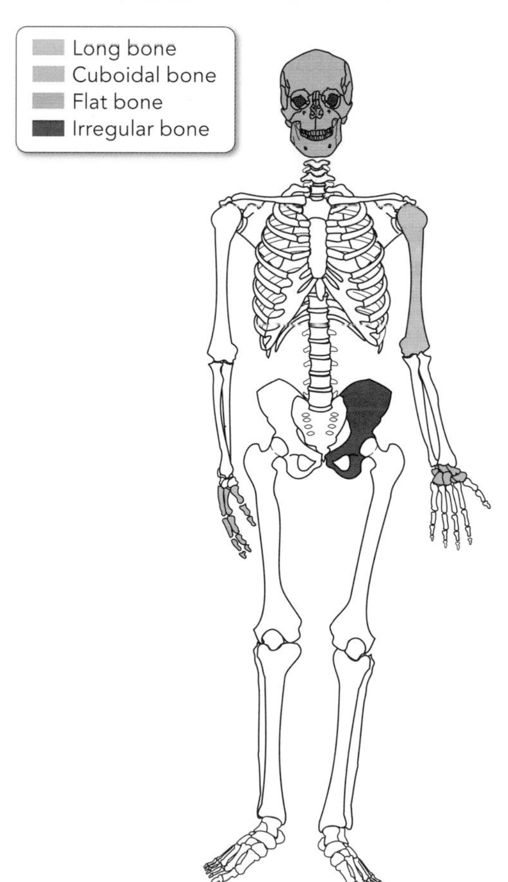

Remember that bone is not just for support, protection, and movement. Bone also stores calcium and phosphate, helping to maintain homeostasis in the concentration of these ions in the blood. Bone stores energy in the form of fat and is also the site of blood cell formation.

FIGURE 6.9 The Human Skeleton

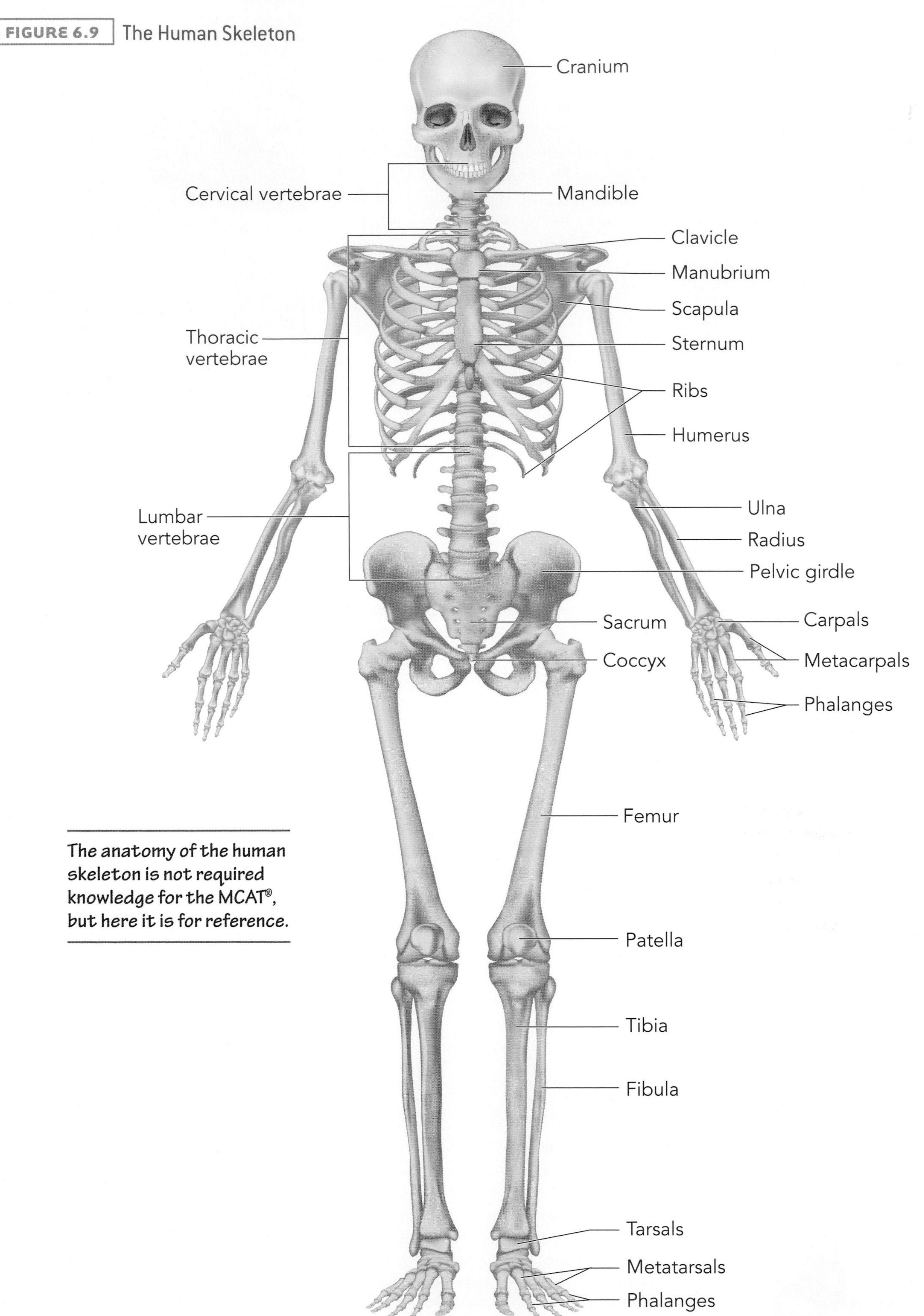

The anatomy of the human skeleton is not required knowledge for the MCAT®, but here it is for reference.

6.7 Cartilage and Joints

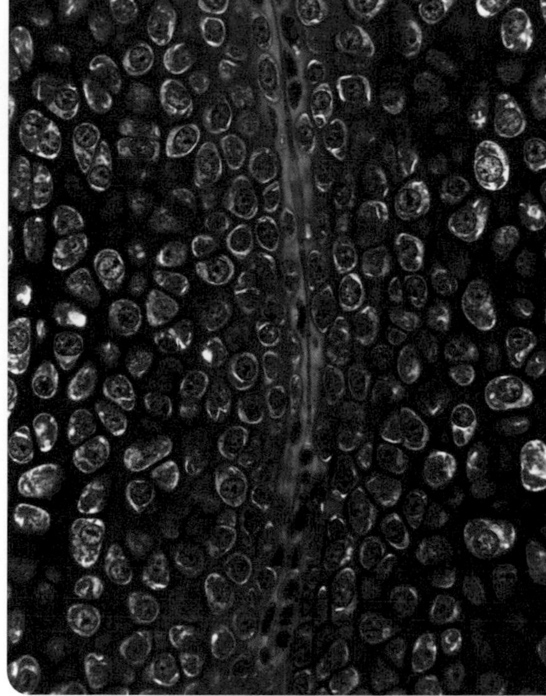

Hyaline cartilage is semi-rigid connective tissue composed of many chondrocytes (cartilage cells, pale purple). These cells synthesize an extracellular matrix (dark purple) of proteoglycans, collagen, and water that keeps them apart from each other in spaces known as lacunae. Hyaline cartilage is strong but compressible due to its high water content. It reduces friction between the bones in the knee joint as they move against each other.

Cartilage

Cartilage is flexible, resilient connective tissue. It is composed primarily of collagen and has great tensile strength. Cartilage gives shape and structure to various body parts (such as the ears) and also serves the important function of providing cushion, connection, and elasticity to the joints of the body. Cartilage has no blood vessels or nerves except in its outside membrane. There are three types of cartilage: 1. *hyaline*, 2. *fibrocartilage*, and 3. *elastic*. Hyaline cartilage, which reduces friction and absorbs shock in joints, is the most common.

Cartilage serves the important function of cushioning and reducing friction at the joints. When cartilage wears away through damage or overuse, one bone will grind against the other, causing the painful condition of osteoarthritis.

Joints

Joints are locations where bones connect in ways that allow for varying amounts of movement, depending on type. Joints are classified into three types according to their structures:

1. **Fibrous joints** occur between two bones held closely and tightly together by fibrous connective tissue, permitting extremely minimal movement. The main purpose of fibrous joints is to maintain a fixed relationship between two bones. Teeth form fibrous joints with the mandible. After early childhood, skull bones form fibrous joints with each other, strengthening and hardening the skull.
2. **Cartilaginous joints** also allow little movement. They occur between two bones tightly connected by cartilage, such as the ribs and the sternum, or the pubic symphysis in the pelvis. The slight flexibility of cartilaginous joints allows them to absorb some energy and can help protect bones in the event of trauma.
3. **Synovial joints** are the type of joint that are most familiar as "joints." Bones joined by synovial joints are not bound directly by the intervening cartilage, so a wide range of movement is possible. Instead, they are separated by a capsule filled with *synovial fluid*. Synovial fluid provides lubrication and nourishment to the cartilage. In addition, the synovial fluid contains phagocytotic cells that remove microbes and particles. The shoulder and knee are two of the many examples of synovial joints in the body.

FIGURE 6.10 Joint Types

Fibrous joint

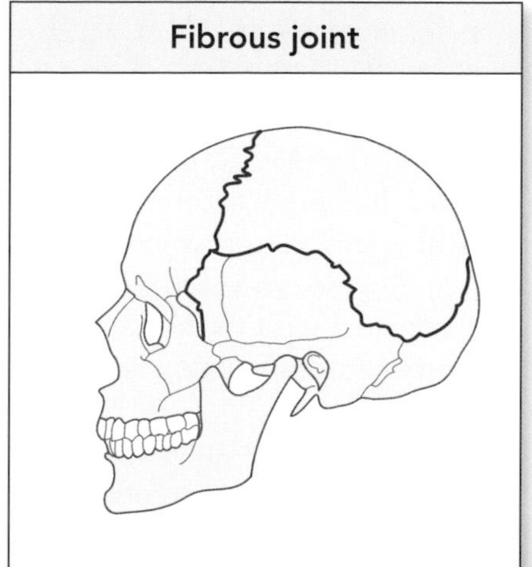

Cartilaginous Joint

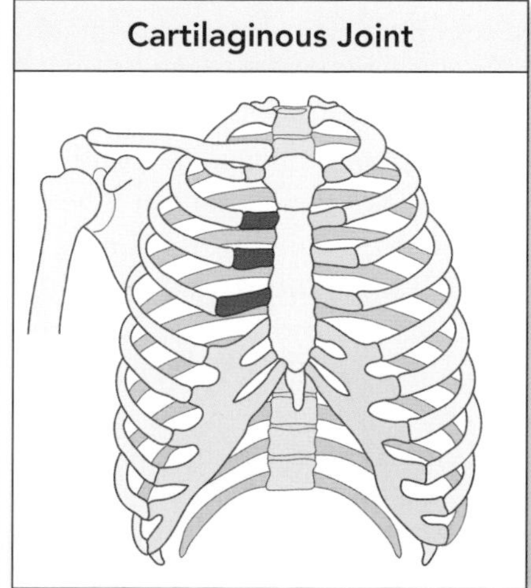

Synovial Joint

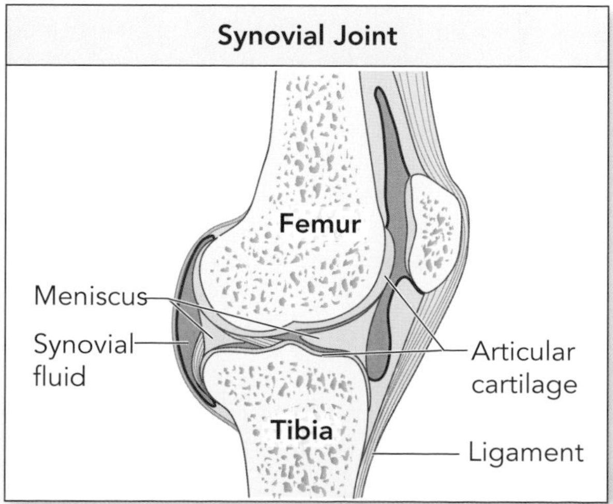

This is a sample of skin from the back of a human hand.

6.8 Skin

Skin may seem different from other organs, but like any other organ, it is a group of tissues working together to perform a specific function. Skin is the largest organ in the body and has a wide variety of functions, many of which contribute to homeostasis. The skin also has specific immune cells and unique patterns of innervation that help protect the body from pathogens, and it transmits information about environmental surroundings to the rest of the body. Some important **functions of the skin** are:

1. **Thermoregulation**: The skin helps to regulate body temperature. Blood conducts heat from the core of the body to skin. Some of this heat can be dissipated by the endothermic evaporation of sweat, but most is dissipated by radiation. Radiation is only effective if the body temperature is higher than room temperature. More blood can be directed to **surface capillaries** through **vasodilation** to allow for greater heat loss, or blood can be shunted away from the capillaries of the skin through the process of **vasoconstriction** to reduce heat loss. Thermoregulation of the body through **sweating**, vasoconstriction, and vasodilation is under the influence of hormonal control. Additionally, hairs can be erected via sympathetic stimulation of associated muscles, insulating warm air next to the skin. Skin has both warmth and cold receptors.
2. **Protection**: The skin is a physical barrier to **abrasion, disease organisms**, many chemicals, and ultraviolet radiation.

Nervous System
BIOLOGY 2

3. **Environmental sensory input**: The skin gathers information from the environment by sensing temperature, pressure, pain, and touch.
4. **Osmoregulation**: Skin is **relatively impermeable to water**, protecting against dehydration. However, some water and salts are excreted through the skin via diffusion. Both **excretion** and sweating are ways that skin contributes to osmoregulation, keeping the concentration of solutes in body fluids at the appropriate level.
5. **Immunity**: Besides being a physical barrier to bacteria, specialized cells of the epidermis are components of the immune system.
6. **Blood reservoir**: Vessels in the dermis hold up to 10% of the blood of a resting adult.
7. **Vitamin D synthesis**: Ultraviolet radiation activates a molecule in the skin that is a precursor to vitamin D. The activated molecule is modified by enzymes in the liver and kidneys to produce vitamin D.

The skin has two principal parts: 1. the superficial **epidermis** and 2. the deeper **dermis**. Beneath both of these layers is a subcutaneous tissue called the *superficial fascia* or *hypodermis*. This subcutaneous **fat layer** is an important heat insulator for the body. It contributes to thermoregulation by helping to maintain normal core body temperatures on cold days when the skin approaches the temperature of the environment.

The epidermis is avascular (no blood vessels) epithelial tissue, which primarily serves the purpose of protection from the environment. It consists of four major cell types. 1. 90% of the epidermis is composed of **keratinocytes**, cells that produce the protein keratin, which helps waterproof the skin. 2. **Melanocytes** transfer *melanin* (skin pigment) to keratinocytes. 3. **Langerhans cells** interact with the helper T-cells of the immune system. 4. **Merkel cells** attach to sensory neurons and function in the sensation of touch.

There are four **layers of the epidermis** (or five in the palms and soles of the feet). With constant exposure to the environment, skin cells must be continually replaced. The deepest layer of the epidermis contains Merkel cells and stem cells.

FIGURE 6.11 Skin

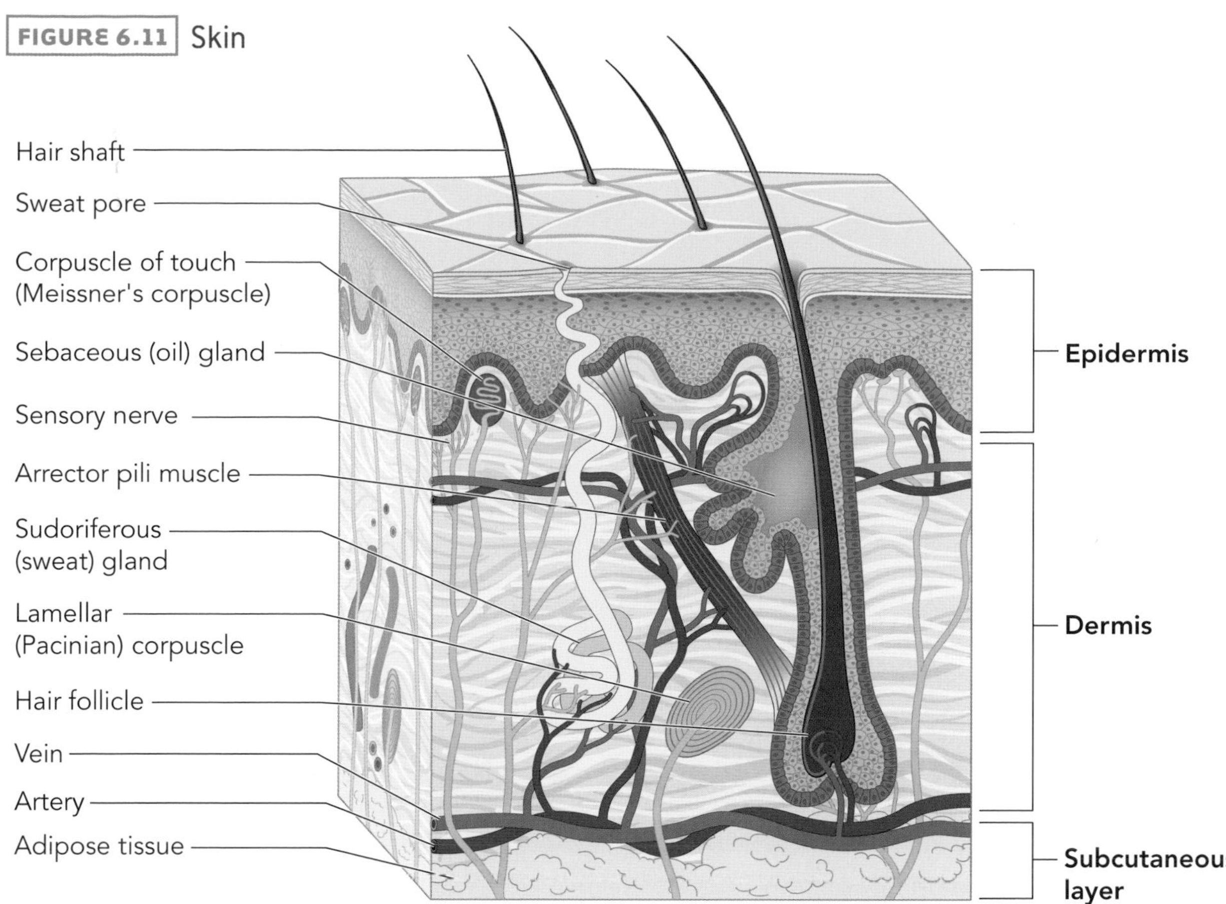

The stem cells continually divide to give rise to keratinocytes and other new, replacement skin cells. Once produced, keratinocytes are pushed to the top layer of the epidermis. As they rise, they accumulate keratin and die, losing their cytoplasm, nucleus, and other organelles. When the cells reach the outermost layer of skin, they slough off the body. The process of keratinization, from birth of a cell to sloughing off, takes two to four weeks. The outermost layer of epidermis consists of 25 to 30 layers of flat, dead cells. Exposure to friction or pressure stimulates the epidermis to thicken, forming a protective **callus**.

The dermis is connective tissue derived from mesodermal cells. This layer serves a variety of functions and is embedded with blood vessels, nerves, glands, and hair follicles. Collagen and elastic fibers in the dermis provide skin with strength, extensibility, and elasticity. The receptors that transmit the sensation of touch, including separate receptors for the sensations of pressure, pain, and temperature, are located in the dermis. *Hair follicles* are embedded in the dermis. **Hair** is a column of keratinized cells held tightly together and plays a role in tactile sensation and thermoregulation. Arrector pili muscles that surround the base of hair follicles allows them to become erect, trapping air close to the skin to help maintain body heat. As new cells are added to its base, the hair grows. **Nails** are also keratinized cells. Hair and nails both contribute to the skin's function of physical protection.

The dermis also contains a wide variety of glands. Most hairs are associated with a *sebaceous (oil) gland* that empties oil directly into the follicle and onto the skin. Two types of **sweat glands** are found in the skin separate from hair follicles. The first type, *eccrine sweat glands*, are found over the entire surface of the skin and produce sweat in response to heat, resulting in the cooling of the body. The second type, *apocrine sweat glands*, are congregated in certain regions of the dermis and produce an acrid-smelling sweat in response to stress.

It is not necessary to know the names of each type of sweat gland, but know that one type is located throughout the skin's surface while the other type is located in certain areas of the dermis.

FIGURE 6.12 The Epidermis

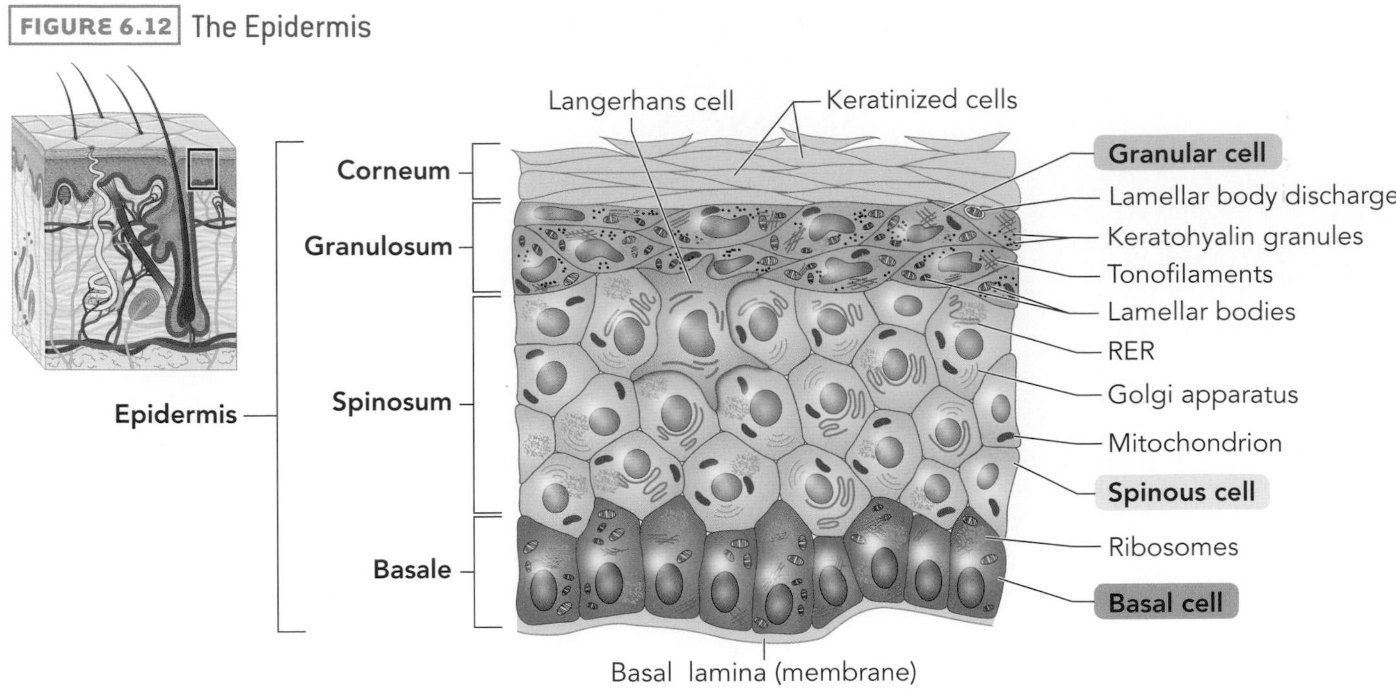

Together, the skin, hair, nails, glands, and some nerve endings function in a variety of roles, primarily concerning the protection of the body and preservation of homeostasis within an organism.

These questions are NOT related to a passage.

Question 137

The production of which of the following cell types would most likely be increased by parathyroid hormone?

- O **A.** Osteoprogenitors
- O **B.** Osteocytes
- O **C.** Osteoclasts
- O **D.** Osteoblasts

Question 138

In a synovial joint, the purpose of synovial fluid is to:

- O **A.** reduce friction between bone ends.
- O **B.** keep bone cells adequately hydrated.
- O **C.** occupy space until the bones of the joint complete their growth.
- O **D.** maintain a rigid connection between two flat bones.

Question 139

Surgical cutting of which of the following tissues would result in the LEAST amount of pain?

- O **A.** Muscle
- O **B.** Bone
- O **C.** Cartilage
- O **D.** Skin

Question 140

In a synovial joint, the connective tissues holding the bones together are called:

- O **A.** ligaments.
- O **B.** tendons.
- O **C.** muscles.
- O **D.** osseous tissues.

Question 141

The top layer of skin cells, or the stratum coronae, is filled with layers of dead cells full of fibrous keratin and connected by glycolipids that seal the space between cells. Which function of the skin does the stratum coronae most likely carry out?

- O **A.** Impermeability to water
- O **B.** Synthesis of vitamin D
- O **C.** Thermoregulation
- O **D.** Osmoregulation

Question 142

All of the following are found in compact bone EXCEPT:

- O **A.** yellow marrow.
- O **B.** Haversian canals.
- O **C.** canaliculi.
- O **D.** Volkmann's canals.

Question 143

Hydroxyapatite, the mineral portion of bone, contains all of the following elements except:

- O **A.** calcium.
- O **B.** sulfur.
- O **C.** phosphate.
- O **D.** hydrogen.

Question 144

Scientists interested in understanding how chemotherapy impacted bone strength treated mice with a cisplatin, a chemotherapeutic that inhibits DNA replication, and measured the tensile strength of the mouse femur.

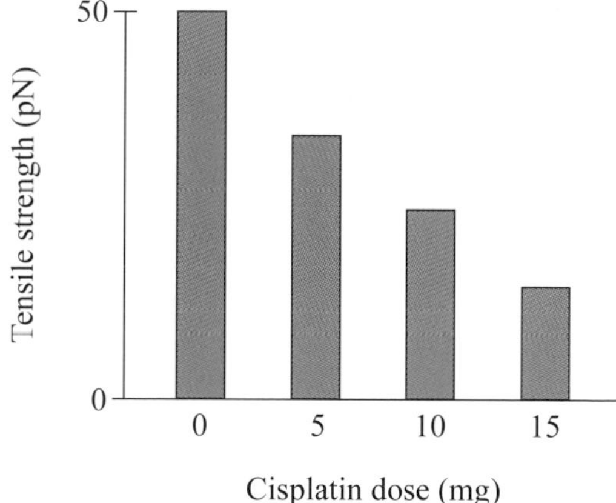

Which of the following cell types would be affected the most by cisplatin treatment?

- O **A.** Osteoblasts
- O **B.** Osteoclasts
- O **C.** Osteocytes
- O **D.** Osteoprogenitor cells

STOP

TERMS YOU NEED TO KNOW

A band
Abrasion
Acetylcholine
Actin
Blood reservoir
Bone
Calcium/protein matrix
Callus
Cardiac muscle
Cartilage
Cartilaginous joints
Compact bone (cortical bone)
Contractile apparatus
Contractile velocity
Cross-bridges
Dermis
Disease organisms
Endocrine control
Endoskeleton
Environmental sensory input
Epidermis
Excretion
Exoskeleton
Fast twitch fibers (type II)
Fat layer
Fibrous joints
Flat bones
Functions of bone
Functions of muscle
Functions of skin
H zone
Hair
I band
Immunity
Intermediate filaments
Involuntary muscle
Irregular bone
Joints
Keratinocytes
Langerhans cells
Layers of the epidermis
Ligament
Long bones
M line
Melanocytes
Merkel cells
Motor end plate
Motor neuron
Muscle fatigue
Myoglobin
Myosin
Nails
Neuromuscular junction
Osmoregulation
Osteoblasts
Osteoclasts
Osteocytes
Oxygen debt
Parasympathetic innervation
Peripheral circulatory assistance
Protection
Red bone marrow
Relatively impermeable to water
Sarcomere
Sarcoplasmic reticulum
Skeletal muscle
Skin
Shivering reflex
Short bones
Sliding filament model
Slow twitch fibers (type I)
Smooth muscle
Specialized bone types
Spongy bone
Striated muscle
Surface capillaries
Sweat glands
Sweating
Sympathetic innervation
Synovial joints
Tendon
Thermoregulation
Thick filaments
Thin filaments
Tropomyosin
Troponin
T-tubules
Vasoconstriction
Vasodilation
Vitamin D synthesis
Voluntary muscle
Z line

DON'T FORGET YOUR KEYS

1. Somatic NS innervates voluntary muscles; autonomic NS innervates involuntary muscle.

2. The energy of an ATP molecule prepares the myosin head so that the muscle is ready and contraction is downhill in energy.

3. See Ca^{2+}, think neuromuscular function and storage in bone.

STOP!

DO NOT LOOK AT THESE EXAMS UNTIL CLASS.

30-MINUTE IN-CLASS EXAM FOR LECTURE 1

Passage I (Questions 1-4)

Septicemia is a serious medical condition where bacteria present in the circulatory system provoke an amplified and dysregulated immune response in the individual. The most common infection sites leading to bacterial entry into the circulatory system are bacterial infections in the lungs, urinary tract, abdominal cavity, and primary infections of the bloodstream. Many bacterial pathogens have become resistant to antibiotic regimens, resulting in an urgent health problem worldwide. One potentially useful method for the treatment of antibiotic resistant bacterial infections employs bacteriophages capable of killing bacteria. Phage-derived enzymes targeting gram-positive bacteria are the most promising candidates to enter the markets for therapeutic use.

Only the most virulent bacterial clones are capable of tissue invasion possibly leading to septicemia, as the bacteria have to overcome anatomical and host immune system barriers to enter the circulatory system. Although there is considerable information regarding prophages and phage-encoded virulence factors in bacterial pathogens, few studies have investigated microbial ecology in clinical bacterial infections. Researchers surveyed the phage ecology in conditions of septicemia. They observed that septicemia-causing bacteria could be induced to produce phages active against other isolates of the same bacterial strain, and that phage production correlated with septicemia. Further characterization of the phage isolates revealed that the virus detected in the blood culture was the same as induced from the bacterium isolated from that particular blood culture sample.

Strain of bacteria	Titer (pFU/ml)
E. coli	
05vv1387	2×10^4
05vv1522	1×10^5
05vv1558	2×10^3
05vv1809	2×10^2
05vv1999	1×10^2
05vv2388	2×10^3
P. aeruginosa	
05vv1315	6×10^3
05vv1400	2×10^3
05vv1973	5×10^4
K. pneumoniae	
05vv2343	1×10^4

Table 1 Phage titer following induction of selected strains of bacteria

Question 1

Which of the following substances would slow the infection and propagation of bacteriophages?

I. Inhibitor of phage enzymes that cut through the bacterial membrane
II. Inhibitor of bacterial genetic machinery
III. Inhibitor of phage binding to a eukaryotic receptor

- **A.** I only
- **B.** III only
- **C.** I and II only
- **D.** I, II, and III

Question 2

Which of the following characteristics of phages would be most important for them to successfully survive within the human body during a bacterial infection?

- **A.** Enveloped and lytic
- **B.** Enveloped and lysogenic
- **C.** Unenveloped and lytic
- **D.** Unenveloped and lysogenic

Question 3

The most virulent bacteria in septicemia bring about their own dominance by the induction of phages. Which of the following statements best explains this phenomenon?

- **A.** A strain releases a phage that it is vulnerable to and this kills off other strains of bacteria.
- **B.** A strain releases a phage that it is vulnerable to and this kills off the strain to conserve resources.
- **C.** A strain releases a phage that it is resistant to and this kills off other strains of bacteria.
- **D.** A strain releases a phage that it is vulnerable to and this strain kills off the strain and nearby strains to conserve resources.

Question 4

If all of the *E. coli* strains in Figure 1 are grown together in a petri dish, which of the following strains is most likely to be dominant?

- **A.** 05vv1387
- **B.** 05vv1522
- **C.** 05vv1558
- **D.** 05vv1315

Passage II (Questions 5-9)

In a healthy cell, smooth endoplasmic reticulum (ER) performs several functions including carbohydrate metabolism, lipid synthesis, and oxidation of foreign substances such as drugs, pesticides, toxins and pollutants.

The smooth ER synthesizes several classes of lipids, including triacylglycerols which are stored in the ER lumen, cholesterol and its steroid hormone derivatives, and phospholipids for incorporation into the various membranous cell structures. Phospholipids are synthesized only along the cytosol side of the ER membrane. They are then selectively flipped to the other side of the cell membrane by phospholipid translocators.

Most detoxification reactions in the smooth ER involve oxidation. Such reactions usually involve the conversion of hydrophobic compounds into hydrophilic compounds, and are governed by a system of enzymes called mixed-function oxidases. Cytochrome P450, a group of iron-containing integral membrane proteins, is a central component of one mixed-function oxidase system. Mixed-function oxidases also govern the oxidation of nonpolar endogenous compounds, such as steroids and fatty acids. Ingestion of the depressant phenobarbital triggers an increase in smooth ER and mixed-function oxidases but not in other ER enzymes.

Previous work has shown that cytochrome P450 activity is enhanced in phospholipid rich environments. This provides a possible mechanism for increasing cytochrome P450 activity in response to toxic foreign substances. Researchers sought to determine what effect a highly toxic substance would have on phospholipid levels by studying the effects of Carbofuran, an organic compound which is used as a pesticide.

Experiment 1

Carbofuran was injected into the abdominal cavity of live catfish. After two days, the catfish were sacrificed. Liver membrane fractions were isolated, and lipid and phospholipid levels were measured. (Figure 1).

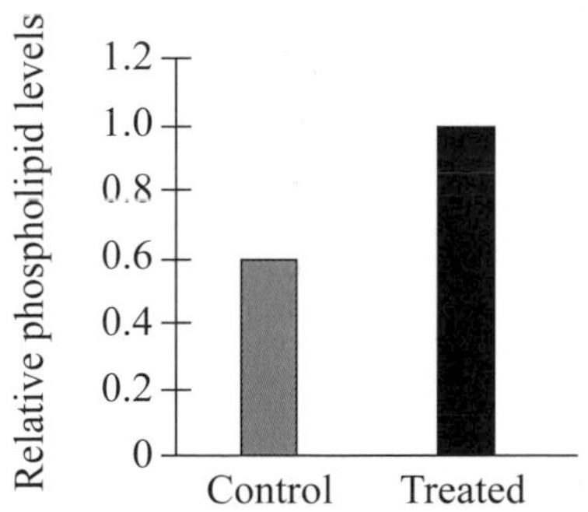

Figure 1 Carbofuran treated fish liver cells

Question 5

Where are many of the enzymes necessary for phospholipid synthesis likely to be located?

- O **A.** ER lumen
- O **B.** Cytosol
- O **C.** Lysosome
- O **D.** Golgi complex

Question 6

How do phospholipids most likely affect cytochrome P450 activity?

- O **A.** By upregulating P450 production
- O **B.** By downregulating P450 degradation
- O **C.** By modulating structural conformation of P450
- O **D.** By providing a pathway for exocytosis

Question 7

The primary structure of mixed-function oxidases is most likely synthesized at the:

- O **A.** smooth ER.
- O **B.** rough ER.
- O **C.** Golgi apparatus.
- O **D.** cellular membrane.

Question 8

Some cells, called adipocytes, specialize in the storage of triacylglycerols synthesized by the smooth ER. The primary function of adipocytes is to:

- O **A.** maintain chemical homeostasis of the body.
- O **B.** filter and remove toxins.
- O **C.** provide for cholesterol synthesis.
- O **D.** serve as a reservoir of stored energy.

Question 9

If a third group of fish was treated with both phenobarbital and Carbofuran, their phospholipid levels would be expected to be:

- O **A.** higher than both the control and experimental group.
- O **B.** higher than the control but lower than the experimental group.
- O **C.** lower than both the control and experimental group.
- O **D.** higher than the experimental but lower than the control group.

Next ▶

Passage III (Questions 10-13)

The cell membrane is generally impermeable to most of the ions necessary for the normal functions of the cell. However, ion channels are embedded into the cell membrane to allow regulated entry and exit of ions. Without these, necessary functions such as metabolism and cell signaling would not be possible.

The selectivity of K^+ channels in cellular membranes allows the passage of K^+ ions while restricting passage of the chemically similar Na^+ ions. This differential permeability maintains the electrical potential across cellular membranes. Using isothermal titration calorimetry (ITC), researchers studied how K^+ channels discriminate between potassium and sodium ions by measuring ion binding to the *Streptomyces lividans* K^+ channel (KcsA) under equilibrium conditions. Two solutions containing equal amounts of KcsA were created. A solution of KCl was titrated into one of the solutions, and a solution of NaCl was titrated into the other. The heat exchange created during the titrations is shown in Figure 1.

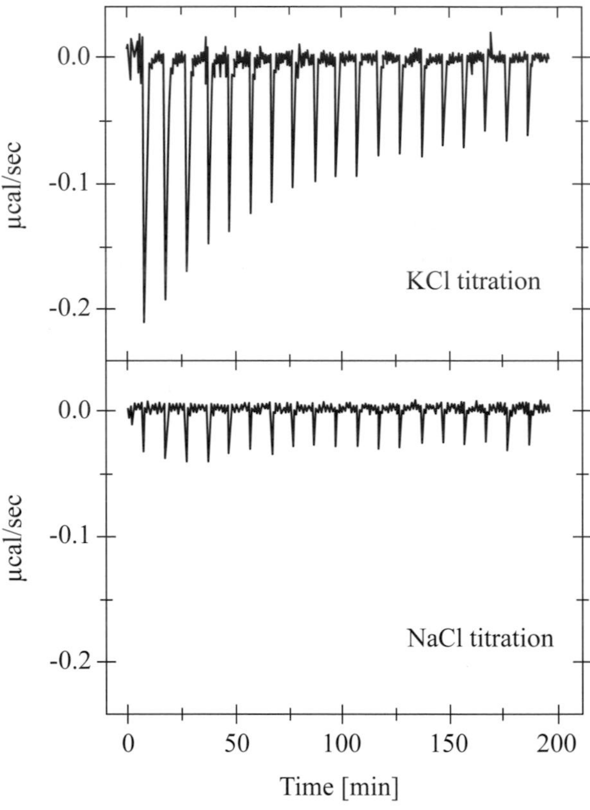

Figure 1 ITC titration of alkali metal cations binding to KcsA

Each downward deflection in the graph represents the heat exchange brought about by an injection of KCl or NaCl. Heat exchange can occur from the heat of diluting the titrant or from the net heat of transfer of ions from solution to the channel, both exothermic processes.

The same procedure was also carried out with other ions. For the ions that exhibited binding to KcsA, the values of ΔG for their reactions with the channel were obtained. The results are shown in Table 1.

Ion	Radius	$\Delta G°$ (kcal M^{-1})
Na^+	0.95 Å	
K^+	1.33 Å	-4.54 ± 0.06
Rb^+	1.48 Å	-5.29 ± 0.29
Cs^+	1.69 Å	-4.53 ± 0.17
Mg^{2+}	0.65 Å	
Ca^{2+}	0.99 Å	
Ba^{2+}	1.35 Å	-5.03 ± 0.15

Table 1 ΔG obtained for ion binding with KcsA

Question 10

According to the data shown in Table 1, which factor was found to be a significant determinant of ion channel permeability?

- O **A.** The charge of the ion
- O **B.** The size of the ion
- O **C.** The concentration of ions in solution
- O **D.** The number of ion channels in solution

Question 11

Ion channels are necessary for ion transport across a cell membrane because:

- O **A.** hydrophobic ions cannot pass through the hydrophilic center of the cellular membrane.
- O **B.** ions are generally too large to pass through the tight connections between cell surface lipids.
- O **C.** the lipid bilayer creates a doubly impenetrable barrier.
- O **D.** the polar ions cannot pass through the non-polar lipid bilayer.

Question 12

Which conclusion is supported by the results shown in Figure 1?

- O **A.** Larger ions pass through ion channels more easily than smaller ions.
- O **B.** The cellular membrane contains surface proteins that allow passage of ions into the cell.
- O **C.** Na^+ ions do not bind to KcsA.
- O **D.** K^+ is an exothermic ion.

Question 13

According to the fluid mosaic model, K^+ channels:

- A. are transported from the nucleus to the outer membrane.
- B. create the electrochemical gradient necessary for cellular functions.
- C. contain a hydrophobic portion that allows them to insert into the membrane.
- D. create energy for the cell.

Passage IV (Questions 14-18)

The propensity to choose risky behaviors in social situations and in economic decisions is known to be influenced by genetic factors. Twin studies have shown that 20% of the variation in financial risk-taking in experimental lotteries is genetic. Recently, particular alleles of several genes involved in the processing of neurotransmitters – monoamine oxidase (MAOA), the serotonin transporter (5-HTTLPR) and a dopamine receptor (DRD4) – have been implicated in risk-taking. In medicine, many decisions involve complex risk-taking where the potential benefits of a treatment may lead patients to assume risks for the possibility of a cure.

Theories of adaptive risk-sensitive foraging have assumed that animals implicitly optimize the average rate of energy intake. If there is a linear relationship between foraging yield and reproductive fitness, then this means that for the effort expended gathering food, there is a reproductive benefit equivalent to that effort. If, however, the payout from foraging could be greater than or less than expected based on the effort invested, then risk-seeking could be selected.

A computational model of selection was used in which two sexually incompatible species "R" (risky) and "A" (averse) compete on equal terms. Each organism begins with the same ability to reproduce, but each "R" organism in each generation experiences a risk event in which it may either have its fecundity boosted by a certain varying reward factor or reduced by a penalty factor. The maximum risk corresponds to the situation in which a loser has no chance to reproduce. The actual, realized number of offspring that each organism contributes to the next generation is then chosen by a statistical distribution with mean equal to that organism's scaled fecundity. A representative diagram of this model is shown below (Figure 1).

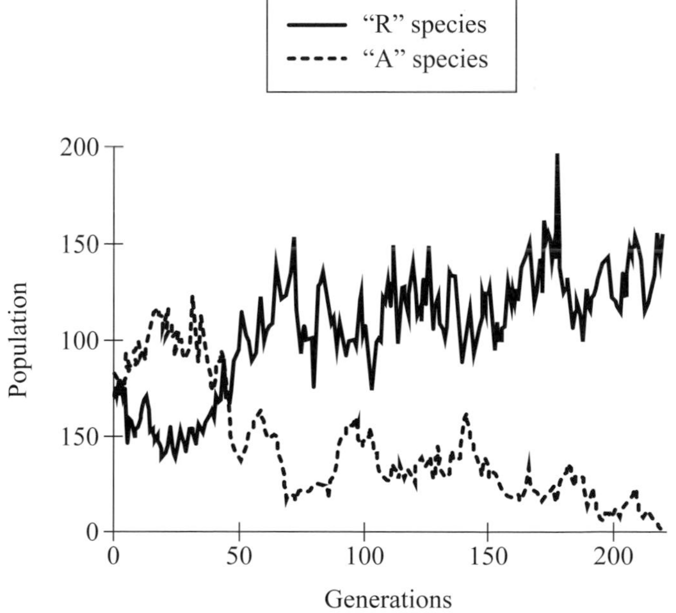

Figure 1 Computational model of theoretical species competing until extinction

Question 14

When a member of the "R" species is faced with the maximum possible risk and loses, how does this event affect the gene pool of the given experimental population?

- O **A.** The gene pool is unchanged due to this individual's inability to mate with other members.
- O **B.** There is speciation where the individual is now part of a new species.
- O **C.** The gene pool of the "A" species is slightly altered due to the addition of a new member.
- O **D.** The gene pool is altered by the loss of the allelic contributions of the individual.

Question 15

Based on the graph below, which of the following explanations would account for the case where the chance of survival is greatest?

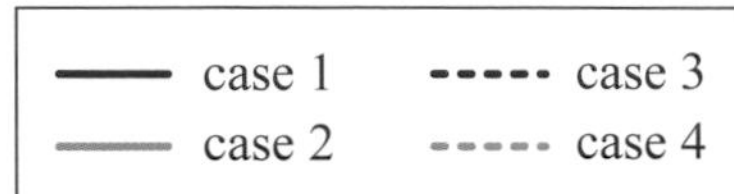

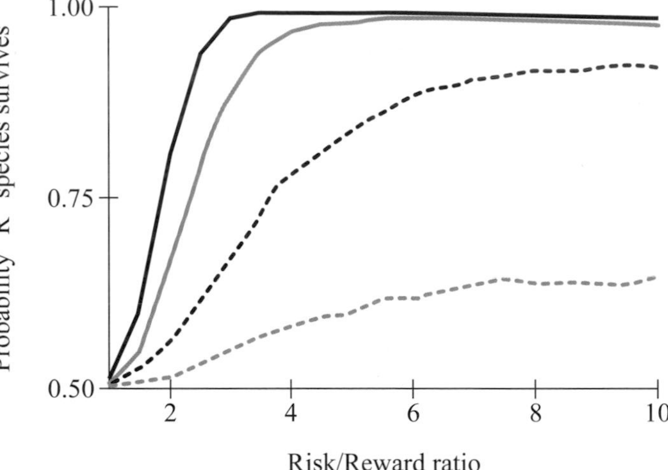

- O **A.** These individuals donate a large share of their rewards or losses to their offspring.
- O **B.** These individuals donate a small share of their rewards or losses to their offspring.
- O **C.** These individuals donate no share of their rewards or losses to their offspring.
- O **D.** There is not enough information to determine the inheritance pattern.

Question 16

The effects of dopamine as a neurotransmitter are expected to be carried out in all of the following ways EXCEPT:

- O **A.** by activating adenylate cyclase to increase cAMP in the cell.
- O **B.** by inhibiting adenylate cyclase to decrease cAMP in the cell.
- O **C.** by opening sodium channels on the plasma membrane.
- O **D.** by binding the DRD4 receptor in the nucleus to activate transcription.

Question 17

Which of the following factors would NOT be required to avoid changes in the gene pool in the "A" species?

- O **A.** A population size that reduces the likelihood of genetic drift
- O **B.** Random mating with different members of the population
- O **C.** A lack of all mutations in the population
- O **D.** A lack of selection on the basis of fitness

Question 18

How was the experimental model set up to create an artificial environment for comparison between the two groups?

- O **A.** The model ignores the contribution of sporadic mutations in the species.
- O **B.** The model only accounts for contributions of three genes in the decision-making process.
- O **C.** The model uses a relatively small sample population size and can be affected by genetic drift.
- O **D.** The model uses two different species that are reproductively isolated from one another.

Next ▶

Questions 19 through 23 are NOT based on a descriptive passage.

Question 19

A cell that is drastically hypertonic to its environment will most likely experience which of the following?

- O **A.** Bursting
- O **B.** Shriveling
- O **C.** Mitosis
- O **D.** Infection

Question 20

Transmission of action potentials in the heart requires a large-scale movement of ions. Which of the following intercellular connections would most likely be enriched in cardiac muscle cells?

- O **A.** Tight junctions
- O **B.** Desmosomes
- O **C.** Intercellular bonds
- O **D.** Gap junctions

Question 21

Streptococcus pyogenes is a common cause of pharyngitis, or inflammation in the throat. How might these bacteria be distinguished in culture?

- O **A.** They will stain pink, indicating they are gram-positive.
- O **B.** They will have a rod-like appearance.
- O **C.** They will stain purple, indicating they are gram-negative.
- O **D.** They will show a growth pattern resembling beads on a chain.

Question 22

Which of the following must be present for a population to be considered in Hardy-Weinberg equilibrium?

- O **A.** Natural selection
- O **B.** Large population size
- O **C.** Non-random mating
- O **D.** Emigration

Question 23

Which of the following does NOT correctly describe a characteristic that is true of eukaryotic cells but not prokaryotic cells?

- O **A.** Eukaryotic cells have a membrane-bound nucleus.
- O **B.** Eukaryotic DNA is coiled with histone proteins.
- O **C.** Eukaryotic cells contain ribosomes.
- O **D.** Eukaryotic flagella are composed of microtubules.

STOP. IF YOU FINISH BEFORE TIME IS CALLED, CHECK YOUR WORK. YOU MAY GO BACK TO ANY QUESTION IN THIS TEST BOOKLET.

30-MINUTE IN-CLASS EXAM FOR LECTURE 2

Passage I (Questions 24-28)

Circadian rhythms, regular cycles of physiological activity throughout the day, are generated by a central clock located in the hypothalamic suprachiasmatic nucleus (SCN) and subordinate clocks in peripheral tissues. The SCN clock responds to external cues, such as light, and drives peripheral clocks via circadian output pathways. Both the central and peripheral clocks are operated by feedback loops of circadian genes, including *Period 1* and *2*. It has been proposed that cell proliferation in vivo is paced by both central and peripheral clocks. Disruption of circadian rhythm, which has been linked with increased cancer risk, would then promote tumor development due to loss of the homeostasis of cell cycle control.

The central clock generates a robust circadian rhythm in sympathetic nervous system (SNS) signaling via direct and indirect targeting of neurons located in the hypothalamic autonomic paraventricular nucleus. In vivo, the SNS controls peripheral tissues by releasing the hormones epinephrine and norepinephrine that target adrenergic receptors (ADRs) on the cell membrane. Norepinephrine is directly released from postganglionic sympathetic neurons, whereas epinephrine is released from chromaffin cells located in the adrenal medulla. Chromaffin cells are controlled by preganglionic sympathetic neurons and are themselves modified sympathetic postganglionic neurons.

Scientists proposed that jet-lag disrupts the homeostasis of multiple circadian output pathways including the SNS. To test this hypothesis, they studied urine epinephrine levels in mice with mutations of the circadian genes *Period1* and *2* (*Per1*$^{-/-}$;*Per2*$^{m/m}$) and weight-matched controls *(Wt)*.

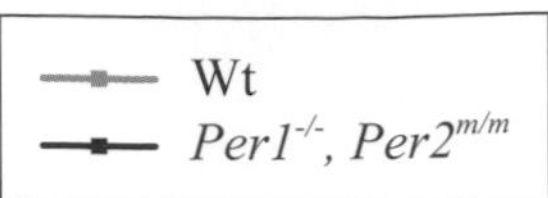

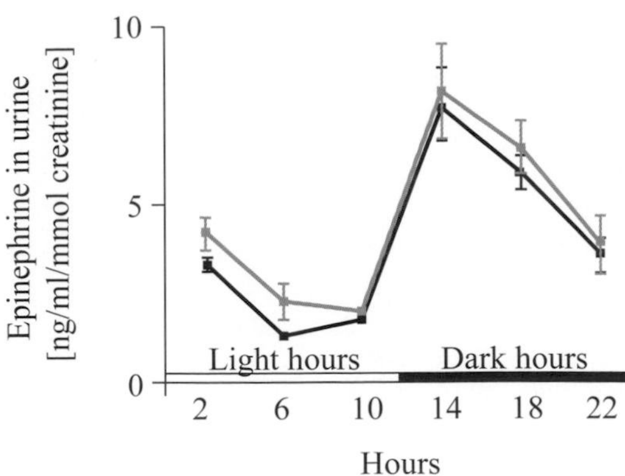

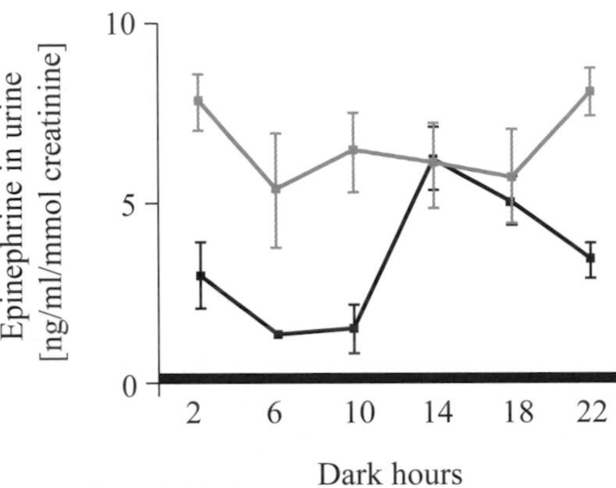

Figure 1 Urine epinephrine levels in *Per*-mutant mice and weight-matched controls during 24 hr light/dark and dark/dark cycles

This passage was adapted from "Disrupting Circadian Homeostasis of Sympathetic Signaling Promotes Tumor Development in Mice." Lee S, Donehower LA, Herron AJ, Moore DD, Fu L. *PLoS ONE*. 2010. 5(6) doi:10.1371/journal.pone.0010995 for use under the terms of the Creative Commons CC BY 3.0 license (http://creativecommons.org/licenses/by/3.0/legalcode).

Question 24

Which of the following cell types most likely contain adrenergic receptors?

- O **A.** Chromaffin cells
- O **B.** Sympathetic postganglionic neurons
- O **C.** Parasympathetic postganglionic neurons
- O **D.** Cardiac muscle cells

Next ▶

Question 25

Which is LEAST likely to be affected by the circadian rhythm of SNS signaling?

- O **A.** The arteries of the heart
- O **B.** The iris musculature of the eye
- O **C.** The diaphragm
- O **D.** Sweat glands

Question 26

Which of the following is most likely NOT associated with activation of chromaffin cells?

- O **A.** Increased heart rate
- O **B.** Elevated blood glucose levels
- O **C.** Increased basal metabolism
- O **D.** Constricted pupils

Question 27

When combined with the data shown in Figure 1, which finding would provide greatest support to the hypothesis that jet lag contributes to tumor formation in Per mutant mice?

- O **A.** The circadian rhythm of SNS signaling encourages activation of the Myc oncogene.
- O **B.** In humans, deregulation of the Per gene is associated with increased risk for a variety of cancers.
- O **C.** Rhythmic SNS signaling plays a role in regulation of cell proliferation.
- O **D.** Previous studies have shown that Per1-/-;Per2m/m mice are deficient in cell cycle regulation and cancer-prone.

Question 28

Which of the following would increase the amount of norepinephrine in the synapse between postganglionic neurons and their effectors?

I. Inhibition of the enzyme that catalyzes norepinephrine breakdown

II. Increased reabsorption by the postganglionic neuron

III. Inhibition of norepinephrine reuptake

- O **A.** I only
- O **B.** I and III only
- O **C.** II and III only
- O **D.** I, II and III

Passage II (Questions 29-32)

Photoreceptors, the sensory receptor cells of the visual system, are a specialized class of neurons found in the retina. Unlike most neurons, photoreceptors are depolarized and release neurotransmitters in their baseline state. When the photoreceptor pigment rhodopsin absorbs a photon, the cell becomes impermeable to cations and hyperpolarizes. This effect is largely due to hydrolysis of the second messenger cyclic GMP (cGMP) and the subsequent closure of cGMP-sensitive cation channels.

To return to a depolarized state after light stimulation, photoreceptors must undergo the process known as recovery, which is largely dependent on the synthesis of new cGMP by retinal membrane guanylyl cyclase (RetGC). RetGCs are activated by guanylyl cyclase activating proteins (GCAPs), whose activity is dependent on free Ca^{2+} concentrations within the cell. In the dark, intracellular Ca^{2+} levels are high, and GCAPs are inhibited. Under illumination, however, Ca^{2+} levels fall, and GCAPs can more easily stimulate RetGCs to produce cGMP.

There are two main GCAPs in mammals, with differing intrinsic sensitivities to Ca^{2+}: GCAP1 and GCAP2. Experimenters explored the differences between the two by generating mutant mice lacking functional GCAP1 (*GCAP1*$^{-/-}$).

Experiment 1

Retinas from wild type (WT) and GCAP1$^{-/-}$ mice were exposed to buffers containing varying Ca^{2+} concentrations, and cGMP levels were measured (Figure 1).

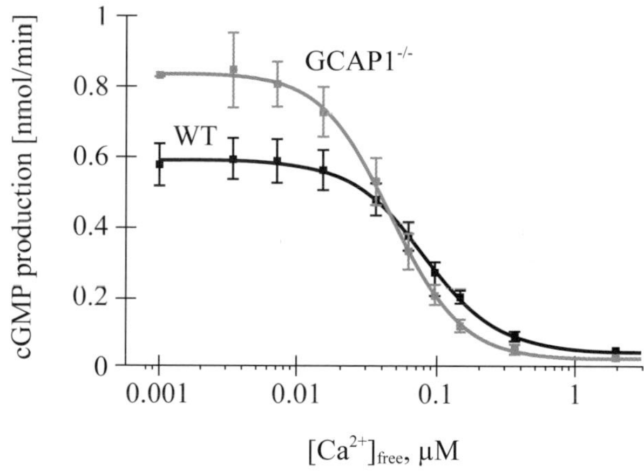

Figure 1 Effects of [Ca^{2+}] on cGMP production in WT and GCAP1$^{-/-}$ retinas

Experiment 2

Experimenters calculated ratios of RetGC activity to baseline (activity in the absence of Ca^{2+}) as Ca^{2+} levels were increased (Figure 2).

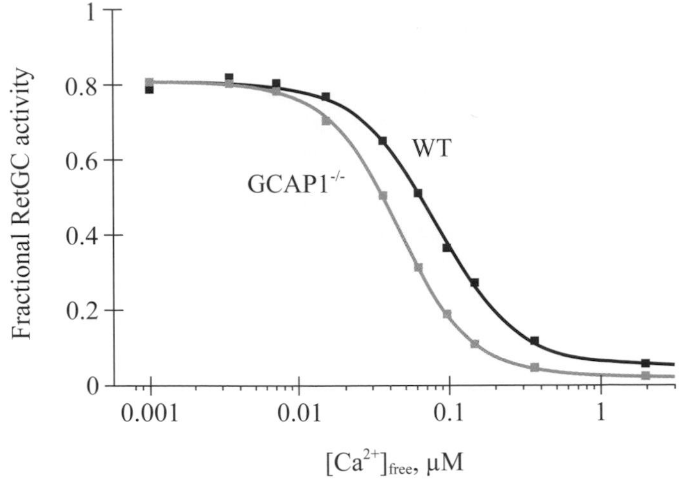

Figure 2 RetGC activity as a fraction of baseline as $[Ca^{2+}]$ increases in WT and GCAP1$^{-/-}$ retinas

This passage is adapted from "Enzymatic Relay Mechanism Stimulates Cyclic GMP Synthesis in Rod Photoresponse: Biochemical and Physiological Study in Guanylyl Cyclase Activating Protein 1 Knockout Mice." Makino CL, Wen XH, Olshevskaya EV, Peshenko IV, Savchenko AB, Dizhoor AM. *PLoS ONE*. 2012. 7(10) doi:10.1371/journal.pone.0047637 for use under the terms of the Creative Commons CC BY 3.0 license (http://creativecommons.org/licenses/by/3.0/legalcode).

Question 29

The experimenters also studied photoreceptor recovery *in vivo* by generating mutant mice that produce fluorescently tagged cGMP and imaging the retina at specific time points after bleaching the photoreceptors with pulses of visible light. This experiment focused only on rod photoreceptors. What adjustments might the researchers make to repeat this study in specific cone cell types?

- O **A.** Use pulses of the entire visible spectrum.
- O **B.** Use pulses of light with wavelengths below 300 nm.
- O **C.** Use pulses of light of only one wavelength.
- O **D.** Use fluorescently tagged cAMP.

Question 30

Given the passage information, which of the following must be FALSE?

I. GCAP1 is necessary for generating new photoreceptor action potentials.

II. GCAP1 is necessary for normal photoreceptor recovery.

III. GCAP2 is necessary for normal photoreceptor recovery.

- O **A.** I only
- O **B.** II only
- O **C.** I and II only
- O **D.** II and III only

Question 31

The amount of light reaching the retina is controlled by the radial and circular muscles of the iris. Would adrenergic signaling onto these muscles be expected to increase or decrease photoreceptor recovery?

- O **A.** Increase due to pupil dilation
- O **B.** Increase due to pupil constriction
- O **C.** Decrease due to pupil dilation
- O **D.** Decrease due to pupil constriction

Question 32

Based on the information in Figure 2, which of the following best explains why cGMP production increased at low Ca^{2+} levels in GCAP1$^{-/-}$ mice in Figure 1?

- O **A.** Compensatory increase in GCAP2
- O **B.** Decreased Ca^{2+} sensitivity of GCAP2
- O **C.** Decreased RetGC activity
- O **D.** Increased photobleaching due to active rhodopsin

Next ▶

Passage III (Questions 33-37)

Cholinergic receptors are subdivided into two broad classifications: nicotinic and muscarinic. The nicotinic acetylcholine receptor (nAChR) is a ligand-gated sodium channel that regulates Na,K-ATPase and is stimulated by nicotine. The nAChR specifically co-immunoprecipitates with both α_1 and α_2 isoforms of the Na,K-ATPase α-subunit and phospholemman (PLM), a muscle-specific auxiliary subunit of Na,K-ATPase which modulates sodium potassium pump activity. Na,K-ATPase is the only known receptor for the poison ouabain.

Active transport by Na,K-ATPase generates a negative membrane potential, V_{pump}, due to the net outward transfer of one positive charge per transport cycle (3 Na^+ out per 2 K^+ in). V_{pump} adds directly to the Nernst potential arising from the ion concentration differences, resulting in the following equation for resting membrane potential (RMP): RMP = $E_{Nernst} + V_{pump}$.

Stimulation of Na,K-ATPase activity by acetylcholine (Ach) produces membrane hyperpolarization, which increases membrane excitability by shifting Na^+ channels from the inactive to available conformation. While nanomolar concentrations of nicotine have been shown to stimulate the Na,K-ATPase α_2 isoform by this mechanism, *in vivo* effects of cigarette smoking on the activity of these nAChRs remains largely uncharacterized.

Nicotine was administered orally to rats for 21-31 days at a dosage shown to produce a plasma nicotine pattern similar to that seen in smokers. The effect on diaphragm muscle removed from euthanized rats is shown in Figure 1.

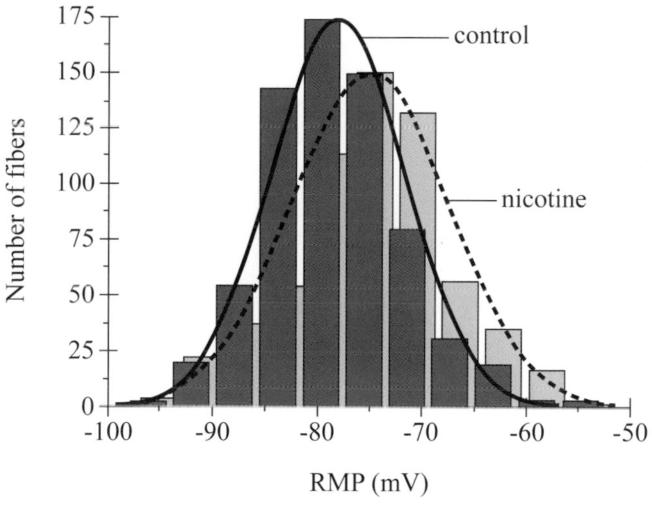

Figure 1 Distribution histogram of resting membrane potentials in the diaphragm of control (black bars) and nicotine-treated rats (light gray bars)

Question 33

According to Figure 1, what was the primary result of nicotine exposure in rats?

- O **A.** Increased muscular excitability
- O **B.** Decreased muscular excitability
- O **C.** Increased neuronal excitability
- O **D.** Decreased neuronal excitability

Question 34

Suppose that repeating the experiment described in the passage with administration of ouabain resulted in equivalent RMPs among both the experimental and control group. What would this suggest about the findings from Experiment 1?

- O **A.** The change in RMP resulted from a change in V_{pump}.
- O **B.** The change in RMP resulted from a change in E_{Nernst}.
- O **C.** The change in RMP was unrelated to changes to E_{Nernst} or V_{pump}.
- O **D.** The change in RMP was due to experimental error.

Question 35

Which conclusions are compatible with the results shown in Figure 1?

I. Nicotine exposure results in overexpression of Na,K-ATPase α_2 isoform.

II. Nicotine exposure results in underexpression of Na,K-ATPase α_2 isoform.

III. Nicotine exposure results in desensitization of the nAChR.

- O **A.** I only
- O **B.** I and III only
- O **C.** II and III only
- O **D.** I, II, and III

Question 36

Based on the passage information, where is phospholemman most likely to be found?

- O **A.** The dendrites
- O **B.** The soma
- O **C.** The presynaptic membrane
- O **D.** The postsynaptic membrane

In-Class Exams

Question 37

Myasthenia gravis is an autoimmune neuromuscular condition that results in production of antibodies that block acetylcholine receptors at the postsynaptic muscular junction. Symptoms include muscle fatigue. Based on the passage, how might cigarette smoking affect sufferers of myasthenia gravis?

- ○ **A.** Symptoms would improve because the resting membrane potential would increase.
- ○ **B.** Symptoms would worsen because the resting membrane potential would increase.
- ○ **C.** Symptoms would improve because the resting membrane potential would decrease.
- ○ **D.** Symptoms would worsen because the resting membrane potential would decrease.

Passage IV (Questions 38-41)

The major histocompatibility complex (MHC) is a cluster of genes that encodes proteins involved in immune responses. Among them, MHC class I (MHC-I) proteins are heterotrimers composed of a transmembrane heavy chain, a non-covalently attached β-2-microglobulin subunit (β2m), and a short peptide comprising 8–15 amino acids derived from self or foreign proteins.

Previous studies have shown that neuronal MHC-I molecules play important roles in synaptic formation, remodeling, and plasticity that are significantly different from their immunological functions. The expression of MHC-I is controlled by the interaction of tissue-specific transcription factors with cognate DNA sequence elements on the extended class I promoter. Scientists interested in the neuronal functioning of MHC-I identified expression in the mouse hippocampus and secondly found that kainic acid (KA), a competitive agonist of glutamate receptors, induced expression of MHC-I in cultured hippocampal neurons.

KA stimulation initiated a calcium-dependent protein kinase C (PKC) phosphorylation cascade that activated IRF-1 and CREB, ultimately resulting in increased expression of MHC-I. The results are shown in Figure 1. In a second experiment, the scientists observed that KA-induced expression of MHC-I can decrease the expression of post-synaptic proteins, which may influence the density of the glutamatergic synapses. This possibility implies that under some pathological conditions, such as epilepsy, dysregulation of neuronal activity might alter the level of neuronal MHC-I, producing changes in synaptic connections and neuronal electrophysiology.

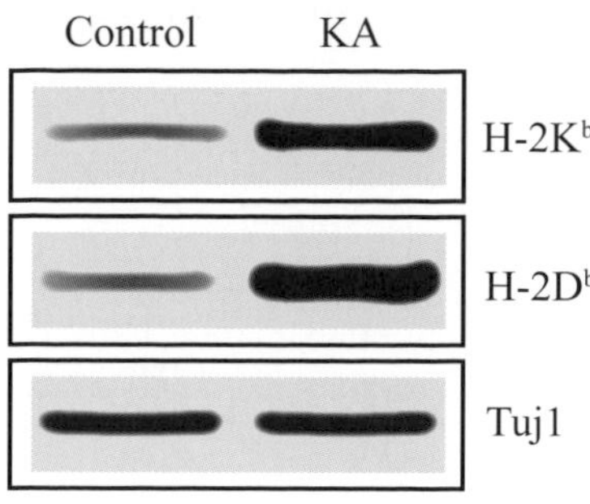

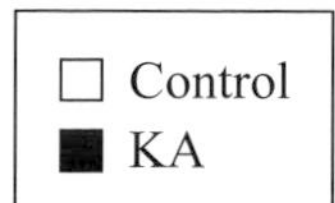

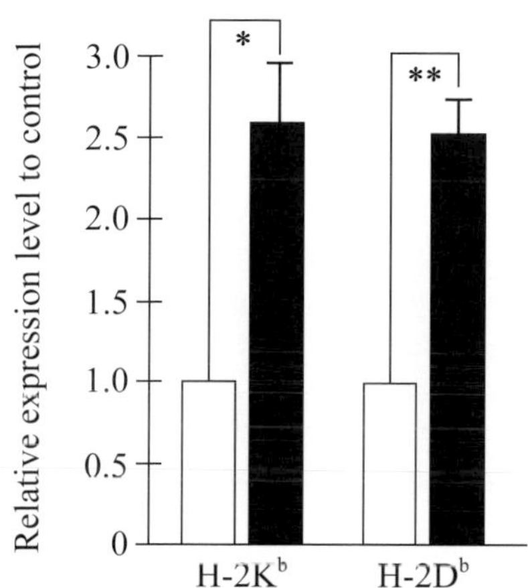

Figure 1 KA-induced expression of MHC-I proteins H-2K^b and H-2D^b by hippocampal neurons

Question 38

A decrease in the number of post-synaptic glutamate channels is LEAST likely to impact the:

- **A.** magnitude of the voltage change of an action potential.
- **B.** probability of an action potential occurring.
- **C.** the formation of long-term memory.
- **D.** rate of neuron firing.

Question 39

In diseases such as multiple sclerosis, neurons in the cortical brain are progressively demyelinated through an autoimmune reaction. Patients with multiple sclerosis are likely to have:

- **A.** decreased threshold for action potential generation.
- **B.** a decreased speed of action potential propagation.
- **C.** fewer sodium/potassium pumps on the neuron plasma membrane.
- **D.** altered concentrations of extracellular and intracellular potassium.

Question 40

Treatment of hippocampal neurons with KA is most likely to:

- **A.** increase the current of calcium across the post-synaptic axon cell membrane.
- **B.** decrease the current of calcium across the pre-synaptic axon cell membrane.
- **C.** decrease the current of calcium across the pre-synaptic Nodes of Ranvier.
- **D.** decrease the current of calcium across the post-synaptic dendrite cell membrane.

Question 41

The current of sodium is likely to be the greatest during which of the following points?

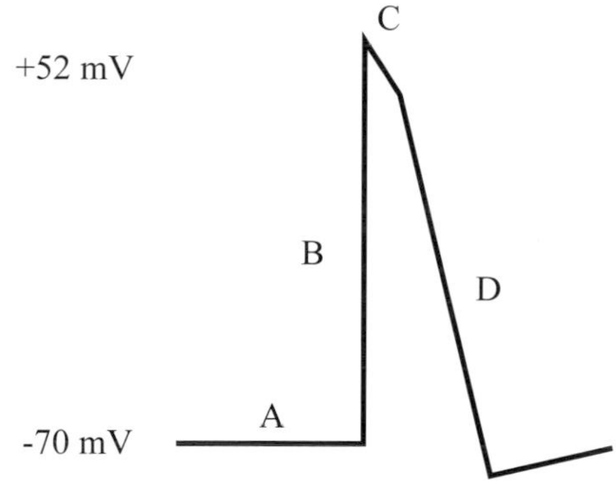

- **A.** A
- **B.** B
- **C.** C
- **D.** D

In-Class Exams

Questions 42 through 46 are NOT based on a descriptive passage.

Question 42

Which of the following statements best describes whether an action potential of a post-synaptic neuron will occur?

- **A.** The electric field of the axon hillock causes voltage-gated ion channels to open.
- **B.** The influx of sodium ions through a glutamate receptor at a dendrite.
- **C.** The release of glutamate from the pre-synaptic axon terminal.
- **D.** The influx of calcium into the pre-synaptic axon terminal .

Question 43

The action potential of the cardiac myocyte is shown below. Which of the regions of the action potential is LEAST likely to be shared by a neuron action potential?

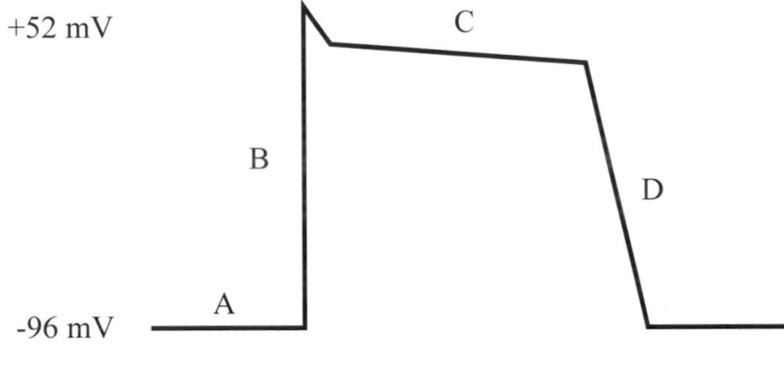

- **A.** A
- **B.** B
- **C.** C
- **D.** D

Question 44

In saltatory conduction:

- **A.** an action potential jumps along a myelinated axon from one node of Ranvier to the next.
- **B.** an action potential moves rapidly along the membrane of a Schwann cell, which is wrapped tightly around an axon.
- **C.** an action potential jumps from the axon of one neuron to the dendrites of the next.
- **D.** ions jump from one node of Ranvier to the next along an unmyelinated axon.

Question 45

All of the following are true concerning a typical motor neuron **EXCEPT**:

- **A.** the K^+ concentration is greater inside the cell than outside the cell.
- **B.** K^+ voltage-gated channels are more sensitive than Na^+ voltage-gated channels to a change in membrane potential.
- **C.** Cl^- concentrations contribute to the membrane resting potential.
- **D.** the action potential begins at the axon hillock.

Question 46

Which of the following statements regarding the thalamus is true?

- **A.** The thalamus maintains homeostasis in many physiological activities by coordinating between nervous and endocrine signaling.
- **B.** The major neural pathways for both visual and auditory information pass through the thalamus.
- **C.** Higher-level processing of the complex features of multiple types of sensory information take place in the thalamus.
- **D.** The thalamus is a critical site of processing for olfactory information.

STOP. IF YOU FINISH BEFORE TIME IS CALLED, CHECK YOUR WORK. YOU MAY GO BACK TO ANY QUESTION IN THIS TEST BOOKLET.

30-MINUTE IN-CLASS EXAM FOR LECTURE 3

Passage I (Questions 47-51)

Peptide hormones and catecholamines, a class of tyrosine derivatives, bind to membrane-bound receptors on their effector cells and act via second messenger systems.

One common second messenger system works as follows: the bound hormone receptor activates a protein inside the cell, called a G-protein. Once activated, the G_S-protein exchanges GDP for GTP, causing a portion of the G-protein to dissociate. The dissociated portion may stimulate or inhibit *adenylyl cyclase*, another membrane bound protein. Inhibitory G-proteins are called G_i-proteins; stimulating G-proteins are called G_S-proteins. After acting on adenylyl cyclase, the dissociated portion of the G-protein hydrolyzes GTP to GDP in order to become inactive and recombine with its other portion. Adenylyl cyclase converts ATP to cyclic AMP (cAMP). cAMP activates *protein kinase A*, which causes changes in protein phosphorylation throughout the cell.

Dopamine (DA) is a catecholamine that can act as a hormone. Depending on the type of receptor that it binds to, hormonal dopamine can have either a stimulatory or inhibitory effect on adenylyl cyclase. *Dopamine receptor D1* (D1R) and *D2* (D2R), for example, act antagonistically to one another.

DA signaling can affect the release of the hormone prolactin (PRL) and the cytokine leptin. Since both of these molecules are secreted by human adipose tissue, researchers investigated the effects of DA on two adipocyte cell lines, SW872 and LS14. Preliminary results indicated that each cell line primarily expressed just one receptor type but that other receptor types might be expressed in limited quantity.

Experiment 1

SW872 cells were treated with either DA, raclopride (RAC), an antagonist of D2R, or both. PRL release and cAMP levels were monitored. (Figure 1)

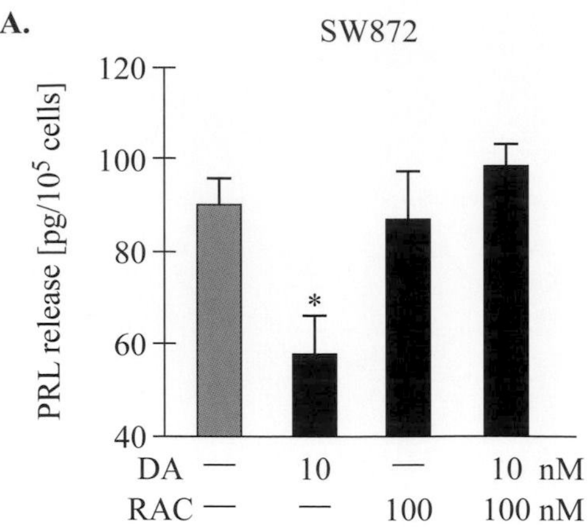

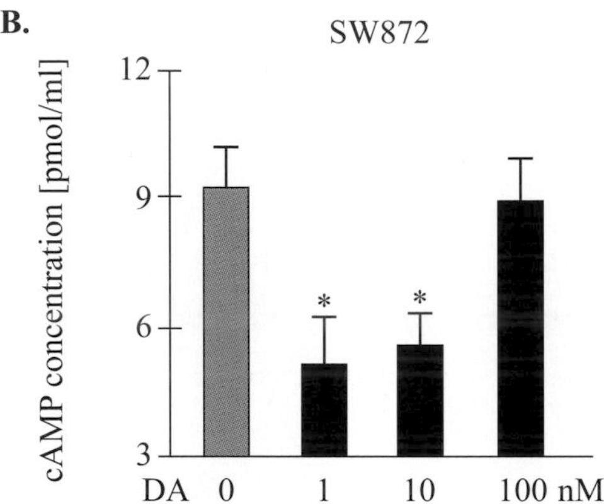

Figure 1 **A.** Prolactin release from SW872 cells treated with DA and D2R antagonist RAC. **B.** cAMP concentration in SW872 cells treated with various concentrations of DA.

Experiment 2

Leptin release was monitored in LS14 cells after treatment with DA or SKF38393 (SKF), a D1R agonist; see Figure 2.

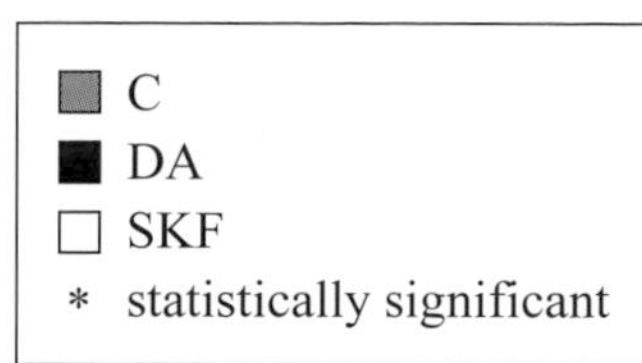

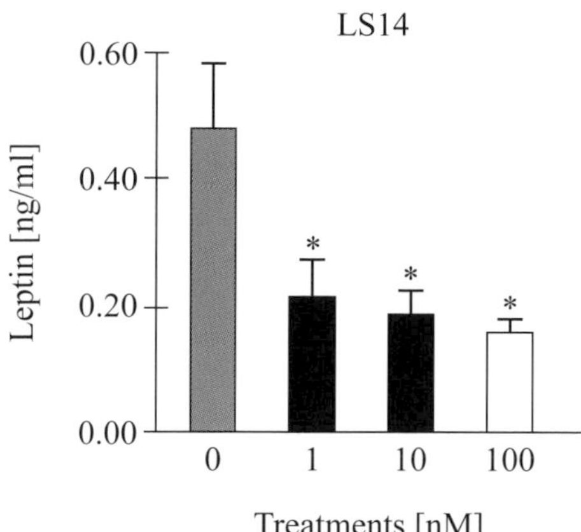

Figure 2 Leptin secretion from LS14 cells after treatment with DA or D1R agonist SKF

Question 47

Cortisol is similar to dopamine in that the changes it causes can vary widely among different types of effector cells. Cortisol most likely binds to a receptor protein:

- O **A.** on the cell membrane.
- O **B.** in the cytosol.
- O **C.** on the nuclear membrane.
- O **D.** just outside the cell.

Question 48

Excess DA is usually converted to the catecholamines epinephrine and norepinephrine. Based on passage information, which of the following would you expect to find in a recently pregnant mother with high prolactin levels?

- O **A.** Low blood DA levels
- O **B.** Increased heart rate
- O **C.** Decreased heart rate
- O **D.** High blood leptin levels

Question 49

What is the most likely explanation for the results shown in Figure 1?

- O **A.** DA binds mainly to D1R and activates G_S-proteins.
- O **B.** DA binds mainly to D2R and activates G_S-proteins.
- O **C.** DA binds mainly to D1R and activates G_i-proteins.
- O **D.** DA binds mainly to D2R and activates G_i-proteins.

Question 50

The catecholamine epinephrine binds to several types of receptors called adrenergic receptors. Heart muscle cells contain β_1-adrenergic receptors, whereas the smooth muscle cells of the gut contain many α_2-adrenergic receptors. Which of the following is most likely true concerning these two types of receptors?

- O **A.** β_1-adrenergic receptors activate G_S-proteins, whereas α_2- adrenergic receptors activate G_i-proteins.
- O **B.** β_1-adrenergic receptors activate G_i-proteins, whereas α_2- adrenergic receptors activate G_S-proteins.
- O **C.** Both β_1-adrenergic receptors and α_2-adrenergic receptors activate G_i-proteins.
- O **D.** Both β_1-adrenergic receptors and α_2-adrenergic receptors activate G_S-proteins.

Question 51

Glucagon works via a G-protein system. Glucagon is broken down by a cAMP- dependent phosphorylase. Which of the following best explains why glucagon stimulates glycogen breakdown in liver cells, but stimulates lipid breakdown in fat cells?

- O **A.** Fat cells contain a G_S-protein while liver cells contain a G_i-protein.
- O **B.** Liver cells do not contain *adenylyl cyclase.*
- O **C.** The two cell types contain a different set of proteins phosphorylated by protein kinase A.
- O **D.** Glycogen and lipid breakdown are not governed by cAMP levels.

Passage II (Questions 52-55)

Leptin is a small, non-glycosylated peptide known to modulate satiety and energy homeostasis. However, leptin has also been implicated in the regulation of embryo implantation and the maintenance of pregnancy. Leptin is synthesized in the placenta, and pregnancy leads to elevated leptin levels. Scientists hypothesize that leptin expression in the placenta is regulated by the peptide hormone human chorionic gonadotropin (hCG).

hCG has many important functions during the course of pregnancy, including stimulation of progesterone production. Progesterone is secreted for the duration of gestation, initially by the corpus luteum, and later by the placenta. hCG is already expressed in eight-cell embryos and is secreted in high local concentrations by the blastula entering the uterine cavity. It can be detected in maternal plasma or urine shortly after trophoblast cells, which form the outermost layer of the blastula, implant in the uterine endometrium.

Binding of hCG to its receptor generates signal transduction through activation of associated G-proteins, causing an increase in cyclic adenosine monophosphate (cAMP) and a consequent activation of protein kinase A (PKA).

In a series of experiments on trophoblast-derived BeWo tumor cells, scientists investigated the regulation of leptin expression in trophoblast cells.

Experiment 1

To measure the effect of hCG on leptin mRNA expression and intracellular cAMP concentration, BeWo cells were treated with hCG. Leptin mRNA expression and intracellular cAMP concentration were measured. The results are shown in Figures 1 and 2.

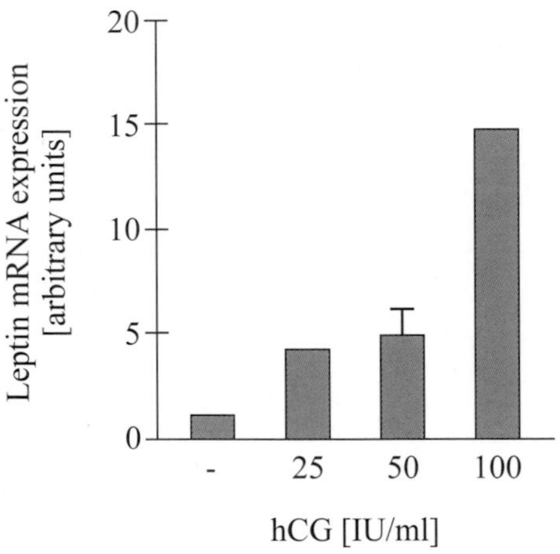

Figure 1 Leptin mRNA expression in BeWo cells treated with hCG

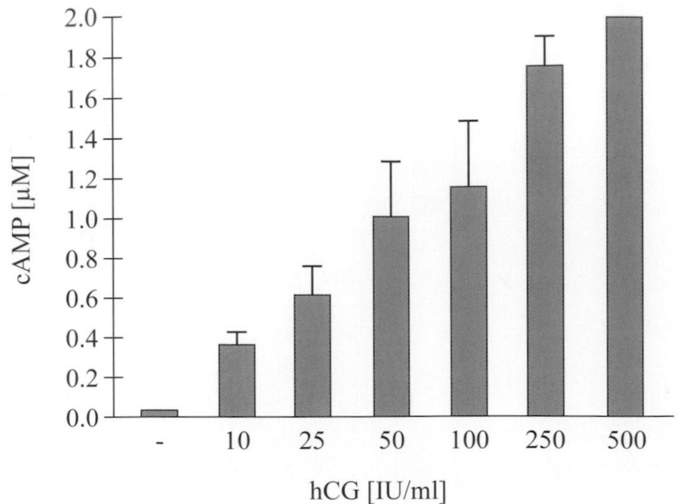

Figure 2 cAMP expression in BeWo cells treated with hCG

Experiment 2

To measure the effect of cAMP on leptin mRNA expression in the presence of high concentrations of hCG, BeWo cells were treated with cAMP and/or high concentrations of hCG. Results are shown in Figure 3.

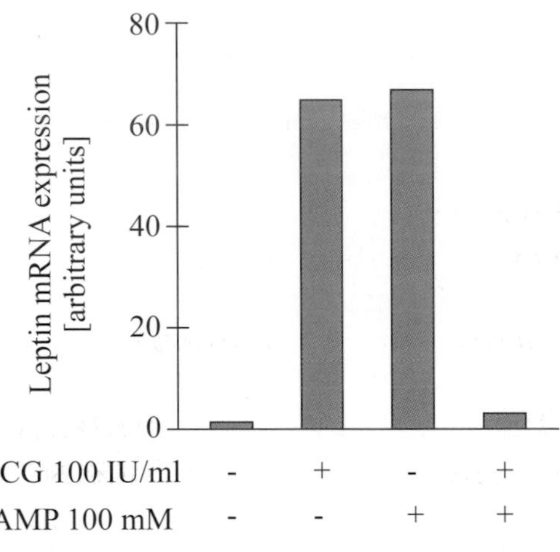

Figure 3 Impact of cAMP on BeWo cell leptin expression under the influence of high concentrations of hCG

Question 52

The data obtained from experiments 1 and 2 best support which of the following conclusions?

- O **A.** At high levels, cAMP interferes with hCG stimulation of leptin expression.
- O **B.** The hCG stimulatory effect on leptin expression is blocked by PKA activation.
- O **C.** cAMP induces leptin stimulation by hCG at high hormone concentrations.
- O **D.** hCG inhibits leptin mRNA expression and suppresses cAMP levels.

Question 53

Syncytiotrophoblast cells in the placenta manufacture and release hCG. Which of the following statements most accurately describes the release and transport of hCG?

- O **A.** hCG is released by exocytosis and dissolves in the blood.
- O **B.** hCG is released by simple diffusion and dissolves in the blood.
- O **C.** hCG is released by exocytosis and binds to carrier proteins.
- O **D.** hCG is released by simple diffusion and binds to carrier proteins.

Question 54

Which of the following is the most likely reason why BeWo tumor cells were used to investigate the regulation of leptin expression in trophoblast cells?

- O **A.** hCG inhibits leptin expression in both trophoblast and BeWo cells.
- O **B.** Leptin deficiency promotes BeWo cell growth.
- O **C.** BeWo cells maintain many characteristics of trophoblast cells.
- O **D.** hCG activates cAMP expression in both trophoblast and BeWo cells.

Question 55

Which of the following hormones is most likely to have an effect similar to that depicted in Figure 2?

- O **A.** Progesterone
- O **B.** LH
- O **C.** Estrogen
- O **D.** Cortisol

Passage III (Questions 56-60)

Gametogenesis is regulated through positive and negative feedback loops that act at the level of the hypothalamus, pituitary, and gonad. Gonadotropin-releasing hormone (GnRH), which is released from the hypothalamus, acts on the anterior pituitary to cause release of the glycoproteins luteinizing hormone (LH) and follicle-stimulating hormone (FSH).

In females, oogenesis occurs in the ovaries. The oocyte, together with surrounding theca and granulosa cells, compose a follicle. LH and FSH act on theca and granulosa cells, respectively, to stimulate steroid production and follicle maturation. When stimulated by LH, theca cells supply androgens to granulosa cells, which convert them to estrogens that exert negative feedback on GnRH, LH, and FSH production.

In males, spermatogenesis occurs in the seminiferous tubules of the testes. LH and FSH act on the Leydig and Sertoli cells of the testis, respectively, to stimulate steroidogenesis and spermatogenesis. Inhibins produced by Sertoli cells and androgens produced by Leydig cells, including testosterone, feedback negatively to reduce GnRH, LH, and FSH levels.

Scientists interested in the effects of gonadotropins and testosterone on mRNA transcript production in Leydig and Sertoli cells performed the following experiment. Mice received an injection of the GnRH antagonist acyline (ACY) every 24 hours for 4 days, followed by a single injection of LH 2 hours after the last ACY injection. Measurements were taken before as well as 1 and 4 hours after LH administration.

In-Class Exams

Blood levels of LH, FSH, and testosterone are shown in Figure 1.

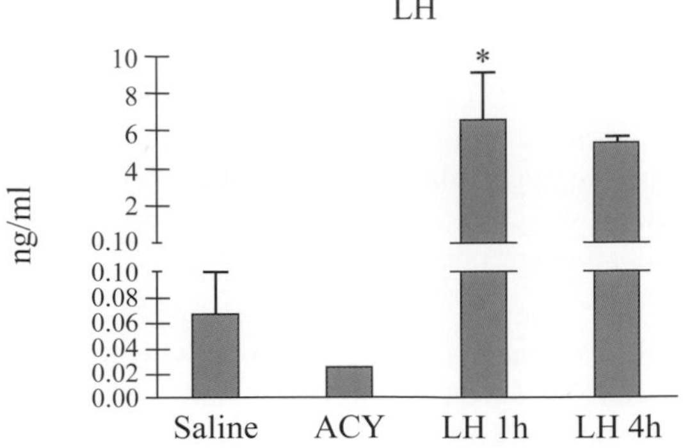

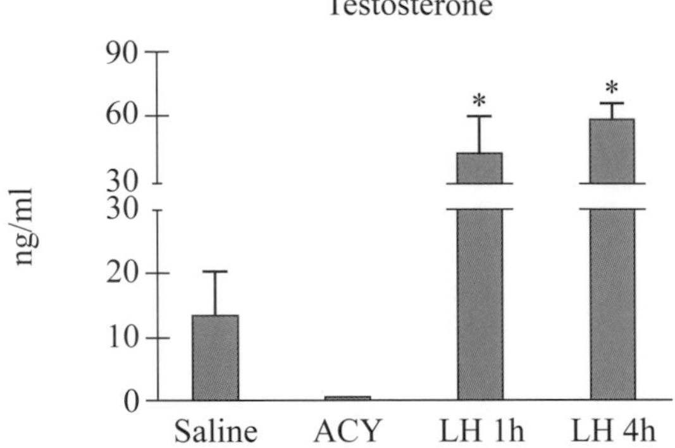

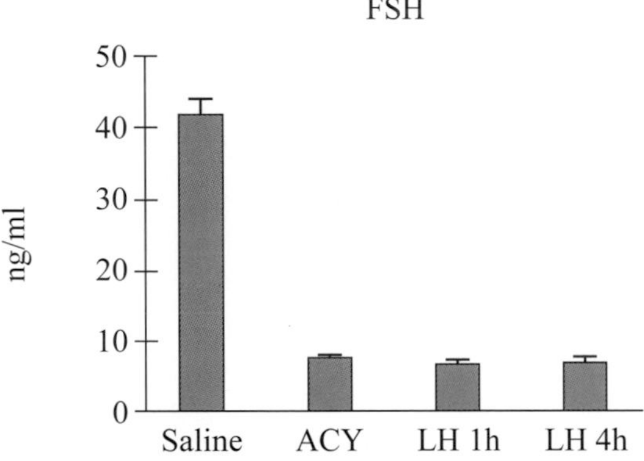

Figure 1 Effect of gonadotropin deprivation and LH restoration on blood levels of LH, FSH, and testosterone

Leydig cells were found to respond robustly to both gonadotropin deprivation and LH restoration with acute changes in mRNA transcript production.

In a second experiment, Sertoli cells were found to respond minimally to either gonadotropin deprivation or FSH and testosterone restoration with minor changes in mRNA transcripts production.

This passage was adapted from "RiboTag Analysis of Actively Translated mRNAs in Sertoli and Leydig Cells In Vivo." Sanz E, Evanoff R, Quintana A, Evans E, Miller JA, Ko C, Amieux PS, Griswold MD, McKnight GS. *PLoS ONE*. 2013. 8(6) doi:10.1371/journal.pone.0066179

Question 56

Androgens are sometimes taken by athletes to improve performance. Which of the following may be a side effect of taking large quantities of androgens?

- O **A.** Infertility due to decreased endogenous testosterone production
- O **B.** Infertility due to decreased secretion of GnRH
- O **C.** Increased fertility due to decreased endogenous testosterone production
- O **D.** Increased fertility due to increased Sertoli cell activity

Question 57

Are the results shown in Figure 1 to be expected given the control pathways described in the passage?

- O **A.** No, because LH stimulates FSH production.
- O **B.** No, because LH inhibits androgen production.
- O **C.** Yes, because LH inhibits FSH production.
- O **D.** Yes, because LH stimulates androgen production.

Question 58

The gonadotropins LH and FSH most likely alter mRNA production by:

- O **A.** acting as a transcription factor.
- O **B.** activating a second messenger system.
- O **C.** binding to a receptor in the cytoplasm.
- O **D.** binding to a receptor in the nucleus.

Question 59

What is the most likely reason that acyline was used in experiments on the effects of gonadotropins and testosterone on mRNA transcript production in Leydig and Sertoli cells?

- O **A.** Acyline is expressed in Leydig and Sertoli cells.
- O **B.** Acyline induces expression of gonadotropins and testosterone.
- O **C.** Acyline inhibits the activity of GnRH.
- O **D.** Acyline is a product of spermatogenesis.

Question 60

A vaccine that stimulates the body to produce antibodies against a hormone has been suggested as a long-term male contraceptive. In order to insure that the vaccine has no adverse effects on androgen production, which hormone should be targeted?

- O **A.** FSH
- O **B.** LH
- O **C.** GnRH
- O **D.** Testosterone

Passage IV (Questions 61-64)

Obesity is an epidemic disease leading to diabetes and metabolic syndrome, and is associated with cardiovascular disorders, especially for women facing the menopause transition. Fluctuations in sex hormones at different stages of reproductive life, such as menarche, pregnancy, and menopause transition, may play a role in the adipose tissue expansion. Notably, menopausal transition is associated with unfavorable changes in body composition, abdominal fat deposition and general health outcomes.

The growth hormone (GH)/insulin-like growth factor-1 (IGF-1) axis regulates growth and development during childhood and adolescence, but also regulates body composition, metabolism and exercise aerobic capacity throughout life. Increased lipolysis and free fatty acids (FFA) mobilization are the main effects of GH in metabolism. These changes in body composition are associated with metabolic derangements including insulin resistance.

In addition to GH, estrogen (E2), and thyroid hormones (TH) are also key hormones that impact body weight. Triiodothyronine (T3) acts virtually in all body tissues. In addition to serum T3, local availability and clearance are well controlled by types I, II, and III iodothyronine deiodinases (D1, D2, and D3, respectively). D1 and D2 catalyze T3 production while D3 inactivates T3. In this way, modulation of deiodinases expression and activity customize T3 biological effects.

Researchers studied the role of estrogen on exercise-induced GH secretion in menopause. Ten days after bilateral removal of ovaries, animals were submitted to 20 min exercise and tissues were harvested immediately or 30 min after exercise. Non-exercised animals were used as controls. Serum GH and T3 levels were measured and this data is shown below.

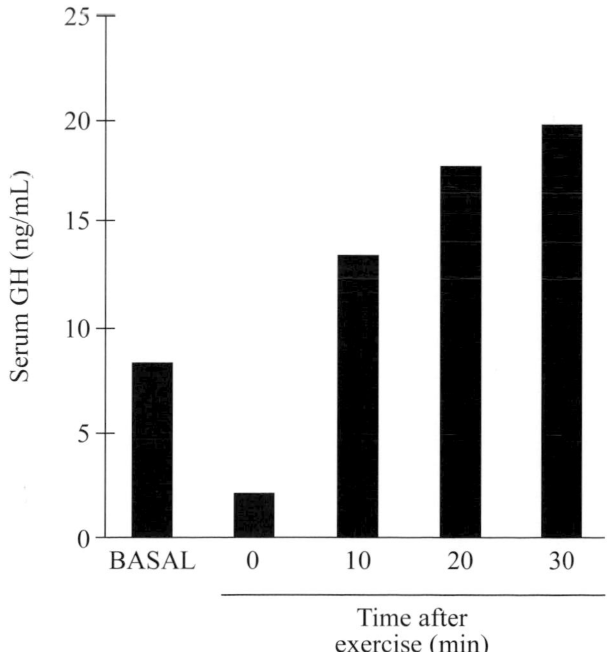

Figure 1 Serum GH levels following exercise

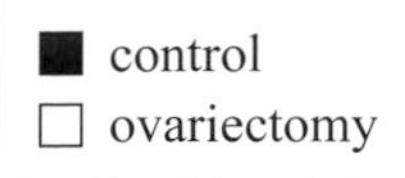

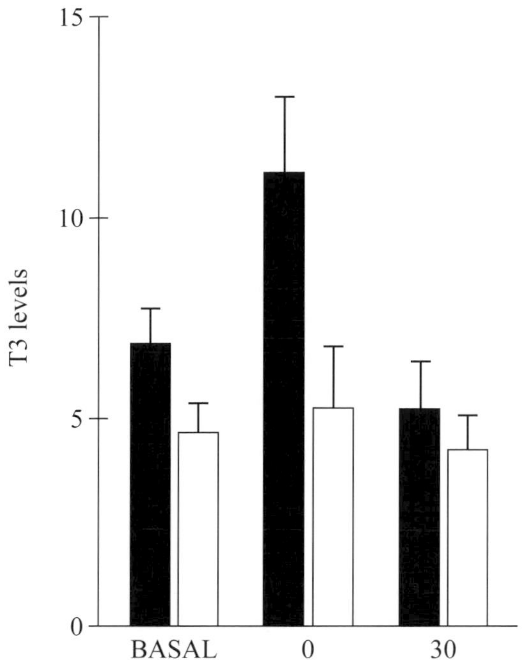

Figure 2 T3 levels following exercise in rats missing ovaries

Question 61

Which of the following statements is true regarding GH and estrogen transport in the blood?

- O **A.** GH and estrogen both use carrier proteins.
- O **B.** GH and estrogen both travel directly through the bloodstream.
- O **C.** GH uses a carrier protein while estrogen travels directly through the bloodstream.
- O **D.** GH travels directly through the bloodstream while estrogen uses a carrier protein.

Question 62

Which of the following effects is likely during exercise in the subjects with ovaries removed?

- O **A.** Reduced D1 activation, which causes the lack of exercised-induced T3 secretion
- O **B.** Reduced D1 activation, which causes the increase in exercised-induced T3 secretion
- O **C.** Increased D1 activation, which causes the lack of exercised-induced T3 secretion
- O **D.** Increased D1 activation, which causes the increase in exercised-induced T3 secretion

Question 63

T3 levels fluctuate depending on level of activity, intake of food, and other factors. When T3 levels are extremely high:

- **A.** D1, D2, and D3 levels increase.
- **B.** D1, D2, and D3 levels decrease.
- **C.** D1 and D2 levels decrease, and D3 levels increase.
- **D.** D1 and D2 levels increase, and D3 levels decrease.

Question 64

Which of the following effects is caused most directly by removal of the ovaries?

- **A.** Increase in luteinizing hormone
- **B.** Reduction in estrogen
- **C.** Increased milk production
- **D.** Decreased milk production

Questions 65 through 69 are NOT based on a descriptive passage.

Question 65

Treatment of a major disease requires removal of the adrenal cortex. Which of these characteristics is most affected by this surgery?

- **A.** Ability to synthesize glycogen
- **B.** Blood calcium control
- **C.** Stimulation of sympathetic actions
- **D.** Ion retention

Question 66

Which of the following is true regarding peptide hormones?

I. Peptide hormones consist entirely of amino acids.
II. Peptide hormones are synthesized by free-floating ribosomes.
III. Peptide hormones act by directly altering transcription levels.

- **A.** I only
- **B.** I and II only
- **C.** III only
- **D.** None of the above

Question 67

A vasectomy is a surgical procedure that severs the vas deferens. What is the effect of this surgery?

- **A.** Inability of sperm to mature
- **B.** Acidification of semen
- **C.** Inability of sperm to penetrate the egg
- **D.** Lack of sperm in semen

Question 68

A competitive inhibitor of TSH binding to TSH receptors on the thyroid would lead to a rise in the blood levels of which of the following?

- **A.** TSH
- **B.** Thyroxine
- **C.** PTH
- **D.** Epinephrine

Question 69

Which of the following hormones most likely acts as the substrate molecule for synthesis of cortisol and aldosterone in the fetal adrenal gland?

- **A.** HCG
- **B.** Estrogen
- **C.** LH
- **D.** FSH

STOP. IF YOU FINISH BEFORE TIME IS CALLED, CHECK YOUR WORK. YOU MAY GO BACK TO ANY QUESTION IN THIS TEST BOOKLET.

30-MINUTE IN-CLASS EXAM FOR LECTURE 4

Passage I (Questions 70-73)

In the 1930s a line of "nude" mice was derived; these mice develop normal numbers of B cells, but lack a thymus and are impaired in producing antibodies. Early models of humanized mice used for human immune system reconstitution were generated in the late 1980s by infusion of human peripheral blood mononuclear cells into a line derived from nude mice. However, the level of reconstitution in such mice was very low and consisted mainly of human T cells. Impairment of human T and B cell function in these humanized mice has been attributed to the lack of expression of Human Leukocyte Antigens (HLA), since HLA molecules are required for development of human T cells that, in turn, are essential for stimulation of B cells. HLA molecules are the human-specific form of major histocompatibility complex (MHC) molecules, which are involved in both the process of clonal selection that selects immature T cells for maturation and the display of foreign particles in the immune system response.

Experiment 1

To test the hypothesis that expression of HLA class II molecules in humanized mice allows development of functioning human CD4 T cells, scientists generated mice expressing HLA-DR4 molecules (DRAG mice). Adult DRAG mice were infused with human hematopoietic stem cells (HSC). Littermates lacking expression of HLA-DR4 molecules were used as controls. The results are shown in Figure 1.

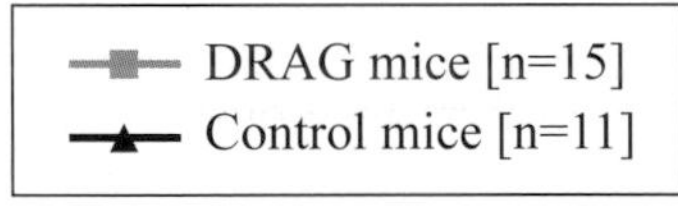

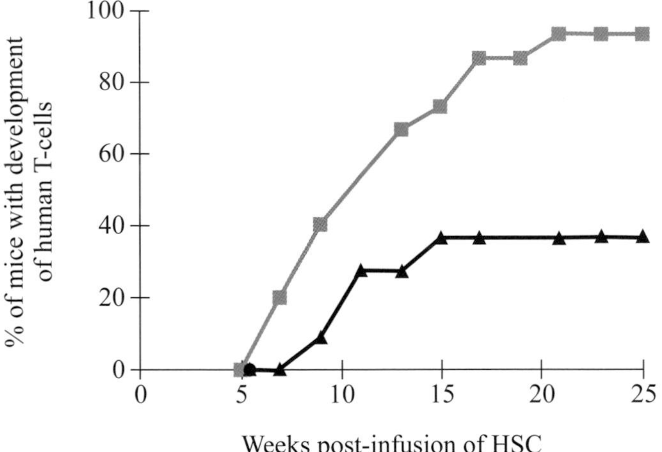

Figure 1 Percent of DRAG and control mice that showed development of human T cells following infusion with human hematopoietic stem cells

Experiment 2

Directly following Experiment 1, the researchers removed human T cells that had developed in the DRAG and control subjects. They then stimulated the T cells with PMA/ionomycin, which leads to the production of cytokines such as interleukins (IL) in functioning T cells. (Interleukins are peptide hormones that stimulate the proliferation of other T cells.) The results are shown in Figure 2.

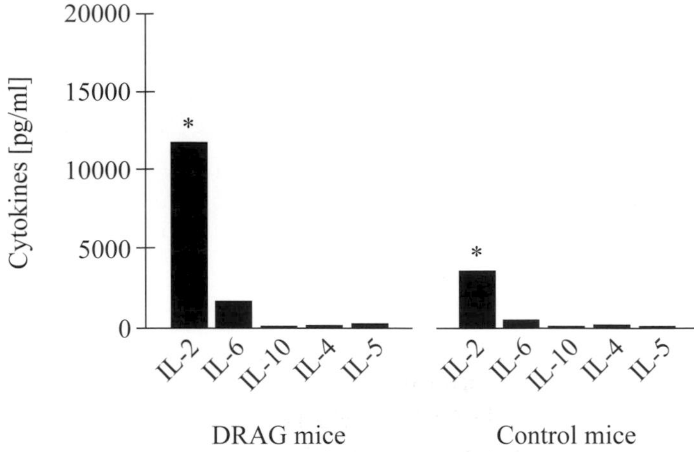

Figure 2 Levels of different types of interleukins secreted by human T cells from DRAG and control mice following stimulation with PMA/ionomycin

Question 70

According to the information given in the passage, interleukins most likely act:

- O **A.** via a second messenger by binding to a membrane bound protein receptor.
- O **B.** by diffusing through the membrane of the helper T cell and binding to a receptor in the cytosol.
- O **C.** at the transcriptional level by binding directly to nuclear DNA.
- O **D.** via the nervous system.

Question 71

All of the following arise from hematopoietic stem cells in bone marrow EXCEPT:

- O **A.** helper T cells.
- O **B.** B lymphocytes.
- O **C.** erythrocytes.
- O **D.** osteoblasts.

Next ▶

Question 72

Which of the following is most likely to create the humoral immune system response mediated by MHC class II molecules?

- A. A bacterial infection
- B. A cell infected by a virus
- C. A tumor
- D. A foreign tissue graft

Question 73

Which of the following is a feature of the study described that may limit the researchers' ability to draw conclusions about the hypothesis of interest?

- A. B cell reconstitution in DRAG mice was not measured.
- B. The researchers did not test the functioning of reconstituted human T cells in DRAG mice.
- C. No wild-type mice were used as controls.
- D. Only control mice that showed reconstitution of human T cells in could be used as controls in the second part of the experiment.

Passage II (Questions 74-78)

Airway hyperreactivity and mucous metaplasia characterize asthma, a chronic inflammatory disease. Mucous metaplasia is an increase in the number of mucus-secreting goblet cells in the epithelium that results in increased mucus synthesis and secretion. Excessive accumulation of airway mucus leads to the formation of mucous plugs that reduce the effective airway diameter and increase airway resistance. Patients who die of severe asthma attacks often exhibit mucus accumulation and large mucus plugs of unusual solidity due to high mucin content in their peripheral airways compared to asthmatic patients who did not die of acute attacks.

Allergic asthma has properties of a type I hypersensitivity, in which T-helper lymphocytes and innate lymphoid cells produce a distinctive set of cytokines in the airways, including IL-5, IL-9 and IL-13. Airway epithelial overexpression of IL-13 is sufficient to induce mucous metaplasia.

The MUC5AC gene encodes the major component of mucin in human airways, and induction of MUC5AC transcription by IL-13 is observed in cultured human airway epithelium. Recent evidence suggests the involvement of β-adrenoceptor (β2AR) signaling in the pathogenesis of asthma. Researchers decided to explore whether β2ARs are also involved in the regulation of MUC5AC transcription.

Human airway epithelial cells were cultured in low concentrations of retinoic acid, conditions where mucin expression is normally minimal. Epinephrine (Epi) was used to activate the β2ARs. The researchers investigated the requirement for β2AR signaling in the transcription of the MUC5AC gene, the expression of MUC5AC protein and intracellular mucin accumulation in response to IL-13. The results are displayed in Figure 1.

In-Class Exams

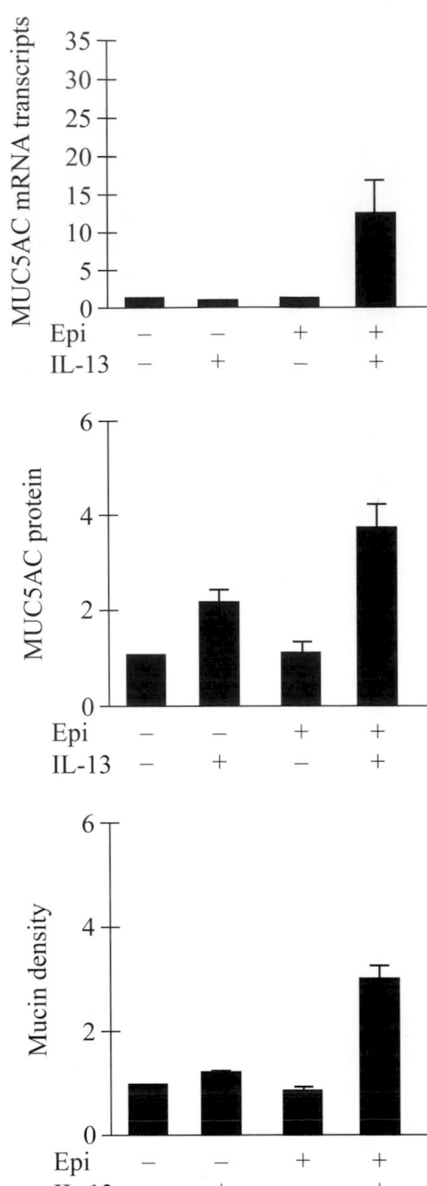

Figure 1 Levels of MUC5AC gene transcription, MUC5AC protein and intracellular mucin accumulation in the presence and/or absence of Epi and/or IL-13

This passage was adapted from "Epinephrine Activation of the β2-Adrenoceptor Is Required for IL-13-Induced Mucin Production in Human Bronchial Epithelial Cells." Al-Sawalha N, Pokkunuri I, Omoluabi O, Kim H, Thanawala VJ et al. *PLoS ONE.* 2015. 10(7) doi: 10.1371/journal.pone.0132559 for use under the terms of the Creative Commons CC BY 4.0 license (https://creativecommons.org/licenses/by/4.0/legalcode).

Question 74

Based on the study described in the passage, which statement is most likely true?

- O **A.** IL-13 is responsible for mucous metaplasia.
- O **B.** Elimination of β2AR signaling would not decrease mucous metaplasia in asthma patients.
- O **C.** Although epinephrine may be used during asthma attacks, it may contribute to the buildup of mucous in allergic asthma patients.
- O **D.** Epinephrine is responsible for mucous metaplasia.

Question 75

A lung biopsy is taken from a patient with allergic asthma. In the tissue, all of the following cells are expected EXCEPT:

- O **A.** mucous secreting goblet cells.
- O **B.** basophils.
- O **C.** mast cells.
- O **D.** skeletal muscle cells.

Question 76

When patients suffer an asthma attack, they are often given oxygen through a mask. Based on Henry's Law, what is the likely reason for this?

- O **A.** Increasing the partial pressure of oxygen inhaled would increase the amount absorbed by the blood.
- O **B.** Pure oxygen will expand the airway more than room air.
- O **C.** Oxygen combats the effects of IL-13 on the airways.
- O **D.** Giving oxygen in a pure form increases its solubility when compared to its solubility in room air.

Question 77

Based on the information in the passage, allergic asthma adversely affects which function of the respiratory system?

- O **A.** Protection against particulate matter
- O **B.** Circulation of oxygen in the body
- O **C.** Alveolar gas exchange
- O **D.** Entry of air through the airways

Question 78

Hypersensitivity and autoimmunity are two ways that the immune system can go awry. Autoimmunity can be a result of:

- O **A.** the body's immune system attacking tissues.
- O **B.** negative selection failure during clonal selection.
- O **C.** positive selection failure during clonal selection.
- O **D.** excessive inflammation.

Passage III (Questions 79-83)

It is well established that low cardiorespiratory and neuromuscular fitness are associated with increased mortality and morbidity in both men and women. In order to counteract an age-related decline in physical fitness, global exercise recommendations suggest a combination of both endurance and strength training. Strength training is especially recommended for post-menopausal women to fortify bones weakened by osteoporosis and generally results in increased muscle mass. While both forms of fitness are important in maintaining general health, many guidelines often overlook the possible complications of concurrent aerobic and strength training, known as the interference effect. The interference effect establishes that muscular and skeletal strengthening may be delayed by the metabolic demands of aerobic exercise. A better understanding of the interference effect will aid in optimizing time-efficient exercise prescription.

Scientists investigated the effects of the endurance and strength training order within the same combined training session and the effects of prolonged concurrent training performed on alternating days on cardiorespiratory fitness. After muscle and fat mass were measured, subjects performed 24 weeks of supervised training in one of the following training groups: i) endurance training immediately followed by strength training (E+S), ii) strength training immediately followed by endurance training (S+E), or iii) endurance and strength training performed on alternating days (AD). Cardiorespiratory fitness was measured by the volume of air consumed during the second minute of exercise (V_{AIR}), which is proportional to the percentage of skeletal muscle in the body (Figure 1). The researchers concluded that endurance and strength training performed on alternating days may optimize cardiovascular fitness.

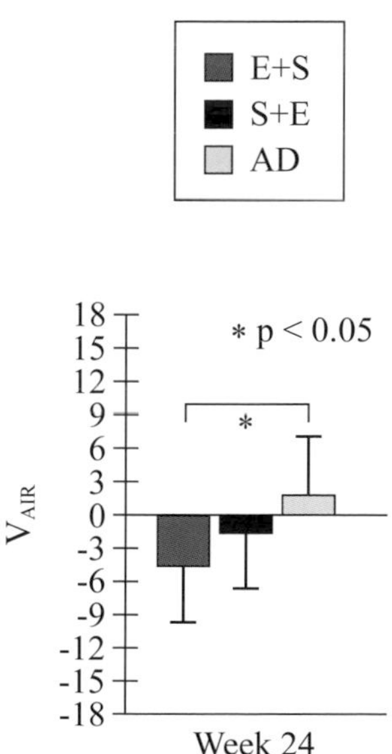

Figure 1 Oxygen consumption of subjects at 24 weeks, as compared to baseline

This passage was adapted from "Cardiorespiratory adaptations during concurrent aerobic and strength training in men and women." Schumann M, Yli-Peltola K, Abbiss C, and Hakkinen K. *PLoS One.* 2015. 10(9) doi:10.1371/journal.pone.0139279 for use under the terms of the Creative Commons CC BY 4.0 license (http://creativecommons.org/licenses/by/4.0/legalcode).

Question 79

Respiratory rate generally increases with increased physical exertion. Which of the following statements best describes how respiration is controlled during exercise?

- O **A.** Expiration is governed by the contraction rate of the external intercostal muscles.
- O **B.** The smooth muscle of the diaphragm contracts quickly to increase the volume of the chest cavity.
- O **C.** The medulla oblongata governs the rate of skeletal muscle contraction.
- O **D.** Inspiration is driven by the contraction of the internal intercostal muscles.

Question 80

Using the V_{AIR} consumed, scientists calculated the oxygen delivered to the tissues. Which of the following findings would most likely result in the scientists overestimating the concentration of oxygen delivered?

- O **A.** 25% of study participants were smokers, which causes increased alveolar wall thickness.
- O **B.** The bronchi of study participants constricted by 30% during intense exercise.
- O **C.** Study participants with increased muscle mass had a higher V_{AIR}.
- O **D.** The concentration of CO_2 expired increased for all study participants during intense exercise.

Question 81

Which of the following figures best depicts the change in oxyhemoglobin binding during exercise?

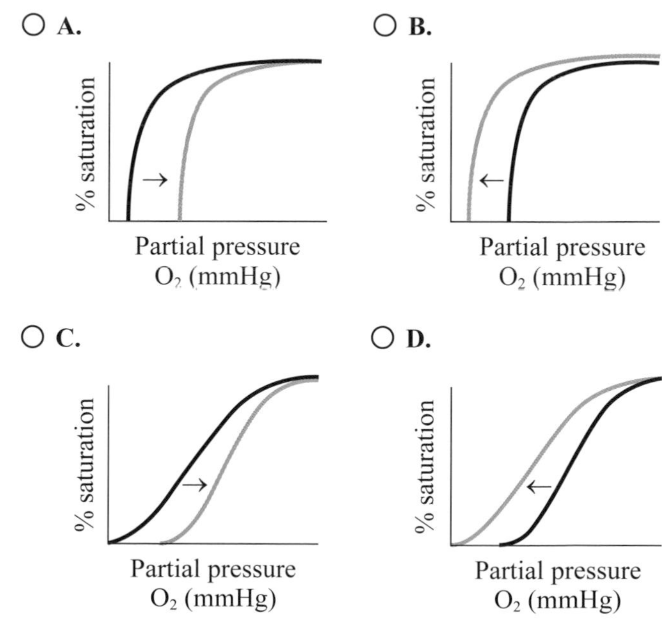

Question 82

Inhibition of carbonic anhydrase with the drug acetazolamide causes:

- ○ **A.** an increase in the concentration of bicarbonate inside the red blood cell.
- ○ **B.** a decrease in the concentration of chloride inside the red blood cell.
- ○ **C.** an increase in the partial pressure of CO_2 in the lungs.
- ○ **D.** a decrease in the intracellular pH of the red blood cell.

Question 83

The interference effect may best be explained by:

- ○ **A.** increased β-oxidation during high intensity aerobic exercise.
- ○ **B.** excess lactic acid fermentation during low intensity aerobic exercise.
- ○ **C.** increased body temperature maintained during high intensity aerobic exercise.
- ○ **D.** increased gluconeogenesis maintained after aerobic exercise.

Passage IV (Questions 84-87)

Recognition of tumor antigens by specific T cells is a prerequisite for the induction of effective anti-tumor immune responses. This is initiated by cross-presentation, a phenomenon where professional antigen presenting cells (APC) such as dendritic cells (DC) capture, process, and present exogenous antigens through the MHC class I pathway. Many of the tumor-associated antigens (TAAs) currently targeted in anti-tumor immunotherapy, such as survivin and PCNA, are predominantly expressed in the nucleus. Therefore, it is important to understand the relative availability for cross-presentation of antigens in the nuclear compartment compared to other cellular compartments.

Scientists hypothesized that nuclear localization of an antigen would limit cross-presentation and activation of naive CD8 T cells. Furthermore, because apoptosis is known to cause nuclear degradation and extracellular release of nuclear contents, they reasoned that apoptosis-inducing chemotherapy would improve the efficiency of cross-presentation. Scientists engineered pancreatic cancer tumors to express a model antigen, an influenza viral surface protein, in the secretory, cytoplasmic, or nuclear compartments, and compared their potential to induce proliferation of killer T cells *in vivo*. In a second experiment, they first treated the cells with a chemotherapy nucleoside analogue, gemcitabine, then exposed them to killer T cells.

The researchers found that the nuclear antigen was not cross-presented as efficiently as its cytoplasmic and secreted counterparts. They also found that treatment with the apoptosis-inducing gemcitabine was able to restore nuclear antigen cross-presentation to levels equivalent to secreted or cytoplasmic compartments. The antigen-dose dependent nature of this improvement suggested that treatment with gemcitabine boosted the amount of antigen available for cross-presentation and thus activation of killer T cells. The researchers reasoned that chemotherapeutic treatment prior to cancer immunotherapy would improve patient outcomes.

Question 84

Which of the following statements best explains why relatively reduced levels of nuclear antigen cross-presentation might correlate with less effective immunosurveillance?

- O **A.** An increased level of survivin but not PCNA is presented on MHC class II molecules.
- O **B.** An increased level of PCNA but not survivin is presented on MHC class I molecules.
- O **C.** A decreased level of PCNA and survivin is presented on MHC class I molecules.
- O **D.** A decreased level of survivin and an increased level of PCNA is presented on MHC class II molecules.

Question 85

Which of the following cells are LEAST likely to cross-present nuclear antigens?

I. Basophils
II. Macrophages
III. Eosinophils

- O **A.** I only
- O **B.** I and II only
- O **C.** I and III only
- O **D.** II and III only

Question 86

According to information contained in the passage, gemcitabine likely acts as:

- O **A.** a competitive inhibitor of DNA polymerase.
- O **B.** a non-competitive inhibitor of DNA topoisomerase.
- O **C.** a non-competitive inhibitor of RNA polymerase.
- O **D.** a competitive inhibitor of MHC Class I.

Question 87

Which of the following adaptive immune responses is LEAST likely to be mediated by antibodies?

- O **A.** Increased phagocytosis of cancer cells with mutation-associated antigens presented on the cell surface
- O **B.** Directly perforating the cell wall of cancer cells, causing lysis
- O **C.** Bind to mast cells to increase the local release of histamine
- O **D.** Cause precipitation of circulating cancer cells in venules

Questions 88 through 92 are NOT based on a descriptive passage.

Question 88

Patients with left sided congestive heart failure have decreased contractile strength in their cardiac muscle leading to fluid accumulation in their lungs. This effect can be explained by which series of events?

- O **A.** Increased stroke volume, blood moving back into the right ventricle, decreased pressure in the pulmonary vessels and fluid moving to the lung interstitium.
- O **B.** Increased stroke volume, blood backing up, increased pressure in the pulmonary vessels and fluid moving to the lung interstitium.
- O **C.** Decreased stroke volume, blood backing up, increased pressure in the pulmonary vessels and fluid moving to the lung interstitium.
- O **D.** Decreased stroke volume, blood moving back into the right atrium, increased pressure in the pulmonary vessels and fluid moving to the lung interstitium.

Question 89

Following injection of an acidic therapeutic agent into the bloodstream of a patient, which of the following statements is true regarding the patient's blood?

- O **A.** Reduced affinity of hemoglobin for oxygen and corresponding leftward shift of the oxygen dissociation curve for hemoglobin
- O **B.** Reduced affinity of hemoglobin for oxygen and corresponding rightward shift of the oxygen dissociation curve for hemoglobin
- O **C.** Increased affinity of hemoglobin for oxygen and corresponding leftward shift of the oxygen dissociation curve for hemoglobin
- O **D.** Increased affinity of hemoglobin for oxygen and corresponding rightward shift of the oxygen dissociation curve for hemoglobin

Question 90

Which of the following is **NOT** true concerning the lymphatic system?

- O **A.** The lymphatic system removes large particles and excess fluid from the interstitial spaces.
- O **B.** The lymphatic system is a closed circulatory system.
- O **C.** The lymphatic system contains lymphocytes that function in the body's immune system.
- O **D.** Most fatty acids in the diet are absorbed by the lymphatic system before entering the bloodstream.

Question 91

DNA sequencing of blood cells and biopsy tissue is used to detect germline mutations that may increase a patient's risk for developing cancer. Mutations in the tumor suppressor gene BRCA1 are LEAST likely to be discovered in:

- O **A.** erythrocytes.
- O **B.** macrophages.
- O **C.** B lymphocytes.
- O **D.** myocytes.

Question 92

Clonal selection is:

- O **A.** the first stage of the innate immune response.
- O **B.** phagocytic destruction of antigens.
- O **C.** the binding to an antigen by an antibody.
- O **D.** the process by which certain immature lymphocytes are selected for maturation.

STOP. IF YOU FINISH BEFORE TIME IS CALLED, CHECK YOUR WORK. YOU MAY GO BACK TO ANY QUESTION IN THIS TEST BOOKLET.

30-MINUTE IN-CLASS EXAM FOR LECTURE 5

Passage I (Questions 93-97)

Since fats are not soluble in the aqueous solution in the small intestine, fat digestion would be very inefficient were it not for bile salts and lecithin, which increase the surface area upon which lipase can act. In addition, bile forms micelles that travel to the brush border of the intestine, where they are absorbed by enterocytes.

Once inside the enterocyte, fatty acids are taken up by the smooth endoplasmic reticulum and new triglycerides are formed. These triglycerides combine with cholesterol and phospholipids to form triglyceride-rich lipoproteins (TRL) called chylomicrons that are secreted through exocytosis to the basolateral side of the enterocyte. From there, the chylomicrons move to the lacteal in the intestinal villus.

Chylomicrons are just one member of the lipoprotein families which transport lipids through the blood. The other members are very low density lipoproteins (VLDL), low density lipoproteins (LDL), and high density lipoproteins (HDL). Lipolysis of chylomicrons by lipoprotein lipase (LPL) results in the formation of smaller remnants that are depleted of triglycerides and enriched in cholesteryl esters. Elevated non-fasting triglyceride (TG) levels and elevated levels of remnants may lead to accelerated atherosclerosis and cardiovascular disease.

The mechanism that leads to high non-fasting TG levels involves decreased hepatic clearance of TRL remnants, which is mediated by syndecan-1 heparan sulfate proteoglycan (HSPG). Dysfunction of syndecan-1 HSPG has been shown to disrupt defective hepatic remnant clearance. One gene responsible for HSPG assembly and disassembly is *SULF2*.

Experiment 1

Scientists tested whether the presence of the G allele in the *SULF2* gene contributes to difficulty in lipid clearance in humans by monitoring blood lipid levels (mmol/L) after standardized consumption of dietary fats. Participants were classified according to genotype (AA, AG, or GG) and blood lipid levels were recorded prior to the administration of the meal and at 3, 4, 6, and 8 hours after eating. Results are shown in Figure 1.

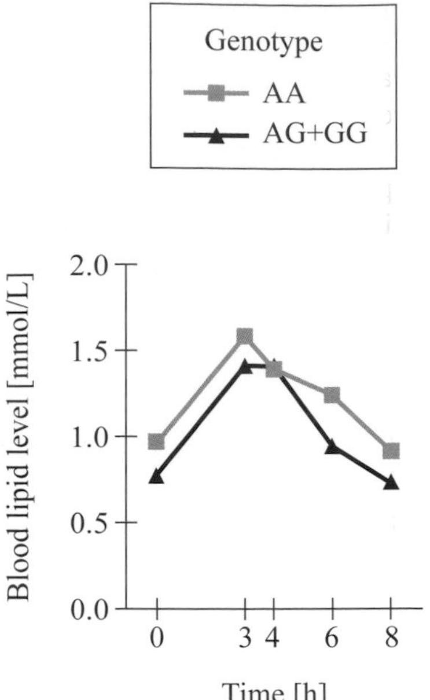

Figure 1 Blood lipid levels 0, 3, 4, 6, and 8 hours after eating

Question 93

The process by which bile increases the surface area of dietary fat is called:

- **A.** lipolysis.
- **B.** adipolysis.
- **C.** malabsorption.
- **D.** emulsification.

Question 94

Most dietary fat first enters the bloodstream:

- **A.** from the right lymphatic duct into arterial circulation.
- **B.** from the thoracic duct into venous circulation.
- **C.** from the small intestine into capillary circulation.
- **D.** from the intestinal enterocyte into the lacteal.

Question 95

Life insurance rates generally increase with increasing total serum cholesterol levels. Which of the following supports the claim that it is important to determine whether HDLs or LDLs are responsible for high serum cholesterol when evaluating a patient's risk for coronary heart disease?

- O **A.** The risk from coronary heart disease doubles from an HDL level of 60mg/100ml to 30mg/100ml.
- O **B.** The incidence of coronary heart disease rises in linear fashion with the level of serum cholesterol.
- O **C.** The optimal serum cholesterol for a middle aged man is probably 200mg/100ml or less.
- O **D.** VLDL is the main source of plasma LDL.

Question 96

Considering the chemical structures of the fats below, which lipid may form part of the surface of a lipoprotein and which will be located in the interior of the lipoprotein?

- O **A.** Triglyceride on the surface; phosphatidylcholine in the interior
- O **B.** Both on the surface
- O **C.** Phosphatidylcholine on the surface; triglyceride in the interior
- O **D.** Both in the interior

Question 97

Results from Experiment 1 suggest that:

- O **A.** polymorphisms in genes responsible for HSPG assembly or disassembly associate with blood lipid clearance ability.
- O **B.** the A allele affects blood lipid clearance, while the G allele has no effect.
- O **C.** blood lipid clearance ability in heterozygotes is intermediate between that of individuals homozygous for AA and those homozygous for GG.
- O **D.** individuals carrying the G allele clear lipids more slowly.

Passage II (Questions 98-101)

The *renal clearance* (C) of a solute is a measure of the rate at which it is cleared from blood plasma by excretion. C is expressed as volume of plasma cleared per unit time, and assumes that 100% of the solute is filtered from the blood into the lumen of the nephron. where U is the urine concentration of the solute, V is the urine flow rate, and P is the plasma concentration of the solute. The clearance rate of any substance that is freely filtered and is not reabsorbed or secreted is equal to the glomerular filtration rate (GFR). Normal GFR is approximately 125 mL/min.

$$C = \frac{U \times V}{P}$$

Renal hyperfiltration, defined as GFR ≥ 135 mL/min, has been associated with the progression of kidney disease, including diabetic nephropathy. Since renal hyperfiltration represents a state of intrarenal *renin-angiotensin-aldosterone system* (RAAS) activation, researchers hypothesized that hyperfiltration in patients with type I diabetes mellitus (DM) would be associated with higher BP and elevated levels of circulating RAAS mediators.

To test this hypothesis, a cohort of patients with type I DM and hyperfiltration (DM-H) was compared to patients with type I DM and normofiltration (DM-N) and healthy control (HC) patients. Researchers determined GFR by measuring the concentration of *inulin*— a nonmetabolizable polysaccharide that is freely filtered and neither reabsorbed nor secreted— in the blood and urine. GFR was found to be higher for DM-H patients than for either DM-N or HC patients. The data are shown by cohort in Table 1.

	HC	DM-N	DM-H
Concentration in plasma (mg/mL)	9.50×10^{-5}	1.05×10^{-4}	1.00×10^{-4}
Urine flow rate (mL/min)	0.75	0.80	1.00
Concentration in urine (mg/mL)	1.45×10^{-2}	1.50×10^{-2}	1.50×10^{-2}

Table 1 Concentrations of inulin (mg/mL) in blood plasma and urine

Researchers then measured systemic blood pressure, renal blood blow, and concentrations of circulating RAAS mediators. The results are shown in Table 2.

	HC	DM-N	DM-H
Systolic BP (mmHg)	109	111	117
Diastolic BP (mmHg)	63	64	67
Effective RBF (mL/minute/1.73m^2)	629	662	891
Renin (pg/mL)	20.0	5.7	5.3
Angiotensin II (pg/mL)	12.5	3.2	3.5
Aldosterone (ng/dL)	276.1	85.9	41.8

Table 2 Blood pressure, effective renal blood flow, and concentrations of circulating RAAS mediators under conditions of normal blood glucose concentration

Question 98

Which of the following must be true of a solute that has a *renal clearance* greater than the GFR?

- O **A.** The solute is being reabsorbed by the nephron.
- O **B.** The solute is being secreted by the nephron.
- O **C.** The plasma concentration of the solute must be greater than the plasma concentration of inulin.
- O **D.** The solute has exceeded its transport maximum.

Question 99

Inulin is neither secreted nor reabsorbed by the kidney. Which of the following must also be true for the clearance of inulin to accurately represent the GFR?

- O **A.** The filtration rate of inulin is less than its excretion rate.
- O **B.** The filtration rate of inulin is greater than its excretion rate.
- O **C.** The concentration of inulin in the filtrate is equal to its concentration in the plasma.
- O **D.** Inulin is completely filtered from the plasma in a single pass through the kidney.

Question 100

Which of the following would LEAST affect the *renal clearance* of a solute?

- O **A.** Solute size and charge
- O **B.** Solute plasma concentration
- O **C.** Glomerular hydrostatic pressure
- O **D.** Inulin plasma concentration

Question 101

The data in Table 2 support the conclusion that hyperfiltration in patients with type I diabetes mellitus is associated with:

- O **A.** higher blood pressure and elevated levels of circulating RAAS mediators.
- O **B.** lower blood pressure and elevated levels of circulating RAAS mediators.
- O **C.** higher blood pressure and suppressed levels of circulating RAAS mediators.
- O **D.** lower blood pressure and suppressed levels of circulating RAAS mediators.

Passage III (Questions 102-105)

The *renin-angiotensin-aldosterone system* (RAAS) regulates renal blood flow (RBF), glomerular filtration rate (GFR), and systemic blood pressure (BP).

Renin, a proteolytic enzyme, is synthesized and stored in the juxtaglomerular granular cells of the afferent arterioles. Each of the stimuli that initiates renin release—including sympathetic stimulation of β-adrenergic receptors on the juxtaglomerular cells, low blood pressure in the renal arterioles, and low rate of NaCl transport across the macula densa of the distal tubule—is either a direct or indirect consequence of low systemic blood pressure.

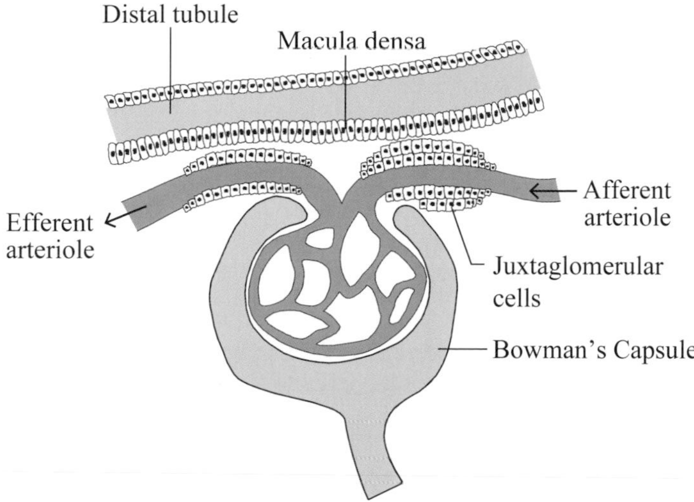

Figure 1 Initiation of renin release involves both the macula densa and juxtaglomerular cells of the nephron

Once in the blood, renin decreases the resistance of the afferent arterioles and converts *angiotensinogen*, an inactive plasma protein, into *angiotensin I*. When angiotensin I (Ang I) encounters *angiotensin converting enzyme* (ACE), it is converted to *angiotensin II* (Ang II). Ang II, a potent vasoconstrictor, increases the resistance of the efferent arterioles and promotes synthesis and release of *aldosterone*, a steroid hormone that increases Na^+ reabsorption in the distal tubule and collecting ducts.

Dysfunction of brain *renin-angiotensin system* (RAS) components is implicated in the development of hypertension. Studies have suggested that Ang II-induced hypertension may be mediated by increased production of proinflvammatory cytokines (PIC), including tumor necrosis factor (TNF).

Researchers experimentally investigated the effects of TNF inhibition on Ang II-induced hypertension. Rats were subjected to intracerebroventricular infusion of entanercept, a TNF inhibitor, with and without concurrent subcutaneous infusion of Ang II. Mean arterial pressure (MAP) and ratio of heart weight to body weight (HW/BW) were measured.

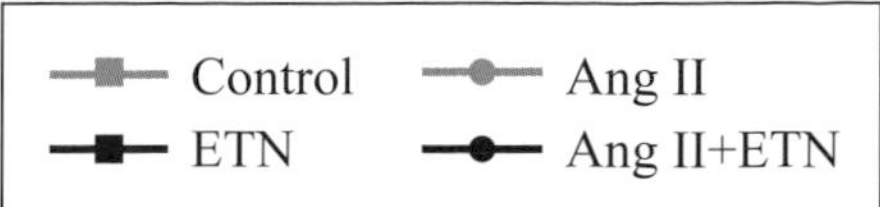

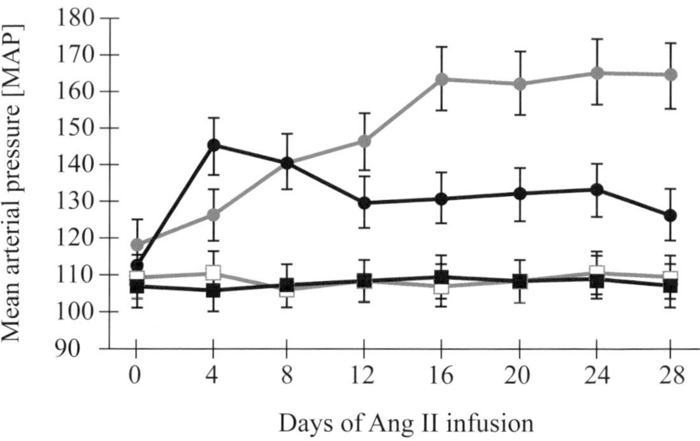

Figure 2 Effects of angiotensin II and central TNF blockade on mean arterial pressure

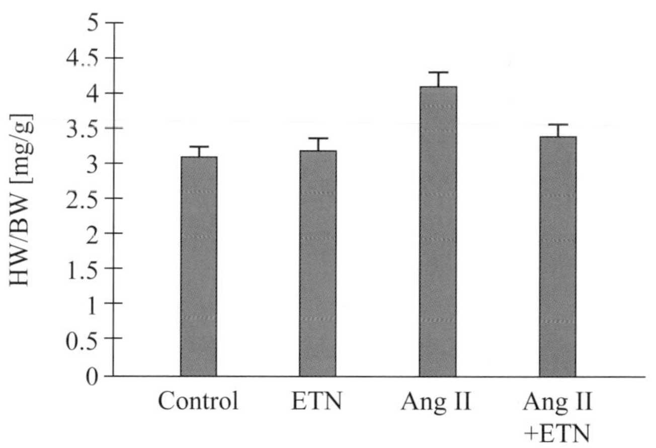

Figure 3 Effects of angiotensin II and central TNF blockade on cardiac hypertrophy

Question 102

The most likely cellular mechanism of action of aldosterone is:

- ○ **A.** modification of existing cellular proteins.
- ○ **B.** initiation of a cAMP second messenger system.
- ○ **C.** diffusion into the nucleus followed by activation or repression of one or more genes.
- ○ **D.** initiation of a signal transduction system.

Question 103

According to Figures 2 and 3, for those rats subjected to Ang II infusion, central TNF blockade tended to:

- **A.** increase mean arterial pressure and increase heart weight to body weight ratio.
- **B.** increase mean arterial pressure and decrease heart weight to body weight ratio.
- **C.** decrease mean arterial pressure and increase heart weight to body weight ratio.
- **D.** decrease mean arterial pressure and decrease heart weight to body weight ratio.

Question 104

The renin-angiotensin-aldosterone system increases all of the following EXCEPT:

- **A.** urine volume.
- **B.** blood pressure.
- **C.** Na^+ reabsorption.
- **D.** K^+ excretion.

Question 105

Amino acids are reabsorbed in the proximal tubule by a symporter driven by the concentration gradient of Na^+. This lowers the concentration of Na^+ that passes the macula densa. A high protein diet would most likely lead to:

I. Decreased resistance of the afferent arterioles
II. Decreased resistance of the efferent arterioles
III. Increased glomerular filtration rate

- **A.** I and II only
- **B.** I and III only
- **C.** III only
- **D.** I, II, and III

Passage IV (Questions 106-110)

Crohn's disease (CD) and ulcerative colitis (UC) are the two main forms of inflammatory bowel diseases (IBD), characterized by intestinal inflammation and ulceration of unknown etiology. Although CD and UC share similar pathophysiological mechanisms, the anatomical localization, histopathological findings, disease progression, and therapeutic responses are unique to each disease. The diagnosis of CD and UC currently relies on a combination of clinical, endoscopic, histological, and imaging parameters. Nevertheless, a subset of patients remains indeterminate in their diagnosis. Because the appropriate treatments differ between the two diseases, it is crucial to accurately diagnose patients.

To elucidate the molecular events involved in IBD pathogenesis, the efforts of some research groups have been focused on the analysis of protein expression. Scientists were interested in understanding the differential expression of the β-catenin and retinoblastoma (Rb) proteins, two key regulators of colonic proliferation, inflammation, and tumorigenesis, in CD and UC. They used colon samples of CD and UC patients to verify by the expression pattern of proteins involved in colonic inflammation by immunohistochemistry and western blot.

β-catenin is mainly detected as part of the adherent junction component, decorating the basolateral membrane of epithelial cells. In the bottom of colonic crypts, however, progenitor cells accumulate cytoplasmic/nuclear β-catenin that binds to transcription factors to drive proliferation. Scientists stained for β- catenin and phospho-Rb in control, CD, and UC tissue samples to determine whether immunolabeling could serve as an adjunct diagnostic tool. The results are shown in Figure 1. Determining the expression status of these molecules could greatly contribute to discriminating between CD and UC.

Next ▶

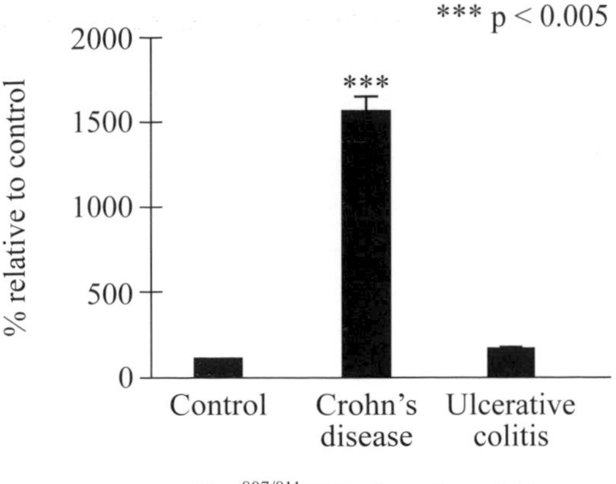

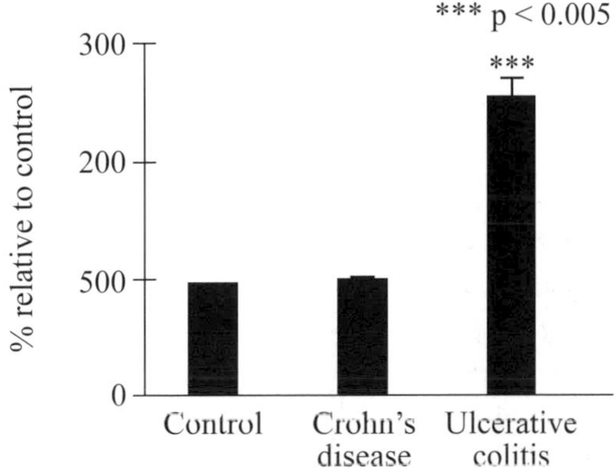

Figure 1 Tissue staining of immunolabeled β-catenin and phospho-Rb

Question 106

Which of the following enzymes is most important for stimulating digestion of complex proteins?

- O **A.** Pepsin
- O **B.** Hormone-sensitive lipase
- O **C.** α-amylase
- O **D.** Chymotrypsinogen

Question 107

Zollinger-Ellison syndrome is characterized by a non-beta cell pancreatic islet tumor that results in significant acidification of the stomach without a proportional increase in protein breakdown. Which of the following molecules is most likely to be secreted by these tumors?

- O **A.** Pepsinogen
- O **B.** Gastrin
- O **C.** Acetylcholine
- O **D.** Insulin

Question 108

Patients with Crohn's Disease often have metabolic abnormalities due to disruptions of bacterial flora in the large intestine. Flux through which of the following metabolic pathways would be decreased in CD patients?

I. β-oxidation
II. Citric acid cycle
III. Electron transport chain

- O **A.** I only
- O **B.** II only
- O **C.** I and II only
- O **D.** I, II, and III

Question 109

Which of the following findings would weaken the assertion by the authors that β-catenin and phospho-Rb could be used as an adjunct diagnostic test?

- O **A.** Increased levels of β-catenin were found in patients with collagenous colitis, a less common form of IBD.
- O **B.** Rb Thr241 was found to be phosphorylated in CD.
- O **C.** Increased levels of β-catenin mRNA were detected in UC patients.
- O **D.** Decreased levels of β-catenin increase water reabsorption in the colon.

Question 110

Which of the following studies would be LEAST likely to provide additional information about how CD and UC affect the normal functions of the large intestine?

- O **A.** Correlation of β-catenin levels and the transcription of Na^+ channels
- O **B.** Co-immunoprecipitation studies of phospho-Rb and aquaporin channels
- O **C.** Correlation of phospho-Rb levels and free Ca^{2+} levels in the blood
- O **D.** Correlation of β-catenin levels and glucose levels in red blood cells

Questions 111 through 115 are NOT based on a descriptive passage.

Question 111

After suffering from chronic inflammation of the pancreas, an individual was advised by his physician that a possible course of treatment could be removal of his entire pancreas. Which of the following would NOT be expected to occur following this procedure?

- O **A.** Difficulty with degradation of proteins into small polypeptides
- O **B.** Limited ability to hydrolyze polysaccharides to disaccharides
- O **C.** A decrease in duodenal pH
- O **D.** Problems with fat emulsification

Question 112

Which of the following cell types is responsible for protecting the brush border from mechanical and chemical damage?

- O **A.** Kupffer cells
- O **B.** Goblet cells
- O **C.** G cells
- O **D.** Parietal cells

Question 113

An individual seeks medical attention because he has been experiencing watery diarrhea for several days. This condition is most likely the result of:

- O **A.** an overactive stomach.
- O **B.** an underactive large intestine.
- O **C.** an overactive small intestine.
- O **D.** an underactive small intestine.

Question 114

Which of the following statements is false regarding hormone interactions with the kidney?

- O **A.** The proximal tubule is sensitive to aldosterone.
- O **B.** The distal tubule is sensitive to ADH.
- O **C.** The collecting tubule is sensitive to ADH.
- O **D.** The distal tubule is sensitive to aldosterone.

Question 115

Which of the following statements about digestion is LEAST likely true?

- O **A.** Carbohydrate metabolism begins in the mouth.
- O **B.** Most dietary protein is absorbed into the body in the stomach.
- O **C.** The large intestine is a major source of water reabsorption.
- O **D.** The liver produces bile for storage in the gallbladder.

STOP. IF YOU FINISH BEFORE TIME IS CALLED, CHECK YOUR WORK. YOU MAY GO BACK TO ANY QUESTION IN THIS TEST BOOKLET.

30-MINUTE IN-CLASS EXAM FOR LECTURE 6

Passage I (Questions 116-120)

The functional unit of a skeletal muscle cell is the sarcomere. The protein polymers actin and myosin lie lengthwise along a sarcomere creating the various regions shown in Figure 1.

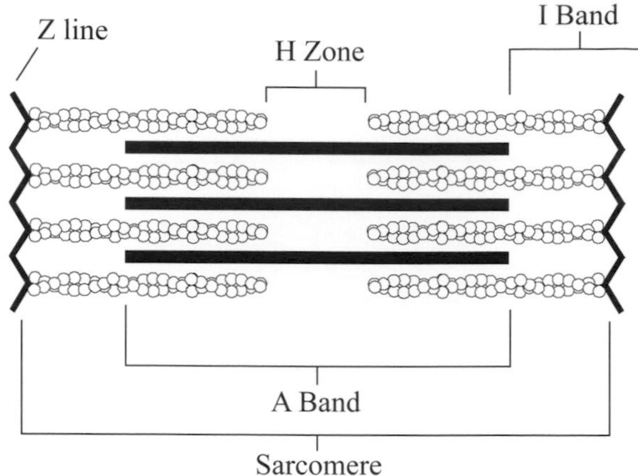

Figure 1 Schematic representation of a sarcomere

In skeletal muscle, the release of calcium by sarcoplasmic reticulum (SR) Ca^{2+} release channels (RyR1s) is the primary determinant of contractile filament activation. The action potential that initiates muscle contraction is delivered deep into each muscle cell via tubular invaginations in the sarcolemma or cell membrane called T-tubules. The change in membrane potential is transferred to the sarcoplasmic reticulum, which causes it to become permeable to Ca^{2+} ions and to release its large stores of calcium into the cytosol, setting off a chain reaction that causes conformational changes in troponin and tropomyosin that allow for the creation of a muscular contraction. The Ca^{2+} ions are removed from the cytosol by extremely efficient calcium pumps via active transport. Once the Ca^{2+} has been sequestered back into the lumen of the sarcoplasmic reticulum, the Ca^{2+} ions are bound by calsequestrin (CASQ1). Much research has been focused on CASQ1 and its role in SR Ca^{2+} buffering as well as its potential for modulating RyR1. Scientists explored the role of CASQ1 by examining muscle contraction in mice with genetic ablation of CASQ1 expression.

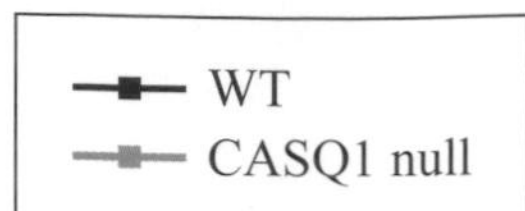

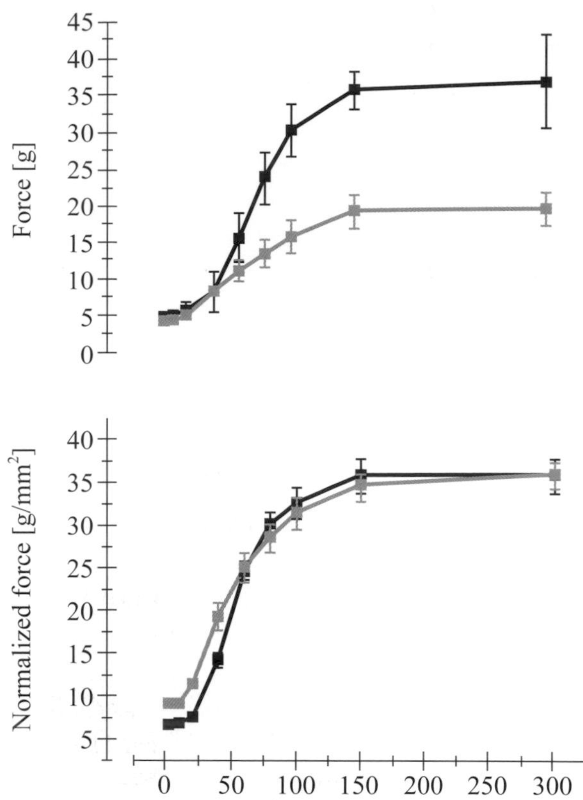

Figure 2 Relationships between force and stimulation frequency in wild-type (n=8) and CASQ1 ablated (CASQ1 null) mice (n=8). **A.** Absolute values of isometric force vs. stimulation frequencies. **B.** Force values normalized by cross-sectional area (CSA) of muscle vs. stimulation frequency.

Although the direct source of energy for muscle contraction comes from ATP, the ATP concentration in actively contracting muscle remains virtually constant. In addition, it has been shown that inhibitors of glycolysis and cellular respiration have no effect on ATP levels in actively contracting muscle over the short term. Instead, phosphocreatine donates its phosphate group to ADP in a reaction catalyzed by creatine kinase. Phosphocreatine levels are replenished via ATP from glycolysis and cellular respiration.

Question 116

Muscle cell T-tubules function to:

- O **A.** create an action potential within a muscle cell.
- O **B.** receive the action potential from the presynaptic neuron.
- O **C.** deliver the action potential directly to the sarcomere.
- O **D.** supply Ca^{2+} to the cytosol during an action potential.

Next ▶

Question 117

Which of the following additions to the study involving CASQ1 ablation and force production would likely be LEAST helpful in increasing the validity of the findings?

- **A.** Age-matching CASQ1 and wild-type mice
- **B.** Including experimental data on higher stimulation frequencies
- **C.** Increasing the sample size of mice
- **D.** Including force data normalized to muscle volume

Question 118

If a creatine kinase inhibitor were administered to an active muscle cell, which of the following would most likely occur?

- **A.** ATP concentrations would diminish while muscle contractions continued.
- **B.** Phosphocreatine concentrations would diminish while muscle contractions continued.
- **C.** ATP concentrations would remain constant while the percent saturation of myoglobin with oxygen would diminish.
- **D.** Cellular respiration and glycolysis would increase to maintain a constant ATP concentration.

Question 119

Which of the following is most likely NOT true concerning the uptake of Ca^{2+} ions from the cytosol during muscle contraction?

- **A.** It requires ATP.
- **B.** The mechanism involves an integral protein of the sarcolemma.
- **C.** It occurs against the concentration gradient of Ca^{2+}.
- **D.** It is rapid and efficient.

Question 120

Which of the following best accounts for the differences between absolute and CSA-normalized force in wild-type and CASQ1 null mice?

- **A.** Wild-type mice are larger than CASQ1 null mice.
- **B.** CASQ1 null mice are larger than wild-type mice.
- **C.** Wild-type mice are able to achieve greater contractile frequency than CASQ1 null mice.
- **D.** CASQ1 null mice are able to achieve greater contractile frequency than wild-type mice.

Passage II (Questions 121-124)

The human fetus has a cartilaginous endoskeleton which is gradually replaced with bone before and after birth until adulthood. Bone forms within and around the periphery of small cartilaginous replicas of adult bones. After it is formed, adult bone is continuously remodeled through the activity of osteoblasts and osteoclasts. *Osteoblasts* differentiate from *fibroblasts* of the perichondrium. *Osteoclasts* differentiate from certain phagocytotic blood cells.

Although studies indicate that both nutrition and amount of load-bearing exercise affect bone density and volume in humans, the full process by which osteoblast and osteoclast activity is regulated in the body is not fully understood. Gastrointestinal peptides are increasingly being linked to processes controlling the regulation of bone mass, particularly under conditions of altered energy balance. Peptide YY (PYY) is a gut-derived peptide of the neuropeptide Y family that is released post-prandially into the bloodstream in proportion to calorie intake. PYY acts to inhibit food intake and increase satiety. It has been shown to be upregulated under some conditions that are also associated with low bone mass. Anorexic conditions and bariatric surgery for obesity both influence circulating levels of PYY and have a negative impact on bone mass.

Experiment 1

To determine the effect of PYY expression on bone mass, scientists compared bone mineral density (BMD) and bone mineral content (BMC) in control and germline PYY knockout ($PYY^{-/-}$) mice. The results for female mice are shown in Figure 1.

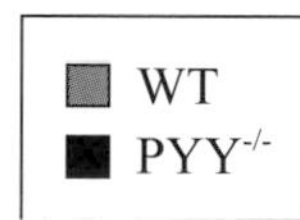

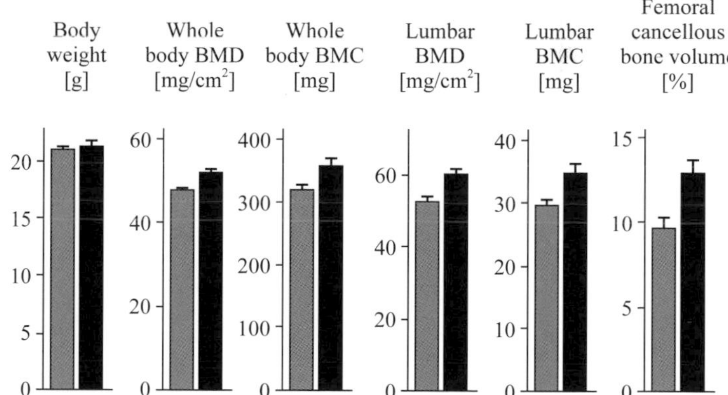

Figure 1 Bone mineral density and bone mineral content in wild-type and PYY knockout mice

Experiment 2

To further evaluate the effects of PYY expression on the regulation of bone mass, scientists examined femoral bone structure in control and conditional adult-onset PYY over-expressing mice (PYYtg). Results for female mice are shown in Figure 2.

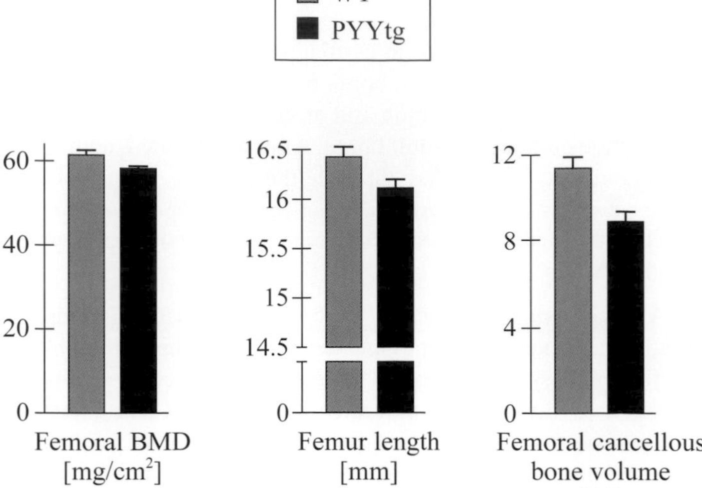

Figure 2 Attributes of femoral bone morphology for wild-type and conditional adult-onset PYY over-expressing mice

This passage was adapted from "Peptide YY Regulates Bone Remodeling in Mice: A Link between Gut and Skeletal Biology." Wong IPL, Driessler F, Khor EC, Shi Y-C, Hörmer B, Nguyen AD, Enriquez RF, Eisman JA, Sainsbury A, Herzog H, Baldock PA. *PLoS ONE*. 2012. 7(7) doi:10.1371/journal.pone.0040038 for use under the terms of the Creative Commons CC BY 3.0 license (http://creativecommons.org/licenses/by/3.0/legalcode).

Question 121

Based on the results in Figure 1, which of the following statements is LEAST likely to be true regarding $PYY^{-/-}$ mice in comparison with wild-type mice?

- **A.** They have increased osteoclast activity and increased osteoblast activity.
- **B.** They have increased osteoclast activity and decreased osteoblast activity.
- **C.** They have decreased osteoclast activity and increased osteoblast activity.
- **D.** They have decreased osteoclast activity and decreased osteoblast activity.

Question 122

Suppose researchers studied bone mass in a mouse genotype they created to be a hybrid of $PYY^{-/-}$ and PYYtg. Which of the following would be true regarding the $PYY^{-/-}$-PYYtg hybrid?

I. Femoral bone would likely decrease, but lumbar bone would likely increase.
II. Deviation from wild-type bone phenotype would likely be age-dependent.
III. It is not possible to predict the net effect on whole body bone phenotype based on the information given.

- **A.** II only
- **B.** III only
- **C.** I and II only
- **D.** II and III only

Question 123

Which of the following hormones most likely stimulates osteoclasts?

- **A.** Parathyroid hormone
- **B.** Calcitonin
- **C.** Epinephrine
- **D.** Prostaglandin

Question 124

Which of the following cells most likely arises from the same stem cell in the bone marrow as an erythrocyte?

- **A.** Osteoblast
- **B.** Fibroblast
- **C.** Osteoclast
- **D.** Chondrocyte

Passage III (Questions 125-129)

The rate at which a muscle can perform work, or the muscle power, varies over time and is given in Table 1.

Time	Power (kg m/min)
First 10 seconds	7000
Next 1.5 minutes	4000
Next 30 minutes	1700

Table 1 Muscle power variance over time

Energy for muscle contraction is derived directly from ATP. However, a muscle cell's original store of ATP is used up in less than 4 seconds by maximum muscle activity. Three systems work to maintain a nearly constant level of ATP in a muscle cell during muscle activity: the phosphagen system, the glycogen-lactic acid system, and the aerobic system. The phosphagen system makes use of phosphocreatine, which contains a high energy phosphate bond that is used to replenish ATP stores from ADP and AMP. This system can sustain peak muscular activity for about 10 seconds. The glycogen-lactic acid system is relied upon for muscular activity lasting beyond 10 seconds but not more than 1.6 minutes. This system produces ATP from glycolysis. Aerobic metabolism of glucose, fatty acids and amino acids can sustain muscular activity for as long as the supply of nutrients lasts.

The recovery of muscle after exercise involves replacement of oxygen. Before exercise, the body contains approximately 2.5 liters of oxygen, but this oxygen store is used up in approximately 1 minute of heavy exercise. After heavy exercise, oxygen uptake increases dramatically at first and then levels back down to normal over a 1 hour period. The extra oxygen taken in is called the oxygen debt, and is about 11.5 liters.

Although the above is true for muscle in general, muscle fiber types vary in endurance, force production, and contractile and metabolic properties. Some of these differences are represented in Table 2.

	Slow Twitch Oxidative (Type I)	Fast Twitch Oxidative (Type IIA)	Fast Twitch Glycolytic (Type IIB)
Time to peak tension	1.0	0.4	0.4
Oxidative potential	1.0	0.7	0.2
Glycolytic potential	1.0	1.5	2.0
[Phosphocreatine]	1.0	1.2	1.2
[Glycogen]	1.0	1.3	1.5
[Triacylglycerol]	1.0	0.4	0.2

Table 2 Contractile and metabolic properties of human skeletal muscle fiber types. All values are expressed as a fold-change relative to slow-twitch oxidative fibers

Proportion of muscle fiber types can influence susceptibility to metabolic diseases such as diabetes. Insulin sensitivity correlates with the proportion of ST oxidative fibers. Specifically, insulin-stimulated glucose transport is greater in skeletal muscle enriched with ST muscle fibers, priming ST muscles for accelerated glucose uptake and metabolism.

Question 125

Based upon the information in Table 1, which of the following are the most likely rates of molar production of ATP for the phosphagen system, the glycogen-lactic acid system, and the aerobic system respectively?

- O **A.** 1.7 *M*/min, 4 *M*/min, 7 *M*/min
- O **B.** 2 *M*/min, 2 *M*/min, 2 *M*/min
- O **C.** 4 *M*/min, 3 *M*/min, 2.5 *M*/min
- O **D.** 4 *M*/min, 2.5 *M*/min, 1 *M*/min

Question 126

Why is the oxygen debt greater than the amount of oxygen stored in the body?

- O **A.** Exercise increases the hemoglobin content of the blood so that it can store more oxygen.
- O **B.** Heavy breathing after exercise takes in more oxygen.
- O **C.** In addition to replenishing the stored oxygen, oxygen is used to reconstitute the phosphagen and lactic acid systems.
- O **D.** In addition to replenishing the stored oxygen, oxygen is used to reconstitute the phosphagen, lactic acid, and aerobic systems.

Question 127

During the aging process, the ratio of fast twitch to slow twitch fibers increases. All of the following might be expected to accompany that transition EXCEPT:

- O **A.** decreased concentration of phosphocreatine.
- O **B.** diminished insulin sensitivity.
- O **C.** decreased oxidative capacity.
- O **D.** increased concentration of glycogen.

Question 128

As shown in Table 2, type I fibers have a greater density of mitochondria than type II fibers. Which of the following helps explain this discrepancy?

I. Type II fibers contract with more force than type I fibers.

II. Type I fibers have greater endurance than type II fibers.

III. Type I fibers make greater use of the aerobic system than type II fibers.

O **A.** II only

O **B.** III only

O **C.** II and III only

O **D.** I, II and III

Question 129

Which of the following is depleted first in the body of an athlete exercising at maximal capacity?

O **A.** Phosphocreatine

O **B.** ATP

O **C.** Glycogen

O **D.** Glucose

Passage IV (Questions 130-133)

The main feature of osteoarthritis (OA) is the progressive degradation and loss of the articular cartilage accompanied by other critical structural changes, such as synovial membrane inflammation and subchondral bone weakening. These structural changes contribute to symptoms of OA, namely severe pain, stiffness, and loss of joint mobility.

Certain cytokines, such as IL-1β and TNF-α produced by activated chondrocytes play a major role in the onset and the progression of OA. These cytokines stimulate their own production in a paracrine or autocrine manner and induce the production of a wide range of other pro-inflammatory mediators, such as nitric oxide (NO). NO promotes cartilage destruction by increasing the production and secretion of matrix metalloproteinases (MMP).

Due to the presence of microcracks, vascular channels link the subchondral bone tissue and cartilage. Scientists hypothesized that cross-talk may occur between osteoblasts and chondrocytes to cause OA. To test this hypothesis, they cultured subchondral osteoblasts with human OA chondrocytes and found an increase in NO, a decrease in type II collagen mRNA levels, and an increase of MMP-3 mRNA in chondrocytes.

Due to the pro-OA effects of NO, the scientists were interested in whether an inhibitor of NO could mitigate damage to cartilage. Carnosol is an anti-inflammatory and anti-oxidant compound from rosemary that has been shown to suppress inducible nitric oxide synthase. In a second experiment, scientists treated the osteoblast/chondrocyte co-culture with carnosol to understand if it could serve as an anti-OA drug by impacting MMP-3 expression (Figure 1).

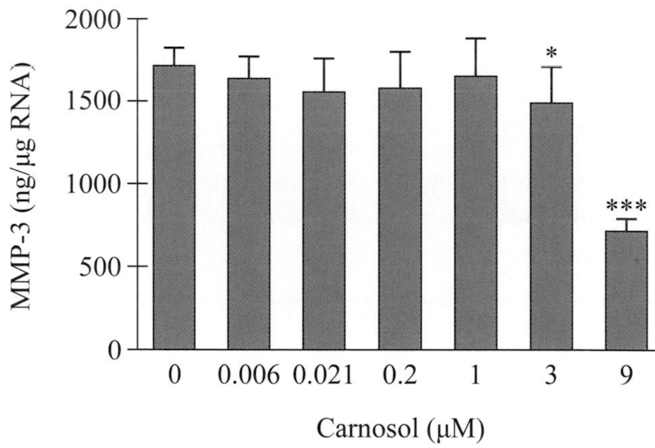

Figure 2 Effect of carnosol on chondrocyte protein production

Question 130

Secretion of which of the following hormones would stimulate osteoprogenitor mitosis in adolescent females?

- A. GHRH
- B. Prolactin
- C. LH
- D. CRH

Question 131

Vascular channels connecting subchondral bone and cartilage function most similarly to:

- A. lamellae.
- B. canaliculi.
- C. Haversian systems.
- D. spongy bone.

Question 132

Which additional experimental finding would be most likely to strengthen the main assertion by the authors?

- A. Carnosol prevents 5' modifications to guanine in the nucleus.
- B. Carnosol prevents protein ubiquitination in the cytosol.
- C. Carnosol weakens disulfide bonds between collagen and aggrecan, a proteoglycan.
- D. Carnosol inhibits acidification of endosomes.

Question 133

OA is most likely to have the greatest effect on which of the following joints?

- A. Joints of the upper rib cage
- B. Joints that connect the parietal and occipital bones of the skull
- C. Joints of the knee and elbow
- D. Joints between jaw and teeth

Questions 134 through 138 are NOT based on a descriptive passage.

Question 134

All of the following are true concerning the musculoskeletal system EXCEPT:

- A. skeletal muscles function by pulling one bone toward another.
- B. tendons connect muscle to muscle.
- C. ligaments connect bone to bone.
- D. the biceps works antagonistically to the triceps.

Question 135

A doctor overprescribes calcium for a patient. Which of the following effects is most likely to occur?

- A. Movement of calcium into blood and maintenance of bone calcium density
- B. Movement of calcium into bone and maintenance of blood calcium concentration
- C. An increase in osteoclast activity
- D. An increase in parathyroid hormone secretion

Question 136

All of the following statements are true concerning bone EXCEPT:

- A. bone is connective tissue.
- B. bone is innervated and has a blood supply.
- C. yellow bone marrow stores triglycerides as a source of energy for the body.
- D. bone is the only nonliving tissue in the body.

Question 137

Dysregulation of which of the following glands is most likely to impact the compressive strength of bone?

- A. Thymus
- B. Parathyroid gland
- C. Adrenal gland
- D. Exocrine pancreas

Question 138

Osteomalacia is a disorder in which bones are softened due to reduced levels of calcium. Following calcium supplementation, an osteomalacia patient shows no improvement. Which of the following conditions could explain this?

I. Low calcitonin secretion
II. Low PTH secretion
III. Lack of vitamin D

- **A.** I only
- **B.** I and II only
- **C.** I and III only
- **D.** II and III only

STOP. IF YOU FINISH BEFORE TIME IS CALLED, CHECK YOUR WORK. YOU MAY GO BACK TO ANY QUESTION IN THIS TEST BOOKLET.

ANSWERS & EXPLANATIONS

FOR
30-MINUTE IN-CLASS EXAMINATIONS

ANSWERS TO THE 30-MINUTE IN-CLASS EXAMS

Lecture 1	Lecture 2	Lecture 3	Lecture 4	Lecture 5	Lecture 6
1. C	24. D	47. B	70. A	93. D	116. C
2. B	25. C	48. B	71. D	94. B	117. B
3. C	26. D	49. D	72. A	95. A	118. A
4. B	27. C	50. A	73. D	96. C	119. B
5. B	28. B	51. C	74. C	97. A	120. A
6. C	29. C	52. A	75. D	98. B	121. B
7. B	30. A	53. A	76. A	99. C	122. D
8. D	31. A	54. C	77. D	100. D	123. A
9. A	32. A	55. B	78. B	101. C	124. C
10. B	33. B	56. B	79. C	102. C	125. D
11. D	34. A	57. D	80. A	103. D	126. C
12. C	35. C	58. B	81. C	104. A	127. A
13. B	36. D	59. C	82. B	105. B	128. C
14. D	37. B	60. A	83. D	106. A	129. A
15. A	38. A	61. D	84. C	107. B	130. A
16. D	39. B	62. A	85. C	108. D	131. C
17. C	40. D	63. C	86. A	109. A	132. A
18. D	41. B	64. B	87. B	110. D	133. C
19. A	42. A	65. D	88. C	111. D	134. B
20. D	43. C	66. D	89. B	112. B	135. B
21. D	44. A	67. D	90. B	113. B	136. D
22. B	45. B	68. A	91. A	114. A	137. B
23. C	46. B	69. B	92. D	115. B	138. C

SCORING

Any attempt we could make at score correlation to the MCAT® would mislead. Unlike the AAMC MCAT®, which includes easy questions, Examkrackers deliberately asks questions only of medium and high difficulty in order to optimize your practice time and maximize the increase to your MCAT® score. To accurately predict your MCAT® score, use an official AAMC MCAT® practice test.

Your goal is to see your raw score improve with each Examkrackers In-Class Exam and full-length EK-Test® you take. Look closely at each question you get wrong to find areas for review and to notice the habits that don't work. As you approach each new In-Class Exam or practice test, make commitments to replace what doesn't work with the approach you need to get questions right. Focus on the questions you get wrong to learn to think like the MCAT® and increase your score.

Toward your success!

EXPLANATIONS TO IN-CLASS EXAM FOR LECTURE 1

Passage I

1. **C is the best answer.** Bacteriophages cut through bacterial membranes and utilize bacterial synthetic machinery to reproduce. Option I would prevent or reduce entry of viruses into bacteria. Viruses must penetrate the cell wall and membrane in order to inject and incorporate their genes into the host genome. This makes option I likely, as it would slow down a vital step in the phage life cycle. Bacterial genetic machinery includes DNA polymerase and RNA polymerase. These enzymes are required for cell division and transcription of genes. Once a bacteriophage is incorporated into the host bacteria, it wants to divide and have its viral genes transcribed. Option II is likely. Next, option III is not a good answer because it references binding to a eukaryotic receptor. Bacteriophages infect prokaryotes, so inhibition of binding to a eukaryotic receptor would not clearly inhibit propagation. Option III can be eliminated and choice C is the best answer.

2. **B is the best answer.** The human body is a foreign and hostile environment to all viruses, including phages. The most successful viruses will be those that evade detection, as being detected will lead to destruction. Viral envelopes are lipid bilayers derived from host cell membranes that cover the protein capsid of viruses. The envelope allows the virus to avoid detection from the immune system because the capsid proteins, which are distinct to viruses, are more difficult for immune cells to detect. An immune cell scanning the area will see a host membrane, and this will not trigger any alarms. Choices A and B are better answers than choices C and D. The lytic viral life cycle involves damage and bursting of cells. This would cause inflammation and attraction of immune cells at the site of cell death. This reasoning makes choice B a better answer than choice A. The lysogenic life cycle is characterized by subtle division within the host without destroying the host cell.

3. **C is the best answer.** The most virulent bacteria – meaning those that produce the most phages – become dominant. This dominance in the population is represented by an increase in proportional representation in the population. Phage-inducing bacteria achieve this by releasing phages that harm other strains of bacteria. Choice A is unlikely because if bacteria release a phage that they themselves are vulnerable to, this would kill the strain itself rather than other strains of bacteria. Choice B is unlikely since a strain that is going to dominate the population is unlikely to kill itself to conserve resources. It is much more likely to kill other strains in order to reserve all of the environmental resources for itself, making choice C the best answer. Choice D does not make sense for the same reason as choice A: a bacteria that will dominate the population would not kill itself off.

4. **B is the best answer.** Figure 1 shows several strains of *E. Coli* and the amount of phages that they produce. Phages are viruses that target bacteria, and each strain releases phages that it can resist. This leads to death of other strains of bacteria while the bacteria that the phages originated from thrive. It is reasonable to conclude that the dominant bacteria would release the highest number of phages. According to Figure 1, 05vv1522 releases more phages than both 05vv1387 and 05vv1558. This makes choice B a better answer than choices A or C. Choice D is misleading because 05vv1315 is not a strain of *E. Coli*. It is a strain of *P. aeruginosa*.

Passage II

5. **B is the best answer.** The passage directly states that phospholipids are produced on the cytosol side of the ER membrane. The lumen is on the opposite side of the ER membrane.

6. **C is the best answer.** Phospholipids are stated to be incorporated into membranous cell structures. Cytochrome P450 is said to be composed of integral proteins, meaning it is in the cell membrane. Since cytochrome P450 is an integral protein, it is likely produced in the rough ER (remember that secretory AND integral proteins are produced in the rough ER). This means that its upregulation and downregulation would probably occur at the level of gene expression or by posttranslational modification, neither of which newly produced phospholipids should affect. Cytochrome P450 is an integral protein, meaning it is not going to be exocytosed.

7. **B is the best answer.** Mixed-function oxidases are proteins, meaning they are synthesized in either the rough ER or on free floating ribosomes in the cytosol. Since cytosol is not a choice here, rough ER is picked by default. Note that even if cytosol were present, you could make an educated guess that rough ER was the right answer by having noticed that the only specific protein listed as being a major part of a mixed-function oxidase is cytochrome P450, which is an integral protein.

8. **D is the best answer.** This is what adipocytes are by definition – adipose tissue, or body fat, is stored energy. The question implies that adipocytes are serving as a reservoir for triacylglycerol, which hints at the best answer. Choices B and C are both stated functions of the smooth ER, not of the adipocytes.

9. **A is the best answer.** Phenobarbital has been stated to increase smooth ER levels and mixed-function oxidase levels. This effect would be compounded when Carbofuran was added, leading to higher levels of phospholipids than either the group which only had Carbofuran added or the control group which had no foreign substances added.

Passage III

10. **B is the best answer.** First of all, Table 1 does not give data regarding concentration of number of ions in solution, so C can be eliminated. D can be eliminated because the passage states that the concentration of the ion channel was the same in each solution. Now you only have to choose between A and B. The passage states that ΔG was only obtained for those ions that exhibited binding with the K^+ channel. Only ions that bind to the channel will be able to pass through it, so identifying the characteristic that determines permeability is as simple as comparing the ions that do and do not have a recorded value of ΔG. According to Table 1, ΔG was obtained for some of the 1+ cations and some of the 2+ cations, so there is no reliable trend for charge. A can be eliminated. Table 1 also shows that only K^+ and ions with larger atomic radii than K^+ have ΔG values recorded. This demonstrates that the ion channel is specifically reacting with ions based on their size, making B the best answer. While the charge of an ion can affect ion channel permeability, this experiment demonstrated that the size of the ion was the important factor in this instance.

11. **D is the best answer.** Choice A can be eliminated because ions are hydrophilic (polar), not hydrophobic (nonpolar), and the center of the lipid bilayer is hydrophobic, not hydrophilic. Choice B is not the best answer because its ions are very small; if they were nonpolar, they would be able to pass through the membrane just as small nonpolar molecules can. Although choice C describes the structure of the cellular membrane accurately, this has nothing to do with the inability of ions to pass through. Choice D is the best answer because it accurately describes the characteristics of the ions and cellular membrane that prevent free passage of ions through the membrane.

12. **C is the best answer.** Figure 1 does not provide any evidence for or against the statements made in A and B, so these answer choices can be eliminated. According to the passage, the downward deflections in Figure 1 represent heat exchange caused by dilution of the titrant or binding of ions to KcsA. The heat exchange associated with dilution of the titrant should be the same for both KCl and NaCl, so any difference between the graphs must reflect a difference in binding of ions to the channel. Therefore, the fact that the deflections in the graph for the KCl titration are greater than those in the NaCl titration must indicate that K^+ binds to the channel while Na^+ does not. This makes choice C the best answer. Choice D is not the best answer because an ion cannot be intrinsically exothermic or endothermic.

13. **B is the best answer.** As described by the fluid mosaic model, proteins inserted into the membrane must have a hydrophobic portion that can interact with the hydrophobic lipids that make up the cell membrane. Choice A is not the best answer because K^+ channels are not synthesized in the nucleus. Just like any other protein, they are made at the rough endoplasmic reticulum. C is not the best answer because the K^+ channels themselves do not "create" the gradient, although they participate in its creation. Also, the fluid mosaic model does not address the electrochemical gradient referred to in choice C. D is not a true statement; the K^+ channels do not create energy for the cell.

Passage IV

14. **D is the best answer.** The gene pool is the collection of all alleles in a population with regards to mating. The passage states that a loss in the case of the maximum possible risk results in a loss of reproductive potential. This would mean that the individual is no longer capable of mating and passing on its genetic components within the population. The alleles of this individual are therefore lost from the pool, so choice A can be eliminated because there is a change. Choice B is misleading since speciation occurs when a new species is formed that is unable to mate with the first species. In this case, the individual is not able to mate at all, so it is not considered to be a separate species. Choice C is inaccurate since there is no indication that the individual is capable of mating with members of the "A" species. Choice D is the best answer since it acknowledges that there is a change to the gene pool due to the loss of reproductive potential of this individual.

15. **A is the best answer.** The graph shows that with a higher risk/reward ratio, the chance for survival is increased. This result is counterintuitive since it would be expected that, with a greater risk, the gamble would be progressively more detrimental. Natural selection would dictate that decreased risk would be favorable in order to minimize losses to reproductive potential. In the top line of the graph, the probability for survival is greatest across the board. The only way to make sense of the findings in the graph is to deduce that there is a heritable component, as hinted in the answer choices. What this means is that, despite the increased risk, there is the possibility of donating the rewards to the offspring to increase their reproductive potential. Choice A is the best answer since this is a case where the inheritance is greatest. Choice B is a case where the inheritance is not as significant, so the payoff of the increased risk is decreased. Choice C is a case without any inheritance, and in such a case, nature would select against increased risk given the unfavorable outcomes for reproductive fitness. Choice D is misleading given the lack of information in the graph, but it mentions inheritance specifically. There is sufficient information in the graph to reason out an inheritance pattern.

16. **D is the best answer.** Dopamine is a neurotransmitter that is released into the extracellular fluid around the cell. Dopamine is expected to be a polar molecule for this reason, and it signals by binding proteins or receptors on the plasma membrane. Only hydrophobic molecules are capable of crossing the membrane to reach intracellular targets. Of all the answer choices, only one is consistent with a hydrophobic signal molecule and not a hydrophilic one. Choices A and B are misleading because they are opposites, making it seem as though they both cannot be true, but recall that with all signaling systems, there is the possibility that a signal can have opposite effects in different cells depending on the receptor it binds. This is in fact the case with dopamine. Choice C is possible for dopamine, where it can affect the opening of channels through interactions along the plasma membrane without having to traverse the membrane. Choice D is tempting because it mentions the receptor name in the passage, indicating that dopamine may signal in this way. However, there is no indication in the passage that the receptor is in the nucleus, so this would be the best answer.

17. **C is the best answer.** This question is asking for the one factor in the list that is not essential to Hardy-Weinberg equilibrium, which is a situation where the gene pool frequencies are fixed. This is a theoretical concept, whereas in reality, changes in the gene pool over time contribute to evolution. The five factors needed to avoid evolution are: a large population size, mutational equilibrium, random mating, no immigration or emigration, and no natural selection. Choice A is discussing a large population size where there is no fear of chance events eliminating alleles, so this would be important. Choice B, or random mating, is also a prerequisite of Hardy-Weinberg equilibrium. Choice C is the best answer choice because it is too extreme in its formulation. Mutational equilibrium is where there can be some forward mutations but also some reverse mutations to balance them out. This means mutations can cause some individuals to have the traits of others in the population, while the reverse also occurs. It does not mean there cannot be any mutations. Choice D describes the condition of the absence of natural selection. Natural selection causes the equilibrium to be disrupted over time, so this choice can be eliminated.

18. **D is the best answer.** This question requires a nuanced understanding of the experimental model. The purpose of the model is to address the question of how natural selection could possibly favor risk-taking behaviors if risks create a situation by which individuals can lose their reproductive fitness. Given certain factors like the ability to confer benefits in the form of inheritance to the offspring, it is possible for such behaviors to be selected for. The question is directly asking for the particular aspect of the model that is used to allow for a comparison between the groups in this artificial setup. Choice A is true of why the setup is artificial but does not address the comparison between the species. Choice B is misleading given the information on the genes mentioned in the first paragraph of the passage. However, there is no indication that the model has anything to do with those genes. The model simply assumes that one species takes risks while the other does not. Choice C is true but is not relevant because the model can simply ignore genetic drift as it is an artificial model. This also says nothing of the comparison that is intended by the experiment. Only choice D provides an explanation of how the lack of mating between the groups sets up a way of comparing the different traits. In reality, this would not be the case since these traits are found within the same species. The model seeks to compare them, so it breaks the traits into two separate species.

Stand-alones

19. **A is the best answer.** When a cell is hypertonic to its environment, it has a higher solute concentration than the environment. This results in an influx of water and potential bursting. Choice A is a likely answer. Choice B is unlikely because shriveling occurs when the environment is hypertonic. The tonicity of the cell is not associated with cell division or presence of pathogens, so choices C and D can be eliminated, and choice A is the best answer.

20. **D is the best answer.** Tight junctions are impermeable connections that do not allow ions to flow. This makes them unlikely to be found in a tissue that requires movement of ions. Desmosomes specialize in cell-to-cell adhesion, making choice B possible but not necessarily related to the ion movement in the question. Intercellular bonds can be ruled out as a distracting answer, as they are not a major type of junction. Choice D, gap junctions, is the best answer because these gap junctions allow the ion flow that occurs in cardiac muscle cells.

21. **D is the best answer.** This question is asking how one might distinguish the bacteria from other bacteria in culture. The hint is that the name ends with "coccus," indicating that these are cocci, or spheres. The answer can be derived by a process of elimination. Choice A is misleading since a purple stain is associated with a positive Gram stain. Choice B is inaccurate since bacilli are rod-shaped. Choice C misleading since a pink stain is associated with a negative Gram stain. Choice D is the best answer since this is true of the bacteria and is also the source of their name. The hint is that they resemble beads, indicating they are spherical.

22. **B is the best answer.** Recall that Hardy-Weinberg equilibrium describes a population in which no evolution is occurring. Natural selection and non-random mating contribute to evolution by the selection and propagation of certain genes at greater frequencies than others. Emigration contributes to evolution by the introduction of new genetic information into a population. Choice B is the best answer because a large population size is necessary to protect the population from evolution caused by genetic drift.

23. **C is the best answer.** Eukaryotic cells have a membrane-bound nucleus and membrane-bound organelles, while prokaryotic cells do not. Eukaryotic DNA is coiled around histone proteins; prokaryotic DNA is "naked" and lacks histone proteins. Eukaryotic flagella are composed of microtubules while those of prokaryotic cells are made of flagellin. Choice C is the best answer because both eukaryotic and prokaryotic cells contain ribosomes (although prokaryotic ribosomes are smaller).

EXPLANATIONS TO IN-CLASS EXAM FOR LECTURE 2

Passage I

24. **D is the best answer.** The passage states that chromaffin cells are modified sympathetic postganglionic neurons, so choices A and B will have similar receptors; this similarity alone is enough to eliminate both answer choices. Also note that postganglionic neurons usually respond to acetylcholine from preganglionic neurons, not epinephrine. Parasympathetic postganglionic neurons will also respond to acetylcholine rather than epinephrine. Process of elimination leaves choice D as the best answer. You could also pick choice D by remembering that cardiac cells are innervated by both branches of the autonomic nervous system, so they must have both cholinergic and adrenergic receptors.

25. **C is the best answer.** This question seems as though it requires detailed knowledge about circadian rhythms, but you only need to know which answer choices are innervated by the sympathetic nervous system. An organ that does not receive signals from the SNS cannot be affected by a circadian rhythm of SNS signaling. Know that the diaphragm is skeletal muscle and is not innervated by the autonomic nervous system. The answer can also be found by process of elimination. Recall that the SNS signals the iris to constrict in dark environments in order to dilate the pupil (allowing you to hunt in the dark); B can be eliminated. Sweat glands and the arteries of the heart also respond to SNS signaling as part of the fight-or-flight response, so choices A and D are also not the best answers.

26. **D is the best answer.** As stated in the passage, chromaffin cells release epinephrine, an SNS neurotransmitter. A, B, and C all describe processes that are associated with the "flight-or-flight" response triggered by SNS activation. Constricted pupils would be associated with PNS activation rather than SNS activation (and therefore would occur in response to acetylcholine, rather than epinephrine). Even if you did not remember that pupil constriction occurs in response to acetylcholine rather than epinephrine, the question can be answered by process of elimination. You should associate increased heart rate, elevated blood glucose, and increased basal metabolism with epinephrine and SNS activation.

27. **C is the best answer.** The data in Figure 1 indicate that jet lag disrupts the circadian rhythm of SNS signaling, since epinephrine urine levels in Per mutant mice experiencing jet lag (the dark/dark cycle) do not follow the regular peak seen in these mice during a normal light/dark cycle. What is needed to support the hypothesis suggested in the question stem is a link between this disruption of the SNS circadian rhythm and tumor growth; the best answer will provide this link. Choice A undermines the hypothesis rather than supporting it. An oncogene promotes tumor formation, so if the normal SNS circadian rhythm activates the Myc oncogene, the disruption associated with jet lag would prevent tumor formation. B might seem tempting, but it does not provide direct evidence of a link between the SNS circadian rhythm and tumor formation (and it also refers to humans rather than the Per mutant mice featured in the experiment and question stem). Choice C is a better answer, since it implies that a disruption of the circadian rhythm of SNS signaling would lead to dysregulated cell proliferation, which is associated with tumor formation. Choice D is not as good an answer: the effects of jet lag found in this experiment could explain the finding described in choice D, but the evidence is less direct than that provided by choice C.

    As this example shows, starting with a loose prediction of what the best answer will look like can be helpful in finding the best answer choice when there are a number of choices that seem like plausible answers.

28. **B is the best answer.** This question requires reasoning about the processes that regulate the amount of neurotransmitter in synapses. It also indicates how important it is to take the time to read the question stem closely: a rushed test-taker might read "decrease" rather than "increase," and you can bet that an answer choice will be included that is opposite to the best one. Choice I is true; notice that the enzyme is associated with norepinephrine breakdown, so inhibition of this enzyme would increase the amount of neurotransmitter available. C does not contain choice I and can be eliminated. Choice II is not true; reabsorption decreases the amount of neurotransmitter in the synapse. D contains choice II, so it can be eliminated. Choice III is true; reuptake is the process of clearing the neurotransmitter from the synapse, so the inhibition of reuptake allows more neurotransmitter to remain in the synapse. A can be eliminated because it does not include choice III. This leaves choice B, the best answer.

Passage II

29. **C is the best answer.** This question is asking you to think about the difference between rods and cones. Know that rods, which are incapable of detecting color, respond to all wavelengths of light in the visible spectrum. Cones, on the other hand, are more finely tuned to smaller wavelength ranges. To properly perform this experiment in a subset of cone cells, you would need to use light with wavelengths corresponding to the range for that particular cell type. Choice A would be true for rods rather than cones, so it can be eliminated. Answer B would have the researchers using a wavelength outside of the visible spectrum. Neither rods nor cones would respond to these wavelengths, so B can also be eliminated. Nothing in the passage or lecture indicates that cones use cAMP for recovery – answer D is also not the best answer choice. This leaves only C. Using light of only one wavelength would make the experiment specific to a specific subset of cones, as long as the cones respond to that wavelength. Although C does not give you all the detail you might have expected from the best answer, it is stronger than any of the other choices.

30. **A is the best answer.** It is important to know that photoreceptors do not send action potentials and instead hyperpolarize when stimulated. This process was referenced at the beginning of the passage, making option I false. Choices B and D can be eliminated because they do not contain option I. To choose between A and C, you have to determine whether option II must be false. Based on the deficits in cGMP production seen in the passage when GCAP1 is knocked out, there is no reason to believe it is not necessary for normal recovery. The key word here is normal; i.e., the wild type phenotype. For this reason, option II is true, and the best answer is choice A. Always use this process of elimination when presented with questions in this format, but to double check yourself, quickly look at option III as well. The passage gives no indication that GCAP2 is not also necessary. It is *possible* that GCAP2 is totally redundant and unnecessary, but remember that the question is asking you for an answer that *must* be false.

31. **A is the best answer.** First you need to think about adrenergic signaling onto the iris. What is adrenergic signaling (a.k.a. epinephrine/norepinephrine signaling) associated with? The sympathetic nervous system. Recall that sympathetic activation leads to the "fight or flight" response, which includes pupil dilation. This narrows the potential best answers down to A and C. The next step is to think about what happens to the retina as the pupil dilates. More light should enter the eye, leading to more hyperpolarization of photoreceptors. This should in turn lead to more photoreceptor recovery. The passage also tells us that Ca^{2+} levels decrease under greater illumination, which should lead to more activation of RetGC and increased recovery.

32. **A is the best answer.** This is a particularly difficult question. Start by eliminating answers. Choice C tells you that the effect is due to decreased RetGC activity, but this would lead to a *decrease* in cGMP production, not an increase. C can be eliminated. Answer D is a distractor answer choice that is not relevant, as the experiment in question does not involve light or rhodopsin, and can also be eliminated. This leaves only A and B. Note that the passage told us that GCAP1 and 2 have different *intrinsic* sensitivities to calcium. In other words, the sensitivity of GCAP2 does not depend on the presence or absence of GCAP1, and should not have changed in this experiment. For this reason, B is also not the best answer. This leaves choice A. An increase in production of GCAP2 to replace the missing GCAP1 would explain an increase in cGMP production at low calcium levels. At higher concentrations of calcium the increase should be negated, since experiment 2 shows that GCAP2 is inhibited at lower calcium levels than GCAP1.

Passage III

33. **B is the best answer.** First, eliminate choices C and D because Figure 1 represents the effect of nicotine exposure on muscle fibers, not neurons. Then examine Figure 1 to determine whether muscular excitability increased or decreased. The figure shows that the experimental condition (exposure to nicotine) causes the distribution to shift to the right, while the shape of the distribution remains the same. This indicates that there is a mean increase in the resting membrane potential among the experimental group. Therefore, exposure to nicotine results in a net membrane depolarization. Although it may seem that depolarization would increase excitability, since it brings the membrane closer to the threshold potential that triggers an action potential, the passage states that hyperpolarization increases excitability. Depolarization will have the opposite effect. Choice A can be eliminated and choice B is the best answer.

34. **A is the best answer.** The passage states that the sodium-potassium pump is the only known receptor for ouabain, that Na/K-ATPase generates V_{pump}, and that the RMP is the sum of E_{Nernst} and V_{pump}. C can immediately be eliminated: by definition, the change in RMP must be due to a change in of E_{Nernst} and/or V_{pump}. If administration of ouabain eliminates any difference in RMP, the difference between the experimental and control groups in RMP seen in Experiment 1 cannot be due to a difference in the Nernst potential, which is unaffected by ouabain. B can be eliminated. The differences in RMP between the experimental and control group must result from differences in Na/K-ATPase activity, making A the best answer. You might be tempted by D, but there is no information given to indicate that experimental error is involved.

35. **C is the best answer.** Notice to start that I and II are opposite from each other, so both cannot be included in the best answer. Even if you did not understand the question, you could eliminate D. The experimental findings are an increased RMP in skeletal muscle among nicotine-exposed rats, so the conclusions must be compatible with this result. The passage states that nicotine exposure results in stimulation of the α_2 isoform of Na/K-ATPase and also states that stimulation of the sodium-potassium pump results in hyperpolarization. Since the results indicate depolarization, stimulation by nicotine must cause the isoform to be underexpressed rather than overexpressed, making option II a strong answer and eliminating option I. Since choice C is the only one that does not contain choice I, it must be the strongest answer. However, as usual, it is best to check yourself by quickly considering all of the choices. Considering option III, the sodium-potassium pump could have simply become desensitized to the stimulus due to constant stimulation. As a result, a given amount of nicotine would produce less stimulation of the enzyme and less hyperpolarization would occur, resulting in a lower potential. This too fits with the experimental findings.

36. **D is the best answer.** As the nAChR immunoprecipitates with phospholemman, we can assume that they are structurally connected. Therefore, we can conclude that phospholemman is present wherever nicotinic acetylcholine receptors are. We know that acetylcholine is a neurotransmitter and neurotransmitters are released across the synapse, from the presynaptic neuron to the postsynaptic membrane.

37. **B is the best answer.** Blocking acetylcholine receptors would result in inhibited stimulation of nAChRs. As stated in the text, stimulation of nAChRs results in hyperpolarization, so the inhibition seen in myasthenia gravis (MG) sufferers would instead lead to depolarization at the neuromuscular synapse. Depolarization is an increase in resting membrane potential, so C and D can be eliminated. Since exposure to nicotine results in an increased RMP and myasthenia gravis also results in an increased RMP, smoking would exacerbate the symptoms of myasthenia gravis rather than alleviating them. A can be eliminated.

Passage IV

38. **A is the best answer.** The number of post-synaptic glutamate channels that can open upon binding of a neurotransmitter would determine the number of ions that could flow into dendrite per unit time, meaning this would control the voltage change in the neuron. Action potentials are an "all-or-none" phenomenon, meaning that when the voltage change reaches the threshold, an action potential will occur. How depolarized a neuron becomes every time post-synaptic channels open is dependent on ion concentrations inside and outside the neuron and how many channels are open to allow the ions to flow. Changing the number of channels could change the probability of an action potential occurring, eliminating choice B. The same logic holds true for choice D. If the probability of an action potential occurring is decreased, the rate of neuron firing will also be lower, eliminating choice D. Memory is formed by a network of neurons that weave together, strengthening connections through constant signaling. A decrease in the number of action potentials between neuron pairs could have implications for memory formation, eliminating choice C. Because an action potential is an "all-or-none" phenomenon, the magnitude of the voltage change is set by the electrochemical gradient of ions, namely sodium, potassium, and to some degree calcium. If an action potential occurs, regardless of the number of receptors, the magnitude of the change will always be the same, making choice A the best answer choice.

39. **B is the best answer.** The threshold of the action potential to fire is a function of ion concentrations, primarily sodium and potassium. The threshold is set and does not depend on the myelination of the axon, eliminating choice A. Ion channels do not exist in regions that are myelinated along an axon. A return of transmembrane proteins to those regions in progressively demyelinated axons would help maintain the negative intracellular voltage. The number would be more likely to increase than decrease, eliminating choice C. Concentrations of extracellular and intracellular potassium also do not depend on the myelination levels of the axon, eliminating choice D. Myelin serves as an insulator for neuron axons, allowing the action potential to rapidly propagate down the length of an axon. In demyelinating diseases, such as multiple sclerosis, decreased myelin would decrease the speed at which an action potential could propagate down an axon, making choice B the best answer.

40. **D is the best answer.** According to Figure 1, treating cells with KA is likely to increase the expression of MHC Class I, which, according to the passage, results in a decrease in post-synaptic glutamate channels. The decrease in post-synaptic glutamate channels would decrease the flow of ions across the post-synaptic membrane. Voltage-gated calcium channels exist alongside voltage-gated sodium and potassium channels in the neuron cell membrane. With fewer glutamate channels to help depolarize the post-synaptic neuron, the current of all ions moving into the cell should decrease, not increase, eliminating choice A. Treatment of KA does not appear to alter the pre-synaptic neuron in a neuron pair, according to the passage information. Therefore, choices B and C, which describe the pre-synaptic neuron, are unrelated to KA treatment, eliminating them as possible answer choices. A decrease in glutamate channels would make it harder to depolarize the post-synaptic neuron, meaning that fewer calcium molecules would move into the neuron over time. This reasoning makes choice D the best answer.

41. **B is the best answer.** Current is defined as the movement of charge per unit time. In the case of a neuron, it is the movement of primarily sodium and potassium across the plasma membrane. Choice A is the resting membrane potential, where the sodium/potassium pump is operational. Some sodium is being pumped out of the neuron, helping to maintain the negative intracellular voltage. However, the amount of sodium moving at rest is small, making choice A less likely to be the best answer. Choice C describes the closing of the voltage-gated sodium channels and the opening of the voltage-gated potassium channels. That means no sodium is flowing, eliminating choice C. As with choice C, the voltage-gated sodium channels are closed during the time interval indicated by choice D, eliminating it as a possible answer. Choice B corresponds to the depolarization of the neurons, where the voltage-gated sodium channels are open, allowing sodium to flow down its concentration gradient into the interior of the neuron. The current is the greatest during this time, making choice B the best answer.

Stand-alones

42. **A is the best answer.** While the influx of sodium ions through a glutamate receptor contributes to the probability of the neuron firing, one local change in voltage does not always mean an action potential is created. Typically, many channels on the post-synaptic dendrite need to open simultaneously, causing a large change in voltage. Because one channel is only open in choice B, it is less likely to be the best answer. As with choice B, the release of a neurotransmitter from the pre-synaptic terminal does not guarantee that the voltage change will be large enough in the post-synaptic dendrite to cause an action potential to form, making choice C unlikely to be the best answer. Influx of calcium into the pre-synaptic terminal does not guarantee that the pre-synaptic terminal will release enough neurotransmitter to cause an action potential in the post-synaptic cell, eliminating choice D. The electric field of the axon hillock determines whether voltage-gated sodium channels will open. The individual movement of ions at different points of the dendrite are summed here, making choice A the best answer choice.

43. **C is the best answer.** Choice A corresponds to the resting membrane portion of the action potential, where sodium/potassium pumps are operating to maintain a negative intracellular voltage. Neurons also share this property, though their potential is around -70 mV due to differences in ion concentrations compared to myocytes. This eliminates choice A. Choice B is the action potential generated by sodium influx into the cell and is shared between neurons and myocytes, as this is the depolarization of the cell. Choice B can be eliminated. Choice D represents the repolarization of the membrane by the opening of potassium channels and the flow of potassium out of the cell. Both neurons and myocytes repolarize, eliminating choice D. Choice C corresponds to the plateau phase of the cardiac action potential, where sustained calcium influx maintains a positive potential. Calcium influx is significant in the myocyte compared to the neuron, as the myocyte uses calcium to trigger muscle contractions. This phase is not found in the neuron action potential, making choice C the best answer choice.

44. **A is the best answer.** Choice A describes saltatory conduction accurately. If you did not remember the exact mechanism of saltatory conduction, at least know that it allows faster movement of an action potential down the axon of a single neuron. C can be eliminated, since it refers to movement from one neuron to another. You should also remember that saltatory conduction has to do with a "jump," making A or D the most likely answer. Furthermore, saltatory conduction occurs only in myelinated axons. D can be eliminated, and A is the best answer.

45. **B is the best answer.** You should know that Na^+ channels are more sensitive than K^+ channels. That is why they open first after the threshold potential is reached, causing the influx of Na^+ ions that depolarizes the neuronal membrane. A is a true statement. Recall that when voltage-gated K^+ channels open, K^+ diffuses *out* of the cell, moving it back towards resting potential; this would not occur if K^+ concentration were not greater inside than outside the cell. C is also true. Although K^+ and Na^+ are by far the major determinants of the resting membrane potential, Cl^- has a small contribution. Finally, D is the best answer. You should know that the action potential starts at the junction between the soma and axon, which is called the axon hillock.

46. **B is the best answer.** Think of the thalamus as a "way station" for most types of sensory information, with the notable exception of the olfactory system. D can be eliminated. The thalamus does not carry out the higher-level processing described in choice C. Instead, the thalamus preserves and communicates sensory information to the cerebral cortex, which is where this type of processing does take place. C can be eliminated. Choice A applies to the *hypo*thalamus, not the thalamus. For the MCAT®, you must be aware of the major functions of both the thalamus and hypothalamus.

EXPLANATIONS TO IN-CLASS EXAM FOR LECTURE 3

Passage I

47. **B is the best answer.** Recognize cortisol as a steroid hormone. Remember that steroid hormones act by diffusing through the effector cell membrane and binding with a receptor in the cytosol. No hormones bind with receptors attached to the nuclear membrane or unattached outside of the cell, so C and D are not the best answer choices. Choice A is true for peptide and catecholamine hormones, and is also not the best answer.

48. **B is the best answer.** This is a feedback loop question – a common MCAT® question type for the endocrine system. Remember that the test usually gives you some condition (in this case, high prolactin levels) and expects you to understand how it will affect the feedback loop, not the other way around. Since the passage indicates that dopamine inhibits the release of prolactin, high prolactin levels will lead to high dopamine levels; A is not the best answer choice. The question stem explains that excess dopamine is converted into epinephrine and norepinephrine, so high dopamine levels will yield high levels of those hormones. Remember that epinephrine and norepinephrine have similar effects to the sympathetic nervous system as hormones, including raising heart rate. Answer B is the best answer choice, and C is not the best. According to the passage, another consequence of high dopamine levels should be decreased secretion of leptin, so D is also not the best answer.

49. **D is the best answer.** Figure 1B indicates that dopamine, except in high concentration, inhibits production of cAMP. From this it is reasonable to conclude that, in SW872 cells, dopamine activates G_i-proteins. Choices A and B, therefore, should be eliminated. Figure 1A indicates that dopamine inhibits prolactin release, except in the presence of a D2R antagonist. Recall that an antagonist is a substance that opposes the action of another. This suggests that dopamine typically inhibits prolactin release by binding to D2R. Choice C should be eliminated. Together, Figures 1A and 1B indicate that DA binds to D2R and activates G_i-proteins, making D the best answer choice.

50. **A is the best answer.** When you see epinephrine, heart, and gut, you should immediately think of the sympathetic nervous system. Remember that epinephrine has similar effects as a hormone. Sympathetic signaling stimulates heart muscle cells and inhibits the smooth muscle cells of the gut. Since all of the answer choices involve differences in G_s- and G_i-protein expression, the strongest answer is the one that has epinephrine stimulating heart cells through G_s-proteins and inhibiting gut cells through G_i-proteins. This is answer A. Answer B is not the best answer choice because it describes the opposite scenario. Neither C nor D makes a distinction between the two cell types in terms of G-protein expression, making them not the best answer choices. It is possible that choices B, C, or D could be strong choices with more information about downstream signaling molecules, but they are not the best answers given the available information. This is a common occurrence on the MCAT®.

51. **C is the best answer.** Remember that different effector cells respond differently to the same hormone because of variability in receptors and other molecules further down signaling cascades. The two answer choices closest to this definition are A and C. The question stem states that both liver phosphorylases are stimulated by cAMP, indicating the G_s-proteins are necessary. Choice A, which says liver cells express Gi-proteins, can be eliminated, and choice C is the best answer. Choice B is not the best answer choice because the question tells us that glucagon works on liver cells through a cAMP-dependent mechanism, which requires adenylyl cyclase. Choice D directly contradicts information in the question stem, so it cannot be the best answer.

Passage II

52. **A is the best answer.** Experiment 1 simply demonstrates that increased hCG is associated with increased expression of both leptin and cAMP in BeWo cells. D contradicts this finding and can be eliminated. The results of Experiment 1 suggest that hCG acts through a cAMP second messenger system to increase leptin expression. Experiment 2 complicates things by suggesting that at high levels, cAMP interferes with hCG stimulation of leptin expression in BeWo cells. A fits both of these findings. B can be eliminated because neither experiment 1 nor experiment 2 examines the effect of PKA activation on leptin expression. C can be eliminated because it contradicts the results of experiment 2.

53. **A is the best answer.** Use your knowledge of hormone polarity to answer this question. The passage introduces hCG as a peptide hormone. Peptide hormones are polar, and so they are unlikely to diffuse through the non-polar interior of the cellular membrane. B and D can be eliminated. Because peptide hormones are polar, they can dissolve in the polar blood and do not require protein carriers. Therefore, C can be eliminated. The best choice is A. Even if you did not notice that hCG was referred to as a peptide hormone or did not remember that peptides are polar, you could at least eliminate B. A substance that is able to diffuse through the cellular membrane must be nonpolar and therefore would not be able to dissolve in the blood.

54. **C is the best answer.** Trophoblast-derived BeWo tumor cells were most likely used because they were similar to trophoblast cells but easier to obtain and culture. A can be eliminated because hCG promotes rather than inhibits leptin expression in trophoblast and BeWo cells. B can be eliminated because the passage gives no indication that leptin deficiency promotes BeWo cell growth. D, while true, is not the best answer. D simply states that hCG activates the cAMP second messenger system in trophoblast and BeWo cells. This is the effect that hCG has on every cell to which it binds. C is a better answer because it explains why BeWo cells could be used in place of trophoblast cells.

55. **B is the best answer.** To answer this question, first refer back to Figure 2 and note that it depicts cAMP expression as a function of increasing hormone concentration. Recall that cAMP is a second messenger system commonly activated by the binding of a peptide hormone to its receptor. The best answer will be a peptide hormone. Progesterone, estrogen and cortisol are all steroid hormones. Only LH is a peptide hormone.

Passage III

56. **B is the best answer.** Androgens are male hormones like testosterone. As the passage states, testosterone inhibits GnRH. GnRH stimulates FSH and LH which are required for gamete production. Choice A is unlikely to be the best answer because, although there would be a decrease in testosterone production, the intake of exogenous androgens more than makes up for this decrease. Otherwise, there would be no point in taking androgens.

57. **D is the best answer.** The results presented in Figure 1 are to be expected given the control pathways presented in the passage. According to the passage, GnRH stimulates release of both LH and FSH. Therefore, A, which states that LH stimulates FSH production can be eliminated. LH, the passage explains, stimulates the Leydig cells to produce androgens, including testosterone, which exert negative feedback on gonadotropin release. Therefore, B, which states that LH inhibits androgen and C, which states that LH—rather than testosterone—inhibits FSH production can be eliminated. D, which states that LH stimulates androgen production, is the best choice.

58. **B is the best answer.** The passage states that LH and FSH are glycoproteins. In general, polar peptide hormones cannot pass through the non-polar interiors of cellular and nuclear membranes and, instead, bind to receptors on the cell membrane. For this reason, A, C, and D, all of which suggest that LH and FSH can pass through cellular and/or nuclear membranes, can be eliminated. By binding to a cell membrane receptor, peptide hormones can activate second messengers that translate hormone signals into intracellular activity. Therefore, B, which states that LH and FSH activate second messenger systems, is the best response.

59. **C is the best answer.** The passage states that acyline is a GnRH antagonist. An antagonist is a substance that opposes the action of another. Therefore, C, which states that acyline inhibits the activity of GnRH is the best choice. A, B, and D are false statements, and can, therefore, be eliminated.

60. **A is the best answer.** FSH blockage would prevent spermatogenesis by interfering with Sertoli cells and would not interfere with Leydig cells, which produce androgens. The others would affect testosterone production.

Passage IV

61. **D is the best answer.** Hydrophobic hormones use carrier proteins, hydrophilic are free in plasma and soluble in water. Steroid hormones are lipid based, and tend to be hydrophobic. Meanwhile, peptide hormones are made of proteins and tend to be hydrophilic. GH is a peptide hormone, while estrogen is a steroid. This means that they will most likely not use the same type of transport, ruling out choices A and B. Choice D is the best answer since it describes a steroid using a carrier protein and a peptide hormone traveling directly through the bloodstream. Choice C is the opposite of this answer, so it is unlikely.

62. **A is the best answer.** This question requires analysis of Figure 1 and noticing a few details in the passage. In the subjects that had an ovariectomy, no increase in T3 was observed as indicated by the lack of change in bar height through each time point. Exercise does not cause a change in T3 levels. According to the passage, D1 activity leads to T3 activation and secretion. There is a lack of exercised-induced T3 secretion, and a potential cause could be reduced D1 activity. Therefore, choice A is a likely answer. Choice B is not the best answer because there is not an increase in T3 secretion. Choices C and D are unlikely because the results shown in Figure 1 (lack of increased T3) could not be related to increased D1 activation.

63. **C is the best answer.** The endocrine system controls levels of hormones typically through negative feedback. When levels of a hormone are high, then the substances that cause it to increase will be less active, and the substances that cause it to decrease will be more active. According to the passage, D1 and D2 lead to increased T3 and D3 leads to decreased T3. It is unlikely that D1, D2, and D3 would change in the same way given their opposite actions. This makes choices A and B unlikely. Choice C is the best answer because in response to high T3, D1 and D2 will decrease to produce less T3, and D3 will increase to inactivate more T3. Choice D is the opposite of this answer and is unlikely.

64. **B is the best answer.** The ovaries primarily secrete estrogen and progesterone. Any choices regarding these hormones and effects that follow closely after are the best answers. LH is not made by the ovaries so choice A is not the best answer. Choice B is a better answer since estrogen would be reduced if ovaries were removed. Choices C and D are unlikely because milk production is dependent on prolactin levels. Though this may be impacted by ovarian hormones, prolactin is not made by the ovaries so choice D is not the best answer.

Stand-alones

65. **D is the best answer.** Glycogen is synthesized when glucose and therefore insulin concentrations are high. The adrenal cortex contributes to sugar metabolism with cortisol which promotes glycogenolysis, not glycogenesis. Choice A is not the best answer. Calcitonin and parathyroid hormone decrease and increase blood calcium, respectively. These are made by the thyroid and parathyroid hormone, so choice B is unlikely. The main stimulator of sympathetic actions is epinephrine. The adrenal medulla is the source of epinephrine, not the cortex. This makes choice C unlikely. Choice D is the best answer as it describes the effects of aldosterone. This hormone is secreted by the adrenal cortex and regulates sodium and potassium excretion.

66. **D is the best answer.** Peptide hormones are hormones that consist primarily of amino acids. Know that they often include carbohydrate portions, which makes option I unlikely. These hormones are secreted, and as with all secreted proteins, they are synthesized in the rough ER. Option II is false. Peptide hormones primarily bind to membrane receptors. These receptors then activate enzymes or second messengers. On the other hand, steroids can enter the cell through the membrane and then act on cytosolic or nuclear receptors. These can act directly on transcription. Option III is also unlikely, making choice D the best answer.

67. **D is the best answer.** Sperm maturation occurs in the epididymis. As this is unaffected by the surgery, sperm will still be able to mature, making choice A unlikely. The vas deferens is not involved in the pH of semen. The prostate is the organ most involved in making semen alkaline, so choice B is not the best answer. Sperm count most likely depends on the epididymis as well, as that is where the majority of growth and maturation occurs. Penetration of the egg depends on acrosomal enzymes, making choice C unlikely. Choice D is the best answer because the vas deferens is the path by which sperm is propelled into the urethra.

68. **A is the best answer.** TSH acts on the thyroid to promote synthesis of T_3 and T_4, also known as thyroxine. T_3 and thyroxine, in turn, inhibit release of TSH from the anterior pituitary. If a competitive inhibitor were to bind to TSH receptors on the thyroid, it would prevent TSH from promoting synthesis of T_3 and thyroxine. In the absence of negative feedback from T_3 and thyroxine, the anterior pituitary would continue to produce TSH, causing a rise in TSH blood levels.

69. **B is the best answer.** To answer this question, recognize that cortisol and aldosterone are both steroid hormones. Most likely, another steroid hormone will act as a substrate molecule for their synthesis. Estrogen is the only steroid hormone listed; hCG, LH, and FSH are all peptide hormones. Therefore, A, C, and D can be eliminated. B is the best answer.

EXPLANATIONS TO IN-CLASS EXAM FOR LECTURE 4

Passage I

70. **A is the best answer.** To answer this question, you must recognize that it is testing your knowledge of how different types of hormones exert their effects, rather than requiring specific knowledge about interleukins. The passage stated that interleukins are peptide hormones, so this question is really asking: "How do peptide hormones act on their effectors?" Choice A describes the mechanism used by peptide hormones, which cannot diffuse through the membrane because they are large and polar. Choice B describes the actions of steroid hormones. Choice C is not true for any type of hormone; steroid hormones can affect DNA at the transcriptional level but must first bind to a receptor in the cytosol. D is not the best answer because hormones are part of the endocrine system, not the nervous system.

71. **D is the best answer.** You should know that all blood cells arise from the same stem cells in the bone marrow. The thymus is required for maturation of T cells, but is not where they are originally produced. D is the only answer choice that is not a type of blood cell.

72. **A is the best answer.** Humoral immunity is directed against an exogenous antigen (one found outside the cell) such as fungi, bacteria, viruses, protozoans, and toxins. Cell-mediated immunity (T cells) works against infected cells, cancerous cells, skin grafts, and tissue transplants.

73. **D is the best answer.** The best answer must 1) be a feature of the study, 2) potentially limit the ability to draw conclusions, and 3) relate to the hypothesis of interest. Any answer choice that does not meet these three requirements can be eliminated. The hypothesis, as stated in the passage, is that expression of HLA in humanized mice will allow these mice to develop functioning T cells. A can be eliminated because it does not relate to this hypothesis. B can be eliminated because it is not a feature of the study described: the researchers did test the functioning of T cells in DRAG mice in the second part of the experiment. C is a feature of the experimental design, but the littermates of the DRAG mice that failed to express HLA-DR4 are more similar to the DRAG mice and thus are better suited for the control group than wild-type mice. D describes a feature of the study design: to compare the functioning of reconstituted T cells between DRAG mice and control mice, only the 40% of control mice that actually produced T cells could be used. The sample sizes were already fairly small, so decreasing the number of subjects even further limits the ability to draw conclusions by statistical analysis.

Passage II

74. **C is the best answer.** Epinephrine and norepinephrine are the principal neurotransmitters for the sympathetic nervous system, which controls the fight-or-flight response. Once triggered, the body has several physiological responses: heart rate increases, pupils dilate and bronchioles dilate, to name a few. Think about the state a person's body might be in if he or she were faced with a lion. Epinephrine or other β2AR agonists can be used during an asthma attack to dilate the bronchioles and rescue the person's ability to breathe. The data in Figure 1 shows that the presence of epinephrine in combination with IL-13 increases levels of MUC5AC gene transcription, MUC5AC protein, and intracellular mucin accumulation. Choice A is true because IL-13 can cause mucous metaplasia, as stated in the passage. It is a possible answer choice. Choice B contradicts the data shown in Figure 1. If there was a method to eliminate β2AR signaling, it would likely decrease levels of mucous metaplasia. Choice C refers to the idea that epinephrine can be used as a bronchodilator and can rescue a person during an allergic asthma attack, and it suggests that epinephrine can also be detrimental to the patient by increasing mucous metaplasia. Choice C is a strong answer choice. Choice D is not supported by the data. Without IL-13, epinephrine cannot cause much mucous metaplasia to occur. Between choice A and choice C, choice C is better supported by the data. This is because in all three parameters used to measure mucous metaplasia, the combination of epinephrine and IL-13 cause significantly higher levels of mucous metaplasia. In MUC5AC gene transcription for example, the levels with only IL-13 are almost identical to levels with neither IL-13 nor epinephrine and to levels with only epinephrine. Because the question is asking what statement it most likely true based on the study findings, choice C is the best answer.

75. **D is the best answer.** As stated in the passage, allergic asthma is a chronic inflammatory disease. A lung biopsy, or tissue removed from a patient to determine the state of their disease, would show signs of inflammation and mucous metaplasia. Choice A, mucous-secreting goblet cells, would be expected to be present in increased numbers due to mucous metaplasia. Recall that the innate immune system is involved in tissue inflammation. Both choice B and choice C are cells of the innate immune system, and their presence would be expected in a patient suffering from a chronic inflammatory disease like allergic asthma. Skeletal muscle is a voluntary muscle. Do not confuse this with smooth muscle, which is involuntary and surrounds the lung airways to facilitate bronchodilation and bronchoconstriction. Choice D would not likely be present in a lung biopsy and is the best answer.

76. **A is the best answer.** A person suffering an asthma attack will have difficulty breathing because his or her airway will be constricted. The patient would need a way to increase the amount of oxygen their body takes in through the narrowed airway. One way to increase the amount of oxygen absorbed by the blood is by giving the person oxygen. This is true due to Henry's Law. Henry's Law states that the amount of a gas that can be dissolved in solution is directly proportional to the partial pressure of gas in equilibrium with the liquid: $C = P \times$ solubility (where C is the concentration of the dissolved gas and P is the partial pressure). As the partial pressure of a gas increases, the concentration of the gas dissolved in solution also increases. Air is a mixture of gases and the partial pressure of oxygen is not as high as it would be if a person is given oxygen through a mask. Choice A is a strong answer choice because it describes how Henry's Law explains the utility of oxygen therapy. Choice B implies that oxygen through a mask would exert a force to expand the airways in a way that room air could not. Since there is no way that Henry's Law could explain how pure oxygen might exert a greater force than room air, choice B can be eliminated. Choice C also provides an explanation that cannot be considered through the concept of Henry's Law. Notice that there is no basis for this statement anywhere in the passage. Choice C can be eliminated. Choice D is tempting because it brings up the concept of solubility, which is a variable in the equation that describes Henry's Law. Recall that the solubility of a gas in a solution is an intrinsic property of the gas in relation to that liquid. In other words, oxygen will always have the same solubility in blood whether it exists as part of a mixture of air or in its pure form. What determines the concentration of oxygen in blood is the partial pressure of oxygen. Choice D can be eliminated, leaving choice A as the best answer.

77. **D is the best answer.** As described in the passage, allergic asthma results in mucous metaplasia, which ultimately decreases the effective diameter of the airways and increases resistance. The lung protects the body from particulate matter with mucous and cilia, which capture particulate matter and then propel it back up the airway, respectively. Excessive mucous would capture more particulate matter, but it could also impair the beating of the cilia. Choice A is a possible answer choice. Choice B, circulation of oxygen in the body, describes the function of the circulatory system, not a function of the lung, so it is a weak answer choice. Because the mucous builds up in the airways, the alveoli themselves are not affected. As long as the air can reach the gas exchange interface, this mechanism is unaffected by allergic asthma. Choice C is not a strong answer choice. Choice D describes the essential problem of allergic asthma. The decreased diameter of the airway prevents sufficient air from reaching the alveoli. While choice A is a possible answer choice, choice D is more closely tied to allergic asthma and more supported by the passage. For this reason, choice D is the best answer.

78. **B is the best answer.** Autoimmunity occurs when a hyperactive immune system attacks the body's own tissues. Choice A describes what autoimmunity is, not what it is a result of. Clonal selection refers to the process by which only certain immature B and T lymphocytes are allowed to mature. Clonal selection has both negative and positive selection. If a lymphocyte reacts too strongly to self-antigens displayed by MHC molecules, it undergoes apoptosis. This is called negative selection. Failure of this process would create mature lymphocytes that would be activated by self-antigens. Those lymphocytes would ultimately attack the body's own tissues. Choice B is a strong explanation for the cause of autoimmunity. If a lymphocyte cannot recognize antigens displayed by MHC molecules, it undergoes apoptosis. This is called positive selection. Failure of this process would create mature, nonreactive lymphocytes to circulate in the body. Choice C does not give a mechanism for how the immune system would begin to attack itself. Choice D, excessive inflammation, describes a symptom of what would occur in an autoimmune reaction. It does not explain how the autoimmunity began. Choice B is the best explanation for why autoimmunity would occur.

Passage III

79. **C is the best answer.** The rate of respiration is proportional to how quickly the chest wall can expand and contract. Expansion of the chest wall lowers the pressure inside the lungs, causing air to flow into the lungs and oxygenate the blood. The chest wall has a high degree of elasticity and will recoil nearly on its own to increase the pressure inside the lungs, driving out waste air with carbon dioxide. Expiration is primarily driven by elastic recoil of the chest wall. Additionally, the internal intercostal muscles cause the chest wall to decrease in volume, not the external intercostal muscles, eliminating choice A. The diaphragm is skeletal muscle, not smooth muscle. Remember that smooth muscle is entirely under the control of the autonomic nervous system. Skeletal muscle may be somatically controlled as well, as is evidenced by forced rapid or slow breathing. While the diaphragm does contract to increase the volume of the chest cavity to cause inspiration, it is not smooth muscle, eliminating choice B. Inspiration is driven by the contraction of the external intercostal, not internal intercostal muscles. The internal intercostal muscles are responsible for expiration and work to decrease the volume of the chest cavity, eliminating choice D. The medulla oblongata, via the phrenic nerve, is responsible for innervating the skeletal muscle of the diaphragm. When the diaphragm contracts, it increases the intrathoracic volume, lowering the pressure and allowing air to flow in. Thus, the breathing rate is partially determined by how quickly the medulla oblongata can control the contraction of the skeletal muscle of the diaphragm, making choice C the best answer choice.

80. **A is the best answer.** According to the question stem, the scientists used the volume of oxygen consumed during exercise to predict the concentration of oxygen in the blood that could be delivered to tissues. Overestimation of the number would mean that scientists predicted that more oxygen would be delivered than was actually transported to the tissues. The bronchi are responsible for bringing air from the outside environment and warming and humidifying it on the way to the lungs. During intense exercise, the body's demand for oxygen increases and the diameter of the bronchi would be expected to increase to allow for more air to flow into the lungs, rather than decrease. Even if the bronchi did constrict during exercise, the scientists used V_{AIR} consumed to estimate oxygen delivery, and a change in the bronchi diameter would have little effect on their estimation since gas exchange does not occur in the bronchi. Choice B can be eliminated. According to paragraph two, increased muscle mass correlates with a higher V_{AIR} consumed. More muscle mass would require a proportional increase in oxygen delivered. If scientists had neglected this information, they would have underestimated the amount of oxygen delivered. Choice C does not provide additional information that helps account for either an overestimation or underestimation of oxygen delivered, making it less likely to be the best answer. Choice D also does not provide any information that helps predict whether an over- or underestimation of oxygen concentration would have occurred. Based on the increased metabolic demands of exercise, cells would be expected to increase their rates of ATP production via the citric acid cycle and the electron transport chain. These processes result in the generation of CO_2. During exercise, it is expected that CO_2 levels would increase but this does not inform about estimated oxygen delivery, eliminating choice D. The amount of oxygen that is able to diffuse and be picked up by hemoglobin in red blood cells is determined by Fick's Law, which states that the rate of diffusion is directly proportional to the surface area and differential partial pressure across a membrane and inversely proportional to the thickness of the membrane. If the patients had increased alveolar wall thickness, diffusion would be reduced and the blood would be less oxygenated. However, the scientists predicted blood oxygen concentrations based on the volume of oxygen moved. If they failed to account for the diffusion deficit, an overestimation of the concentration of oxygen delivered would occur. Choice A is the best answer.

81. **C is the best answer.** The figures above are for two different molecules, myoglobin in choices A and B, and hemoglobin in choices C and D. The best way to distinguish between these curves is to note that the myoglobin graph does not show cooperativity and does not have a sigmoidal shape. Cooperativity occurs when the four different proteins that make up the hemoglobin molecule work together to help make loading or unloading of oxygen easier. This gives the sigmoidal shape seen in choices C and D. Additionally, myoglobin binding to oxygen is not altered in exercise, for the purposes of the MCAT®, eliminating choice A and B. Choices C and D present two different shifts of the graph, a right shift in choice C and a left shift in choice D. A left shift occurs in decreased temperatures, low partial pressures of CO_2, and increased pH. Exercise creates high temperatures, increased CO_2 due to additional ATP generation, and decreased pH, due to lactic acid build up. Choice D represents the opposite case, eliminating it as a possible answer choice. Choice C displays the right shift of the oxyhemoglobin curve with increased temperature, low pH, and a higher partial pressure of CO_2. The graph shows that it is easier to deliver oxygen to tissues, even at higher partial pressures of oxygen. Choice C is the best answer.

82. **B is the best answer.** Carbonic anhydrase is an enzyme that is responsible for catalyzing the conversion of carbon dioxide and water to bicarbonate and a hydrogen ion. Bicarbonate is soluble in the blood and serves as the main transporter of carbon dioxide from the cells to the lungs for excretion. CO_2 diffuses out of tissues and into red blood cells, where it is converted. If carbonic anhydrase were inhibited, a decrease, not an increase in the concentration of bicarbonate would be seen inside the red blood cell, eliminating choice A. Because CO_2 could not be effectively converted to bicarbonate, less bicarbonate would travel to the lungs to be excreted during expiration. This would lead to a decrease, not increase, in the partial pressure of CO_2 in the lungs, eliminating choice C. Similar to choice A, inhibition of carbonic anhydrase prevents the formation of the hydrogen ion. A decreased concentration of protons leads to an increased, not decreased intracellular pH, eliminating choice D. Remember that a higher concentration of protons gives a lower pH. The Haldane effect is when high concentrations of bicarbonate in the red blood cell are exchanged for chloride ions outside the red blood cell. The Cl^- helps balance out the H^+ to maintain electric neutrality, while dissipating the bicarbonate gradient. If less bicarbonate were generated, less chloride would need to be exchanged, leading to a decrease in the concentration of chloride inside the red blood cell. Choice B is the best answer.

83. **D is the best answer.** According to paragraph one, the interference effect occurs when aerobic training prevents effective strength training, defined by increased muscle mass. The best answer to the question will be the one that prevents the buildup of muscle. β-oxidation occurs when fatty acids are broken down to provide acetyl-CoA for the citric acid cycle to fuel the creation of ATP. While muscle can use fatty acids, β-oxidation itself is unlikely to prevent the buildup of muscle fibers, as these are made up proteins, eliminating choice A. Low intensity exercise is likely to be aerobic. During aerobic metabolism, lactic acid does not build up, eliminating choice B. While an increase in body temperature would occur during high intensity exercise, choice C does not provide a rationale as to why a delay in muscle formation would occur, making it less likely to be the best answer. Gluconeogenesis is the formation of new glucose from precursor molecules. These carbon molecules for gluconeogenesis may be derived from the breakdown of proteins into amino acids, which are them metabolized to give intermediates in gluconeogenesis. If muscle were being broken down to help fuel gluconeogenesis, it would help explain the delay in strength training. According to Figure 1, the greatest increase in muscle mass, as exemplified by an increase in V_{AIR}, is when aerobic metabolism and strength training alternate, preventing the body from needing to use proteins as a carbon source for gluconeogenesis. Choice D is the best answer choice.

Passage IV

84. **C is the best answer.** The first key to answering this question is to determine whether proteins such as PCNA and survivin would be presented on MHC class I or class II molecules. Exogenous proteins that are phagocytized by APCs are traditionally presented on MHC class II molecules. However, paragraph one states that cross-presentation occurs when these antigens are instead presented on class I molecules. The best answer will have proteins presented on a class I, not class II molecule, eliminating choices A and D. The next step is to determine whether an increased or decreased level of protein would account for reduced immune activation. T cells are activated when their T cell receptor (TCR) binds an antigen on an MHC receptor on an APC. Increasing the amount of protein available to activate the TCR would increase immune activation. However, the question stem describes the opposite. The best answer will be the one that has decreased protein, eliminating choice B and making choice C the best answer choice.

85. **C is the best answer.** While this question appears to test specific knowledge about antigen cross-presentation, it really draws on a distinction between the innate and adaptive immune systems. The innate immune system is non-specific and arrives early on, while the adaptive immune system is specific to a given antigen and mounts a sustained response. Paragraph one of the passage describes that antigen cross-presentation occurs in antigen presenting cells, such as dendritic cells. Dendritic cells phagocytize dead cells and process these proteins for display on MHC Class I in the cross-presentation pathway. Basophils are innate immune cells that mainly release chemicals like histamine to induce local inflammation. They are not responsible for antigen presentation and direct activation of T cells. Similarly, eosinophils are responsible for defending against parasitic infections and are also innate immune cells that do not directly activate T cells. Macrophages, by contrast, are phagocytic cells that ingest dead cells and present antigens both on MHC Class I and II molecules. They are the most likely to have cross-presentation and can directly activate adaptive immune system cells. The best answer will be the one that contains the innate immune system cells basophils and eosinophils, or options I and III. Option II should not be in the best answer. Choice A does not contain option III and can be eliminated, while choice B contains option II, which corresponds to macrophages that are likely to cross-present antigens. Choice D also contains option II for macrophages, eliminating it as the best answer choice. Choice C contains only the innate immune cells that are LEAST likely to cross-present, making it the best answer choice.

86. **A is the best answer.** According to paragraph two, gemcitabine is a nucleoside analog, meaning it most likely resembles DNA bases that would be incorporated into the growing DNA strand during DNA replication. Compounds that resemble the biological substrate are typically competitive antagonists, as they bind the active site of the enzyme similar to the normal biological molecule. The best answer will be the one that is a competitive inhibitor, eliminating choices B and C. MHC molecules do not display nucleotides but instead display short protein sequences. Choice D can be eliminated because gemcitabine would not be found on an MHC molecule. Competitive inhibition of DNA polymerase would prevent successful DNA replication, likely killing the cancer cells. Because it resembles a nucleoside, it likely binds in the active site, making it a competitive inhibitor instead of a non-competitive inhibitor. Choice A is the best answer.

87. **B is the best answer.** Antibodies, produced by specialized B cells called plasma cells, serve four main functions. By binding to antigens on the surface of cells, they make it easier for macrophages and natural killer cells to phagocytize foreign cells. Therefore, choice A can be eliminated. In allergies, antibodies on the surfaces of mast cells can link together, causing the local release of histamine, eliminating choice C. Agglutination is the clumping of cells together that have all been bound by antibodies. If multiple antibodies bound the surfaces of cancer cells, these collections could precipitate in the venules of the blood, eliminating choice D. While antibodies help activate the complement cascade that ultimately results in cell lysis, they are not directly responsible for perforating cancer cell walls. Choice B is the best answer.

Stand-alones

88. **C is the best answer.** Stroke volume is the volume of blood that is pumped out of the left ventricle and into the body with each heartbeat. Congestive heart failure patients experience decreased contractility of their heart muscle, which results in decreased stroke volume. Choice A and choice B can be eliminated. Because less blood is moving forward from the heart to the body, blood begins to back up. Consider the structure of the heart to determine where it backs up. Moving backwards, the blood builds up in the left atrium and then the pulmonary veins. This would increase the pressure in the pulmonary veins to the point that the pressure inside the pulmonary capillaries exceeds the fluid pressure in the lung interstitium. This pressure differential causes fluid to seep out of the capillaries and fill the lungs. Choice C describes this sequence of events. Choice D skips past the lungs and describes the blood as accumulating in the right atrium. As time passes, this may occur for patients with congestive heart failure, but fluid backed up in the right atrium does not explain why fluid would accumulate in the lungs. For this reason, choice C is the best answer.

89. **B is the best answer.** This question concerns the impact of pH on hemoglobin's capability to carry oxygen. Consider an oxygen dissociation curve to immediately rule out some answer choices. An oxygen dissociation curve plots quantity of oxygen on the x-axis and hemoglobin saturation on the y-axis. If the curve shifts left, this means that less oxygen is required to reach the same saturation as before, indicating an increase in affinity. If the curve shifts right, more oxygen is required to reach the same saturation and affinity is decreased. Choices A and D are unlikely since the two half-statements do not match each other. Injecting an acidic compound decreases the pH of the blood. The increase in concentration of protons in the blood causes a rightward shift on the curve through changes like protonation of certain amino acids on the hemoglobin. There is a rightward shift that corresponds to reduced affinity for oxygen. Choice B is the best answer and choice C is unlikely.

90. **B is the best answer.** The lymphatic system is an open system, meaning something goes in one end and out the other. In contrast, the blood flows in a closed system. Choices A, B and D all accurately describe major functions of the lymphatic system.

91. **A is the best answer.** According to the question stem, isolation and sequencing of nuclear DNA may be used to detect mutations in genes that increase a patient's risk of developing cancer. One class of genes that helps prevents cancer is tumor suppressor genes, such as *BRCA1*, which greatly increase the risk of breast and ovarian cancer in women when mutated. Macrophages are immune system cells that are responsible for clearing away dying cells and phagocytizing pathogens like bacteria and viruses. They have nuclei and likely also contain DNA from cells they have engulfed. Mutations in inherited genes would be found in their nuclear DNA, eliminating choice B. B lymphocytes are adaptive immune system cells that help produce antibodies to protect the body against viruses and bacteria. They need nuclear DNA to transcribe to create the antibody proteins, meaning their DNA would also contain germline mutations, eliminating choice C. For the MCAT®, remember that myocytes are muscle cells, do not reproduce, and may contain multiple nuclei per cell. Since they contain nuclei that could be sequenced, choice D can be eliminated. Circulating erythrocytes, also known as red blood cells, do not contain nuclei or mitochondria. Their main purpose is to transport oxygen to tissues with hemoglobin. Because they do not contain DNA that could be sequenced, erythrocytes are the LEAST likely to provide information about germline DNA mutations, making choice A the best answer.

92. **D is the best answer.** Clonal selection allows only those lymphocytes that demonstrate the ability to recognize foreign antigens and avoid responding to self-antigens to mature; those that violate either of these conditions are destroyed.

EXPLANATIONS TO IN-CLASS EXAM FOR LECTURE 5

Passage I

93. **D is the best answer.** You should know the word emulsification.

94. **B is the best answer.** You should narrow this down to A or B from the passage. The thoracic duct delivers lymph to the venous circulation from the lower part of the body and the left arm. This question requires that you know either that the thoracic duct delivers the fat to the blood, or that the lymphatic ducts empty their contents into the veins, not the arteries. Even if you haven't memorized the fact that lymphatic ducts empty into the venous system, you can figure it out by remembering that the function of the lymph system is generally to clear things out from arterial circulation, so it would not make sense for the lymph system to dump directly back in to the arteries. Choice D states how dietary fat enters the lymphatic system, not the bloodstream. This question is on the trivial side for an MCAT® question, but it is not impossible that they would ask it.

95. **A is the best answer.** Only choice A shows that serum cholesterol level alone might not indicate a health risk to the patient. The HDLs might cause a high serum cholesterol but indicate a healthy patient.

96. **C is the best answer.** Remember that the purpose of chylomicrons (and other lipoproteins) is to transport water-insoluble fats through the aqueous environment of the bloodstream. This means that chylomicrons must consist of polar fat molecules surrounded and protected by amphipathic molecules. Triglycerides are nonpolar fats and are transported in the interior of the chylomicron. Phosphatidylcholine is amphipathic, so it can contribute to the surface of the chylomicron. Phosphatidylcholine's nonpolar end faces the interior of the chylomicron while its polar end faces the aqueous environment of the bloodstream.

97. **A is the best answer.** The blood lipid levels differ between genotype categories at every time point. Answer B is unlikely to be the best answer because the experimental setup does not allow for the results to draw this conclusion. Drawing this conclusion would require a comparison of lipid levels with a control group who have neither the A nor the G allele. The present experimental setup can only answer the question of whether there is a difference between genotypes studied. For answer C, again the experimental setup does not allow this conclusion because heterozygotes are lumped in with the AG + GG group rather than being examined separately. The graph indicates that individuals carrying the G allele clear lipids more quickly, making answer choice D less likely to be the best answer.

Passage II

98. **B is the best answer.** The passage states that renal clearance is equal to GFR for a substance that is not reabsorbed or secreted. Since the question refers to a scenario in which GFR and renal clearance do not have the same value, either reabsorption of secretion must be occurring, making A or B the best answer. If clearance is greater than filtration, then the urine must be receiving an additional supply of the substance from some other source; this is consistent with secretion. Reabsorption would instead cause clearance to be lower than filtration. A can be eliminated, and B is the best answer.

99. **C is the best answer.** The passage states that clearance rate is equal to GFR only if there is no reabsorption or secretion, and the question stem reminds you that inulin does not undergo either of these processes. A suggests that inulin is secreted and B suggests that it is reabsorbed, so both A and B can be eliminated. The glomerular filtration rate (GFR) is the volume of fluid that filters into Bowman's capsule per unit time. To select between C and D, look for a restatement of the other requirement described in the passage: that the substance be freely filtered. Freely filtered does not mean completely filtered; it just means that the kidney's filtration barriers do not select for or against the substrate, so its filtrate concentration equals its plasma concentration. D states that inulin must be completely filtered, so D can be eliminated. C is the best answer, because C says that inulin's filtrate concentration must equal its plasma concentration.

100. **D is the best answer.** Since the formula for clearance includes solute concentration in the urine, urine flow rate, and solute concentration in the plasma, you can eliminate any answer choices that involve these variables. For this reason, B can be eliminated. The rate at which a substance is cleared depends, in part, on the rate at which it flows out of the glomerulus and into Bowman's capsule. Since kidney filtration barriers select against substances of a certain size and charge, you can eliminate A. Because pressure differences drive fluid flow, C can be eliminated. D is the best answer choice; the plasma concentration of a substance that is neither secreted nor absorbed should not appreciably affect the clearance rate of another solute.

101. **C is the best answer.** In Table 2, the column labeled DM-H presents the data for patients with type I diabetes mellitus and hyperfiltration. Average systolic and diastolic pressure is higher for DM-H patients than for either DM-N or HC patients, so B and D can be eliminated. By contrast, renin and ALD concentrations are lower for DM-H patients than for either DM-N or HC patients. In other words, levels of circulating RAAS mediators are suppressed. Therefore, A can be eliminated and C is the best answer choice.

Passage III

102. **C is the best answer.** Aldosterone is a steroid hormone. Steroid receptor-hormone complexes enter the nucleus and act as transcription factors that activate or repress one or more genes. In this way, steroid hormones induce new protein synthesis. Only peptide and amine hormones induce modification of existing proteins. While some steroid hormones do bind to membrane receptors that use second messenger systems, the use of signal transduction pathways, such as cAMP second messenger systems, is more commonly ascribed to peptide hormones. In general, steroid hormones are associated with slower genomic rather than rapid non-genomic responses.

103. **D is the best answer.** Intracerebroventricular infusion of entanercept, a TNF inhibitor, results in central TNF blockade. Figures 2 and 3 show that, compared to Ang II rats, Ang II + ENT rats tended to have decreased mean arterial pressure and decreased heart weight to body weight ratio.

104. **A is the best answer.** Ang II increases blood pressure by increasing vasopressin secretion, stimulating thirst, constricting blood vessels, and increasing sympathetic output to the heart and blood vessels. Because aldosterone increases the activity of the Na^+-K^+-ATPase pump, it increases both Na^+ reabsorption and K^+ excretion, and, thereby, indirectly increases blood pressure. Neither renin nor angiotensin nor aldosterone increases urine volume.

105. **B is the best answer.** Decreased delivery of sodium past the macula densa stimulates renin release by the juxtaglomerular granular cells. Renin decreases the resistance of the afferent arterioles and eventually converts angiotensinogen to angiotensin II, which, in turn, *increases* the resistance of the efferent arterioles. Both decreasing the resistance of the afferent arterioles and increasing the resistance of the efferent arterioles increases the glomerular filtration rate. The former increases both renal blood flow and glomerular filtration rate, whereas the latter decreases renal blood blow but increases glomerular filtration rate.

Passage IV

106. A is the best answer. The question asks about the digestion of proteins. The best answer will contain an enzyme that can directly or indirectly catalyze digestion of complex peptides into amino acids. Hormone-sensitive lipase is responsible for breaking down fats, especially in response to hormonal signaling like glucagon or ACTH. Because the question specifically asks about proteins, choice B can be eliminated. α-amylase breaks down sugars from polysaccharides into monosaccharides. However, the question asks about proteins, making choice C less likely to be the best answer. Chymotrypsinogen is the zymogen form of chymotrypsin and is not able to process any proteins until activated by pepsin, making choice D less likely to be the best answer. In the stomach, pepsinogen is released from chief cells. Pepsinogen is a zymogen, meaning that it is released in inactive form. The acidic environment of the stomach catalyzes the activation of pepsinogen to pepsin, which can then begin protein digestion. Other enzymes are released from the pancreas in zymogen form and require the activity of pepsin to be transformed from zymogens to active enzymes. In this way, pepsin regulates both of the breakdown proteins by activating precursor zymogens to active enzymes, making choice A the best answer.

107. B is the best answer. In Zollinger-Ellison (Z-E) syndrome, the stomach becomes increasingly acidic. The best answer is likely a molecule that stimulates acid secretion but does not stimulate other cells. Pepsinogen is the zymogen form of pepsin that is released by chief cells of the stomach. Pepsin is responsible for breaking down proteins. The question notes that Z-E syndrome does not show increased protein breakdown, eliminating choice A. Acetylcholine is responsible for increasing the secretion of all types of cells in the stomach. While this would further acidify the stomach by stimulating parietal cells, it would also stimulate chief cells, leading to a rise in protein breakdown. Because this is not seen in Z-E syndrome, choice C is less likely to be the best answer. Insulin is released from β-cells of the pancreas and helps regulate blood glucose levels. The question stem states that Z-E syndrome is a non-β-cell tumor, eliminating choice D. Gastrin is responsible for stimulating parietal cells to secrete HCl into the lumen of the stomach. Z-E syndrome patients have pancreatic tumors that secrete gastrin, which then flows through the blood to activate the parietal cells, making choice B the best answer choice. It is also important to know for the MCAT® that histamine also stimulates parietal cells to secrete HCl.

108. D is the best answer. The bacterial flora of the large intestine maintain a symbiotic relationship with the human body and are responsible for breaking down some macromolecules that human enzymes are incapable of processing. They produce vitamin K, thiamine, B12, and riboflavin, a key component in molecules like FAD. While this question appears to require in-depth knowledge of each biochemical pathway, it is best answered by considering where co-factors, such as FAD, are present in reactions. β-oxidation is the breakdown of fatty acids into acetyl-CoA, NADH, and $FADH_2$. Because a product of β-oxidation uses $FADH_2$, reduced concentrations of FAD that would result from disruption of the bacterial flora in the gut could decrease flux through the pathway. In this way, option I should be contained in the best answer choice. Similar to β-oxidation, the citric acid cycle produces an $FADH_2$ per turn of the cycle, making option II a component of the best answer choice. The electron transport chain oxidizes $FADH_2$ to FAD in complex II, meaning that a ribonucleotide is a component of this pathway and making option III a component of the best answer choice. The best answer choice will contain all options, I, II, and III, making choice D the best answer.

109. A is the best answer. The higher the bar along the Y-axis of the figure, the greater the amount of the indicated protein in the tissue sample. The phosphorylation status of Thr241 is not established in the passage, so no information is available to weigh whether or not this impacts the authors' claims, making choice B less likely to be the best answer. In most cases, increased mRNA levels correspond to increased protein levels. β-catenin levels are increased in UC patients, so it is reasonable to assume that mRNA levels may be increased as well. Choice C supports, not weakens, the authors' findings, eliminating choice C. The passage does not address how these proteins modulate the functions of the large intestine, which include water and electrolyte reabsorption. While β-catenin levels could impact water reabsorption, this does not impact the use of immunolabeling as a diagnostic tool, eliminating choice D. A requirement of the diagnostic test is that there are detectable differences among diseases that can be identified by immunolabeling. If the profile of protein expression of UC and CD is not unique, then the authors' assertion would be weakened. Choice A introduces the possibility that using the limited expression pattern may lead to a misdiagnosis of another related disease. While not directly diminishing the ability of the test to distinguish between CD and UC, the fact that an increased level of β-catenin is not unique to UC weakens the assertion that it can reliably be used as a diagnostic test. Choice A is the best answer.

110. **D is the best answer.** The large intestine is responsible for water and electrolyte absorption, in addition to supporting commensal bacterial flora and serving as a storage site for feces before elimination. The question asks for the experiment that would be LEAST likely to help elucidate how CD and UC impact large intestine functioning. Because the large intestine helps with electrolyte absorption, it would be a reasonable experiment to determine whether β-catenin would increase the transcription of channels involved in electrolyte absorption from the large intestine, eliminating choice A. Aquaporin channels are responsible for reabsorbing water. They exist in the large intestine and kidney collecting duct, among other sites, to transport water from the lumen into the epithelial cells. Choice B represents an experiment that would be reasonable to determine whether Rb impacted water reabsorption. Similar to choice A, calcium is an electrolyte that is also reabsorbed in the large intestine. An experiment that looked at Rb levels and circulating calcium could inform how well the large intestine absorbed calcium, eliminating choice C. Glucose is almost entirely absorbed in the small intestine, not the large intestine. Choice D represents an experiment that would correlate β-catenin levels with a function of the small, not large intestine, making choice D the least likely to provide information about large intestine functioning. Choice D is the best answer.

Stand-alones

111. **D is the best answer.** The pancreas serves many functions within the digestive system. The pancreas is responsible for synthesizing and releasing enzymes that aid in digestion. Trypsin and chymotrypsin are enzymes released from the pancreas that break down proteins into smaller polypeptides. Removal of the pancreas could lead to difficulty with degradation of proteins into smaller polypeptides, so choice A can be eliminated. Amylases are synthesized and released from the pancreas, as well as in the saliva, and break down polysaccharides into disaccharides. Removal of the pancreas could lead to difficulty with carbohydrate breakdown, so choice B can be eliminated. The pancreas is also responsible for secreting bicarbonate ions into the duodenum in order to neutralize the acid from the stomach. Bicarbonate works to raise the duodenal pH, so removal of the pancreas would lead to a decrease in the duodenal pH, eliminating choice C. The pancreas is responsible for releasing lipases which contribute to the degradation of fats. Lipases degrade fat, but they rely on bile to emulsify it. Bile is synthesized in the liver and released into the duodenum from the gallbladder. Assuming that the bile can still reach the duodenum, removal of the pancreas should not result in problems with fat emulsification, making choice D the best answer.

112. **B is the best answer.** The brush border, the microvilli that line the intestine, is subject to mechanical and chemical damage. Lubricating mucous can help protect the brush border from damage, and there are certain epithelial cells that serve the function of secreting this mucous. Kupffer cells serve immune functions in the liver, phagocytizing bacteria and destroying irregular erythrocytes. Since Kupffer cells do not secrete mucus, choice A is not the best answer. Goblet cells in the small intestine secrete a lubricating mucous in order to help maintain the heath of the brush border, so choice B is a strong answer. Choice C can be eliminated because G cells are sensory cells that know when the stomach is full and secrete gastrin in order to promote acid production. Parietal cells secrete hydrochloric acid in order to increase the acidity of the stomach. Hydrochloric acid would not protect the brush border from mechanical or chemical injury, eliminating choice D and leaving choice B as the best answer.

113. **B is the best answer.** This question tests the function of the stomach, as well as the functions of both the small and large intestines. Within the stomach, many major digestive processes begin to take place. The environment is highly acidic to help aid in the breakdown of macromolecules into their constituents. The stomach also serves as an area in which to mix and store food. An overactive stomach might lead to increased acid secretion or an increased amount of food mixing, but it would not directly explain why an individual may be experiencing watery diarrhea, so choice A is not the best answer. The small intestine is typically associated with the function of digestion and absorption, while the large intestine is primarily responsible for water absorption. An underactive large intestine would not absorb enough water, which could result in watery diarrhea. Over- or under-activity of the small intestine could result in a variety of nutritional problems, but neither would explain excess loss of water. Since the small intestine would not be implicated in watery diarrhea, choices C and D can be eliminated, leaving choice B as the best answer.

114. **A is the best answer.** The hormones ADH and aldosterone both work to modulate the water and salt concentrations within the filtrate and body. Aldosterone works on the distal tubule and collecting duct to promote salt transport into the body. ADH works to increase the cellular permeability to water, increasing water flow out of the tubule and back into the body. Aldosterone works on the distal tubule, not the proximal tubule, so choice A is not a strong answer. ADH works on the distal tubule, specifically on the terminal portion known as the collecting tubule. Choice B is true and can be eliminated. The collecting tubule is also sensitive to ADH, eliminating choice C. Aldosterone works on the distal tubule, so choice D is true and can be eliminated, leaving choice A as the best answer.

115. B is the best answer. Carbohydrate metabolism begins in the mouth with the breakdown of sugars by the enzyme amylase. Water reabsorption is a major function of the large intestine. The gallbladder is a storage receptacle for the bile produced by the liver. B is the only false choice because, while protein DIGESTION begins in the stomach, the small intestine is the major site for ABSORPTION of nutrients.

EXPLANATIONS TO IN-CLASS EXAM FOR LECTURE 6

Passage I

116. C is the best answer. T-tubules are invaginations of the sarcolemma that deliver the action potential directly to the sarcoplasmic reticulum along the center of each sarcomere. The function of T-tubules is implied by the passage assertion that T-tubules allow action potentials to be "delivered deep into each muscle cell."

117. B is the best answer. The patterns of force production as stimulation frequencies increase in graphs 2A and 2B are regular and non-erratic. While there is always a chance that the pattern could change if data from a larger range of frequencies were collected, extrapolating from the present data, there is no reason to believe that measurements at higher frequencies would be more informative; B is the best answer choice. Regarding answer choice A, it is reasonable to expect that young mice might be stronger than old mice. Hence, age-matching mice in different experimental groups would help prevent confounding factors (such as differential age-related strength or other age-related health problems) from influencing the outcomes of absolute force production. For choice C, both CASQ1 null and wild-type mice were studied with a small sample size (n = 8), so increasing sample size would help scientists know if their results were more generally applicable to a larger population of mice. Choice D would be helpful because the current normalization procedure (cross-sectional area normalized force) accounts for muscle thickness in force production but does not take into account muscle length, or total size.

Exam Explanations

118. A is the best answer. From the passage, the function of creatine kinase is to allow phosphocreatine to give or receive a phosphate group to or from ATP. This is what maintains ATP levels during muscle contraction. The passage states that cellular respiration is not fast enough to maintain ATP levels at a constant level (so D can be eliminated). However, muscular contractions would continue because there would still be a store of ATP available. B can be eliminated because the inhibition of the enzyme that breaks down phosphocreatine would cause a buildup of phosphocreatine rather than a decrease. The passage does not suggest any mechanism that would support answer choice C.

119. B is the best answer. The mechanism involves an integral protein of the sarcoplasmic reticulum, not the sarcolemma. The passage says that uptake is active, so it requires ATP and can occur against the concentration gradient of Ca^{2+}, meaning that A and C can be eliminated. Notice that since A and C must both be true if either is true, both could be eliminated even if you did not remember that Ca^{2+} uptake is an active process.

120. A is the best answer. Graph 2A shows that wild-type mice, in general, produce greater amounts of force than CASQ1 mice, but in graph 2B when force is normalized by cross-sectional area of the muscle, those differences mostly disappear. This is due to the fact that wild-type mice have larger muscle mass which contributes to their greater force production. Choice A could help explain why wild-type mice would have larger muscle mass. If choice B were true, the opposite effect would occur (an exaggeration of differences in the graph of normalized force). Choices C and D do not address how normalization by cross-sectional area affects force.

Passage II

121. B is the best answer. The graphs in Figure 1 and the language of the passage indicate that, in comparison to wild-type mice, $PYY^{-/-}$ mice have increased bone mineral density and bone mineral content. You should know that osteoclasts resorb bone while osteoblasts build new bone. These two processes happen simultaneously. In order to have the possibility of net bone gain in comparison to wild-type mice, $PYY^{-/-}$ mice must have decreased osteoclast activity (less breakdown), increased osteoblast activity (more build-up) or both of these conditions. The only choice in which neither condition is met is choice B – a situation in which no net bone gain is possible.

122. D is the best answer. Option I may be tempting because femoral bone mass decreases in PYYtg mice while lumbar bone increases in $PYY^{-/-}$ mice. However, there is nothing to suggest that the effects of these genotypes are sampling site-specific. In fact, whole body measurements indicate that PYY expression has a whole body effect. Therefore option I can be discarded and answer choice C can be eliminated. Option II is likely true because the passage states that the PYYtg genotype has a conditional, adult-onset phenotype while the knockout genotype presumably affects expression right away. The experiments show that, in general, $PYY^{-/-}$ mice have increased bone while PYYtg mice have decreased bone. Presumably, each genotype might cancel out the effect of the other to some degree. However, it is not possible from the information given to know which effect might be stronger or whether there are unexpected phenotypic outcomes from the interaction of the two genotypes. Option III is also true, making D the best answer choice.

123. A is the best answer. There are two pieces of MCAT®-required knowledge that will help you answer this question: 1. Parathyroid hormone is released in order to increase blood calcium levels and 2. Bone tissue is the body's major repository for storing calcium and other minerals. When osteoclasts break down bone, the calcium that was contained in the bone becomes available to the circulatory system. Parathyroid hormone increases blood calcium by breaking down bone via stimulation of osteoclasts.

124. C is the best answer. The passage states that osteoclasts are differentiated from phagocytic blood cells. Erythrocytes are also a type of blood cell and all blood cells differentiate from the same precursor. The fact that osteoblasts and osteoclasts were the focus of the passage while the other two answer choices were not discussed should have made you give careful consideration to choices A and C. Be careful, though, the best answer choice is not necessarily one that was discussed in the passage.

Passage III

125. D is the best answer. The phosphagen system, glycogen-lactic acid system, and aerobic system take place primarily in the first, second, and third time categories laid out in Table 1, so that each time category may be considered to represent a system. The values in answer choice D are in the same ratio as the values of power given in the table. Since ATP concentration remains relatively constant within the cell, it makes sense that ATP production rate would mirror muscle power for these systems.

126. C is the best answer. The body must reconstitute the phosphagen system, which requires ATP to make phosphocreatine. Oxygen is required to make ATP. The heavier than normal breathing that follows exercise for approximately 1 hour takes in oxygen to convert most of the lactic acid produced by the glycogen-lactic acid system back into glucose. (This is done principally in the liver.) Choice A can be eliminated because the time frame is too short to make new hemoglobin. B is a true statement, but it does not answer the question. D can be eliminated because the passage states that aerobic metabolism can continue indefinitely or as long as nutrients last. Nutrients are not replenished with oxygen, but with food.

127. A is the best answer. The last paragraph of the passage states that slow twitch fibers are associated with increased insulin sensitivity, so the change to fast twitch fibers would decrease insulin sensitivity, eliminating answer choice B. Answering this question requires an accurate interpretation of Table 2. Fast twitch fibers have INCREASED (not decreased) phosphocreatine concentration, decreased oxidative potential, and decreased glycogen concentration compared to slow twitch fibers. Therefore, as the ratio of fast twitch fibers increases, these traits would accompany that change.

128. C is the best answer. All three statements are true. Fast twitch (type II) fibers do contract with more force than slow twitch (type I) fibers, while slow twitch fibers have greater endurance and therefore require the replacement of more ATP through aerobic respiration during sustained periods of contraction. However, option I does not help explain why type I fibers have a greater density of mitochondria, so C is the best answer. Remember that mitochondria are involved in energy production. Option III is valid because of the connection between mitochondria and energy production through aerobic respiration. Option II is also about type I fibers requiring more energy. However, option I is about type II fibers requiring more energy and, taken out of context, would actually suggest that type II fibers would have more mitochondria than type I fibers; therefore, II does not answer the question and is not included in the best answer.

129. **A is the best answer.** The passage states that ATP levels remain nearly constant during muscle activity, so choice B can be eliminated. While glycogen can be converted to make more glucose, and glucose can be metabolized to make more ATP, the passage also states that phosphocreatine is the first substance used to maintain those stable ATP levels and is therefore the first to be depleted. Notice that you could eliminate C immediately even in the absence of other information, since glucose (choice D) would be depleted before glycogen, which is a storage molecule.

Passage IV

130. **A is the best answer.** Osteoprogenitor cells give rise to osteoblasts that build bone during puberty. Mitosis of osteoprogenitor cells would create the populations needed to allow for the lengthening of bones and can be stimulated by some hormones. Prolactin is a hormone secreted to induce the production of milk from breast tissue in women after childbirth. It does not serve to stimulate bone growth, eliminating choice B. Luteinizing Hormone (LH) would begin to be released in cycles during puberty for females. However, it serves to induce ovulation, not control bone growth, eliminating choice C. Corticotrophin Releasing Hormone (CRH) is released from the hypothalamus, causing the release of ACTH from the pituitary. ACTH has many functions in glucocorticoid homeostasis, including regulating cortisol levels, but it does not directly induce bone formation or maturation, eliminating choice D. Growth Hormone Releasing Hormone (GHRH) is released from the hypothalamus, causing release of Growth Hormone (GH) from the pituitary. GH is responsible for inducing the growth of bone and would be the most likely to stimulate mitosis of osteoprogenitor cell populations that would create the osteoblasts needed for bone elongation. Choice A is the best answer choice.

131. **C is the best answer.** The passage describes that vascular channels connect the subchondral bone and the cartilage and may facilitate the transfer of cytokines that induce OA. Lamellae are rings of mineralized bone that serve to support the strength of bones, not to directly transport nutrients, eliminating choice A. Canaliculi are small openings in compact bone that allow for the exchange of nutrients between the larger vascular systems and individual osteocytes in bone. While they do allow some exchange functioning, Haversian systems are large systems of blood vessels and lymphatic vessels that transport most nutrients and waste throughout the bone microenvironment. Choice C more similar to vascular channels than choice B is. Spongy bone is where red blood cells develop. It is not directly responsible for delivery or transport of nutrients or waste, eliminating choice D. Haversian systems deliver oxygen and other nutrients to cell populations in the bone. Additionally, they remove waste products via venules and lymphatics. They serve as the transport system of the bone, much like the vascular channels described between subchondral bone and cartilage, making choice C the best answer choice.

132. **A is the best answer.** According to Figure 1, as the dose of carnosol increases, the mRNA levels of MMP-3 decrease. Presumably, protein levels would also decrease as mRNA levels decreased. The best answer to the question will justify why expression of MMP-3 would fall in the presence of carnosol. Protein ubiquitination targets proteins for degradation by the proteasome. If carnosol prevented ubiquitination, less MMP-3 would be targeted for degradation and the levels would not decrease. This is in contrast to the information provided by the passage, eliminating choice B. Weakening of disulfide bonds between two components of the cartilage extracellular matrix, collagen and proteoglycans, would lead to more OA and is contradicted by information in the passage, eliminating choice C. Similar to the proteasome, endosomes and lysosomes are responsible for breaking down proteins, especially those that are extracellular. Endosomes and lysosomes are acidic, allowing them to cleave peptide bonds. If carnosol inhibited this process, then less MMP-3 would be broken down. This would maintain higher levels of MMP-3, as opposed to reduced levels, eliminating choice D. In order to exit the nucleus, mRNA must have a 5′ cap and a 3′ poly-A tail. The 5′ cap is on a guanine residue. If carnosol prevented formation of the 5′ cap, the levels of MMP-3 mRNA would decrease due to increased degradation of the non-capped transcript. This would support the finding seen in Figure 1, making choice A the best answer choice.

133. **C is the best answer.** Three types of joints exist and include fibrous joints that occur between bones held closely together, allowing no motion, cartilaginous joints that occur between bones held closely together by cartilage, and synovial joints that occur between bones that move relative to cartilage on the end of long bones. The best answer will be the joint that experiences the most friction and would need constant repair of cartilage. Joints of the upper rib cage are cartilaginous joints that also do not move relative to each other, eliminating choice A. Joints of the skull, teeth, and jaw are all fibrous joints and do not move relative to each other. These joints do not experience frictional forces due to their lack of motion, eliminating choices B and D. The joints of motion, including the knee and elbow, are synovial joints that allows the cartilage at the end of bones to slide past one another. Cartilage that is damaged by immune cells and the process of OA would be affected the most, due to the frictional forces generated at synovial joints. Choice C is the best answer choice.

Stand-alones

134. B is the best answer. Tendons connect muscle to bone. Muscle pulls on bone, not on other muscles.

135. B is the best answer. This question tests homeostasis as well as the effect of the endocrine system on bone and calcium. When calcium levels are too high or low in the blood, the body acts to keep it at an approximately constant level. Remember that the focus of homeostasis is often on the bloodstream. Overprescribing calcium leads to an influx of extra calcium into the body. The body will seek to maintain blood calcium concentration, not bone density, making choice A unlikely. Choice B is a good answer because the extra blood calcium will move into bone, leading to a more constant blood calcium concentration. Choice C is not the best answer because osteoclasts break down bone to release minerals and calcium into the bloodstream. Choice D is also unlikely since parathyroid hormone increases blood calcium levels. Calcitonin is its antagonistic hormone, as it decreases blood calcium concentration.

136. D is the best answer. There are several types of connective tissue in the body, of which bone and cartilage are two. Bone is living tissue containing blood and nerves. One of its main functions is the storage of yellow fat within the bone marrow. Even without all of this background knowledge, the answer choices can be narrowed to B and D. If bone were a nonliving tissue, it would not need its own blood and nerve supply, indicating that either B or D will be false.

137. B is the best answer. The compressive strength of bone comes from the hydroxyapatite crystals of calcium and phosphate that lie alongside collagen fibers. Osteoblasts synthesize this matrix, while osteoclasts break this down. The thymus is responsible for helping to mature T cells and does release a hormone thymogen to help stimulate T cell development. This does not contribute to bone formation, eliminating choice A. The adrenal glands release epinephrine, mineralocorticoids, and glucocorticoids. These hormones do not directly impact the activation of osteoblasts and osteoclasts, making choice C less likely to be the best answer. The exocrine pancreas releases enzymes that break down sugars, fats, and proteins in digested food. These enzymes do not enter the systemic circulation and would be unlikely to impact bone formation, eliminating choice D. The parathyroid gland synthesizes and secretes PTH, which indirectly activates osteoclasts to break down bone and return calcium to the circulation. Because the compressive strength of bone is determined by the amount of calcium, reducing the calcium content by breaking down bone would decrease the compressive strength, making choice B the best answer.

138. C is the best answer. Osteomalacia could potentially be caused by low calcium. The body will prioritize maintaining blood calcium at the expense of bone mineral density. Consider what could prevent the supplemented calcium from being added to bones. Calcitonin is a hormone that leads to addition of calcium to bones, making option I very likely. Options I and II are unlikely to be in the same answer since PTH and calcitonin have opposite effects. Parathyroid hormone moves calcium from bone to the bloodstream, so low levels would actually potentially increase bone calcium density. Option III is also likely because the role of vitamin D is proper calcium absorption. If calcium cannot be absorbed, then it cannot move through the bloodstream and eventually reach bone. The best answer is choice C, option I and III.

ANSWERS & EXPLANATIONS

FOR

QUESTIONS IN THE LECTURES

ANSWERS TO THE LECTURE QUESTIONS

Lecture 1	Lecture 2	Lecture 3	Lecture 4	Lecture 5	Lecture 6
1. A	25. B	49. D	73. B	97. C	121. B
2. B	26. D	50. C	74. C	98. D	122. C
3. A	27. A	51. B	75. A	99. B	123. C
4. D	28. C	52. C	76. B	100. A	124. D
5. D	29. A	53. D	77. C	101. B	125. A
6. C	30. D	54. B	78. B	102. C	126. C
7. C	31. A	55. B	79. A	103. A	127. C
8. B	32. B	56. B	80. D	104. C	128. D
9. D	33. D	57. C	81. D	105. C	129. B
10. B	34. A	58. C	82. C	106. D	130. C
11. D	35. D	59. D	83. A	107. D	131. A
12. A	36. B	60. A	84. A	108. A	132. C
13. B	37. A	61. D	85. B	109. B	133. D
14. A	38. A	62. A	86. A	110. B	134. D
15. A	39. D	63. C	87. A	111. B	135. A
16. A	40. A	64. D	88. B	112. A	136. A
17. C	41. C	65. C	89. D	113. C	137. C
18. B	42. A	66. A	90. D	114. D	138. A
19. D	43. B	67. C	91. D	115. B	139. C
20. A	44. B	68. C	92. D	116. D	140. A
21. B	45. D	69. D	93. D	117. D	141. A
22. C	46. B	70. C	94. C	118. B	142. A
23. A	47. C	71. C	95. D	119. B	143. B
24. B	48. C	72. D	96. D	120. A	144. D

EXPLANATIONS TO QUESTIONS IN LECTURE 1

1. **A is the best answer.** Retroviruses inject viral RNA into host cells. This RNA is then reverse transcribed into DNA. In DNA form, it can then be integrated into the host genome, and replicated along with host DNA, especially in the lysogenic life cycle. Once the virus is in the host, the reverse transcriptase gene is transcribed, and the enzyme can then convert viral RNA into DNA. The DNA can then be transcribed by the cell, and the resulting mRNA is translated by the host's ribosomes.

2. **B is the best answer**. The best way to answer this question is to understand the difference between positive and negative sense viruses, and the way it affects their replication cycles. Positive sense viruses have coding RNA that can immediately be transcribed by the host ribosomes. They do not require RNA-dependent RNA polymerase (RdRP) because these viruses can simply carry the gene for it and have the host translate new RdRP. This makes option II a weak answer and eliminates choices C and D. Negative sense viruses contain the complement of the coding strand. Since human cells do not normally make RNA from other RNA, there is no way for these viruses to begin replication. This makes option III a strong choice and makes choice B the best answer. Retroviruses are viruses that use an enzyme called reverse transcriptase to insert themselves into the host genome. Note that knowledge of their polarity is not required since it is only an option by itself and with option II, which can be eliminated.

3. **A is the best answer.** Most animal viruses enter host cells via receptor-mediated endocytosis. Because of this, a receptor is required to recognize the specific appropriate host cell. Unlike bacteriophages, animal viruses do not leave capsids outside host cells. Finally, no virus can reproduce independently, as a defining characteristic of viruses is reproduction within a host cell.

4. **D is the best answer.** Eubacteria have cell walls containing peptidoglycan. Bacterial cell membranes do not contain cholesterol. Finally, ribosomes are present in all cells, eukaryotic or prokaryotic.

5. **D is the best answer.** Bacteriophages can only infect those bacteria who have the specific receptor that the virus can attach to and use to enter the cell body. Viruses do not reproduce until they enter a cell. Integration of the viral genome also does not occur until after infection. Finally, bacterial cell wall penetration is the first step of infection, right after attachment to the receptor. As the question asks what occurs before infection, it is presumably also asking for the step that occurs first, and that is receptor attachment.

6. **C is the best answer.** Transformation is the process of picking up stray DNA from the environment and incorporating it. As the DNA is already on the agar, the originally sensitive bacteria most likely underwent transformation and incorporated the resistance genes. Transduction is transfer of genes via a viral vector, and while that could potentially occur here, transformation would definitely have to occur first. Sexual reproduction and conjugation involve transfer and swapping of genes between multiple bacteria, and the resistance genes are on the extracted DNA in the environment, not in a bacterium that could conjugate or do sexual reproduction.

7. **C is the best answer**. Catastrophic events will not cause significant genetic drift to a large, homogeneous (well-mixed) population. These kinds of events only significantly disrupt small and/or non-homogeneous populations. Emigration, selection, and mutation all affect the HW equilibrium, regardless of population size.

8. **B is the best answer.** Archaea are distinct from bacteria in a number of important ways. Their cell walls do not have peptidoglycan, one of many characteristics they share with eukaryotes. As prokaryotes, they still lack membrane-bound organelles. Choice A is inaccurate since these organisms are roughly the same size as bacteria and would not be expected to be any smaller. Since penicillin acts on the peptidoglycan layer of bacteria, these organisms would not be affected by the drug. Choice B is a strong answer. While it is true that archaea typically live in harsh environments, they lack mitochondria, which are membrane-bound organelles not found in prokaryotic cells. Choice C is a weak answer. While archaea may be rod-like in some cases, they are prokaryotic cells, so they lack nuclei. Their genetic material is wound together in the nucleoid region within the cytoplasm as it is for bacteria. Choice D is misleading and not a strong answer.

9. **D is the best answer.** The flagella of bacteria are made from the protein flagellin. Microtubules are a main building block of the other three organelles.

10. **B is the best answer.** The nucleolus disappearing is one of the defining events of prophase. The nucleolus is the site of rRNA transcription, not translation. It is not membrane bound and is not to be confused with the nucleoid of prokaryotes.

11. **D is the best answer.** This toxin binds to and stops the activity of ribosomes. Because ribosomes are enzymes present in both prokaryotes and eukaryotes, choices A and B are unlikely. Ribosomes in eukaryotes are found in the cytoplasm, on the rough ER, and in the mitochondria. Choice D is a better answer than choice C because it includes the rough ER instead of the smooth ER, and also includes the mitochondria instead of the peroxisome. The peroxisome is responsible for breaking down long-chain fatty acids in eukaryotic cells.

12. **A is the best answer.** The smooth ER plays a major role in detoxification, and the liver detoxifies the body, so it is most likely a liver cell. The other options do not have a significant role that would suggest more than normal smooth ER.

13. **B is the best answer.** Examine each organelle and its function to answer this question. The rough ER is involved in translation of secreted proteins and is not related to processing of alcohol. Choice A is unlikely. The smooth ER is heavily involved in detoxification, so activity in this organelle would be increased to deal with the alcohol intake. Smooth ERs in the liver process and break down ethanol. Choice B is a strong answer. The nucleolus is the site of ribosome creation. Extra ribosomes and increased capacity to generate proteins are not necessarily required for detoxification and alcohol processing. Choice C is a weak answer. Lysosomes break down old and damaged cell parts. It would not have increased activity due to alcohol presence, so choice D is unlikely to be true.

14. **A is the best answer.** The smooth endoplasmic reticulum helps to detoxify alcohol in the liver. The nucleus runs the cell and makes nucleic acids; the Golgi body packages materials for transport. The rough endoplasmic reticulum makes proteins for use outside the cell.

15. **A is the best answer.** The hydrolytic enzymes of lysosomes are activated by a low pH achieved by pumping protons into the interior. Choices B and C can be ruled out immediately, as the pH change described is the opposite of what actually occurs based on the proton movement. Finally, the enzymes in lysosomes require a low pH to be active, so choice D is unlikely.

16. **A is the best answer.** Peptide hormones are synthesized in the rough ER. Insulin and glucagon are two major peptide hormones synthesized in the pancreas, so a pancreatic cell is most likely. The other cell types do not perform any major functions that would require more than normal amounts of rough ER. Neurons do use amino acids to transmit signals. However, the amino acids are not linked together as proteins, so the rough ER is not directly involved in their production. Choice D is not the best answer.

17. **C is the best answer.** The sodium-potassium pump transports both ions against their concentration gradient. If they were moving along the gradient, a pump, and energy input, would not be necessary. The sodium-potassium pump is also a form of primary active transport. Energy is used to pump ions against their gradients.

18. **B is the best answer.** In secondary active transport, one molecule diffuses down its gradient, and this favorable change is paired with the unfavorable movement of another molecule against its gradient. This is happening here with sodium and glucose. Primary active transport uses energy, usually ATP, to pump something against its gradient. Passive and facilitated diffusion both involve molecules only going down their gradient.

19. **D is the best answer.** The main characteristics that need to be considered here are size and polarity. Carbon dioxide is a small molecule, and is nonpolar and hydrophobic. Therefore, it can go straight through the hydrophobic cell membrane. All other molecules are polar, and larger, and would most likely require some sort of transporter to enter a cell.

20. **A is the best answer.** Protein hormones bind to structures on the plasma membrane, as proteins are often large and hydrophilic, so they cannot go through the membrane easily. The only option within the cell is binding to a nuclear receptor. Typically, only steroid hormones will bind to a nuclear receptor. All other options are located on the cell membrane, and protein hormones are likely to use them.

21. **B is the best answer.** I and II are true, but desmosomes are anchored to the cytoskeleton and are stronger than tight junctions.

22. **C is the best answer.** The red blood cells would burst. The osmolarity in the cells is much higher than the pure water, with no solutes. For this reason, water would continuously flow into the cells until they burst. Cells stay the same size in an isotonic solution. They would expand in a slightly hypotonic solution, but not in a solution as hypotonic as pure water. Finally, shrinking occurs in a hypertonic solution, which is not the case for pure water.

23. **A is the best answer.** Adult neurons are not preparing for, or in the process of, cell division. Cells in the G_0 phase cease to divide, describing these neurons perfectly. All other phases listed are preparing for cell division, producing DNA, organelles, and other components necessary for mitosis.

24. **B is the best answer.** Desmosomes are used purely for cell-to-cell adhesion, and not so much for communication. Gap junctions allow molecules and ions to flow freely between the two cells, and are used for communication. Membrane and nuclear receptors also bind signaling molecules in cell-to-cell communication.

EXPLANATIONS TO QUESTIONS IN LECTURE 2

25. **B is the best answer.** A signal which is being received from another neuron is typically transmitted to the dendrites of a postsynaptic neuron. It then travels to the cell body and finally down the axon; however, synapses are found all along the neuron and a signal may begin anywhere on the neuron. Although an action potential moves in all directions along an axon, the cell body and dendrites do not normally contain enough sodium channels to conduct the action potential for any length.

26. **D is the best answer.** This question is testing your knowledge of an action potential. The major ions involved in the action potential are sodium and potassium. An action potential is an all or nothing response which is determined by depolarization caused by the opening of sodium channels. Blocking sodium channels is the only way given that would block an action potential.

27. **A is the best answer.** The sodium/potassium pump moves potassium inside the membrane. Potassium is positively charged making the inside of the membrane more positive. The resting potential is measured with respect to the inside.

28. **C is the best answer.** The Na^+/K^+ pump uses 1 ATP to move 3 sodium ions out of the cell and 2 potassium ions into the cell. Because all of these ions have a +1 charge, the net change in membrane potential is negative. This makes sense functionally, as the pump is used to maintain a negative resting membrane potential. There is no way that the pump can lead to an increase in membrane potential, making choice A unlikely. As 3 sodium ions flow out, and none ever flow in from the pump, choice B is unlikely. When only the sodium/potassium pump operates, the membrane potential will drop slowly so long as ions are available to be moved. Choice C is a strong answer. Choice D is the opposite of that answer and can be eliminated. Another potential answer is a lack of change due to deficient ATP concentration, as ATP is used to power the sodium/potassium pump. However, since this is not an answer, it is only an idea to consider, and choice C is the best out of the existing answers.

29. **A is the best answer.** Although the nervous and endocrine systems are both involved in communication throughout the body, they differ in speed and specificity. The nervous system provides fast, specific signals. Since the endocrine system involves the release of hormones into the bloodstream, its effects are both slower to occur and more generalized than those of the nervous system, usually affecting a wider range of cells.

30. **D is the best answer.** Choice A must be wrong because the resting membrane potential of a neuron is primarily dependent on the electrochemical gradient of K^+ because the cell membrane is much more permeable to it than Na^+. A change in the balance of intracellular and extracellular potassium ion concentrations must cause a change in the resting membrane potential. Choice B can be eliminated because the new resting membrane potential, and therefore the change caused by a change in intracellular K^+, can be predicted by using the Nernst equation, as mentioned in the question stem. The easiest way to choose between C and D is to recall that under normal conditions the intracellular K^+ concentration is greater than the extracellular concentration, causing a negative resting membrane potential. If the intracellular K^+ concentration increases, the resting membrane potential must become even more negative.

31. **A is the best answer.** The electrochemical gradient of K^+ is the most significant contributor to the resting membrane potential, so choices C and D can be eliminated. Note that if the membrane became impermeable to potassium, the electrochemical gradient of potassium would be disrupted; in other words, choice C would have to be the strongest answer if choice D is true. This is another reason to eliminate both answers. Leakage of Na^+ down its electrochemical gradient also contributes to the resting membrane potential, making it slightly less negative than the equilibrium potential of potassium, ruling out choice B . The electrochemical gradient of Cl^- makes only a minor contribution to the resting membrane potential compared to those of K^+ and Na^+, making choice A the strongest answer.

32. **B is the best answer.** Action potentials start with an upswing in membrane potential as a result of the opening of voltage-gated sodium channels. Excess sodium channels outside the cell flow in. Near the peak, potassium channels open, letting potassium ions out of the cell and causing the membrane potential to drop back to resting levels. Blocking potassium channels will not affect the initial jump in the action potential, making choice A unlikely. Potassium channels are heavily involved in dropping back to resting potential, so choice B is very likely. Na^+/K^+ pumps maintain resting potential by moving more sodium ions out than potassium ions in. Potassium channels play a much smaller role in this, so choice C is not likely. Saltatory conduction refers to the jumping of the action potential from one node of Ranvier to the next. The presence of myelination is the key to this process, not voltage-gated potassium channels, so choice D is unlikely and choice B is the best answer.

33. **D is the best answer.** Acetylcholinesterase is an enzyme that degrades acetylcholine. Remember that enzymes are often named by their substrate followed by the suffix '-ase'. Since acetylcholine is the substrate of this enzyme, not the product, choice C can be eliminated. Choice B can also be eliminated because acetylcholine, not the enzyme acetylcholinesterase, is what binds to the acetylcholine receptor. If this enzyme is inhibited, then acetylcholine will not be catabolized as quickly, and it will bind and release repeatedly with postsynaptic receptors.

34. **A is the best answer.** Sensory processing occurs throughout the lobes of the brain. Visual information is mainly processed in the occipital lobe. Meanwhile, the temporal lobe is primarily concerned with auditory and olfactory function. This makes option I likely to be in the best answer, so choices A and C are the strongest candidates. It makes sense that a stroke in a particular lobe would damage those senses processed in that lobe. Options II and III are both unlikely to be in the best answer, making choice A the strongest answer. The sense of touch is processed in the somatosensory cortex in the parietal lobe.

35. **D is the best answer.** All neurons only have one cell body with a single nucleus. In fact, most cells in the human body other than a few types of muscle cells only have a single nucleus. Choice A is unlikely. White matter and grey matter are distinct types of tissue. White matter consists of the myelinated axons and other processes that branch off of cell bodies. Grey matter consists of clumps of cell bodies. Neurons consist of both processes and central soma, so neither choice B nor choice C is true. Choice D is the best answer because most neurons are interneurons, or neurons that communicate from one neuron to another. These are the intermediate signal transmitters of the nervous system.

36. **B is the best answer.** The cerebellum controls finely coordinated muscular movements, such as those that occur during a dance routine. Involuntary breathing movements are controlled by the medulla oblongata. The knee-jerk reflex is governed by the spinal cord.

37. **A is the best answer.** Every type of synapse in the peripheral nervous system uses acetylcholine as its neurotransmitter except the second (the neuroeffector or postganglionic) synapse in the sympathetic nervous system. As the "effector," it can be reasoned that this synapse is the end organ synapse. An effector is an organ or a muscle, something that responds to neural innervation by making something happen in the body.

38. **A is the best answer.** Parasympathetic stimulation results in "rest and digest" responses, or responses that are not involved in immediate survival or stress. Choice B is a sympathetic response, as is choice C. Choice D is mediated by skeletal muscles, which do not receive autonomic innervation.

39. **D is the best answer.** In order to prevent conflicting contractions by antagonistic muscle groups, reflexes will often cause one muscle group to contract while also sending an inhibitory signal to its antagonistic muscle group. Motor neurons exit ventrally from the spinal cord, not dorsally, so choice A is out. Reflex arcs (at least somatic ones) are usually confined to the spinal cord; they do not require fine control by the cerebral cortex. This eliminates choice B. Reflex arcs may be integrated by an interneuron in the spinal cord. Choice C is out as well.

40. **A is the best answer.** The central nervous system is comprised of the brain and spinal cord. An effector is organ or tissue affected by a nervous impulse.

41. **C is the best answer.** Pressure waves, or sound, are converted to neural signals by hair cells in the organ of Corti in the cochlea. The retina is related to the eye and vision, and is a distractor.

42. **A is the best answer.** All of the answer choices can be converted to electrical signals by the human body that can then be transmitted to the central nervous system, but choices B, C, and D are converted by chemoreceptors, while choice A, light, is converted by electromagnetic receptors.

43. **B is the best answer.** As light enters the eye, it first passes through the clear cornea and then passes through the aqueous humor that fills the anterior chamber of the eye. The lens divides the anterior segment from the posterior segment. After the light passes through the lens, it enters the posterior segment, which is filled with vitreous fluid. After passing through the vitreous fluid, the light strikes the posterior wall of the eye, where the retina is located. The photoreceptors (rods and cones) of the retina work to convert the stimulus into an electrical signal to be sent to the CNS.

44. **B is the best answer.** One of the significant features of the olfactory system is that information about olfactory stimuli has a direct route to the amygdala and hippocampus, rather than first being processed in the cortex. Choice B is not a true statement and is the best answer. The other answer choices accurately describe features of the olfactory system. Note that, as stated in choice D, olfactory sensory information bypasses the thalamus in its path to the cortex. This aspect of the olfactory system is unique among the sensory systems and must be known for the MCAT®.

45. **D is the best answer.** This question is a straightforward test of knowledge about the different types of sensory receptors. Nociceptors contribute to the sensation of pain, making choice D the best answer. Chemoreceptors detect chemical stimuli, as in the olfactory and taste systems; mechanoreceptors detect touch and pressure; and photoreceptors respond to photons.

46. **B is the best answer.** The vestibular system contributes to the maintenance of balance, as described in choice B. Although the structures of the vestibular system are located in the inner ear, which contains structures involved in hearing, the vestibular and auditory systems have separate functions, and choice A can be eliminated. Choice C describes the functions of the somatosensory system. The detection of pheromones occurs in the olfactory system.

47. **C is the best answer.** Choice A is a weak answer because the optic nerve conveys visual information to the brain for advanced processing, rather than carrying out such processing itself. Processing of "what" and "where" information takes place in the temporal and parietal lobe, respectively. Choice B can be eliminated because the lateral geniculate nucleus is in the brain (in the thalamus) and therefore cannot be said to convey information "to the brain," although it does send information to another part of the brain (the primary visual cortex in the occipital lobe). This statement describes the optic nerve, rather than the LGN. Choice C accurately describes the role that photoreceptors play in the visual system. Choice D is wrong because hair cells are part of the auditory system, not the visual system. The function that is inaccurately assigned to hair cells in choice D is actually carried out by cones in the retina.

48. **C is the best answer.** Sensory adaptation is the adaptive phenomenon where sensory receptors become less sensitive to a repeated stimulus, facilitating the ability to detect new stimuli. Choice C accurately describes sensory adaptation. Choice A is a distractor, intended to be appealing simply because it includes the word "adaptation." Similarly, choice D misuses the word; the phenomenon described in choice D is transduction, not adaptation. Choice B describes summation, a different feature of the nervous system.

EXPLANATIONS TO QUESTIONS IN LECTURE 3

49. **D is the best answer.** Aldosterone is a steroid, just as any hormone whose name ends in "-sterone" or something similar is a steroid. This reasoning eliminates choice A, because steroid hormones do not need cell membrane receptors or second-messenger systems. They simply diffuse across the cell membrane. Choice B can be eliminated because the adrenal cortex is aldosterone's source, not its target tissue. Choice C describes how neurotransmitters act at a synapse. For the most part, endocrine hormones act by being released into the blood stream and having effects at target tissues. Epinephrine, also produced in the adrenal gland, is the exception. It has roles as both a hormone and neurotransmitter. Aldosterone exerts its effect as is written in choice D, increasing the production of sodium-potassium pump proteins.

50. **C is the best answer.** This is a negative feedback question. If another source of aldosterone exists in the body besides the adrenal cortex, negative feedback (through the renin-angiotensin system and increased blood pressure) would suppress the level of aldosterone secreted by the adrenal cortex. Answer choice A is out because the levels of renin in the blood would decrease, not increase; aldosterone release would increase blood pressure, and renin is released in response to low blood pressure. Oxytocin plays no role in blood pressure (vasopressin does) and would not be affected by this tumor. Choices C and D are both possibilities, but the adrenal cortex would definitely respond to negative feedback, while the tumor is a wild card. At this point, choice C is the best answer. In fact, this is a good choice because normally, hormone-secreting tumors will not respond to negative feedback. Because the behavior of the tumor is unpredictable, choice D can be ruled out.

51. **B is the best answer.** All hormones bind to a protein receptor, whether at the cell membrane, in the cytoplasm, or in the nucleus of the cell. Steroids and thyroxine require a transport protein to dissolve in the aqueous solution of the blood. However, protein hormones do not. Steroids are derived from cholesterol, not protein precursors. Steroids, unlike protein hormones, do not act through a second messenger system.

52. **C is the best answer.** Acetylcholine acts through a second messenger system, and is not a second messenger itself. The other choices, cAMP, cGMP, and calmodulin are all examples of second messengers.

53. **D is the best answer.** Steroids are lipid-soluble. Different steroids may have different target cells. For instance, estrogen is very selective while testosterone affects every, or nearly every, cell in the body. Steroids act at the transcription level in the nucleus, and are synthesized by the smooth endoplasmic reticulum.

54. **B is the best answer.** Exocrine function refers to enzyme delivery through a duct. Answer choices C and D refer to hormones, which are always part of the *endocrine* function of a gland, so they can be eliminated. Releasing digestive enzymes straight into the blood would be a very bad idea. The same enzymes that help digest food would have no problem digesting the cells of the body.

55. **B is the best answer.** Steroids act at the level of transcription by regulating the amount of mRNA transcribed. Because steroids diffuse into the cell and head for the nucleus, choice C can be eliminated – translation happens in the cytoplasm, not the nucleus. Although some steroid hormones have an effect on cellular growth, they do not all control replication. In a cell, reactions are typically regulated by controlling relative concentrations of the enzyme and/or inhibitors (i.e. through regulation of transcription or translation).

56. **B is the best answer.** This is another negative feedback question. T_3 and T_4 (thyroxine) production is controlled by a negative feedback mechanism involving TSH (thyroid stimulating hormone) from the anterior pituitary. If there is an exogenous source of thyroxine the body will sense the higher level and react by decreasing levels of TSH. If there is enough thyroxine around the anterior pituitary does not need to produce TSH to tell the thyroid to produce more thyroxine. Parathyroid hormone production is not be affected by thyroxine levels.

57. **C is the best answer.** Epinephrine release leads to "fight or flight" responses, as does sympathetic stimulation. Choice A is out because insulin causes cells to take up glucose. In fact, in the "flight or fight" response, it is advantageous for blood glucose levels to be higher and insulin secretion is suppressed. Choice B is out because acetylcholine is a neurotransmitter; it has few, if any, known hormonal actions. Choice D is out because aldosterone is involved in sodium reabsorption by the kidney; it has no role in "fight or flight" responses.

58. **C is the best answer.** The nervous and endocrine systems are the two systems that respond to changes in the environment. In general, the endocrine system's responses are slower to occur but last longer. Think about the difference in your reaction when you stub your toe – a nervous system response- and when you get stressed – and endocrine response.

59. **D is the best answer.** The only important thing to recognize from the question is that high insulin levels exist. Then go to the basics; insulin decreases blood glucose. Telling the liver to make more glucose through gluconeogenesis will raise the blood glucose and is the job of glucagon. Although glucagon will definitely be released in response to the dangerously low blood glucose levels, it is not the direct result of high insulin.

60. **A is the best answer.** This is an important distinction to be made. The hormones of the posterior pituitary are synthesized in the bodies of neurons in the hypothalamus, and transported down the axons of these nerves to the posterior pituitary. Remember the hormones produced in the anterior pituitary are FLAT PG. The hormones produced in the hypothalamus are regulating hormones used to control the release of other hormones from the pituitary. The kidney is the target organ of vasopressin, not where it is produced.

61. **D is the best answer.** Calcitonin builds bone mass – it "tones the bone". If there were excess levels of calcitonin this would increase bone tissue mass. Parathyroid hormone increases blood calcium levels in part by breaking down bone. Bring more sensitive to parathyroid hormone would certainly lead to a decrease in bone tissue mass. Defective intestinal calcium absorption would lead to low blood calcium levels and activation of parathyroid hormone in an attempt to restore blood calcium. Parathyroid hormone would break down bone to release the calcium into the blood. Menopause contributes to osteoporosis by reducing estrogen levels leading to diminished osteoblastic activity. Choice D has the opposite effect of any of the other choices, so it can also be picked in this way. The answer that is different from the rest will often be a strong candidate.

62. **A is the best answer.** Thyroxine (T_4) is produced by the thyroid gland. The anterior pituitary hormones are FLAT PG (FSH, LH, ACTH, TSH, Prolactin, GH). Choice A would result from mixing up thyroxine, T_4, with thyroid stimulating hormone, or TSH.

63. **C is the best answer.** Glucagon increases blood sugar, a good thing while running a marathon. An increased heart rate would be expected in someone who had just run 25 miles. Blood flow to the small intestine would be rerouted to where it was need – the heart, lungs, and leg muscles. Remember, rest and digest, the body does not want to waste blood flow for digestion while finishing the last mile of a marathon. ACTH regulates cortisol levels and the stress response, which would be expected to be higher near the end of a marathon.

64. **D is the best answer.** Parathyroid hormone is all about raising blood calcium levels. It stimulates osteoclast (bone resorption) activity, decreasing bone density. It also works in the kidney to slow calcium lost in urine. It controls blood calcium levels via these two mechanisms. Because parathyroid hormone has to do with blood calcium levels, choices A and B can be eliminated because they do not contain option III. Option I is in both remaining answers, so it does not need to be carefully considered. Afterwards, determine if parathyroid hormone impacts renal calcium reabsorption, which it does.

65. **C is the best answer.** Choice A looks good (testosterone does stimulate the testes to descend) until the question states that the person is a physically mature male. The testes normally descend during late fetal development. Choice B is out because increased testosterone would cause puberty to occur early, and would not change the timing of puberty if it has already happened (dealing with a physically mature male). Choice D is out because it is unknown whether any direct mechanism by which testosterone increases body temperature exists. More testosterone does enhance secondary sex characteristics even in a physically mature male. Think about athletes using steroids (often exogenous testosterone) to have bigger muscles.

66. **A is the best answer.** Increased secretion of estrogen sets off the luteal surge, which involves increased secretion of LH and leads to ovulation. Choice B is close, but estrogen stimulates a surge in LH, not FSH. Choice C describes what takes place when there are decreasing levels of estrogen. Progesterone is not secreted by the anterior pituitary: it is secreted by the placenta or the corpus luteum.

67. **C is the best answer.** Programmed cell death, or apoptosis, is indeed a normal part of development. Many tissues that are no longer required are destroyed throughout development. Option I is likely to be in the answer. Option II is false because the germ layers actually form during gastrulation, a process that occurs after implantation. Option III is true since HCG is the hormone that prevents degeneration of the corpus luteum. This makes choice C the best answer.

68. **C is the best answer.** Decreased progesterone secretion results from the degeneration of the corpus luteum, which occurs because fertilization of the egg and implantation did not happen. Choice A is out because thickening of the endometrial lining occurs while estrogen and progesterone levels are high, not while progesterone secretion is decreasing. Choice B can be eliminated because increased estrogen secretion causes the luteal surge, and because the luteal surge occurs earlier in the cycle. Choice D is out because while the flow phase does follow decreased progesterone secretion, it does not occur as a result of increased estrogen secretion.

69. **D is the best answer.** The layer of cilia along the inner lining of the Fallopian tubes serves to help the egg cell move towards the uterus, where it will implant if it has been fertilized. (Fertilization usually happens in the Fallopian tubes.) Choice A describes what the ciliary lining in the respiratory tract does. Choice B may sound good, but the Fallopian tubes are far enough away from the external environment that protection from its temperature fluctuations is not an issue. Choice C would seem to prevent continuation of the human race. The cilia would most likely try to improve the odds of fertilization rather than decrease it.

70. **C is the best answer.** The adrenal cortex makes many other steroid-based hormones, as well as testosterone. Again, the anterior pituitary hormones are FLAT PG. LH does stimulate the production of testosterone, but testosterone is not produced in the anterior pituitary. The pancreas produces insulin and glucagon. The adrenal medulla produces epinephrine and norepinephrine.

71. **C is the best answer.** Mammalian eggs undergo holoblastic cleavage where division occurs throughout the whole egg. At a first glance, this question appears to ask for somewhat obscure knowledge about meroblastic cleavage. However, choices A, B, and D can be eliminated because they are part of human embryonic cleavage, so it is unnecessary to know meroblastic or holoblastic cleavage.

72. **D is the best answer.** Generally, the inner lining of the respiratory and digestive tracts, and associated organs, come from the endoderm. The skin, hair, nails, eyes and central nervous system come from ectoderm. Everything else comes from the mesoderm. The gastrula is not a germ layer.

EXPLANATIONS TO QUESTIONS IN LECTURE 4

73. **B is the best answer.** Hyperventilation results in loss of CO_2, leading to lower concentrations of carbonic acid in the blood, and an increase in pH. Hypoventilation would result in the reverse. Breathing into a bag would increase the CO_2 content of the air and lead to acidosis. Excess aldosterone may lead to metabolic alkalosis due to hydrogen ion exchange in the kidney, so an insufficiency in aldosterone might lead to acidosis.

74. **C is the best answer.** The rate of gas exchange in the lungs would decrease, as carbonic anhydrase catalyzes the reaction involving carbonic acid and carbon dioxide. The carbon dioxide in the blood would not be expelled as quickly with lowered carbonic anhydrase activity. Catalysts increase the rate of a reaction. If the catalyst is inhibited, the rate decreases. Since the reaction moves in one direction in the lungs, and the opposite direction in the tissues, choice A is ambiguous. Unless one believes that a carbonic anhydrase inhibitor will affect transcription or degradation of hemoglobin, choices B and D are equivalent. If one were true, the other should also be true, so both can be ruled out.

75. **A is the best answer.** Cellular respiration produces carbon dioxide, which, in turn, lowers blood pH. Increased cardiac contractions and blood velocity would likely increase oxygen delivery to muscle tissue. During heavy exercise, capillaries dilate in order to deliver more oxygen to the active tissues. Nitrogen is irrelevant to respiration.

76. **B is the best answer.** The increased red blood cell count and hemoglobin concentration in the blood after reinjection increase the blood's ability to deliver oxygen to the tissues, often a limiting factor in endurance competitions. Choice A is wrong because, while this may happen, it is not the primary benefit of blood doping in the context of sports. Choice C is wrong because only blood cells are reinjected, so the body's hydration status is unaffected. Choice D is wrong because, while blood is less viscous with a lower concentration of red blood cells, the whole blood is being removed, not just red blood cells are being taken out. The blood viscosity does not change.

77. **C is the best answer.** Carbon dioxide is produced in the tissues. It is transported by the blood to the lungs, where it is expelled by diffusing into the alveoli. Since the concentration gradient carries CO_2 into the capillaries from the tissues and from the blood into the alveoli, it can be reasoned that there is a higher concentration of CO_2 in the tissues than in the alveoli. Choice A is wrong because blood CO_2 concentration will not change in the veins; it has nowhere to go. Choice B is wrong because CO_2 is expelled into the lungs at the pulmonary capillaries, so CO_2 that was present in the pulmonary arteries (before the capillaries) will largely be gone in the pulmonary veins (after the capillaries). Choice D describes the opposite of the concentration gradient that actually exists for CO_2 between the systemic tissues and the systemic capillaries.

78. **B is the best answer.** The person would need increased vascularity to deliver more blood to the tissues because the blood would carry less oxygen. Red blood cells carry oxygen, and an increase would be helpful in a low oxygen environment. Increased pulmonary ventilation would continuously introduce fresh oxygen into the alveoli and circulation. Finally, increased diffusing capacity increases the efficiency of gas exchange, helping to compensate for the lower oxygen content at high altitude.

79. **A is the best answer.** Constricted air passages is the key clue. The bronchioles are surrounded by smooth muscle and small enough to constrict. Cartilage does not constrict, while muscle does. The skeletal muscle in the thorax does not constrict the air passages. The alveoli are not part of the air passages.

80. **D is the best answer.** Heavy exercise manifests increased carbon dioxide production that leads to increased carboxyhemoglobin. Heavy exercise generates hydrogen ions in many ways, including by lactic acid fermentation and increased carbon dioxide in the blood. Increased metabolism contributes to increased temperature during exercise.

81. **D is the best answer.** The atrioventricular node, which sits at the junction between the atria and the ventricles, pauses for a fraction of a second before passing an impulse to the ventricles.

82. **C is the best answer.** Stroke volume must be the same for both ventricles. If it weren't, there would be a never-ending backlog of blood in one or the other circulations, ending with the faster circulation running dry. To keep the whole system running smoothly, both halves of the circulation must pump the same quantity of blood with each stroke.

83. **A is the best answer.** The cardiac action potential is spread from one cardiac muscle cell to the next via ion movement through gap junctions. Desmosomes are primarily for cell adhesion and are not as involved with signaling. Tight junctions do not allow the same movement of ions that gap junctions do. Acetylcholine is a signaling molecule, but it is not the one used to spread the cardiac action potential.

84. **A is the best answer.** Less oxygenated blood will reach the systemic system because some oxygenated blood will be shunted from the aorta to the lower pressure pulmonary arteries. The pulmonary circulation will carry blood that is more oxygenated than normal, since highly oxygenated blood from the aorta is mixing with deoxygenated blood on its way to the lungs. The entire heart will pump harder in order to compensate for less efficient oxygen.

85. **B is the best answer.** This question tests knowledge of blood flow and oxygenation. The aorta carries oxygenated blood from the left ventricle to the rest of the body, while the pulmonary artery takes deoxygenated blood from the right ventricle to the lungs. A connection between these vessels leads to a mixing of oxygenated and deoxygenated blood. Blood going to the lungs will be slightly more oxygenated than it normally would be due to aortic blood entering pulmonary circulation. Blood that enters the lungs typically has very low oxygen content so that it can maximize the gradient and uptake a lot of oxygen from alveoli. This gradient is reduced in the mixed blood, resulting in reduced oxygen uptake. Choices A and C can be eliminated and choice B is the best answer. The gradient is reduced, not increased, and oxygen uptake is subsequently reduced. There is not as clear of an effect on blood volume, so choice D is not a likely answer. In fact, red blood cell counts and blood volume could potentially increase to compensate for reduced oxygen carrying capacities.

86. **A is the best answer.** Blood loss is likely to be more rapid during arterial bleeding due the greater blood pressure in the arteries. Low oxygen intake would not affect blood volume. Lastly, excess sodium consumption would increase osmolarity of blood, increasing blood volume.

87. **A is the best answer.** The not-so-subtle physics reference is a trick. Bernoulli's equation, which would indicate a greater pressure at the greater cross-sectional area, does not work here. The blood pressure in a human is greatest in the aorta and drops until the blood gets back to the heart. The longer the blood has travelled since leaving the heart, the lower the pressure at that point.

88. **B is the best answer.** All of these factors affect the dissociation curve of O_2 from hemoglobin. Regulation of peripheral hemoglobin activity is almost always focused on lowering affinity, that is, shifting the curve to the right. This facilitates the unloading of O_2 at specific targets and tissues within the body. Choice A, the elevated concentration of CO_2 in the blood, has an allosteric effect on the hemoglobin molecule where it interacts with the N-terminal residue. It leads to increased offloading of O_2, which would shift the curve to the right, eliminating it as a possibility. Choice B, elevated pH, makes the blood more basic. Elevated pH is the opposite situation from the usual increased CO_2 leading to increased $[H^+]$, leading to increased O_2 dissociation. This lowers pH and shifts the curve to the left. Choice B presents the opposite stimulus, which leads to the opposite effect. This makes choice B a strong possibility. Choice C, 2,3-bisphosphoglycerate, is a product of metabolism and allows for more O_2 to be released in response to metabolic need. This also shifts the curve to the right, eliminating it as a possibility. Choice D, temperature, is not as commonly discussed. It makes sense that areas of the body with increased temperature would require more oxygen. Heat is produced by active muscles so heat would signal offloading of oxygen which corresponds to a right shift so choice D is not a good answer. As choices A, C, and D would shift the oxygen dissociation curve to the right, choice B remains as the only option to shift it to the left. This makes choice B the best answer.

89. **D is the best answer.** The key effect of the cancer in this question is the depletion and loss of stem cells in the bone marrow. Stem cells in the bone marrow are known to divide into red blood cells, megakaryocytes, and a variety of immune cells including neutrophils, eosinophils, and others. A reduced ability to replenish red blood cells would result in anemia, which by definition is a reduced hematocrit or red blood cell count. Option I is likely to be in the best answer. The loss of immune cell differentiation and production would lead to reduced immune function and impaired ability to combat pathogens. Option II is also likely, making choices C and D better answers than choices A and B. Megakaryocytes are gigantic cells that fragment into platelets, which are essential in clotting. This makes option III likely and choice D the best answer.

90. **D is the best answer.** Immunoglobulins, or antibodies, are involved in the humoral immune system, or the B cell system. Choice A is out because cytotoxic T cells work in cell mediated immunity. Choice B is out because stomach acid plays a role in the nonspecific innate immunity; humoral immunity is specific and acquired. Choice C is unrelated.

91. **D is the best answer.** Antibodies bind to antigens through interactions between the antibody's variable region and the antigen. Antibodies do not phagocytize anything, so choice A is out. Antibodies are produced by plasma cells, they do not normally bind to them; choice B is out. Plasma cells are derived from stem cells in the bone marrow. Antibodies will not usually prevent their production. Choice C is out.

92. **D is the best answer.** Fluid that is picked up by the lymphatic tissues is returned to the circulation at the right and left lymphatic ducts, which feed into veins in the upper portion of the chest. The rest of the options are false.

93. **D is the best answer.** T cells interact with both categories of MHCs, which like all lymphocytes undergo both positive and negative selection. Positive selection determines whether T cells can recognize MHCs and loaded peptides. Negative selection determines whether there is any association with self-peptide when presented in MHCs. T cells that have a strong interaction are directed down the apoptotic pathway. Because all T cells undergo negative selection, peptide would need to be presented in the context of both MHC classes to prevent the development of an immune response against self-antigen. This means choices A and B are both weaker than choice D because they are not as inclusive. Choice C refers to non-self antigen, which is not a component of negative selection. The recognition of non-self antigen is the function of mature, circulating T cells, eliminating this option. This makes choice D the best answer.

94. **C is the best answer.** The innate immune response does not involve humoral immunity (B cells) or cell-mediated immunity (T cells). The innate immune system responds to any and every foreign invader with the white blood cells called granulocytes, as well as with inflammation and other actions.

95. **D is the best answer.** The structure presented in the question is an antibody, a specialized protein produced by a subset of B cells called plasma cells. Antibodies serve a few functions in the adaptive immune system, including binding to antigens on the surface of cells, making it easier for macrophages and natural killer cells to phagocytize foreign cells. Antigen presenting cells process exogenous proteins and present them on MHC Class II molecules. Viral proteins found extracellularly, including capsid proteins, may be endocytosed by binding to antibodies and processed and presented on MHC Class II molecules. Option I is a potential answer choice. Antibodies also cover the surface of mast cells, specialized cells that are part of the response to allergens. Allergens induce histamine release from mast cells, which causes the endothelial cells of blood vessels to become leaky. Option II is also mediated by antibodies. Agglutination is the clumping of molecules together that have all been bound by antibodies. Clumping increases the efficiency with which macrophages and other phagocytic cells can pick up and eliminate pathogens, making option III also a function facilitated by antibodies. The best answer will contain all three options, making choice D the best answer choice. Eliminate choices A, B, and C, which do not contain all applicable descriptions of antibodies.

96. **D is the best answer.** Positive and negative selection are the two ways that T cells are selected. In positive selection, only T cells that can successfully interact with MHCs survive. In negative selection, potentially auto-immune, self-reactive T cells are selected against. Type I diabetes results from self-reactive T cells, so negative selection has failed. Choices A and B are unrelated to T cell selection.

EXPLANATIONS TO QUESTIONS IN LECTURE 5

97. **C is the best answer.** Pepsin works primarily in the stomach, and has an optimum pH around 2.0. It denatures in the environment of the small intestine, where the pH is between 6 and 7. Choice A is wrong because pepsin is not working at all in the small intestine; it won't be working synergistically with trypsin. Choice B is wrong because pepsinogen is activated in the stomach by low pH. Choice D is wrong because pepsin is a catalyst, which makes it a protein; amylase digests starch.

98. **D is the best answer.** The best answer is choice D because only choice D is both true and reveals a benefit for enzymes to be inactive while in the pancreas. Choice A may seem logical but would only apply to lipase. Choice B is false; zymogens are in inactivated precursor form. Choice C is true but is not an adequate explanation.

99. **B is the best answer.** The increase in HCl secretion following a large meal is countered by the alkaline tide, which occurs via the release of bicarbonate ions into the bloodstream. An increase in bicarbonate should decrease the concentration of free hydrogen ions in solution, eliminating choices A and C. A decrease in free hydrogen ions means the blood becomes more basic and pH increases. For this reason, choice D can be eliminated and choice B is the best answer.

100. **A is the best answer.** All macronutrients are digested through hydrolysis, or the breaking of bonds by adding water. Reduction does not necessarily take place. Glycolysis and phosphorylation are generally processes relating to metabolic, not digestive, processes.

101. **B is the best answer.** If food is moving too rapidly through the digestive tract, there will be insufficient time for proper digestion of absorption. This would lead to a lack of energy and nutrient uptake, likely causing these patients to become underweight and malnourished. This condition could lead to vitamin deficiencies in these individuals. That is consistent with the hypothesis presented in choice A, so this choice can be eliminated. The hypothesis presented in choice B directly contradicts the logic used to support choice A, and this hypothesis would be inconsistent with what is known about digestion. The rate that food moves through the digestive system is largely reliant upon hormonal communication. When the duodenum senses food, it sends hormonal signals to the stomach to prevent more food from being passed through. This allows the duodenum and the rest of the small intestine time to digest and absorb. If these hormones were down-regulated, one could imagine that there would be food passing through this system too quickly. The hypothesis in choice C is consistent with this logic and can be eliminated. The communication between stomach and duodenum is largely hormonal, so choice D presents a feasible hypothesis and can also be eliminated along with choice C. This leaves choice B as the best answer, as this hypothesis would be least consistent with what is known about digestive processes.

102. **C is the best answer.** Disaccharide digestion and absorption occurs at the microvilli (known as a "brush border") of the small intestine. Enzymes that complete digestion are incorporated into the brush border. For instance, maltase hydrolyzes maltose into two molecules of glucose; sucrose hydrolyzes sucrose into glucose and a fructose; and lactase hydrolyzes lactose into glucose and a galactose. Carbohydrate breakdown begins in the mouth with the amylase breakdown of long chain carbohydrates, so choice D can be eliminated. Proteins break down in the stomach due to the enzyme pepsin. Disaccharides are not digested until chyme (food mass leaving the stomach) enters the small intestine. Choice A can be eliminated. Finally, the large intestine (colon) is mostly responsible for water reuptake, so choice B can be eliminated.

103. **A is the best answer.** Lipid digestion occurs in the small intestine. Specifically, the pancreatic enzyme, lipase, is secreted into the duodenum. Lipid digestion occurs in the duodenum. Only proteins are digested in the stomach by the enzyme pepsin, eliminating choice B. Water and electrolyte reabsorption occurs in the large intestine, ruling out choice C. The ileum is where absorption of nutrients primarily occurs, eliminating choice D.

104. **C is the best answer.** Chemical digestion of carbohydrates begins in the mouth and chemical digestion of proteins begins in the stomach. Both the stomach and the mouth are also large contributors to physical breakdown of food. However, most chemical digestion occurs in the first part of the small intestine, the duodenum, making C the best answer. By the time food reaches the ileum, it is mostly digested. The primary function of the ileum is the absorption of nutrients from the broken-down food.

105. **C is the best answer.** While approaching this question, keep in mind that the structure of a triglyceride consists of a glycerol backbone that is linked by ester bonds to three fatty acid chains. Now start by imagining the steps that take place during lipid digestion and the order in which they occur. Upon entering the small intestine, triglycerides are exposed to bile acid and pancreatic lipase. The function of the bile is to emulsify these lipids in order to expose the ester bonds for hydrolysis by lipase. This enzyme hydrolyzes a triglyceride to produce a monoglyceride and two free fatty acid chains. The resulting monoglycerides and fatty acids aggregate to form a structure known as a micelle in order to facilitate their absorption by enterocytes. Note that one of the ways in which micelles differ from chylomicrons is where they are formed. Whereas micelles are formed in the small intestine prior to absorption, chylomicrons are assembled within the enterocytes and then secreted into the lymphatic system. It is important to remember that during the digestive process, lipids are initially transported to the lymphatic system prior to entering the bloodstream.

106. **D is the best answer.** Gluconeogenesis is the production of glycogen from non-carbohydrate precursors. This function is performed mainly in the liver. Glycolysis can be performed by any cell, eliminating choice C. Fat storage takes place in adipocytes which can be found throughout the body, making choice A a weak answer. Protein degradation occurs in all cells, which eliminates choice B.

107. **D is the best answer.** Choice A is a weak answer because fat digestates are shuttled into intestinal epithelial cells by micelles, not chylomicrons (which primarily transport fats through the lymph and blood). Most fat digestates first enter the lymph as chylomicrons via lacteals, which eliminates choice B and makes choice D the best answer. In choice C, it is true that triglycerides are degraded into fatty acids during digestion, but the second part of the answer choice is not true; smooth endoplasmic reticulum synthesizes triglycerides.

108. A is the best answer. 'Essential' means that the body cannot synthesize them. Nonessential amino acids are synthesized by the liver. Choices B and C are true because amino acids are absorbed through both facilitated and active transport. Choice D is true; urea is the end product of amino acid deamination in the liver.

109. B is the best answer. Glucose is not an electrolyte (eliminating choice C), but sodium is an electrolyte and the absorption of glucose increases the absorption of sodium because glucose is absorbed in a symport mechanism with sodium; choice B is the best answer. The focus of the question is on reestablishing water and electrolyte balance. Choices A and D are irrelevant to this goal and can be eliminated.

110. B is the best answer. The primary function of aldosterone is in the regulation of salt, so choice D is less relevant and can be eliminated. Glycogenolysis converts stored glycogen into products ready for metabolism. Cortisol and glucagon are energy mobilizers and work to increase blood glucose levels. Insulin has the opposite effect and decreases blood sugar levels in several ways. One of the ways that it accomplishes this goal is by inhibiting glycogenolysis, making choice B the best answer.

111. B is the best answer. This is a knowledge-based question. The knowledge that blood is an aqueous solution is required. This eliminates choice D and makes it apparent that either choice A or B must be the best answer. Since hydrophilic compounds easily dissolve in aqueous solutions while hydrophobic ones do not, choice A is eliminated and choice B is the best answer. To double-check, choice C can be eliminated because chylomicrons are a combination of proteins and fats and, therefore, DO bind fatty acids.

112. A is the best answer. Single amino acids are the most basic building blocks, while dipeptides, polypeptides and proteins represent increasingly large and complex structures composed of amino acids. During digestion, macromolecules are broken down into their basic nutrients, making choice A the best answer.

113. C is the best answer. The only process available for the removal of wastes by the Bowman's capsule is diffusion, aided by the hydrostatic pressure of the blood. The mechanisms mentioned in choices A, B, and D do not take place.

114. D is the best answer. Glucose is normally completely reabsorbed from the filtrate and thus does not appear in the urine, making choice B a weak answer. When glucose does appear in the urine, the glucose transporters in the PCT (not in the loop of Henle, as stated in choice A) are unable to reabsorb all of the glucose from the filtrate. Choice C is wrong because the proximal tubule does not secrete glucose.

115. B is the best answer. Choice A can be eliminated because villi do not increase the volume of filtrate that reaches the loop of Henle. The purpose of the brush border is to increase the surface area available to reabsorb solutes from the filtrate. The brush border is made from villi, not cilia, and so has no effect on the direction or rate of fluid movement, making choice C and D irrelevant. The question suggests a similar function to the brush border in the small intestine, further confirming choice B.

116. D is the best answer. Renin secretion catalyzes the conversion of angiotensin I to angiotensin II, which increases the secretion of aldosterone. If renin is blocked, then aldosterone cannot cause increased synthesis of sodium absorbing proteins, and sodium absorption decreases, making D the best answer. Under these circumstances, blood pressure would decrease, not increase; choice A can be eliminated. Platelets are irrelevant, eliminating choice B. Choice C is a weak answer because the effect of inactivating renin primarily impacts sodium reabsorption, not rate of flow through the nephron.

117. **D is the best answer.** The collecting duct is a structure which lies within the concentrated environment of the renal medulla. Due to the highly concentrated environment, the collecting duct, in response to antidiuretic hormone, allows for passive reabsorption of water. For mammals that reside in water deprived habitats, longer collecting ducts would allow for better reabsorption of water and a subsequently increased osmolarity of the excreted urine. Both reabsorption and secretion take place in the proximal tubule. This allows the kidneys to retain valuable nutrients that were inadvertently filtered out. Reabsorption in the proximal tubule can occur via passive or active transport. Water is reabsorbed into the renal interstitium of the proximal tubules across relatively permeable tight junctions by osmosis. However, the overall osmolarity of the filtrate is unchanged as both solutes and water are reabsorbed. For a mammal in a water-deprived habitat, reabsorbing more water and increasing the osmolarity of the filtrate would be required. The same principle applies to the distal convoluted tubule. The Bowman's capsule plays no role in secretion or reabsorption.

118. **B is the best answer.** Vasopressin is antidiuretic hormone, which will rise in response to dehydration increasing water retention. Aldosterone is a mineralocorticoid released by the adrenal cortex in response to low blood pressure. In a severely dehydrated person, blood volume would be low, likely resulting in diminished blood pressure. Aldosterone levels would rise in response to low blood pressure.

119. **B is the best answer.** Under the circumstances described, high blood pressure would result in more fluid being forced into Bowman's capsule, making choice B the best answer. There is no reason to assume that the increased pressure would result in constriction of the fenestrations; choice A is not a strong answer. Filtrate volume could be expected to be larger (not smaller) due to increased fluid pressure, so choice C is false and can be eliminated. The question stem does not provide insight into the solute concentration of the person in question, so choice D can be eliminated.

120. **A is the best answer.** While approaching this question, keep in mind that even though aldosterone achieves various functions by regulating fluid volume, this question focuses on the specific role that aldosterone plays at the level of nephron. Aldosterone, which is released from the adrenal cortex, acts on the nephrons of the kidneys to increase the reabsorption of sodium and excretion of potassium. In the situation described in the question stem, aldosterone levels are elevated. With more aldosterone acting on the nephrons to reabsorb sodium and excrete potassium, plasma sodium levels will be expected to increase and potassium levels to decrease. Since aldosterone has an opposite effect on sodium and potassium transport, it would not simultaneously increase (or, alternatively, decrease) the levels of both sodium and potassium.

EXPLANATIONS TO QUESTIONS IN LECTURE 6

121. **B is the best answer.** This question requires an understanding of sarcomeres and skeletal muscle contraction. A sarcomere is defined as the section between two Z-lines. The Z-line is marked by a dark stain of proteins that bisects the I-band of skeletal muscle. The I-band is the area of thin filaments that is not superimposed by thick filaments. The A-band contains the length of the entire thick filament in the sarcomere. The H-zone is the area of thick filaments that are not superimposed by thin filaments. The M-line is the dark band in the center of the H band. During contraction, the I-band and H-zone shorten, causing Z-lines to come closer to one another. Options III and IV can be eliminated. The A-band's length stays constant throughout contraction, ruling out option I. Option II is the only accurate statement, making choice B the best answer.

122. **C is the best answer.** Permanent sequestering of calcium in the sarcoplasmic reticulum would prevent calcium from binding to troponin, which is what causes the conformational change that moves tropomyosin away from the myosin binding sites on actin. Choice A would occur if calcium were present and ATP were not. Loss of ATP would prevent the myosin from releasing from actin, resulting in rigor mortis-like conditions. Choice B can be eliminated because resorption of calcium from bone would result in a decrease in bone density, not an increase. Choice D is a weak answer because loss of calcium would not cause depolymerization of actin filaments. If this actually occurred, it would pose a serious problem every time calcium was re-sequestered into the SR after the completion of a contraction.

123. C is the best answer. The question requires an understanding of excitation-contraction coupling. Choice C is not accurate because calcium binds to troponin, not tropomyosin, in order to allow the myosin cross-bridges to bind to the actin filaments. Choice A can be eliminated because the T-tubules transfer the depolarization to the sarcoplasmic reticulum. Choice B can be ruled out because the sarcoplasmic reticulum serves as a storage reservoir for calcium. Choice D is a weak answer because excitation-contraction coupling utilizes voltage-gated calcium channels.

124. D is the best answer. Muscles cause movement by contracting, not lengthening, eliminating choices B and C. The contraction brings the origin and insertion closer together, usually by moving the insertion. Neurons cause contractions in muscles, not tendons, and the neural signals are not initiated by the muscle. This eliminates choice A.

125. A is the best answer. Slow oxidative fibers fatigue slowly because they undergo aerobic respiration and while they do not contract quickly, they are the last type of muscles to fatigue. That combined with the large amount of myoglobin (making this fiber type red) makes this type of muscle ideal for long-term running. Fast oxidative fibers definitely contribute to running, but they fatigue faster than slow fibers and would not be the best fiber suited for the job. Slow glycolytic fibers are not one of the three most common muscle fiber types. Fast glycolytic fibers would be a poor choice since they contract and fatigue quickly.

126. C is the best answer. All muscles will need more energy, which can be obtained through the metabolic processes of glycolysis and the citric acid cycle, as well as increased protein if they are being used rigorously, making choices A, B, and D accurate. However, in humans, mature skeletal muscle cells usually do not divide, so no mitosis occurs.

127. C is the best answer. Choices A, B, and D are true: third class levers require more force to perform a certain amount of work than if no lever were present due to the fact that the force of the muscle acts between the fulcrum and the load force. The need for greater force to perform smaller movements increases the body's capacity for precision movement (as opposed to the opposite – small forces producing huge movements). The arrangement of muscles in order to act as third class levers actually reduces the bulk of the body making choice C false and the best answer.

128. D is the best answer. Choices A, B, and C are all functions of skeletal muscle. Muscular contraction can assist in venous blood movement and lymph fluid movement. Shivering is an example of temperature regulation by skeletal muscle. Skeletal muscles help to hold the body erect and maintain posture. Peristalsis refers to the muscular contraction which pushes partially digested food through the digestive tract; it is accomplished through the action of smooth (not skeletal) muscle.

129. B is the best answer. Gap junctions provide direct contact between adjacent cardiac muscle cells and allow for the rapid spread of the action potential throughout the heart. The gap junctions do not anchor muscle fibers together (a task which is accomplished by other types of cardiac cell-cell junctions); nor do they control capillaries or the release of calcium.

130. C is the best answer. The key to answering this question is the knowledge that skeletal muscle is innervated by the somatic nervous system while smooth muscle is innervated by the autonomic nervous system. The muscles of the leg for the knee-jerk reaction and the diaphragm are all skeletal muscles. Therefore, choices A and D can be eliminated outright. An action potential in cardiac muscle is conducted from cell to cell by gap junctions, making choice B a weak answer. Peristalsis is a smooth-muscle activity, and is, therefore, controlled by the autonomic nervous system, making C the best answer.

131. A is the best answer. The heart requires long, steady contractions in order to pump blood effectively, so the longer amount of time spent in a state of depolarization prevents the occurrence of contractions in quick succession. Adjacent cardiac cells must contract together to produced coordinated and effective pumping: choice B can be eliminated. Neurons are not the focus of the question, and the answer choice about neurons is not logical. Faster repolarization of the neuron would allow for more rapid firing so choice C is eliminated. Sodium flows into the cell, not out. Therefore, choice D can be eliminated as well.

132. **C is the best answer.** Smooth muscle contains thick and thin filaments, so it requires calcium to contract, making choice C the best answer. Choices A, B, and D are all true regarding smooth muscle, which tends to have slower, lengthier contractions. Smooth muscle is also under involuntary control through the autonomic nervous system. Under some conditions, changes in the chemical environment can result in contraction of smooth muscles cells.

133. **D is the best answer.** This question requires an understanding of acetylcholine and its effects on different types of tissues. Acetylcholine is a neurotransmitter used in the peripheral and central nervous systems. In the peripheral nervous system, acetylcholine serves as a neurotransmitter between motor neurons and skeletal muscle. In the central nervous system, acetylcholine is found in interneurons. Choice A is true and can be eliminated. In smooth and cardiac muscle, it binds to muscarinic receptors, which form G-protein receptor complexes. Choice B is true and can be eliminated. In skeletal muscle, acetylcholine activates the postsynaptic cell. Choice C is true and can be eliminated. In cardiac muscle, acetylcholine serves to decrease heart rate because it creates an inhibitory postsynaptic potential. Choice D is false and is the best answer.

134. **D is the best answer.** Choice A is a tempting answer because cardiac muscle is under the control of the autonomic nervous system and parasympathetic innervation is part of the autonomic nervous system. However parasympathetic input does not excite the cardiac muscle, but rather represses heart rate. T-tubules are invaginations of the sarcolemma and their constriction does not result in cardiac muscle excitation. All muscles contract in response to increased cytosolic calcium concentration, which eliminates choice C and makes choice D the best answer.

135. **A is the best answer.** Choices B and C are not logical and can be eliminated. To decide between choices A and D, it is necessary to know that the vagus nerve is a parasympathetic nerve, which slows the heart rate from the pace naturally set by the SA node. Without inhibitory input from the vagus nerve, the heart would normally beat at 100 – 120 beats per minute.

136. **A is the best answer.** The word dilation here is the give-away. Smooth muscle in the walls of blood vessels must be relaxing in order to allow dilation. While some of the other answer choices could result in greater blood flow reaching the extremities, none of the other choices (paralysis of skeletal muscle, tachycardia, or blood shunting) necessarily result in vasodilation as indicated in the question stem.

137. **C is the best answer.** Parathyroid hormone is released when blood calcium levels are low and need to be increased. Choice C is the best answer because osteoclast activity breaks down bone tissue and releases the stored calcium back into the blood. Therefore, need for higher blood calcium levels increases osteoclast activity and osteoclast count. An increased number of osteoblast would have the opposite of the desired effect, making choice D a weak answer. Increased osteocyte and osteoprogenitor numbers would have minimal effects on blood calcium concentrations, making choice C the best answer.

138. **A is the best answer.** Synovial fluid acts as a lubricant, decreasing friction between the ends of the bones as they move. Bone cells receive circulation to keep them adequately hydrated; they do not need synovial fluid for this purpose, eliminating choice B. Synovial fluid persists in adults, after bones have stopped growing, so choice C must be wrong. Synovial fluid is found in synovial joints, which allow movement. Its purpose is not to prevent movement. Choice D can be eliminated.

139. **C is the best answer.** Of the tissues listed, only cartilage does not contain nerves. Therefore, the cutting of cartilage would result in much less pain than cutting skin, muscle, or bone.

140. **A is the best answer.** Bands of connective tissue attaching bone to bone are called ligaments. Tendons attach bone to muscle, eliminated choice B. Choice C can be eliminated because muscles are not a type of connective tissue. Choice D can be ruled out because osseous tissues make up bone and would contribute to the formation of one continuous bone rather than the connection between two bones.

141. **A is the best answer.** To answer this question, visualize what the question stem is describing. These dead cells are sealed together and have tough keratin molecules everywhere. The fact that there are many layers is important, it seems like this part of the integumentary system is designed for protection or insulation so choice A is the best answer. Dead cells are unlikely to help synthesize vitamin D and would be unable to transport it anyway, so choice B is not a good answer. This layer could conceivably help in insulating against heat loss, but as dead cells, it seems unlikely that they could participate in regulation itself. This is more a function of the capillaries in the dermis, so choice C is not a strong answer. Just like thermoregulation, dead cells are unlikely to play a role in active osmoregulation so choice D is not a strong answer. Notice that this description could be true for physical insulation against abrasion, but this is not an answer choice.

142. **A is the best answer.** Haversian canals, canaliculi, and Volkmann's canals are some of the structural features that are found within compact bone on a microscopic level. Yellow bone marrow is usually found in the medullary cavity of long bones.

143. **B is the best answer.** Hydroxyapatite is made up of calcium and phosphate in a compound that includes hydroxyl groups as well. This question can be answer by process of elimination. The fact that bone acts as a storage place for phosphate and calcium is required knowledge and eliminates choices A and C. Ascertaining that the compound contains hydroxyl can be inferred from the name hydroxyapatite, which eliminates choice D and leaves B as the best answer.

144. **D is the best answer.** DNA replication is necessary for cell division and the production of new daughter cells during mitosis. Cisplatin treatment impairs the ability of cells to undergo mitosis successfully and would impact cells that divide quickly the most. Osteoblasts, osteoclasts, and osteocytes all derive from osteoprogenitor stem cells in the bone. In addition, these three cell types do not undergo mitosis, so they would be unaffected by cisplatin treatment, eliminating choice A, B, and C. Rapidly dividing cells, like osteoprogenitor cells, would no longer be able to undergo mitosis upon cisplatin treatment, making choice D the best answer choice.

Photo Credits

Covers

Front cover, The upper body muscles: © Eraxion/ iStockphoto.com; The abdominal anatomy: © Eraxion/ iStockphoto.com

Lecture 1

Pg. 1, Bacteriophages: © Science Picture Co/Getty Images

Pg. 2, Zebra-horse hybrid: © Mary Beth Angelo/Science Photo Library

Pg. 4, One cheetah sitting: Stolz, Gary M/U.S. Fish and Wildlife Service, Washington, D.C. Library

Pg. 5, Influenza Virus: Frederick Murphy/CDC Public Health Image Library

Pg. 9, E. coli bacteria: © Andrew Syred/Photo Researchers, Inc.

Pg. 11, Cholera bacteria: © Juergen Berger/Photo Researchers, Inc.

Pg. 11, Gut bacterium reproducing: © Hazel Appleton, Health Protection Agency Centre for Infections/Photo Researchers, Inc.

Pg. 13, T4 bacteriophages: © Russell Kightley/Science Source

Pg. 19, Mitochondria: © Don W. Fawcett/Science Photo Library

Pg. 21, Lung and trachea epithelium: Charles Daghlian/Dartmouth Electron MIcroscopy Center

Pg. 21, Human Sperm: Michael Crammer/Cell Image Library

Pg. 25, Diffusion: © Charles D. Winters/Photo Researchers, Inc.

Pg. 33, Los Angeles from Space: NASA, courtesy photo

Lecture 2

Pg. 37, Male nervous system: © Sciepro/Science Photo Library

Pg. 37, Hippocampus: Margaret I. Davis/Cell Image Library

Pg. 70, Inner Ear organ of corti: © Dr Goran Bredberg/Science Photo Library

Lecture 3

Pg. 97, Sperm production: © Susumu Nishinaga/Photo Researchers, Inc.

Pg. 99, Ovulation: © Profs. P.M. Motta & J. Van Blerkom/Photo Researchers, Inc.

Pg. 100, Sperm cell fertilizing an egg cell, colored ESEM (environmental scanning electron micrograph): © Thierry Berrod, Mona Lisa Production/Photo Researchers, Inc.

Lecture 4

Pg. 107, Male vascular system: © Sciepro/Science Photo Library

Pg. 107, Lungs anatomy, artwork: © Henning Dalhoff/Science Photo Library

Pg. 109, X-ray of a lung cancer: © muratseyit/iStockphoto.com

Pg. 119, Blood clot: © Steve Gschmeissner/Photo Researchers, Inc.

Pg. 122, Color enhanced scanning electron micrograph (SEM) of a single red blood cell in a capillary: © Dr. Cecil H. Fox/Photo Researchers, Inc.

Pg. 126, Doctor checking patient blood pressure: © Rudyanto Wijaya/iStockphoto.com

Pg. 132, Histology - Immune System: Human macrophage ingesting pseudomonas: © David M. Phillips/Photo Researchers, Inc.

Pg. 135, Scanning electron micrograph of HIV-1 virions budding from a cultured lymphocyte: C. Goldsmith/CDC Public Health Image Library

Pg. 140, Blood storage: © Tek Image/Photo Researchers, Inc.

Lecture 5

Pg. 143, Male anatomy: © Sciepro/Science Photo Library

Pg. 143, Stomach interior walls © Garry DeLong/Photo Researchers, Inc.

Pg.145, Stomach lining: © Steve Gschmeissner/Photo Researchers, Inc.

Pg. 148, False-color scanning electron micrograph of a section through the wall of the human duodenum: © David Scharf/ Photo Researchers, Inc.

Pg. 149, Pancreas surface: © Susumu Nishinaga/Photo Researchers, Inc.

Pg. 151, Radiologic evaluation of the large intestine often involves performing a barium enema: © Medical Body Scans/Photo Researchers, Inc.

Pg. 155, Pasta: © Akihito Fujii/Flickr, adapted for use under the terms of the Creative Commons CC BY 2.0 license (http://creativecommons.org/licenses/by/2.0/legalcode)

Pg. 157, The normal surface pattern of the jejunal (small intestine) mucosa: © Biophoto Associates/Photo Researchers, Inc.

Pg. 157, Celiac Disease: © Biophoto Associates/Photo Researchers, Inc.

Pg. 158, Nutrition Label: © Brandon Laufenberg/iStockphoto.com

Pg. 167, LM of Simple Cuboidal Epithelium: © Robert Knauft/Biology Pics/Science Source

Lecture 6

Pg. 175, Anatomical Overlays - Man Running Front View: © LindaMarieB/iStockphoto.com

Pg. 178, Light micrograph of a section through human skeletal (striated) muscle: © Manfred Kage/Photo Researchers, Inc.

Pg. 185, Heart muscle: © Manfred Kage/Photo Researchers, Inc.

Pg. 187, Light micrograph of a section of human smooth muscle: © SPL/Photo Researchers, Inc.

Pg. 189, Transmission electron micrograph (TEM) of human bone showing an osteocyte: © Biophoto Associates/Photo Researchers, Inc.

Pg. 189, Osteoblast: © CNRI/Photo Researchers, Inc.

Pg. 191, Computer-generated image of multi-nucleated osteoclasts etching away trabecular bone in a process called bone resorption: © Gary Carlson/Photo Researchers, Inc.

Pg. 193, Hyaline cartilage: © Innerspace Imaging/Photo Researchers, Inc.

Pg. 194, Sample of skin from the back of a human hand: © Eye of Science/Photo Researchers, Inc.

Index

Symbols

A

B

C

G

H

M

N

R

S

T

U

V

W

Z

About the Author

Jonathan Orsay is uniquely qualified to write an MCAT® preparation book. He graduated on the Dean's list with a B.A. in History from Columbia University. While considering medical school, he sat for the real MCAT® three times from 1989 to 1996. He scored above the 95th percentile on all sections before becoming an MCAT® instructor. He has lectured in MCAT® test preparation for thousands of hours and across the country. He has taught premeds from such prestigious universities as Harvard and Columbia. He has written and published the following books and audio products in MCAT® preparation: "Examkrackers MCAT® Physics", "Examkrackers MCAT® Chemistry", "Examkrackers MCAT® Organic Chemistry", "Examkrackers MCAT® Biology", "Examkrackers MCAT® Verbal Reasoning & Math", "Examkrackers 1001 questions in MCAT® Physics", "Examkrackers MCAT® Audio Osmosis with Jordan and Jon", all of which have evolved when the MCAT® has changed, and which continue to be the number one bestselling MCAT® materials available.

A Student Review of This Book

The following review of this book was written by Teri from New York.

"The Examkrackers MCAT® books are the best MCAT® prep materials I've seen-and I looked at many before deciding. The worst part about studying for the MCAT® is figuring out what you need to cover and getting the material organized. These books do all that for you so that you can spend your time learning. The books are well and carefully written, with great diagrams and really useful mnemonic tricks, so you don't waste time trying to figure out what the book is saying. They are concise enough that you can get through all of the subjects without cramming unnecessary details, and they really give you a strategy for the exam. The study questions in each section cover all the important concepts, and let you check your learning after each section. Alternating between reading and answering questions in MCAT® format really helps make the material stick, and means there are no surprises on the day of the exam-the exam format seems really familiar and this helps enormously with the anxiety. Basically, these books make it clear what you need to do to be completely prepared for the MCAT® and deliver it to you in a straightforward and easy-to-follow form. The mass of material you could study is overwhelming, so I decided to trust these books—I used nothing but the Examkrackers books in all subjects and scored in the 99th percentile in all sections. Thanks to Jonathan Orsay and Examkrackers, I was admitted to all of my top-choice schools (Columbia, Cornell, Stanford, and UCSF). I will always be grateful. I could not recommend the Examkrackers books more strongly. Please contact me if you have any questions."

BIOLOGICAL SCIENCES

DIRECTIONS. Most questions in the Biological Sciences test are organized into groups, each preceded by a descriptive passage. After studying the passage, select the one best answer to each question in the group. Some questions are not based on a descriptive passage and are also independent of each other. You must also select the one best answer to these questions. If you are not certain of an answer, eliminate the alternatives that you know to be incorrect and then select an answer from the remaining alternatives. A periodic table is provided for your use. You may consult it whenever you wish.

PERIODIC TABLE OF THE ELEMENTS

1 **H** 1.0																	2 **He** 4.0
3 **Li** 6.9	4 **Be** 9.0											5 **B** 10.8	6 **C** 12.0	7 **N** 14.0	8 **O** 16.0	9 **F** 19.0	10 **Ne** 20.2
11 **Na** 23.0	12 **Mg** 24.3											13 **Al** 27.0	14 **Si** 28.1	15 **P** 31.0	16 **S** 32.1	17 **Cl** 35.5	18 **Ar** 39.9
19 **K** 39.1	20 **Ca** 40.1	21 **Sc** 45.0	22 **Ti** 47.9	23 **V** 50.9	24 **Cr** 52.0	25 **Mn** 54.9	26 **Fe** 55.8	27 **Co** 58.9	28 **Ni** 58.7	29 **Cu** 63.5	30 **Zn** 65.4	31 **Ga** 69.7	32 **Ge** 72.6	33 **As** 74.9	34 **Se** 79.0	35 **Br** 79.9	36 **Kr** 83.8
37 **Rb** 85.5	38 **Sr** 87.6	39 **Y** 88.9	40 **Zr** 91.2	41 **Nb** 92.9	42 **Mo** 95.9	43 **Tc** (98)	44 **Ru** 101.1	45 **Rh** 102.9	46 **Pd** 106.4	47 **Ag** 107.9	48 **Cd** 112.4	49 **In** 114.8	50 **Sn** 118.7	51 **Sb** 121.8	52 **Te** 127.6	53 **I** 126.9	54 **Xe** 131.3
55 **Cs** 132.9	56 **Ba** 137.3	57 **La*** 138.9	72 **Hf** 178.5	73 **Ta** 180.9	74 **W** 183.9	75 **Re** 186.2	76 **Os** 190.2	77 **Ir** 192.2	78 **Pt** 195.1	79 **Au** 197.0	80 **Hg** 200.6	81 **Tl** 204.4	82 **Pb** 207.2	83 **Bi** 209.0	84 **Po** (209)	85 **At** (210)	86 **Rn** (222)
87 **Fr** (223)	88 **Ra** 226.0	89 **Ac**= 227.0	104 **Unq** (261)	105 **Unp** (262)	106 **Unh** (263)	107 **Uns** (262)	108 **Uno** (265)	109 **Une** (267)									

*	58 **Ce** 140.1	59 **Pr** 140.9	60 **Nd** 144.2	61 **Pm** (145)	62 **Sm** 150.4	63 **Eu** 152.0	64 **Gd** 157.3	65 **Tb** 158.9	66 **Dy** 162.5	67 **Ho** 164.9	68 **Er** 167.3	69 **Tm** 168.9	70 **Yb** 173.0	71 **Lu** 175.0
=	90 **Th** 232.0	91 **Pa** (231)	92 **U** 238.0	93 **Np** (237)	94 **Pu** (244)	95 **Am** (243)	96 **Cm** (247)	97 **Bk** (247)	98 **Cf** (251)	99 **Es** (252)	100 **Fm** (257)	101 **Md** (258)	102 **No** (259)	103 **Lr** (260)